ENVIRONMENTAL ENGINEERING V

PROCEEDINGS OF THE FIFTH NATIONAL CONGRESS OF ENVIRONMENTAL ENGINEERING, LUBLIN, POLAND, 29 MAY–1 JUNE, 2016

Environmental Engineering V

Editors

Małgorzata Pawłowska & Lucjan Pawłowski
Lublin University of Technology, Lublin, Poland

CRC Press is an imprint of the
Taylor & Francis Group, an **informa** business

A BALKEMA BOOK

CRC Press/Balkema is an imprint of the Taylor & Francis Group, an informa business

© 2017 Taylor & Francis Group, London, UK

Typeset by V Publishing Solutions Pvt Ltd., Chennai, India
Printed and bound in Great Britain by CPI Group (UK) Ltd, Croydon, CR0 4YY

All rights reserved. No part of this publication or the information contained herein may be reproduced, stored in a retrieval system, or transmitted in any form or by any means, electronic, mechanical, by photocopying, recording or otherwise, without written prior permission from the publisher.

Although all care is taken to ensure integrity and the quality of this publication and the information herein, no responsibility is assumed by the publishers nor the author for any damage to the property or persons as a result of operation or use of this publication and/or the information contained herein.

Published by: CRC Press/Balkema
              P.O. Box 11320, 2301 EH Leiden, The Netherlands
              e-mail: Pub.NL@taylorandfrancis.com
              www.crcpress.com – www.taylorandfrancis.com

ISBN: 978-1-138-03163-0 (Hbk)
ISBN: 978-1-315-28197-1 (eBook)

Environmental Engineering V – Pawłowska & Pawłowski (Eds)
© 2017 Taylor & Francis Group, London, ISBN 978-1-138-03163-0

Table of contents

Preface	ix
About the editors	xi

Safety analysis of water supply to water treatment plant
B. Tchórzewska-Cieślak, D. Papciak, P. Koszelnik, J. Kaleta, A. Puszkarewicz & M. Kida ... 1

The use of geographical information system in the analysis of risk of failure
of water supply network
I. Piegdoń, B. Tchórzewska-Cieślak & D. Szpak ... 7

Method for forecasting the failure rate index of water pipelines
K. Boryczko & A. Pasierb ... 15

Ground water levels of a developing wetland—implications for water management goals
A. Brandyk, G. Majewski, A. Kiczko, A. Boczoń, M. Wróbel & P. Porretta-Tomaszewska ... 25

Detection of potential anomalies in flood embankments
M. Chuchro, M. Lupa, K. Szostek, B. Bukowska-Belniak & A. Leśniak ... 33

Stochastic model for estimating the annual number of storm overflow discharges
B. Szeląg ... 43

Assessment of water supply diversification using the Pielou index
J. Rak & K. Boryczko ... 53

Reduction of water losses through metering of water supply network districts
T. Cichoń & J. Królikowska ... 59

Potential DBD-jet applications for preservation of nutritive compounds on the example
of vitamin C in water solutions
D. Bozkurt, M. Kwiatkowski, P. Terebun, J. Diatczyk & J. Pawłat ... 65

Modeling and predicting the concentration of volatile organic chlorination by-products
in Krakow drinking water
A. Włodyka-Bergier, T. Bergier, Z. Kowalewski & S. Gruszczyński ... 71

The carbon and nitrogen stable isotopes content in sediments as an indicator of the trophic
status of artificial water reservoirs
L. Bartoszek, P. Koszelnik & R. Gruca-Rokosz ... 83

Design errors in working water and sludge installations in industrial plants
T. Piecuch, J. Piekarski & A. Kowalczyk ... 89

Impact of UV disinfection on microbial growth in drinking water
A. Włodyka-Bergier & T. Bergier ... 95

Electron microscopy assessment of the chemical composition of sediments from water supply pipes
J. Bąk, J. Królikowska, A. Wassilkowska & T. Żaba ... 101

Effect of chemical coagulants on the sedimentation properties of activated sludge
A. Masłoń & J.A. Tomaszek ... 109

Bacteria from on-site wastewater treatment facilities as enzymes producers for applications
in environmental technologies
Ł. Jałowiecki, J. Chojniak, G. Płaza, E. Dorgeloh, B. Hegedusova & H. Ejhed ... 115

v

Evaluation of the possibilities of water and sewage sludge disposal 123
J. Górka & M. Cimochowicz-Rybicka

Application of iron sludge from water purification plant for pretreatment of reject water
from sewage sludge treatment 131
K. Piaskowski & R. Świderska-Dąbrowska

Magnetic separation of submicron particles from aerosol phase 137
A. Jaworek, A. Marchewicz, T. Czech, A. Krupa, A.T. Sobczyk & K. Adamiak

Treatment of wastewater from textile industry in biological aerated filters 145
J. Wrębiak, K. Paździor, A. Klepacz-Smółka & S. Ledakowicz

Hydroecological investigations of water objects located on urban areas 155
W. Wójcik, V.P. Osypenko, V.V. Osypenko, V.I. Lytvynenko, N. Askarova & M. Zhassandykyzy

An influence of sludge compost application on heavy metals concentration in willow biomass 161
D. Fijałkowska, L. Styszko & B. Janowska

An influence of municipal sewage sludge and mineral wool application on sorption properties
of coarse-grained soil 169
*S. Wesołowska, S. Baran, G. Żukowska, M. Myszura, M. Bik-Małodzińska,
A. Pawłowski & M. Pawłowska*

Nitrate monoionic form of anion exchanger as a means for enhanced nitrogen fertilization
of degraded soils 181
M. Chomczyńska

The process generation of WWTP models for optimization of activated sludge systems 187
J. Drewnowski, K. Wiśniewski, A. Szaja, G. Łagód & C. Hernandez De Vega

Oxidation of organic pollutants in photo-Fenton process in presence of humic substances 197
R. Świderska-Dąbrowska, K. Piaskowski & R. Schmidt

Gas sensors array as a device to classify mold threat of the buildings 203
Z. Suchorab, H. Sobczuk, Ł. Guz & G. Łagód

Operation of detention pond in urban area—example of Wyścigi Pond in Warsaw 211
A. Krajewski, M. Wasilewicz, K. Banasik & A.E. Sikorska

Hydraulic equations for vortex separators dimensioning 217
M.A. Gronowska-Szneler & J.M. Sawicki

Gaseous fuel production from biomass using gasification process with CO_2 emission reduction 225
S. Werle

Methane emissions and the possibility of its mitigation 231
A. Czechowska-Kosacka & W. Cel

Mitigation of pollutant migration from landfill to underground water and air 239
K. Szymański, B. Janowska, A. Czechowska-Kosacka & W. Cel

Assessment of odour nuisance of wastewater treatment plant 249
Ł. Guz, A. Piotrowicz & E. Guz

Air pollution in Poland in relation to European Union 257
W. Cel, Z. Lenik & A. Duda

The influence of external conditions on the photovoltaic modules performance 261
A. Zdyb & E. Krawczak

Operational characteristics of the heat and cold storage in traction vehicles 267
D. Zieliński, K. Przytuła & K. Fatyga

Parameterisation of an electric vehicle drive for maximising the energy recovery factor 273
K. Kolano & M. Litwin

Coal and biomass co-combustion process characterization using frequency analysis
of flame flicker signals 279
A. Kotyra, W. Wójcik, D. Sawicki, K. Gromaszek, A. Asembay,
A. Sagymbekova & A. Kozbakova

Battery-supported trolleybus traction network—a component of the municipal smart grid 287
W. Jarzyna, D. Zieliński & P. Hołyszko

Analysis of energy consumption of public transport in Lublin 293
M. Dziubiński, E. Siemionek, A. Drozd, S. Kołodziej & W. Jarzyna

Ecological aspect of electronic ignition and electronic injection system 299
M. Dziubiński

Testing of exhaust emissions of vehicles combustion engines 305
M. Dziubiński

Artificial intelligence methods in diagnostics of coal-biomass blends co-combustion
in pulverised coal burners 311
A. Smolarz, W. Wójcik, K. Gromaszek, P. Komada, V.I. Lytvynenko, N. Mussabekov,
L. Yesmakhanova & A. Toigozhinova

Combustion process diagnosis and control using optical methods 319
K. Gromaszek, A. Kotyra, W. Wójcik, B. Imanbek, A. Asembay, Y. Orakbayev & A. Kalizhanova

Author index 325

Environmental Engineering V – Pawłowska & Pawłowski (Eds)
© 2017 Taylor & Francis Group, London, ISBN 978-1-138-03163-0

Preface

Environmental Engineering V summarizes research carried out in Poland in the area of environmental engineering. The main goal of the book is to improve technology transfer and scientific dialogue in the time of economic transformation from a planned to a free market economy, thereby leading to a better comprehension of solutions to a broad spectrum of environmentally related problems.

Increased use of motor vehicles is one of the most serious problems in Poland today. No incentives or economic stimulation for buying pro-ecological cars have yet been introduced. In order to mitigate unfavorable environmental changes, especially one of the highest levels of air pollution, the new Polish government has taken actions to begin the production of electric cars in Poland.

Therefore, a presentation of scientific findings and technical solutions created by the Polish research community ought to be of the interest not only for Polish institutions, but also for international specialists, searching for solutions for environmental problems in new emerging democracies, especially those who plan to participate in numerous projects sponsored by the European Union. Finally, we would like to express our appreciation to all who have helped to prepare this book: Mr. Szymon Obrusiewicz for improving the linguistic side of the papers. Anonymous reviewers who not only evaluated papers, but very often made valuable suggestion helping authors and editors to improve the scientific standard of this book. And finally, last but definitely not least Mrs. Katarzyna Wójcik Oliveira for her invaluable help in preparing a layout of all papers.

Lucjan Pawłowski
Małgorzata Pawłowska
Lublin, September 2016

About the editors

MAŁGORZATA PAWŁOWSKA

Małgorzata Pawłowska, Ph.D., Sc.D. (habilitation) was born in 1969 in Sanok, Poland. In 1993 she received M.Sc of the protection of the environment at the Catholic University of Lublin. Since that time she has been working in the Lublin University of Technology, Faculty of Environmental Engineering. In 1999 she defended Ph.D. in the Institute of Agrophysics of the Polish Academy of Science and in 2010 she defended D.Sc. thesis at the Technical University of Wroclaw and was appointed as associate professor and head of Engineering of Alternative Fuels Department at Faculty of Environmental Engineering Lublin University of Technology. Now she is working on biomethanization processes and application of selected wastes for remediation of degraded land.

She has publishes 59 papers, 2 books and is co-author of 7 polish patents, 1 European patent, 28 polish patent applications and 25 European patent applications.

LUCJAN PAWŁOWSKI

Lucjan Pawłowski, was born In Poland, 1946. Director of the Institute of Environmental Protection Engineering of the Lublin University of Technology, Member of the European Academy of Science and Arts, Member of the Polish Academy of Science, Deputy President of the Engineering Science Division of the Polish Academy of Science, honorary professor of China Academy of Science. He received his Ph.D. in 1976, and D.Sc. (habilitation) in 1980, both at the Wrocław University of Technology. He started research on the application of ion exchange for water and wastewater treatment. As a result, together with B. Bolto from CSIRO Australia, he has published a book "Wastewater Treatment by Ion Exchange" in which they summarized their own results and experience of the ion exchange area. In 1980 L. Pawłowski was elected President of International Committee "Chemistry for Protection of the Environment". He was Chairman of the Environmental Chemistry Division of the Polish Chemical Society from 1980–1984. In 1994 he was elected the Deputy President of the Polish Chemical Society and in the same year, the Deputy President of the Presidium Polish Academy of Science Committee "Men and Biosphere". In 1999 he was elected President of the Committee "Environmental Engineering" of the Polish Academy of Science. In 1991 he was elected the Deputy Reactor of the Lublin University of Technology, and held this post for two terms (1991–1996). He has published 19 books, over 128 papers, and authored 98 patents, and is a member of the editorial board of numerous international and national scientific and technical journals.

Environmental Engineering V – Pawłowska & Pawłowski (Eds)
© 2017 Taylor & Francis Group, London, ISBN 978-1-138-03163-0

Safety analysis of water supply to water treatment plant

B. Tchórzewska-Cieślak, D. Papciak, P. Koszelnik, J. Kaleta, A. Puszkarewicz & M. Kida
Faculty of Civil and Environmental Engineering and Architecture, Rzeszow University of Technology, Rzeszow, Poland

ABSTRACT: Collective water supply system safety analysis should include risk assessment of threats at each stage of water production, from the water intake through treatment, distribution to the point of delivery to consumers. The proposed method, using the definition of the expected value of water scarcity, enables to conduct the analysis and assessment of the risk of lack of water supply to the distribution subsystem, taking into account different combinations of reliability states. The paper analyses the operation of a water treatment plant. The analysis used data operational work of the water intake. The expected value is used as a measure of the risk of the lack of water supply. Calculations were carried out in two variants: for a nominal value, and for the theoretical increase in the demand for water, in which the nominal demand for water is two times higher than the current maximum water production.

Keywords: water supply system, safety analysis, water treatment plant

1 INTRODUCTION

State, level and area of population using Collective Water Supply System (CWSS) are a measure of the level of civilization. Water quality, quantity and availability should be a priority for the sustainable development of a local community. In accordance with The European Programme for Critical Infrastructure Protection, collective water supply system belongs to critical infrastructure. The identified critical infrastructure must have a protection plan including identification of key resources, risk analysis based on major threats scenarios and vulnerabilities of resources, as well as identification, selection and procedures prioritization and protection measures. CWSS safety has its own international legal regulations, the source of which are primarily the guidelines of the World Health Organization (WHO). The primary and basic subject the notion of water safety is concerned with is a consumer (Tchórzewska-Cieślak 2007, Tchórzewska-Cieślak 2009, Boryczko & Tchórzewska-Cieślak 2015, Olkiewicz et al. 2015, Mrozik et al. 2015).

The secondary subject is a supplier, i.e. a manufacturer of water. In this respect, one can consider the risk of the consumer and the producer. The important elements in this regard also include the environmental aspects and the principles of sustainable development in water management. The environmental aspect and the principles of sustainable development in the widely understood water management are relevant as well.

Operational reliability of the CWSS is the ability to supply a constant flow of water for various groups of consumers, with a specific quality and specific pressure, according to the users' requirements in the specific operational conditions, at any time (in which case we use the readiness indicator K) or in the specific time range (then we use reliability function R(t)) (Rak 1993, Rak 2005, Kwietniewski et al. 1993, Tchórzewska-Cieślak & Rak 2010).

The definition of the CWSS safety, including technical, economic and environmental aspects, is the following: "safe CWSS operation means ensuring continuity of water supply to the consumer while the following criteria are met" (Rak 2005, Tchórzewska-Cieślak 2011, Nowacka et al. 2016):

- system reliability (in terms of quantity and quantity),
- socially acceptable level of prices per m^3 of delivered water, taking into account aspects arising from the requirements for public safety, natural aquatic environment protection and standard of life quality.

CWSS safety analysis should include an analysis and risk assessment of threats at each stage of water production, from the water intake through treatment, distribution to the point of delivery to consumers (Hrudey & Hrudey 2004). In most studies regarding the analysis and assessment of CWSS safety it is assumed that the measure of CWSS safety is risk associated with the possibility of different threats occurrence (Haimes 2009).

One of the most common ways to conduct threat analysis is the study of threats using data from:

- previous safety analyses, conclusions—drawn from undesirable events and their causes,
- the experience of experts from—the operation of existing systems of water-pipelines.

Risk identification involves selecting the representative emergency events that may occur during the operation of the system, including the initiating events, which may cause the so-called domino effect (Pollard et al. 2004, Rak 2009, Rak 2003, Rak & Tchórzewska-Cieślak 2013).

Risk assessment is the process of qualitative and quantitative analysis using methods adequate for the given type of risk, with the determination of the criteria value for the assumed risk scale (Zio 2007) or in the form of failure prediction (Tchorzewska-Cieslak et al. 2016). For example, a three-step scale distinguishing tolerated risk, controlled risk and unacceptable risk, or a five-step scale, in which the area of neglected risk and absolutely unacceptable risk is also distinguished (Apostolakis & Kaplan 1981, Michaud & Apostolakis 2006).

The determination of risk acceptability criteria should primarily take into account the aspect of water consumer safety, as well as technical and economic analysis. The risk acceptability criteria are used in making decisions concerning the operation of the system (e.g. repairs, modernization and allowing to be used). The system can be considered safe if the risk level created during system operation does not exceed the limit values.

It is very important for waterworks to identify risk correctly and to divide it into consumer risk and water producer risk (Boryczko & Tchórzewska-Cieślak 2015). It allows to choose the right method for calculating different types of risks. The correct WSS risk management process should contain suitable organizational procedures within the framework of regular waterworks activity, the WSS operation technical control and supervisory system, a system of automatic transfer and data processing about WSS elements operation (Zimoch 2007). The key role in this process is played by a system operator, whose main purpose is (Ezell, Farr & Wiese 2000, Iwanejko 2009, Królikowska 2011, Rak 2005, Sadiq, Kleiner & Rajani 2004):

- to implement the reliability and safety management system,
- to operate the WSS according to valid regulations and in a way which ensures its long and reliable operation,
- to execute a program of undesirable events prevention,
- to develop failure scenarios for water supply in emergency situations,

- to develop a complex system of information about the possible threats for water consumers.

In practice, the WHO recommends the development of the so called Water Safety Plans (WSP) and the new Water Cycle Safety Plan (WCSP) approach, based on analyses and risk assessment. It is important for the entire water cycle in urban catchments and the impact of rainfall on the functioning of water supply system (Kazmierczak & Kotowski 2015). At the same time there are still risks such as floods, droughts, failures of electrical power, accidental pollution of water sources, and even terrorist and cyber terrorist attacks, which are often the cause of serious disruption in CWSS subsystems functioning and thus contribute to the loss of water consumers' safety.

2 MATERIAL AND METHODS

The risk assessment is concerned with what can go wrong, as well as its likelihood and consequences. Risk is a measure of the probability and severity of adverse effect (Haimes 2009).

Kaplan and Garrick introduced the theory of scenario and the triplet questions in the risk assessment process (Kaplan & Garrick 1981, Kaplan 1997):

- what can go wrong?
- what is the likelihood?
- what are the consequences?

They introduced that the risk is the function: $r = f(S_i, L_i, C_i)$, where S_i = denotes the i-th risk scenario (initiating events scenarios); L_i = denotes the likelihood of that scenario; C_i = denotes resulting consequences.

In order to perform effective risk assessment and management, the analyst must understand the system and its interactions with its environment, and this understanding is a requisite to modeling the behaviour of the state of the system under varied probabilistic conditions (Boryczko & Tchórzewska-Cieślak 2015).

A factor determining the risk of lack of water supply may be a scarcity of water production during the failure of the particular subsystems of water production. It is assumed that the absolute risk of lack of water supply is a product of the probability of water scarcity and the losses related to it.

This relation can be determined using the so-called expected value of water shortage $E(\Delta Q)$ (Rak 2005, Tchórzewska-Cieślak 2011; Kwietniewski et al. 1993):

$$r_a = E(\Delta Q) \tag{1}$$

and

$$E(\Delta Q) = \int_0^\infty \Delta Q f(\Delta Q) d(\Delta Q) = \sum_{i=0}^{i=n} \Delta Q \cdot P_i \qquad (2)$$

where r_a = the absolute risk of lack of water supply (m³/d); $E(\Delta Q)$ = the expected value of water shortage; i = a number of operating state; n = a maximum number of the possible states of reliability; n = 2^m (m = a number of all water sources); ΔQ_i = deficiency of water sources in the given state of unreliability; P_i = probability of the i-th state of the water production subsystem operation.

The value of the deficiency of water sources ΔQ is calculated as the difference between the required capacity of water sources and the capacity of sources in the i-th state, according to the (Rak 2005, Tchórzewska-Cieślak 2011, Kwietniewski et al. 1993):

$$\Delta Q = Q_n - \sum_{i=1}^{k_i} Q_{ik} \qquad (3)$$

where Q_n = the required water demand, the required system capacity during the normal operation (usually the value of Q_n is assumed to be the maximum daily demand for water Q_{maxd} or the design value of water production); k_i = a number of faulty water sources in the i-th state; Q_{ik} = production of the particular water sources in the i-th state with k_i failures.

Probability value P_i is determined by the following formula (Kwietniewski M. et al. 1993):

$$P_i = \prod_{j \in S} K_j \cdot \prod_{j \in N} (1 - K_j) \qquad (4)$$

where K_j = the availability index of the j-th water supply subsystem; $j \in S$ = the set of those subsystems of delivery (or their components) that are efficient in the i-th state, marked with the symbol (+); $j \in N$ = the set of those subsystems of delivery (or their components) that are inefficient in the i-th state, marked with the symbol (–).

In order to assess the risk of lack of water supply, the relative risk of lack of water supply to the CWSS is defined, referring the expected value of water scarcity (the absolute risk) to the nominal value of water supply, which also allows to determine the criteria values. In order to better illustrate the relative risk, its value can be expressed as a percentage, according to the formula (Tchórzewska-Cieślak 2011):

$$r_r = \frac{r_a}{Q_n} \qquad (5)$$

Table 1. The safety criteria values for levels of the relative risk of lack of water supply.

Water supply system	r_r [%]	Safety level
large supply system,	≤2	TSL
number of inhabitants	(2÷5)	CSL
>500 000	≥5	USL
medium supply system,	≤ 4	TSL
number of inhabitants	(4÷9)	CSL
50 000 ÷ 500 000	≥9	USL
small supply system,	≤5	TSL
number of inhabitants	(5÷9)	CSL
<50 000	≥9	USL

where r_r = the relative risk of lack of water supply [%]; Q_n = the nominal value of the water demand (consumption) [m³/d].

If the sum of the capacity of all sources is higher than the required capacity, then so called water reserve occurs (scarcity is equal to zero).

Criteria for the assessment of the safety of CWSS on absolute risk are the following:

- Tolerable Safety Level (TSL),
- Controlled Safety Level (CSL),
- Unacceptable Safety Level (USL), and are presented in Table 1 (Tchórzewska-Cieślak 2011).

If the calculated values indicate that safety level is:

- TSL—one can assume that the subsystems of water production fulfils its functions in the satisfying way,
- CLS—an improvement in the work of some elements of water production subsystems or alternative water sources should be considered,
- USL—the subsystems of water production does not fulfil its functions and should undergo a complete modernization and alternative sources of water are needed.

3 RESULTS AND DISCUSSION

The paper analyses the functioning of the Water Treatment Plant (WTP) for a town located in the eastern Poland and supplying water for about 80 000 residents. Water Treatment Plant takes water from the unconfined quaternary aquifer from a depth of about 15 m.

The quaternary aquifer level within which there is a water intake is part of the Main Groundwater Reservoir. Wells, due to their location, are divided into two intakes I and II. Water is directed to the collective well (retention time 24 hours, depending on the current water production). The weak side of supply is connecting the intake wells I and II by means of one transmission pipeline with WTP.

It is compensated by a store of water in the collective well; however, it is a weak point of the intake.

Water facilities for water intake are pump intakes:

- I—which includes 5 pieces of drilled wells (Q_{emax} 183 m³/h).
- II—which includes 22 pieces of drilled wells (Q_{emax} = 715 m³/h).
- the maximum daily water consumption (2013/2014) is on average Q_{maxd} = 6 958.5 m³/d.

Table 2 lists the values of water production for intakes I and II and adopted values of reliability indexes K (Kwietniewski M. et al. 1993; Rak J. 1993).

Calculations were carried out in two variants:

- for a nominal value, accepted on the base of the average maximum water production, Q_n = 7 000 m³/d, was assumed.
- for the theoretical increase in the demand for water, in which the nominal demand for water is two times higher than the current maximum water production, $Q_n = Q_{maxd}$ = 14 000 m³/d, was assumed.

Table 3 presents the results of calculations of the absolute risk according to the formulas (1) ÷ (4) for the nominal value of the water demand, adopted on the base of the average maximum water production: Q_n = 7 000 m³/d and we have to assume:

- The state of reliability is marked with "1", the state of unreliability with "0".

For the two water sources the number of the possible states is: $2^2 = 4$, i = 1,2,3,4.

Table 2. The reliability indexes K for water intake I and II.

Intake	Q [m³/d]	K
I	2 976	0.984
II	15 797	0.995

The absolute risk of lack of water supply:

r_a = 20.35 m³/d.

The relative risk was calculated using the formula (5) for the demand for water: $Q_n = Q_{maxd}$ = 7 000 m³/d (adopted for the current maximum water consumption), which is approximately 40% of the maximum operational capacity of intakes I and II.

It is stated that the system has excess production capacity. The relative risk for the lack of water supply is:

r_r = 0.3%.

According to the data contained in table 1, the relative risk of lack of water supply to the water supply system of the city, for a variant with an excess (real state) is at a tolerable level—TSL.

In order to analyse the risk of lack of water supply in case of a theoretical increase in the demand for water, the risk of lack of water supply for a hypothetical state in which the nominal demand for water is two times higher than the current maximum water production, was calculated.

$Q_n = Q_{maxd}$ = 14 000 m³/d was adopted, which represents approximately 75% of the maximum operational capacity of intakes I and II.

Table 4 summarizes the results of calculation of the absolute risk according to formulas: (1) ÷ (4) for a nominal value adopted for the hypothetical variant.

The absolute risk of lack of water supply:

r_a = 55.36 m³/day.

The relative risk was calculated using the formula (5) for the demand for water:

$Q_n = Q_{maxd}$ = 14 000 m³/d

The relative risk for the lack of water supply is:

r_r = 0.4%.

Table 3. Analysis of the risk of lack of water supply (I and II) for the existing state.

	Characteristics of operating states		Capacity m³/d]		Total m³/d]	Deficiency m³/d]	Probability of i state	
i	I	II	Q_I	Q_{II}	Q	ΔQ	P_i	$P_i \cdot \Delta Q$
1	1	1	2976	15797	18773	0	0.9791	0
2	1	0	2976	0	2976	4024	0.0049	19.79
3	0	1	0	15797	15796	0	0.0159	0
4	0	0	0	0	0	7000	0.00008	0.56
							Σ	20.35

Table 4. Analysis of the risk of lack of water supply for a hypothetical state.

i	Characteristics of operating states		Capacity [m³/d]		Total [m³/d]	Deficiency [m³/d]	Probability of i state	
	I	II	Q_I	Q_{II}	Q	ΔQ	P_i	$P_i \cdot \Delta Q$
1	1	1	2976	15797	18773	0	0.9791	0
2	1	0	2976	0	2976	11024	0.0049	54.24
3	0	1	0	15797	15796	0	0.0159	0
4	0	0	0	0	0	14000	0.00008	1.12
				Σ				55,36

According to the data contained in Table 1, the relative risk of lack of water supply to the water supply system of the city, for a variant with an excess (real state) is on a TSL too.

In the analysed case it was found that:

- the risk of lack of water delivery from the intake to the water treatment plant is at a tolerable level.
- the production capacity of the water intake is much bigger than the current demand for water, which allows the optimal exploitation of wells adapted to the current demand for water and raw water quality needs.
- the weak side of supply is connecting wells I and II with the water treatment plant by one pipeline, which is compensated by water reserve in a collective well, however, the modernization of this element should be considered.

4 CONCLUSION

- Water supply system belongs to the so called critical infrastructure, and it should be a priority task for waterworks to ensure the suitable level of its safety.
- The management of risk connected with the CWSS can be defined as a process of coordination of the operation of the CWSS elements and its operators, using available means, in order to obtain the tolerable risk level in the most efficient way, as far as technology, economic and reliability are concerned. The exploitation of urban CWSS should take into account the minimization of water losses, operational and safety reliability. The main purpose of the decision-maker is to make the right choice that means to choose the best alternative which will assure them the best results in their economical activity
- Currently growing stream of research on water supply system operation, which is derived directly from the reliability theory, is the analysis con-

cerning water consumers safety. To adopt risk as a measure of safety has become the paradigm.
- The proposed method of risk analysis, using the definition of the expected value of water scarcity, allows the analysis and assessment of the risk of lack of water supply to the distribution subsystem taking into account different combinations of reliability states of water production subsystem (water sources, treatment plant and pumping stations).
- The results can be used to analyse the degree of safety of supply and the degree of water sources diversification. Reserve of water sources is the main factor minimizing the risk of lack of water supply.
- The paper proposes the determination of the absolute risk and the relative risk. The proposals of criteria values for the assessment of risk, depending on the size of the city, were given. The given values can be modified and adjusted to the existing conditions.
- The limitation of using this method is the necessity to have a reliability index K for subsystem, which is not always available.
- The method can be used in making decisions, such as the extension or modernization of existing water sources.

REFERENCES

Apostolakis, G. & Kaplan, S. 1981. Pitfallsin risk calculations. *Reliability Engineering and System Safety*. 2: 135–145.

Boryczko K. & Tchórzewska-Cieślak B. 2015. Analysis of risk of failure in water main pipe network and of delivering poor quality water. *Environment Protection Engineering*. 40(4): 77–92.

Ezell, B., Farr, J. & Wiese, I. 2000. Infrastructure risk analysis of municipal water distribution system. *Journal of Infrastructure Systems, ASCE*. 6(3): 118–122.

Haimes, Y.Y. 2009. On the complex definition of risk: a systems-based approach. *Risk Analysis*. 29(12): 1647–1654.

Hastak, H. & Baim, E. 2001. Risk factors affecting management and maintenance cost of urban infrastructure. *Journal of Infrastructure Systems, ASCE.* 7(2), 67–75.

Hrudey, S.E. & Hrudey, E.J. 2004. Safe drinking water. *Lessons from recent outbreaks in affluent nations.* New York: IWA Publishing.

Iwanejko R. 2009. Preliminary analysis of risks attributed to operation of small surface water intakes. *Water Supply and Water Quality.* 1: 229–239.

Kaplan S. & Garrick B.J. 1981. On the quantitative definition of risk. *Risk Analysis.* 1(1): 11–27.

Kaplan S. 1997. The words of risk analysis. *Risk Analysis,* 7(4):407–417.

Kazmierczak, B. & Kotowski, A. 2015. The suitability assessment of a generalized exponential distribution for the description of maximum precipitation amounts. *Journal of Hydrology.* 525. 345–351.

Królikowska J. 2011. Damage evaluation of a town's sewage system in southern Poland by the preliminary hazard analysis method. *Environment Protection Engineering.* 37(4): 131–142.

Kwietniewski M., Roman M. & Kłos-Trębaczkiewicz H. 1993. *Niezawodność wodociągów i kanalizacji.* Warszawa: Arkady.

Michaud, D. & Apostolakis, G. 2006. Methodology for ranking elements of water-supply networks. *Journal of Infrastructure Systems. ASCE.* 12(4): 230–242.

Mrozik K., Przybyła C., Pyszny K. 2015. Problems of the Integrated Urban Water Management. The Case of the Poznań Metropolitan Area (Poland). *Rocznik Ochrona Środowiska/Annual Set the Environment Protection,* 17(1), 230–245.

Nowacka A., Wlodarczyk-Makula M. & Tchorzewska-Cieślak B. & Rak J. 2016. The ability to remove the priority PAHs from water during coagulation process including risk assessment. *Desalination and Water Treatment.* 57(3): 1297–1309.

Olkiewicz, M., Bober B.& Majchrzak-Lepczyk J. 2015. Instrumenty zarządzania w ochronie środowiskowej, *Rocznik Ochrona Środowiska/Annual Set the Environment Protection,* 17(1): 710–725.

Pollard, S.J.T., Strutt, J.E., Macgillivray, B.H., Hamilton, P.D. & Hrudey S.E. 2004. Risk analysis and management in the water utility sector—a review of drivers, tools and techniques. *Process Safety and Environmental Protection.* 82 (6): 1–10.

Rak J. Tchórzewska-Cieślak B. 2013. *Ryzyko w eksploatacji systemów zbiorowego zaopatrzenia w wodę.* Warszawa: Seidel-Przywecki.

Rak J. 1993. *Niezawodność systemu uzdatniania wód powierzchniowych. Rzeszów*: Rzeszow University of Technology.

Rak J. 2003. A study of the qualitative methods for risk assessment in water supply systems. *Environment Protection Engineering.* 3(4): 123–134.

Rak J. 2005. *Podstawy bezpieczeństwa systemów zaopatrzenia w wodę.* Lublin: PAN.

Rak, J. 2009. Selected problems of water supply safety. *Environmental Protection Engineering.* 35(2): 23–28.

Sadiq, R. & Kleiner, Y. & Rajani B. 2004. Aggregative risk analysis for water quality failure in distribution networks. *Journal of Water Supply: Research & Technology—AQUA.* 53 (4): 241–261.

Tchórzewska-Cieślak B. 2009. Risk Management in Water Safety Plans. *Ochrona Środowiska.* 31(4): 57–60.

Tchórzewska-Cieślak B. 2011. *Metody analizy i oceny ryzyka awarii podsystemu dystrybucji wody.* Rzeszow: Oficyna Wydawnicza Politechniki Rzeszowskiej.

Tchorzewska-Cieslak B. 2011. A Fuzzy Model for Failure Risk in Water-pipe Networks Analysis. *Ochrona Środowiska* 33 (1). 35–40.

Tchórzewska-Cieślak, B. & Rak, J. 2010. Method of identification of operational states of water supply system. *Envronmental Engineering III.* Sound Parkway Nw: Crc Press-Taylor & Francis Group. 521–526.

Tchórzewska-Cieślak, B. 2007. Method of assessing of risk of failure in water supply system. *European Safety and Reliability Conference—ESREL 2007. Risk, reliability and societal safety.* Taylor & Francis, 2: 1535–1539.

Tchórzewska-Cieślak, B., Pietrucha-Urbanik, K. & Urbanik M. 2015. Analysis of the gas network failure and failure prediction using the Monte Carlo simulation method. *Eksploat. Niezawodn.* 18: 254–259. http://dx.doi.org/10.17531/ein.2016.2.13.

Zimoch I. 2009. Pressure Control as Part of Risk Management for a Water-pipe Network in Service. *Ochrona Środowiska.* 34(4): 57–62.

Zio, E. 2007. *An introduction to the Basics of Reliability and Risk Analysis.* Singapore: World Scientific Publishing.

Environmental Engineering V – Pawłowska & Pawłowski (Eds)
© 2017 Taylor & Francis Group, London, ISBN 978-1-138-03163-0

The use of geographical information system in the analysis of risk of failure of water supply network

I. Piegdoń, B. Tchórzewska-Cieślak & D. Szpak
Department of Water Supply and Sewage Systems, Rzeszow University of Technology, Rzeszow, Poland

ABSTRACT: The possibilities offered by the GIS in the management of critical infrastructure are enormous. The most important of these possibilities are documenting, processing, storing, analysing data and decision-making. GIS is an excellent tool for database analysis, which has a significant role in the decision making process associated with the water supply network renovation process. The main aim of this work is to present the possibility of the use of information systems and the GIS databases in the risk analysis of failures of water pipes in water distribution subsystem, in selecting pipes with the highest number of failures which create the greatest risk of lack of water supply. The aim of the analysis and presentation of the results is to support the decision making process related to the improvement of activities of repair teams and to eliminate potential water pipes posing risk of further damage.

Keywords: failure, water network, GIS, spatial analysis

1 INTRODUCTION

Water supply systems and wastewater disposal systems are the oldest systems used in any urban area. In the modern era of civilization skilful management of network assets in any waterworks company greatly affects the quality of decisions and influences the increased competitiveness in the market (Kaźmierczak 2016). Decision-makers at every company should have full information about the network and managed assets, which affects not only the efficient operation of the company but also the level of provided services. For each inhabitant of the city or village—every customer—it is extremely important to feel stability and safety in the supply of water (WHO 2005, WHO 2013). Safety in the field of water supply is defined as the state of water management that allows to cover the current and prospective demand for water, in a technically and economically justified way, by the requirements of environmental protection. For the realization of tasks connected with the safety of water supply it is therefore necessary to have a lot of information about the operating system, as well as statistical details regarding undesirable events.

These data should also allow for accurate identification of the analysed object, the event that occurred, as well as its cause and effect analysis. Completeness and details of the databases determines the correct analysis and risk assessment and thus the process of its management.

Safety is inherently connected with the responsibility during the decision-making by the operator of the Collective Water Supply System (CWSS) (Jaźwiński 2000, Lewandowski 2000, Rak 2011, Rak 2013, Tchórzewska-Cieślak 2009, Tchórzewska-Cieślak 2010). Therefore, in order to avoid the negative consequences of making wrong decisions concerning the operation of the network we must learn how to judge the risk of undesirable events and inform others of its size, as well as gain the knowledge necessary to conduct proper activity when the risk occurs.

The use of the Geographic Information System in the supply of water to consumers is an important part of the reliability and safety management of the network. Applications of the Geographical Information System support decision making process in every waterworks company. The available literature focuses mainly on the GIS applications in planning, development and generally on the management of company assets (Gonzalez 2000, Kwietniewski 2008, Michael 2009, Shamsi 2002, Shamsi 2005, Sherer Phebey 1995, Vaughan & Kirby 1988, Vemulapally 2010, Zhang 2006).

The system includes functions of acquiring and processing data about the failure frequency of water supply network and allows to visualize data spatially. The GIS programs allow a complete visualization of the various components of the critical infrastructure and tracking the factors influencing the increase in risk (Kwietniewski 2008, Shamsi 2005, Vemulapally 2010).

The GIS applications in enterprises are a convenient tool used in risk analysis. Using data visualization we can create maps based on the failure frequency of water supply network, the age of pipes and their current technical condition. Modern geoinformation systems are often complex systems, integrated with information tools such as water supply system monitoring or mathematical models (Studziński 2006, Studziński 2010, Zimoch 2006, Zimoch 2013).

The main aim of this work is to present the possibility of the use of information systems and the GIS databases in the risk analysis of failures of water pipes in water distribution subsystem, in selecting pipes with the highest number of failures which create the greatest risk of lack of water supply. The aim of the analysis and presentation of the results is to support the decision making process related to the improvement of activities of repair teams and to eliminate potential water pipes posing risk of further damage.

For the purpose of ensuring the safety of the water supply system the analysis of undesirable events that can result in a lack or reduction in the supply of water to consumers should be performed. In the analyses of the risk of failure of water pipes most commonly are used parametric matrices of risk assessment (Piegdoń 2012, Rak 2009, Tchórzewska-Cieślak 2011). Most often this process comprises the following steps (Tchórzewska-Cieślak 2011, Tchórzewska-Cieślak 2015):

– determination of the size of resources by determining the number of residents using public water supply,
– determination of the system susceptibility to undesirable events,
– the impact of threats on water consumers safety,
– determination of levels of risk and the analysis of tolerated, controlled and unacceptable risk.

Currently, an important element in the analysis and assessment of risk of failure is the possibility of using the GIS databases which significantly affect the degree of detail of the carried out calculations (Piegdoń 2013, Piegdoń 2014, Tchórzewska Cieślak 2014).

Opportunities that the GIS gives in the management of water supply infrastructure are huge. The use of databases in the systems of supplying water to consumers refers to the support investment and modernization processes and the use and operation of these systems. An important area of application in the process of operation is to assess the reliability and safety and the failure frequency of facilities and water supply network. The specificity of waterworks companies using the network infrastructure requires that the GIS not only archives and updates data on the status of the property but also shares functions and information related to the exclusion of the sections of the water supply network where the failure is being removed, the location of the failure with a description of its cause, location and the nature of the carried out repair, data on customers who may be deprived of water supply for the duration of the failure or even information about the repair team removing the failure. Dispersion of water supply infrastructure in the city or village causes that the question *where* the failure occurred is very important for the repair team. Information about the location of sections of the network, water supply connections and fittings is particularly important in the operation and effectiveness of the removal of the negative consequences of undesirable events. Equally important information is the data on the location of other networks (sewage, gas, etc.), which are valuable clue during repair. The GIS is often used to set priorities in the field of renovation (replacement, restoration, reconstruction) of water supply infrastructure.

There are few works on the use of the GIS to assess the risk of failure of water pipes and thus to ensure the required degree of safety of water supplies. Most often, databases are used as a tool to locate the failure. This action allows to perform further analysis related to the functioning and operation of the network, e.g. estimating the failure frequency of pipes working in the area of water supply and developing the strategies for network renovation (Choi et al. 1997, Zhang 2006).

The GIS users can get the information they require by asking simple attribute questions or complex spatial questions. Simple questions can concern e.g. indicating on the map the location of the failure which took place on a particular day at a particular time and showing appropriate attributes for the object (parameters description). It can be done by selecting in the created database the object we are interested in or its characteristic attributes e.g. identification number, diameter, etc. and the system will find the location of the object on the map. The user can also get information related to the analysed network area, for example, mark on the map those parts of the network made of cast iron in which there were failures in the particular year of operating.

More advanced level of the utility of the GIS applications can be obtained by performing a spatial analysis in the program, making assessment and presentation of results. It allows to obtain answers to such questions as, for example, which recipients will be deprived of water in case of failure of a particular section of the pipe.

After receiving the graphical presentation of the results of failure frequency analyses of water pipeline, the next step is to analyse and assess

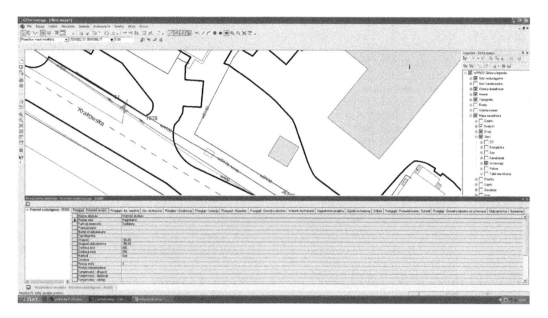

Figure 1. An exemplary fragment of a digital map with the locations of failures together with the characteristics of a water pipeline on which the failure occurred.

Figure 2. Layer of the main network with an exemplary site of the failure and buildings situated 250 m away (failure on the mains Ø400, material cast iron).

the risk of these pipes. In the quantitative matrix methods for all risk parameters are assigned the appropriate weight points (Boryczko 2013, Boryczko 2014, Piegdoń 2012, Pietrucha-Urbanik 2015, Rak 2006, Studziński 2014, Tchórzewska-Cieślak 2009, Tchórzewska-Cieślak 2010, Zimoch 2011). In order to estimate the level of risk of failure of water pipes we can use the three parametric risk matrix. On the basis of the available literature, the individual parameters of risk can be attributed to the criteria point scale (Rak 2009). The parameter of water pipe failure probability may be estimated based on the number of failures of the given pipe or on the basis of its failure rate. The parameter concerning the consequences/losses and susceptibility to the failure can be determined on the basis of the criteria of descriptive point scale which are available, among others, in the literature (Rak 2013, Tchórzewska-Cieślak 2011).

2 MATERIAL AND METHODS

The object of the research was the main water network and distribution water network, for which the analysis of the failure frequency was made and the necessary database for the visualization in the selected GIS program was created. Inhabitants of the city are supplied with water by boundary-chamber water intake with a capacity of 84000 m^3/d. The total length of the water supply system is 903.8 km (data from 2015). The main network, 49.8 km long, is made of iron and steel pipes. The distribution pipes are made of cast iron, steel, PE and PVC, with a total length of 530 kilometres. The water supply connections, with a length of 324 km, are mainly made of steel, cast iron, PE and PVC. Figure 3 shows the structure of the water supply system with division into water pipes functions (purple—main network, green—distribution network).

A proposal for a graphical visualization of the analysis of the failure frequency of water pipes means to create a database of failures and an orderly set of technical parameters of the water supply network. Due to the laborious process, the graphical presentation of the results was made for 2014.

The scope of spatial and text data describing the failure that had to be put in the GIS database included:

– location of failure,
– cause of failure,

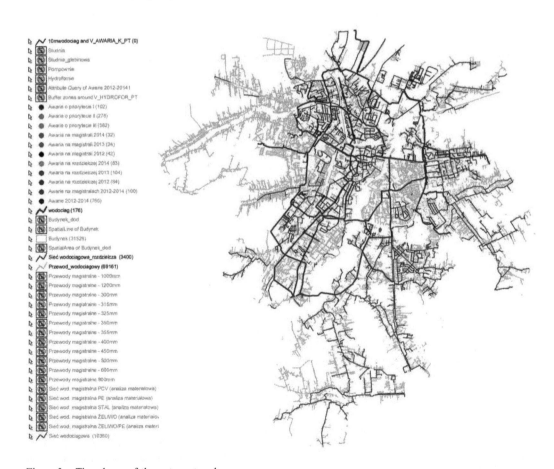

Figure 3. The scheme of the water network.

- type, diameter and material of pipe where failure occurred,
- the number of brigades that services given failure,
- the date of report about failure,
- assigning the identification number to the failure.

3 RESULTS

3.1 Pipeline failure analysis

In order to present the analysis of failure frequency of water supply pipelines using the GIS application in the first place the analysis of failure frequency of water supply network was made. The main source of data was the operating documentation of waterworks concerning the inventory, recorded failures, repairs and overhaul of damaged pipes.

Analysis of water pipes failure was made on the basis of actual operational data (10 years of exploitation of water network). The analysis required to collect, sort out and verify the data. Water supply network with water connections was analysed due to the type of water network (Table 1) and due to the reason of failure (Table 2).

Before conducting the failure pipes analysis, it is necessary to collect the data, sort them out and verify the data.

Currently, in many water companies the available data relating to failure of water pipes do not provide sufficient possibility of determining the exact parameters of reliable operation of pipes. These data are hardly legible, written in old books or kept only in the memory of workers. Such situation makes the analysis of the technical condition of the network on the basis of failure frequency

Table 1. Failure frequency for water supply network for failure data from 2005 to 2014.

	Type of water network			
	Main		Distribution	
Years	Length (km)	Number of failures	Length (km)	Number of failures
2005	49.5	48	350.5	87
2006	49.5	35	384.4	114
2007	49.5	40	443.5	90
2008	49.5	21	447.7	81
2009	49.8	32	468	72
2010	49.8	35	490.5	96
2011	49.8	45	504.1	92
2012	49.8	44	520.5	95
2013	49.8	24	524.8	103
2014	49.8	33	530	69

Table 2. The failure data of water supply network sorted out according to failure modes from 2005 to 2014.

	Type of damage				
Years	Unsealing	Fracture	Corrosion	Crack	Mechanical damage
2005	105	34	96	10	0
2006	107	47	118	24	2
2007	82	28	127	18	0
2008	71	30	102	15	0
2009	75	40	88	12	2
2010	76	32	127	19	1
2011	107	31	145	16	0
2012	95	32	145	11	0
2013	65	37	136	20	0
2014	69	25	119	11	0

inaccurate. As a result, actions aimed at maintenance planning, decision-making or any investment in water supply system are difficult. Therefore, it is necessary to present available data and to complete them in the GIS database.

3.2 Analysis presentation

Creating attribute and spatial questions in the program provides the possibility to present the results of the analysis in a graphical way. Figure 4 and 5 show the analysis of the failure frequency conducted in the GIS. Presentation of the results of the analysis of failure frequency of water supply network is possible to perform for every analysed year, as well as the presentation of the number of failures related to cause, diameter or pipe material. Visualization of the results is possible thanks to the formatting of size and colour of the "Failure" in accordance with the needs of the user. You can influence the size, colour and shape depending on the value of the descriptive attribute of "Failure".

Data entered in the GIS can be presented spatially (against a basic map) on the screen, in the form of a printout from the program or in the Internet program GoogleEarth.

Presented results of the analyses provide the possibility of a graphical presentation of analysis of the risk of failure of water supply pipes. On the basis of the results of analysis and assessment of the risk of failure of water supply pipes, it is possible to graphically select those pipes for which the risk of failure is at an unacceptable level and to create for them the priority service at the time of failure or priority for the renovation. The generated maps of the risk of failure analysis can be made for each type of pipe. It is possible to enrich them with reports and charts that—in an easy and clear way—would make water consumers aware of the technical condition of the water supply network, of pipes exposed to the great-

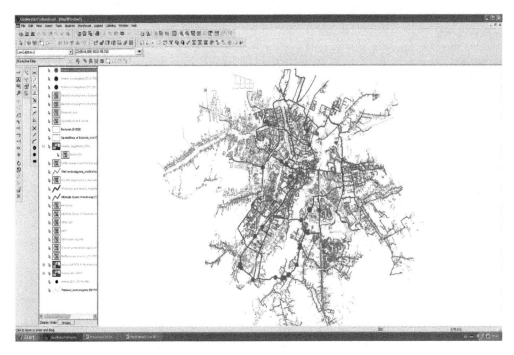

Figure 4. Presentation of the analysis of the failure frequency of the main water pipeline recorded in 2014.

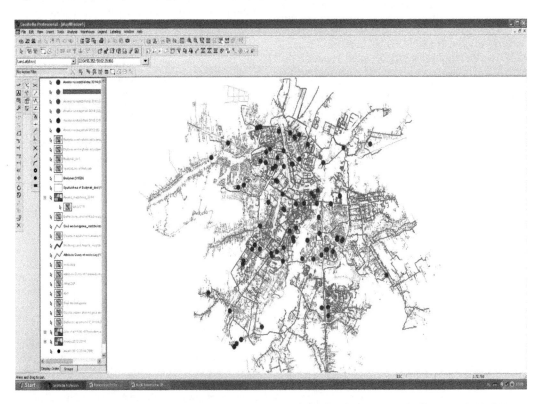

Figure 5. Presentation of the analysis of the failure frequency of the distribution water pipeline recorded in 2014.

Figure 6. Spatial visualization of the fragment of the main pipeline with failure (presentation in GoogleEarth).

est risk of failure and the buildings and the number of people exposed to a lack or reduction in water supply due to the failure in a specific pipe (Dadic et al. 2010, Kwietniewski 2008, Piegdoń 2014).

4 CONCLUSION

Implementation of the GIS allows to store increasing amounts of data, while providing them with the required level of safety. Accurate and complete databases are the basis for a more or less complex risk analysis in relation to the lack of water supply to consumers. Continuously updated data are the basis for the GIS applications.

Analysing the risk of failure of water pipes greatly increases the safety of water supply in the context of a water producer and water recipient. Graphical presentation of risk analysis offers the possibility to classify water pipes which are marked to be repaired or replaced as the first and whose failure would endanger consumers to long lack of water.

The system allows to make different kinds of risk analysis and statistics aimed at better planning of the investments and making interruptions in water supply caused by failures shorter. Thanks to the fast transfer of current and consistent data, it is possible to shorten the time of decision making and to make right decisions.

The GIS systems enable the effective management of many areas of activity in the company, among others, inventory of assets, planning of infrastructure development and management of work of repair teams. The use of information technology in the management of water supply network contributes significantly to the improvement of operational reliability and safety of water supply systems. In addition, monitoring of the network combined with the GIS software should be an important element in the analysis and assessment of risk of failure of water distribution subsystem.

REFERENCES

Boryczko, K., Piegdoń, I. & Eid, M. 2014. Collective water supply systems risk analysis model by means of RENO software, In P.H.A.J.M. Van Gelder, R.D.J.M. Steenbergen, S., Miraglia & A.C.W.M. Vrouwenvelder (eds), Safety, Reliability and Risk Analysis: Beyond the Horizon: 1987–1992. London: Taylor & Francis Group.

Boryczko K. & Tchórzewska-Cieślak A. 2013. Analysis and assessment of the risk of lack of water supply using the EPANET program. *Environmental Engineering IV*. D. M. R. Pawłowski L., Pawłowski A. London, Taylor & Francis Group: 63–68.

Choi, D. Y., Kim, J. H., et al. 1997. Pipeline planning in multi-regional water supply system—Focusing on the application GIS and MCA technique. *Proceedings of the Seventh International Conference on Computing in Civil and Building Engineering*, 1–4: 1555–1560.

Dadic, Z., Ujevic, M., et al. 2010. Integral Management of Water Resources in Croatia: Step Towards Water Security and Safety for All. *Threats to Food and Water Chain Infrastructure*: 131–140.

Gonzalez, F.C. 2000. Managing Anomalies and Customer Care with a GIS. *Water Supply* 18(4): 62–76.

Jaźwiński, J. & Smalko, Z. 2000. Problemy decyzyjne w inżynierii niezawodności. *XXVIII Zimowa Szkoła*

Niezawodności—Problemy decyzyjne w inżynierii niezawodności, Szczyrk.

Kaźmierczak, B. & Wdowikowski, M. 2016. Maximum Rainfall Model Based on Archival Pluviographic Records—Case Study for Legnica (Poland). *Periodica Polytechnica-Civil Engineering* 60(2): 305–312.

Kwietniewski, M. 2008. GIS w wodociągach i kanalizacji. Warszawa, Wyd. Naukowe PWN.

Kwietniewski, M., Miszta-Kruk K. & Wróbel K. 2008. Możliwości zastosowania GIS w wodociągach na przykładzie wybranego systemu dystrybucji wody. *Ochrona Środowiska* 29(3): 73–76.

Lewandowski, J. 2000. Aspekty decyzyjne zarządzania środkami trwałymi w przedsiębiorstwie. *XXVIII Zimowa Szkoła Niezawodności—Problemy decyzyjne w inżynierii niezawodności*, Szczyrk.

Michael, G. & Zhang, J. 2009. Simplified GIS for Water Pipeline Management. *International Conference Pipelines: Infrastructure's Hidden Assets*, San Diego, American Society of Civil Engineers.

Piegdoń I. & Tchórzewska-Cieślak B. 2012. Matrix analysis of risk of interruptions in water supply in terms of consumer safety. *Journal of KONBiN*. Warszawa, Wydawnictwo Instytutu Technicznego Wojsk Lotniczych. 4: 125–140.

Piegdoń I. & Tchórzewska-Cieślak B. 2013. Zarządzanie ryzykiem w przedsiębiorstwach wodociągowych z wykorzystaniem nowoczesnych systemów informatycznych. *Gaz, Woda i Technika Sanitarna* 10: 401–404.

Piegdoń I. & Tchórzewska-Cieślak B. 2014. Methods of visualizing the risk of lack of water supply. *Proceedings of the European Safety and Reliability Conference, ESREL 2014*, Taylor & Francis Group.

Pietrucha-Urbanik, K. 2015. Failure analysis and assessment on the exemplary water supply network. *Engineering Failure Analysis* 57: 137–142.

Rak, J. 2009. Bezpieczna woda wodociągowa. Zarządzanie ryzykiem w systemie zaopatrzenia w wodę. Rzeszów, Oficyna Wydawnicza Politechniki Rzeszowskiej.

Rak, J. 2009. Selected problems of water supply safety. *Environment Protection Engineering* 35(2): 23–28.

Rak, J. & Kwietniewski, M. 2011. Bezpieczeństwo i zagrożenia systemów zbiorowego zaopatrzenia w wodę. Rzeszów, Oficyna Wydawnicza Politechniki Rzeszowskiej.

Rak, J., Kwietniewski, M., et al. 2013. Metody oceny niezawodności i bezpieczeństwa dostawy wody do odbiorców. Rzeszów, Oficyna Wydawnicza Politechniki Rzeszowskiej.

Rak, J. & Tchórzewska-Cieślak, B. 2013. Ryzko w eksploatacji systemów zbiorowego zaopatrzenia w wodę. Warszawa, Wydawnictwo Seidel-Przywecki Sp. z o.o.

Shamsi, U.M. 2002. GIS Tools for Water, Wastewater, and Stormwater Systems. New York, American Society of Civil Engineers.

Shamsi, U.M. 2005. GIS Applications for Water, Wastewater, and Stormwater Systems. London, CRC Press.

Sherer P. T. 1995. Geographical Information System. *Journal of Water Supply Research and Technology: AQUA* 44(3): 118–124.

Studziński, A. 2014. Amount of labour of water conduit repair, In P.H.A.J.M. Van Gelder, R.D.J.M. Steenbergen, S. Miraglia, & A.C.W.M. Vrouwenvelder (eds), Safety, Reliability and Risk Analysis: Beyond the Horizon: 2081–2084. London: Taylor & Francis Group.

Studziński, J. 2010. Narzędzia informatyzacji miejskich sieci wodociągowych. *Wodociągi Kanalizacja* 7(75): 34–37.

Studziński, J. & Bogdan, L. 2006. Informatyczny system wspomagania decyzji do zarządzania, sterowania operacyjnego i planowania miejskiego systemu wodno-ściekowego. *Badania Systemowe* 49: 149–157.

Tchórzewska-Cieślak, B. 2009. Water supply system reliability management. *Environment Protection Engineering* 2: 29–35.

Tchórzewska-Cieślak, B. 2010. Water consumer safety in water distribution system. *Environmental Engineering III*. D. M. R. Pawłowski L., Pawłowski A. London, Taylor & Francis Group: 527–532.

Tchórzewska-Cieślak, B. 2011. Matrix method for estimating the risk of failure in the collective water supply system using fuzzy logic. *Environment Protection Engineering* 37: 111–118.

Tchórzewska-Cieślak, B. 2011. Metody analizy i oceny ryzyka awarii podsystemu dystrybucji wody. Rzeszów, Oficyna Wydawnicza Politechniki Rzeszowskiej.

Tchórzewska Cieślak, B., Piegdoń, I. & Boryczko, K. 2014. Wykorzystanie nowoczesnych technik informatycznych oraz baz danych w analizach ryzyka awarii podsystemu dystrybucji wody. *INSTAL*(6): 76–79.

Tchórzewska-Cieślak, B. & Szpak, D. 2015, A Proposal of a Method for Water Supply Safety Analysis and Assessment. *Ochrona Środowiska* 37(3): 43–47.

Vaughan, R. A. & Kirby, Roger P. 1988. Geographical information system and remote sensing for local resource planning: an introduction for potential users in local authorities and other public sector bodies. Dundee, Remote Sensing Products and Publications.

Vemulapally, R. 2010. Development of Standard Geodatabase Model and its Applications for Municipal Water and Sewer Infrastructure. Blacksburg, Faculty of the Virginia Polytechnic Institute and State University.

World Health Organization, 2005. Water Safety Plans. Managing drinking-water quality from catchment to consumer. Geneva, Water, Sanitation and Health. Protection and the Human Environment World Health Organization.

World Health Organization, 2013. Water Quality and Health Strategy 2013–2020. Geneva, WHO Press.

Zhang, T. 2006. Application of GIS and CARE-W Systems on Water Distribution Networks, Skärholmen. Sztokholm, Royal Institute of Technology.

Zimoch, I. 2006. Zastosowanie technik komputerowych w analizie niezawodnościowej funkcjonowania systemu zaopatrzenia w wodę w sytuacjach awaryjnych. *XIX Krajowa Konferencja VII Międzynarodowa Konferencja „Zaopatrzenie w wodę, jakość i ochrona wód"*, Poznań—Zakopane, PZITS O/Wielkopolski.

Zimoch, I. 2011. Zintegrowana metoda analizy niezawodności funkcjonowania i bezpieczeństwa systemów zaopatrzenia w wodę. Gliwice, Wydawnictwo Politechniki Śląskiej.

Zimoch, I. & Paciej, J. 2013. Zastosowanie Geograficznych Systemów Informacyjnych w zarządzaniu oraz prowadzeniu kontroli jakości wody przeznaczonej do spożycia przez ludzi. *INSTAL* 1: 38–41.

Environmental Engineering V – Pawłowska & Pawłowski (Eds)
© 2017 Taylor & Francis Group, London, ISBN 978-1-138-03163-0

Method for forecasting the failure rate index of water pipelines

K. Boryczko & A. Pasierb
Faculty of Civil and Environmental Engineering and Architecture, Rzeszow University of Technology, Rzeszow, Poland

ABSTRACT: The paper presents a method of forecasting the failure rate index of water pipelines. The proposed method uses moving average, seasonal index and the centered average value to the forecast. By using the actual operating database, the proposed method makes it possible to estimate the prospective demand for materials and equipment for removal of damage to the water supply network, as well as to plan the work schedule repair teams. Weather can also be helpful when planning the exchange of parts of the water supply network.

Keywords: forecasting, failure rate, water supply system, pipelines

1 INTRODUCTION

Failure frequency of water distribution subsystem is one of the major operational problems in delivering water to consumers. Reducing water losses caused by failures of water supply is an important issue that should be solved effectively because of higher and higher cost of water intake, water treatment and delivery to consumers. The treated water must comply with precise conditions of water quality. It should also be delivered to the users in a specific quantity and under appropriate pressure. The fulfilment of quantitative objectives can be achieved by identifying customers' needs for long-time operation of water supply. The necessary pressure, however, may be ensured by appropriate facilities that are adapted to the expected situation. In addition to the two above-mentioned requirements which should be met by the Collective Water Supply System (CWSS), yet another element is increasingly often enumerated, that is the need to ensure the optimal delivery costs, which can be done more effectively by ensuring the lowest possible failure rate index of water pipeline (Biedgunis et al. 2007; Budziło 2010; Gandy 1997; Kleiner et al. 2006; Kwietniewski 2006; Kwietniewski 2004; Lindhe et al. 2009; Magelky 2009; Rak 2007; Rak & Pietrucha 2008; Sadiq et al. 2004; Sadiq et al. 2008, Szkarowski & Janta-Upińska 2015).

The analysis of the CWSS safety and reliability remains therefore invariably, for more than 30 years, the subject of research (Kowalski 2011; Rak et al. 2013; Zimoch 2009). Reliability analysis is also directly related to the detailed research on the failure rate of water distribution subsystem, in order to find the most common causes of failures and determine the most effective methods to eliminate them. Despite extensive studies conducted so far, the need to continue these studies in the future because of the constant changes in the structure and age of water pipes, is also stressed. The necessity of conducting research on the reliability and failure rate results from the fact that they are a key indicator of proper operation of water supply systems (Bajer 2007; Bajer 2013; Bergel & Pawełek 2008; Boryczko et al. 2014; Boryczko & Tchórzewska-Cieślak 2014; Dąbrowski 2006; Iwanejko & Bajer 2009; Ke et al. 2016; Kowalski & Miszta-Kruk 2013; Kruszyński & Dzienis 2008; Li et al. 2016; Miksch et al. 2015; Pietrucha-Urbanik 2013; Pietrucha-Urbanik 2014; Rabczak 2016; Studziński 2014; Szymura & Zimoch 2014; Tchórzewska-Cieślak 2012; Tchórzewska-Cieślak et al. 2015; Tchórzewska-Cieślak & Szpak 2015, Czapczuk et al. 2015).

In connection with the different failures that occur during the operation of the network and cause more or fewer limitations in water supply, the operator managing proper operation of the network should take a number of steps to reduce the impact of the undesirable event (Iwanek et al. 2014; Iwanek et al. 2016; Zimoch 2012). These steps, however, can be taken only after the development of appropriate failure analysis and on the basis of extensive knowledge of the given water supply network. An additional tool to support CWSS operators may be forecasts of failure frequency of water supply network, through which they can predict the required future state of spare parts in a warehouse and create a general schedule of work for a renovation and repair team (Kleiner & Rajani 2001).

The aim of the study is to assess the failure frequency of water supply network located in a city with 34 thousand inhabitants and to forecast the failure rate index using method of centred moving average and seasonal indicator with irregularities.

2 FORECASTING

The term "forecasting" is defined as the ability to predict the future in a rational way, based on the scientific methods. The process of forecasting is based on theoretical studies, analytical considerations, logical assumptions and experiences that become the basis for more and more dynamically developing statistical prediction theory (Birek et al. 2014; Chen et al. 2005; Kutyłowska 2015; Lajevardi & Minaei-Bidgoli 2008; Taigbenu & Ilemobade 2007; Yang et al. 2006).

Classification of predictions can be performed in view of the period for which it was built, that is the so-called prediction horizon, according to which we distinguish:

- Direct forecast not exceeding 1 month
- Short-term forecast covering the period from 1 to 3 months
- Medium-term forecast not exceeding two years
- Long-term forecast built for a period covering more than two years.

Forecasts can also be divided into qualitative and quantitative. We speak about the quantitative forecast when the state of forecast variable was expressed by the number, while the qualitative forecast is the one whose state is a single value or a vector of numbers. Moreover, one can distinguish the disposable and repeatable, comprehensive and sequential, and self-fulfilling and destructive forecasts.

Some prediction postulates mentioned in the literature are:

- Constructing forecasts along with the measure of the order of accuracy
- Striving for highly efficient prediction, which would allow to achieve possibly the most accurate value of the measure of the order of prediction accuracy.

The choice of forecasting model should be based firstly on a thorough analysis of statistical material, which must be subjected to thorough graphical and quantitative processing with the use of the measures adapting the model to the interpreted statistical data (Zeliaś & Wanat 2003).

The first method used in the course of forecasting is the one based on classical trend models. The trend is referred to as a long term tendency to changes going in one direction. In this method it is necessary to accurately determine the form of the function that describes the trend and agree the way in which separated components of the time series overlap each other. In the model of development tendency, the explanatory variable is a time variable which is the number of the period or time moment "t" (Kiełtyka & Nazarko 2005). This method, in addition to determining the trend, also needs to take into account random fluctuations and periodic fluctuations. For this purpose, the absolute seasonal fluctuations or seasonal indicators should be determined for the individual phases of the cycle. After determining the impact of a seasonal factor, taking into account separated regularities one can start building the forecast for the course of the examined process in the future.

Another method is forecasting based on the adaptation models. Among them, the method of moving averages used for short-term forecasting should be mentioned at first. In this method, the forecast is calculated as the arithmetic mean of the real values of the variable from a given time interval, which is called the smoothing interval. There are two types of the moving average: Simple and weighted. The simple moving average is calculated according to the formula:

$$y = \frac{1}{k} \sum_{t=n-k+1}^{n} y_t \qquad (1)$$

where k = the size of the interval for which the moving average is calculated [–]; y = forecast for variable determined at time t [–]; t = number of the period or moment in time [–]; n = number of terms of time series [–]; y_t = consecutive term of the time series in period t [–].

The weighted moving average method is expressed by the formula:

$$y = \sum_{t=n-k+1}^{n} y_t \cdot w_t \qquad (2)$$

where w_t = the weight given by the investigator [–].

Another adaptive method is the exponential equalization method which is used when the values of dependent variable are constant or change on a regular basis. The essence of this method is to put out a time series using the weighted moving average, whose weights were determined under the law of exponential. The next adaptive method is the naïve method that, similarly to the moving average method, is used in short-term forecasting. It is based on the assumption that the forecast takes the value of the last known realization of the analysed variable in the given one nominal period. It is used most often to compare the accuracy of the forecasts which were built using a different, more complex method. Adaptive methods are also the exponential autoregressive smoothing method, the Holt method (for the time series without seasonal fluctuations), the Winters method (for time series superimposed on the trend in a multiplicative or additive way) and the forecasting method based on creeping trend with harmonic weights (Zieliaś & Wanat 2003).

The method commonly used for the prediction is the prediction based on the linear econometric models. The linear model is an example of an econometric single equation action which can be used to provide links between economic phenomena that have been analysed.

Among the forecasting methods, an important role is also played by the autoregressive models. These models can define the relationship between the values of forecast variable at time t and the values of this variable at preceding moments, for example, t-1, t-2, etc.

The last important method for predicting phenomena are forecasts prepared by analogy, among which we distinguish:

- The biological analogy—involving the transfer of biological knowledge on other objects,
- The spatial analogy—that allows to predict the appearance of new events that relate to the existing events,
- The historical analogy,
- The spatial analogy—temporal analogy utilizing knowledge of changes in the analysed process in a long-term time intervals.

The analog methods are mainly used to predict changes in the relationship between variables, to predict the form and turning points of the trend, as well as to predict the possibility of new events in the object. The essence of forecasting using the analog methods lies in predicting the analysed variable which course at the preceding period is similar to the course of the variable during the period for which is the forecast.

2 BRIEF CHARACTERISTICS OF THE ANALYSED COLLECTIVE WATER SUPPLY SYSTEM

Water for the city is taken from the river thanks to the waterside pipeline intake with a capacity of 17,280 m^3/d. Its integral part is a weir through which water is piled up to a height of 1 m and which provides a guaranteed depth of water for the Water Treatment Plant. In the city there is also a reserve underground intake with a capacity of 348 m^3/d.

There are two tanks located at the end of the network, each with a capacity of 900 m^3 and two initial tanks, the total capacity of which is 5000 m^3.

The primary task of the tanks located at the beginning of the network is to collect treated clean water. These tanks, together with the tanks located at the end of the network, provide daily water supply to the city. Moreover, the initial tanks stabilize the operation of water treatment plant and water supply system and are a reservoir in case of temporary threats caused by water pollution. Large capacity of the tanks also allows alignment of the daily operation of the water treatment plant while keeping the possibility of unlimited work of high pressure water pumping station.

Water treated during the process described above flows through pipelines DN 300 and is then sucked by pumps in the third stage pumping station and pumped to the municipal network.

The water supply network has been made in the mixed system. It consists only of the distribution pipes and water supply connections (there is no main network in the city). It is estimated that at the end of 2015, the network delivered water to 34000 residents of the city. The length of the water supply network in the last five years is shown in Table 1.

The water supply network is made of, among others, steel, cast iron, polyethylene, polyvinyl chloride and asbestos (until 2013).

The owner of the water pipeline does not have data on the exact material structure of the water supply connections. However, a material structure of the distribution network has been documented:

- Steel 19.5 km
- Cast iron 6.1 km
- PE-HD 80.2 km
- PVC-U 44.6 km.

Most water pipes are up to 10 years old (81.8 km), however, the decrease in the length of distribution pipelines older than 25 years can be seen, while in 2011–2015 there was the increase (21.5 km) of the length of pipelines with the operating period up to 10 years.

Table 1. The length of water supply network in 2011–2015.

Rok	Podłączenia domowe	Przewody rozdzielcze
	km	km
2011	60	133.9
2012	60	138.4
2013	60	140.6
2014	60	148.6
2015	60	150.4

3 FORECAST OF THE FAILURE RATE INDEX FOR THE SELECTED WATER SUPPLY NETWORK

In order to evaluate the failure frequency of the linear sewage and water supply infrastructure we use the failure rate index which is defined as:

$$\lambda(\Delta t) = \frac{n(\Delta t)}{L \cdot \Delta t} \quad (3)$$

where $\lambda(\Delta t)$ = the unit failure rate (constant within the time interval Δt) [failures/(km·year)]; $n(\Delta t)$ = number of failures within the time interval Δt [–]; L = the length of the examined pipelines within the time interval Δt (the average length of the pipelines in this interval); Δt = the time interval as the reference period for which failure rate was calculated [year].

This parameter indicates the technical state of pipelines, allows for proper planning of investments relating to the repairs and replacement of transmission pipelines. It is assumed that the lower the failure rate index, the higher the system reliability and vice versa. Moreover, the unit failure rate is considered to be very comfortable indicator for comparing failure rate of pipelines made of different materials, pipes with different functions, different age groups, pipes arranged in different grounds and operating at varying pressure and different temperatures of ground and transferred water.

The forecast was carried out according to the following steps. The first step is to present the values of the failure rate index in a graphical way (Fig. 1).

Historical values of the failure rate index were visualized using a line chart with markers. The blue line in the graph clearly shows random and periodic fluctuations of the index. These fluctuations are directly related to repeated changes in the forecast variable which are of similar sizes at regular intervals. Other elements which we can see in the figure are irregular elements, impossible to be adapted to any pattern, generated at random, which are inherent elements in each group of data.

The next step of the forecast is alignment of time series by using the moving average for 12 periods (corresponding to 12 months at each time interval), calculated according to the formula (2).

By using the moving average we obtained values which are the effects only of main factors, while eliminating random variations. Because of an even number of periods, the moving average cannot be related to any moment which occurs in a given time series, but only to an intermediate period (between the central terms of the given alignment interval). Therefore, it is necessary to calculate the centred moving average which directly corresponds to the centre of the period for which the alignment was carried out. In case of an odd number of periods, calculating the centred moving average would be pointless. The centred moving average was determined by averaging the two successive values of the moving average (column MA (12) – Tab. 4) and its values are listed in the column CMA (12). The values of the centred moving average are marked in the Figure 1. The line determined by the centred moving average is devoid of seasonality and other irregularities. The smoothed line of the centred

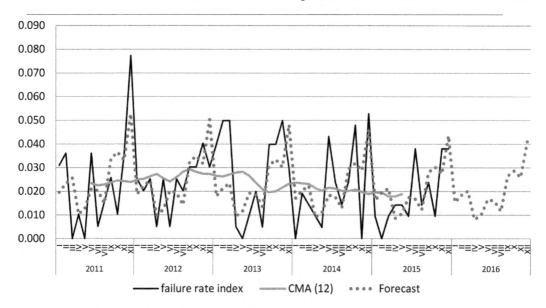

Figure 1. The failure rate index for the analysed city with forecast.

moving average is shorter than the original series by k-1 observations (where k—the size of the interval for which the moving average was calculated).

In connection with the smoothing of data by determining the moving average and the centred moving average, the next step of forecasting will be to extract the seasonality indicator (S_t) and the irregularity indicator (I_t) from the available data. Seasonal and irregular components are two of the three elements of the classical multiplicative model, where the third factor is the trend function. Thus, the multiplicative model can be written with the following formula:

$$Y_t = S_t \cdot I_t \cdot T_t \qquad (4)$$

where Y_t = the real value of the analysed variable, i.e. the value of the failure rate index at time t [−]; S_t = the seasonal index [−]; I_t = the irregular component [−]; T_t = the trend component [−].

The separation of seasonal and irregular components is made by dividing the real value of the analysed variable by the centred moving average (column CMA (12)—Tab. 4). This equation can be written using the following formula:

$$S_t I_t = \frac{Y_t}{CMA(12)} \qquad (5)$$

Where S_t, I_t = the seasonal indicator with irregularities; Y_t = the real value of the analysed variable, i.e. the value of the failure rate index at time t [−]; CMA (12) = the value of the centred moving average [−].

The values in the column $S_t I_t$ being a result of the above actions should be interpreted as follows:

- When the value from the column $S_t I_t$ is greater than unity, e.g. 1.543, it means that the seasonal index along with irregularities were 54.3% higher than the centred moving average.
- When the value from the column $S_t I_t$ is less than unity, e.g. 0.229, it means that the seasonal index along with irregularities was 77.1% below the centred moving average.

In the next stage of forecasting, only the seasonal indicator S_t was separated.

The Table 2 contains data without the irregularity (irregularity was removed by calculating the average for each period i.e. a month, separately, in other words, the mean calculated for one nominal months). For example for month 3 is mean of $S_t I_t$ values: 0.959, 1.841, 0.623, 0.532 equals 0,989.

Interpretation of the seasonal index is analogous as for the situation when the seasonal index and irregularities occur together. The next step of the forecast is the deseasonalization of output data:

Table 2. The seasonal index for the particular months.

Month	S_t
1	0.751
2	0.899
3	0.989
4	0.403
5	0.496
6	0.825
7	0.794
8	0.585
9	1.336
10	1.449
11	1.317
12	2.108

Table 3. The number of failures of water supply network in 2011–2015.

	House connections	Distribution network
Year	Failures	Failures
2011	28	27
2012	18	38
2013	30	38
2014	15	38
2015	14	32

$$Des = \frac{Y_t}{S_t} \qquad (6)$$

where Des = the deseasonalized output data [−].

The last stage before the final forecast is determining the trend component. It can be done by means of the simple linear regression using the deseasonalized data (variable y—the explained variable, dependent) and the variable t. The final value of the forecast (Fore) both for the historical data and for 2015 is the result of the following equation:

$$Fore = S_t \cdot T_t \qquad (7)$$

where $Fore$ = the definitive, final value of the forecast [−].

On the basis of the presented data it can be stated that in 2011–2015 a total of 278 failures (an average of 4.6 per month) were recorded. Within five years in the connections of water supply there were 105 failures, while in the distribution network there were 173 failures, which is illustrated in Table 3.

19

On the basis of the data from Table 2 it can be concluded that the number of failures of the distribution network over the past five years remained stable (38 failures) at a slightly lower number of failures in 2011 (27 failures) and 2015 (32 failures). Larger differences in the number of failures are noticeable in water supply connections, where in 2014 the number of failures decreased by half in comparison with 2013.

Table 4. Forecast of the failure rate index for the analysed city.

Year month	Number of failures	Failure rate index	MA (12)	CMA (12)	S_t, I_t	S_t	Deseasonalized	T_t	Forecast
2011 I	6	0.031				0.751	0.041	0.0259	0.019
II	7	0.036				0.899	0.041	0.0258	0.023
III	0	0.000				0.989	0.000	0.0257	0.025
IV	2	0.010				0.403	0.026	0.0256	0.010
V	0	0.000				0.496	0.000	0.0255	0.013
VI	7	0.036	0.024	0.023	1.543	0.825	0.044	0.0255	0.021
VII	1	0.005	0.023	0.022	0.229	0.794	0.006	0.0254	0.020
VIII	3	0.015	0.022	0.023	0.676	0.585	0.026	0.0253	0.015
IX	5	0.026	0.024	0.024	1.088	1.336	0.019	0.0252	0.034
X	2	0.010	0.023	0.025	0.420	1.449	0.007	0.0251	0.036
XI	7	0.036	0.026	0.024	1.486	1.317	0.027	0.0250	0.033
XII	15	0.077	0.023	0.024	3.245	2.108	0.037	0.0249	0.052
2012 I	5	0.025	0.025	0.025	1.013	0.751	0.034	0.0248	0.019
II	4	0.020	0.025	0.025	0.798	0.899	0.023	0.0247	0.022
III	5	0.025	0.025	0.026	0.959	0.989	0.025	0.0246	0.024
IV	1	0.005	0.027	0.027	0.185	0.403	0.013	0.0245	0.010
V	5	0.025	0.027	0.025	0.989	0.496	0.051	0.0245	0.012
VI	1	0.005	0.024	0.024	0.209	0.825	0.006	0.0244	0.020
VII	5	0.025	0.025	0.026	0.970	0.794	0.032	0.0243	0.019
VIII	4	0.020	0.027	0.028	0.714	0.585	0.034	0.0242	0.014
IX	6	0.030	0.029	0.029	1.033	1.336	0.023	0.0241	0.032
X	6	0.030	0.029	0.028	1.072	1.449	0.021	0.0240	0.035
XI	8	0.040	0.027	0.027	1.473	1.317	0.031	0.0239	0.031
XII	6	0.030	0.028	0.027	1.105	2.108	0.014	0.0238	0.050
2013 I	8	0.040	0.027	0.027	1.504	0.751	0.053	0.0237	0.018
II	10	0.050	0.026	0.026	1.897	0.899	0.057	0.0236	0.021
III	10	0.050	0.027	0.027	1.841	0.989	0.050	0.0235	0.023
IV	1	0.005	0.027	0.028	0.179	0.403	0.012	0.0235	0.009
V	0	0.000	0.028	0.028	0.000	0.496	0.000	0.0234	0.012
VI	2	0.010	0.028	0.027	0.375	0.825	0.012	0.0233	0.019
VII	4	0.020	0.025	0.024	0.843	0.794	0.025	0.0232	0.018
VIII	1	0.005	0.022	0.021	0.239	0.585	0.009	0.0231	0.014
IX	8	0.040	0.019	0.020	2.034	1.336	0.030	0.0230	0.031
X	8	0.040	0.020	0.020	1.994	1.449	0.028	0.0229	0.033
XI	10	0.050	0.020	0.022	2.310	1.317	0.038	0.0228	0.030
XII	6	0.030	0.023	0.023	1.293	2.108	0.014	0.0227	0.048
2014 I	0	0.000	0.023	0.024	0.000	0.751	0.000	0.0229	0.017
II	4	0.019	0.024	0.023	0.819	0.899	0.022	0.0225	0.020
III	3	0.014	0.023	0.023	0.623	0.989	0.015	0.0225	0.022
IV	2	0.010	0.023	0.021	0.449	0.403	0.024	0.0224	0.009
V	1	0.005	0.019	0.020	0.237	0.496	0.010	0.0223	0.011
VI	9	0.043	0.021	0.022	2.000	0.825	0.052	0.0222	0.018
VII	5	0.024	0.022	0.021	1.132	0.794	0.030	0.0221	0.018
VIII	3	0.014	0.020	0.020	0.713	0.585	0.025	0.0220	0.013
IX	5	0.024	0.020	0.020	1.189	1.336	0.018	0.0219	0.029
X	10	0.048	0.020	0.021	2.311	1.449	0.033	0.0218	0.032
XI	0	0.000	0.021	0.020	0.000	1.317	0.000	0.0217	0.029
XII	11	0.053	0.018	0.019	2.787	2.108	0.025	0.0216	0.046

(*Continued*)

Table 4. (*Continued*).

Year month	Number of failures	Failure rate index	MA (12)	CMA (12)	S_t, I_t	S_t	Deseasonalized	T_t	Forecast
2015 I	2	0.010	0.020	0.020	0.487	0.751	0.013	0.0215	0.016
II	0	0.000	0.019	0.019	0.000	0.899	0.000	0.0215	0.019
III	2	0.010	0.019	0.018	0.532	0.989	0.010	0.0214	0.021
IV	3	0.014	0.016	0.018	0.798	0.403	0.035	0.0213	0.009
V	3	0.014	0.019	0.019	0.757	0.496	0.029	0.0212	0.010
VI	2	0.010	0.018			0.825	0.012	0.0211	0.017
VII	8	0.038				0.794	0.048	0.0210	0.017
VIII	3	0.014				0.585	0.024	0.0209	0.012
IX	5	0.024				1.336	0.018	0.0208	0.028
X	2	0.010				1.449	0.007	0.0207	0.030
XI	8	0.038				1.317	0.029	0.0206	0.027
XII	8	0.038				2.108	0.018	0.0205	0.043
2016 I						0.751		0.0204	0.015
II						0.899		0.0204	0.018
III						0.989		0.0203	0.020
IV						0.403		0.0202	0.008
V						0.496		0.0201	0.010
VI						0.825		0.0200	0.017
VII						0.794		0.0199	0.016
VIII						0.585		0.0198	0.012
IX						1.336		0.0197	0.026
X						1.449		0.0196	0.028
XI						1.317		0.0195	0.026
XII						2.108		0.0194	0.041

The deseasonalized output data are listed in Table 4 in the column Deseasonalized. Assuming that in the linear regression the dependence of variables can be described by a straight line and the equation:

$$T_t = \beta_1 \cdot t + \beta_0 \qquad (8)$$

where β_1 = is the directional coefficient; β_0 = the intercept (the point of intersection with the axis of ordinates).

Linear regression, which was determined using the MS Excel software resulted in:

$$T_t = -0,000091 \cdot t + 0,026 \qquad (9)$$

where T_t = the trend component, the explained variable, predicted [−].

The coefficient of determination R^2 is 0.012.

Determined equation (9) Tt is intended only to indicate the general trend of changes in the value of failure rate index. It does not have to describe seasonal changes (for such purpose seasonal index S_t is used).

The predicted values are shown in the column Forecast (Tab. 4). Numeric values of failure rate index and forecast are presented in failures/(km·month).

In the forecast above, along with the determining the impact of seasonal factor on the formation of the unit failure rate index, we used detected regularities to build the forecast for the further course of the analysed process (2016). On the basis of the carried out forecast it must be noted that the failure rate index in 2016 will be the lowest in April (λ = 0.008 failures/(km·month)) and May (λ = 0.010 failures/(km·month)). In October (λ = 0.028 failures/(km·month)) and December (λ = 0.041 failures/(km·month)) the values of the failure rate index will be the highest.

Comparing the forecast with the actual data (2011–2015) must be noted that the forecast satisfactorily indicated periods (months) with lower failure rate index. The forecast indicates that the spring months should be characterized by low failure rate index, and the growth occurs during the early winter.

The verification of the results of the analysis with the real data will be made at the beginning of 2017.

4 CONCLUSIONS

The presented method is a model that allows to forecast the failure rate index on the basis of data that are held by the majority of water companies and therefore may be implied there.

Forecasting involves the continuous and meticulous creation of databases about failures and accurate description of the parameters that may have an impact on the future value of the index.

Method of centred moving average and seasonal indicator with irregularities is simple method that involves only one parameter—time to create forecast. Most of water companies have failure data bases which could be used to create forecast with presented method. Only the study taking into account more factors that may influence the failure rate index and a comparison the forecast for next period (for example year) with the actual data will answer the question if method described in the article is effective through its simplicity.

Authors plan to expand the methods of forecasting taking into account other factors affecting the failure rate of water pipelines like age of pipe, material, ground temperature changes. Only the study taking into account those factors that may influence on failure rate index and a comparison of forecast with the actual data will answer the question if described in the article method is effective through its simplicity.

REFERENCES

Bajer, J. 2007. Reliability analysis of variant solutions for water pumping stations, In M.R. Dudzińska, L. Pawłowski & A. Pawłowski (eds), *Environmental Engineering*: 253–261. New York, Singapore: Taylor & Francis Group.

Bajer, J. 2013. Ecomical and reliability criterion for the optymization of the water supply pumping stations designs, In M. R Dudzińska, L. Pawłowski & A. Pawłowski (eds), *Environmental Engineering IV*: 21–28. London: Taylor & Francis Group.

Bergel, T. & Pawełek, J. 2008, Quantitative and economical aspects of water loss in waterworks systems in rural areas. *Environment Protection Engineering* 3: 59–64.

Biedgunis, S., Smolarkiewicz, M., Podwójci, P. & Czapczuk, A. 2007, Mapa ryzyka funkcjonowania rozległych systemów technicznych. *Rocznik Ochrony Środowiska* 9: 303–311.

Birek, L., Petrovic, D., & Boylan, J. 2014. Water leakage forecasting: the application of a modified fuzzy evolving algorithm. Applied Soft Computing, 14: 305–315.

Boryczko, K. & Tchórzewska-Cieślak, B. 2014, Analysis of risk of failure in water main pipe network and of developing poor quality water. *Environment Protection Engineering*, 40(4): 77–92.

Boryczko, K., Piegdoń, I. & Eid, M. 2014. Collective water supply systems risk analysis model by means of Reno software, In P.H.A.J.M. Van Gelder, R.D.J.M. Steenbergen, S., Miraglia & A.C.W.M. Vrouwenvelder (eds), *Safety, Reliability and Risk Analysis: Beyond the Horizon*: 1987–1992. London: Taylor & Francis Group.

Budziło, B. 2010. *Niezawodność wybranych systemów zaopatrzenia w wodę w południowej Polsce.* Kraków: Wydawnictwo Politechniki Krakowskiej.

Chen, W.X., Nie, B.S. & He, X.Q. 2005. Application of fault tree and artificial neural network on mechanical gearbox gear wheel failure predication of forklift truck. *Progress in Safety Science and Technology, Vol V, Pts a and B*, 5: 586–589.

Czapczuk, A., Dawidowicz, J., Piekarski, J. 2015. Metody sztucznej inteligencji w projektowaniu i eksploatacji systemów zaopatrzenia w wodę. *Rocznik Ochrona Środowiska*, 17(2), 1527–1544.

Dąbrowski, W. 2006. O współczesnych metodach zarządzania rewitalizacją sieci wodociągowych. *Forum Eksploatatora* 1(22): 26–30.

Gandy, M. 1997. The making of a regulatory crisis: restructuring New York City's water supply. *Transactions of the Institute of British Geographers* 22(3): 338–358.

Iwanejko, R. & Bajer, J. 2009. Determination of the optimum number of repair units for water distribution systems. *Archives of Civil Engineering*, 55(1): 87–101.

Iwanek, M., Kowalska, B., Hawryluk, E. & Kondraciuk, K. 2016. Distance and time of water effluence on soil surface after failure of buried water pipe. Laboratory investigations and statistical analysis. *Eksploatacja i niezawodnosc—Maintenance and Reliability*. 18(2): 278–284.

Iwanek, M., Kowalski, D., Kowalska, B., Hawryluk, E. & Kondraciuk, K. 2014. Experimental investigations of zones of leakage from damaged water network pipes. In: C.A. Brebbia, S. Mambretti (eds.): *WIT Transactions on The Built Environment 139, Urban Water II: 257–268. Boston*: WIT Press Southampton.

Ke, W., Leia, Y., Shaa, J., Zhangb, G., Yana, J., Lind, X. & Pana X. 2016. Dynamic simulation of water resource management focused on water allocation and water reclamation in Chinese mining cities. *Water Policy* 18(1): 1–18.

Kiełtyka, L. & Nazarko, J. 2005. *Technologie informatyczne i prognozowanie w zarządzaniu. Wybrane zagadnienia.* Białystok: Wydawnictwo Politechniki Białostockiej.

Kleiner, Y. & Rajani, B. 2001. Comprehensive review of structural deterioration of water mains: statistical models. *Urban Water Journal.* 3: 131–150.

Kleiner, Y., Rajani, B. & Sadiq, R. 2006. Failure risk management of buried infrastructure using fuzzy-based techniques. *Journal of Water Supply Research and Technology—AQUA* 55(2): 81–94.

Kowalski, D. & Miszta-Kruk, K. 2013. Failure of water supply networks in selected Polish towns based on the field reliability tests. *Engineering Failure Analysis* 12(35): 736–742.

Kowalski, D. 2011. *Nowe metody opisu struktur sieci wodociągowych do rozwiązywania problemów ich projektowania i eksploatacji.* Lublin: Komitet Inżynierii Środowiska PAN.

Kruszyński, W. & Dzienis L. 2008. Selected Aspects of Modelling in Water Supply System on the Example of Lapy City. *Rocznik Ochrony Środowiska* 10:605–611.

Kutyłowska, M. 2015 Neural network approach for failure rate prediction. *Engineering Failure Analysis*, 47:41–48.

Kwietniewski, M. 2004. Reliability modeling of Water Distribution System for Operation and Maintrnance Needs. Archives of Hydro-Engineering and Envirnmental Mechanics. *Monografia Instytutu Budownictwa Wodnego PAN*, 51: 23–32.

Kwietniewski, M. 2006. Field reliability tests of water distribution system from the point if view of consumer's needs. *Civil Engineering and Environmental Systems* 23(4): 287–294.

Lajevardi, S.B. & Minaei-Bidgoli, B. 2008. Combination of Time Series, Decision Tree and Clustering: A case study in Aerology Event Prediction. *Iccee 2008: Proceedings of the 2008 International Conference on Computer and Electrical Engineering*. 111–115.

Li, F.,Wang, W. & Ramírez, L.H.G. 2016. The determinants of two-dimensional service quality in the drinking water sector—evidence from Colombia. *Water Policy* 18(1).

Lindhe, A., Rosen, L., Norberg, T., & Bergstedt, O. 2009. Fault tree analysis for integrated and probabilistic risk analysis of drinking water systems. *Water Research* 43(6): 1641–1653.

Magelky, R. 2009. In K. M Kenny & J.J. Jr. Galleher (eds), *Assessing the Risk of Water Utility Pipeline Failures Using Spatial Risk Analysis*, San Diego, International Conference Pipelines: Infrastructure's Hidden Assets: American Society of Civil Engineers.

Miksch, K., Cema, G., Felis E. & Sochacki A. 2015. Novel Methods and Technologies in Environmental Engineering. *Rocznik Ochrony Środowiska* 17: 833–857.

Pietrucha-Urbanik, K. 2013. Multidimensional comparative analysis of water infrastructures differentiation, In M.R. Dudzińska, L. Pawłowski & A. Pawłowski (eds), *Environmental Engineering IV*: 29–34. London: Taylor & Francis Group.

Pietrucha-Urbanik, K. 2014, Assessment model application of water supply system management in crisis situations. *Global Nest Journal* 16(5): 893–900.

Rabczak, S. 2016, Free-Cooling in Seasonal Cold Accumulator. *International Journal of New Technology and Research* 1(8):49–52.

Rak, J. & Pietrucha, K. 2008. Some factors of crisis management in water supply system. *Environment Protection Engineering* 34(2): 57–65.

Rak, J. 2007, Some aspects of risk managementin waterworks. *Ochrona Środowiska* 29(4): 61–64.

Rak, J., Kwietniewski, M., Kowalski, D., Tchórzewska-Cieślak, B., Zimoch, I., Bajer, J., Iwanejko, R., Miszta-Kruk, K., Studziński, A., Boryczko, K., Pietrucha-Urbanik, K. & Piegdoń, I. 2013. *Metody oceny niezawodności i bezpieczeństwa dostawy wody do odbiorców*. Rzeszów: Oficyna Wydawnicza Politechniki Rzeszowskiej.

Sadiq, R., Kleiner, Y. & Rajani, B. 2004. Aggregative risk analysis for water quality failure in distribution networks. *Journal of Water Supply Research and Technology: AQUA* 53(4): 241–261.

Sadiq, R., Saint-Martin, E., & Kleiner, Y. 2008. Predicting risk of water quality failures in distribution networks under uncertainties using fault-tree analysis. *Urban Water Journal* 5(4): 287–304.

Studziński, A. 2014. Amount of labour of water conduit repair, In P.H.A.J.M. Van Gelder, R.D.J.M. Steenbergen, S. Miraglia, & A.C.W.M. Vrouwenvelder (eds), *Safety, Reliability and Risk Analysis: Beyond the Horizon*: 2081–2084. London: Taylor & Francis Group.

Szkarowski, A. & Janta-Lipińska, S. 2015. Badania doświadczalne a dokładność opracowanego modelu. *Rocznik Ochrona Środowiska/Annual Set the Environment Protection*, 17(1), 576–584.

Szymura, E. & Zimoch I. 2014. Operator reliability in risk assessment of industrial systems function. *Przemysł Chemiczny* 93(1): 111–116.

Taigbenu, A.E. & Ilemobade, A.A. 2007. Software development for the water sector. *Advances in Materials and Systems Technologies* 18–19: 543–548.

Tchórzewska-Cieślak, B. & Szpak, D. 2015, A Proposal of a Method for Water Supply Safety Analysis and Assessment. *Ochrona Środowiska* 37(3): 43–47.

Tchórzewska-Cieślak, B. 2012. Urban Water Safety Management, In V. DeRademaeker, E. Cozzani, S. Pierucci & JJ. Klemes (eds), *Chemical Engineering Transactions*: 201–206. Milano: AIDIC SERVIZI SRL.

Tchórzewska-Cieślak, B., Boryczko, K. & Piegdoń, I. 2015. Possibilistic risk analysis of failure in water supply network, In Nowakowski et al. (eds), *Safety and Reliability: Methodology and Applications*: 1473–1480. London: Taylor & Francis Group.

Yang, Z.X., Yuan, X.B., Feng, Z.Q., Suzuki, K. & Inoue, A. 2006. A fault prediction approach for process plants using fault tree analysis in sensor malfunction. *IEEE ICMA 2006: Proceeding of the 2006 IEEE International Conference on Mechatronics and Automation, Vols 1–3, Proceedings*, 2415–2420.

Zeliaś, A. & Wanat, S. 2003. *Prognozowanie ekonomiczne. Teoria, przykłady, zadania*, Warszawa: Wydawnictwo Naukowe PWN.

Zimoch, I. 2009. Bezpieczeństwo działania systemów zaopatrzenia w wodę w warunkach zmian jakości wody w sieci wodociągowej. *Ochrona Środowiska* 31(3): 51–55.

Zimoch, I. 2012. Pressure control as part of risk management for a water-pipe network in service. *Ochrona Środowiska* 34(4): 57–62.

Ground water levels of a developing wetland—implications for water management goals

A. Brandyk, G. Majewski & A. Kiczko
Faculty of Civil and Environmental Engineering, Warsaw, Poland

A. Boczoń & M. Wróbel
Forest Research Institute, Sękocin Stary, Poland

P. Porretta-Tomaszewska
Faculty of Building Services, Hydro and Environmental Engineering, Warsaw, Poland

ABSTRACT: Wetlands, along with their biodiversity, have been subject to major threats worldwide. This was the reason to undertake restoration measures in order to cease their deterioration and loss of natural values. Controlled water table management in irrigation channels was said to be vital for maintenance of proper ground water levels on wetland areas that are favourable for wet, low-productive meadows state. We aimed to demonstrate this on a "developing wetland" area, which is part of Całowanie Peatland, consisting of sandy soils. The possible increase in channel water table caused that on 29% of research area (171.6 ha) the ground water depth ranged from 0.3 to 0.45 m., which was found to be indispensable for the development of wet meadows, according to the adopted hydrologic criteria. The evidence was found through ground water modeling and assuming criteria for so-called "hydrologic uniformity", which were still subject to discussions in the literature and herein.

Keywords: wetlands restoration, modeling, habitat uniformity, hydrological criteria

1 INTRODUCTION

Wetland habitats and their unique vegetation are threatened by human activities all over the world (Schot et al. 2004). Among the many threatened vegetation types, low-productive fen meadows are particularly known for species richness, thus giving it a high priority from the viewpoint of biodiversity conservation (Grygoruk et al. 2015). The species richness of fen meadow vegetation is closely related to the availability of nutrient-poor groundwater (Wheeler & Shaw 1995, Dąbrowski et al. 2015).

Drainage and groundwater outflow through ditches may lead to the development of so-called transitional zones in the upper groundwater of fen meadows, which prevent the upward seeping alkaline groundwater from reaching the root zone, but enhance the presence of infiltrated rainfall water instead (Fig. 1). This threatens the conservation of the species rich vegetation and may bring about changes to soil organic matter (Brandyk & Majewski 2013, Stelmaszczyk et al. 2015, Kasperek et al. 2015).

Decline of water tables allows atmospheric oxygen to enter the soil inducing mineralisation of organic matter and, moreover, the release of large quantities of nutrients with adverse effects on nutrient-poor vegetation of wet, fen meadows (Keizera et al. 2014).

Decreased water tables also enable recharge by local precipitation giving rise to so-called "rainfall-dominant" zones that interfere with the inflowing deeper regional groundwater, as it was already indicated above (Fig. 1). Therefore, the properties of water in the uppermost soil layers may temporarily be rainfall-like (e.g. electrical conductivity and

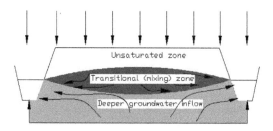

Figure 1. Infiltrated rainfall interference with ground water, mineralization and oxygen supply—transitional zone.

pH), until it is fully mixed or percolated so as to reach ground water properties. This may lead to a precipitation-dependent vegetation type and a loss of biodiversity. Hydrological restoration of degraded fens may entail reduction of drainage in order to re-establish the desired water levels and ground water inflow higher than precipitation recharge (Schot et al. 2004).

In order to conduct fen meadows restoration, one has to consider the historical hydrological conditions. However, as it is commonly known, the rewetting projects of previously drained organic soils cause the enrichment of groundwater in phosphates, for instance. In this regard, although the restoration and maintenance of degraded wetlands by rewetting is generally found to improve their environment quality, the eutrophication threat, which is often related with drainage system malfunctioning, is nowadays a matter of considerable concern (Grootjans & Wołejko 2007). However, large restoration projects in the world,, discuss such use of drainage-irrigation infrastructure, so as to restore proper water levels and minimize contaminant transport (Dietrich et al. 2008). Very often, those problems are approached through hydrological modeling. For instance, Modflow—a common ground water model—is widely used in order to find wetland ground water response to transient boundary conditions. Scenario studies are then envisaged, involving hydrological criteria that quantify soil wetness and potential habitat type versus the achieved ground water levels. This gives the basis to select the "best" or "optimal" scenario that helps to keep natural values of wetlands (Brandyk 2011).

The Authors put emphasis on maintaining the hydrologic properties of a developing wetland with respect to ground water levels. Moreover, the goal of this study was to demonstrate that an artificial stream network may provide for keeping such ground water table position which meets threshold values for the achievement of pre-defined soil moisture with regard to conservation of low-productive, wet meadows.

2 MATERIAL AND METHODS

2.1 Research area

The selected research area is the southern part of Całowanie Petalnd, has already been under protection as habitat for endangered wetland species of flora and fauna for about 10 years. The whole Całowanie Peatland, having the total area of 3500 ha, forms a vast and interrelated complex of wetlands. It is located about 35 km south of the capital city of Warsaw, in the valley of the Vistula River (Fig. 2). The most valuable landscape components include: raised and transitional bogs, as well as single-swath meadows, the biodiversity of which strongly depends on human maintenance and which have been claimed to be one of the most vulnerable habitat mosaics in Europe.

However, the southern part, which has been selected as the research area (588.8 ha), is still a developing wetland on top of a sandy aquifer, and also an area protected by means of a habitat directive.

First signs of high moisture appear there: increased ground water levels (frequently near land surface), and gradual appearance of hydrophilius vegetation (Pierzgalski et al. 2013). Since the future water management and the condition of drainage-irrigation system in that area could most likely affect the habitat quality, it was being resolved to maintain such water levels that help fulfilling environment protection goals. We tried to gain insight into the ground water level changes on account of keeping the assumed water table elevations in the main canals. It was hypothesized that with homogenous geological setting (sandy layers prevailing over the area) by adjusting surface water levels, so –called „hydrologic stability " could be achieved in neighboring wetland areas with comparable ground water depth. This could guarantee proper moisture conditions for the development of plant communities. The induced water level changes and resulting soil moisture are dependent on the existing inflows or outflows from the analysed wetland system, but their significance for possible nature development strongly relies on the applied hydrological criteria which enable to compare the results of hydrological models and quantify the wetness of the restored wetlands.

2.2 Channel network

Before the setup and exploitation of ground water model, an observation network of piezometers and water gauges was prepared within the research

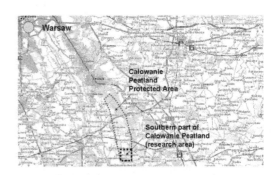

Figure 2. Location of Całowanie Peatland Protected Area.

area (Fig. 3). The channels that surround the area from the south, north and west, are considered to be model boundaries, on which we applied known surface water stages(1st type boundary condition). Those channels are part of a large drainage-irrigation system which in the past served for the supply of water to the meadows located within Całowanie Peatland. In the research area, there is also a network of secondary ditches (Fig. 3) which is out of order nowadays, mostly dried, not supplied with water from the main irrigation channels and overgrown with vegetation. Therefore, they were not considered in the model used here, but their future restoration seems to be vital for environmental protection; however, a very dense and accurate monitoring network as well as adopting different modeling scale are required.

2.3 Ground water model

In order to conduct analysis of ground water level changes versus the assumed ditch water table position, we utilized common and recognized Modflow code, which can simulate ground water flow response to changing boundary conditions. The governing, partial-differential equation of this model can be written in the following form (McDonald & Harbaugh 1988):

$$\frac{d}{dx}\left(T_x \frac{dH}{dx}\right) + \frac{d}{dy}\left(T_y \frac{dH}{dy}\right) + \frac{d}{dz}\left(T_z \frac{dH}{dz}\right) - W = S\frac{dH}{dt} \quad (1)$$

where T_x, T_y, T_z,—aquifer transmissivities in three directions: x, y, z [m²/s]; H—hydraulic head [m]; W—inflow or outflow from internal sources or sinks of water [m/s] (e.g. rainfall, evaporation, channel seepage); S—specific yield [-]; t—time [days, years etc.].

Analytical solutions of equation 1 are possible only in few simple cases. That is why, most often, it should be solved numerically. For the numerical solution, equation 1 is replaced by proper differential equations that describe the mass balance for a set of finite cells (rows, columns and layers), into which the ground water system is discretized (an example is given in Figure 4).

Basic input parameters include values of Transmissivity (T) or saturated hydraulic conductivity (K_a) of main geological layers, the conductivity and the thickness of riverbed materials, Vertical Leakance (Vl) between main hydro-stratigraphic units (vertical flow term), heads in the rivers, heads or fluxes on model boundaries, flows representing rainfall and evapotranspiration and other sources or sinks of water (e.g. drainage, irrigation, flow barriers etc).

A simple geologic model was developed from the materials observed in 20 soil borings taken across the site (Grootjans & Wołejko 2007). The resulting model includes a basal layer of sand and loamy sand, which is about 2.2 m. thick, and is overlain by 3.7 m-thick alluvial sand. Both sand layers are part of a regional aquifer that has a large spatial extent from glacial uplands near the town of Osieck to the Vistula River. At the very bottom of the model, an impermeable base (no flow condition) is assumed underneath the sandy layers.

Model layers of the Modflow program, depending on the adopted scale, modeling goals and expected accuracy, are assigned to individual soil layers, leading to accurate, but also very complex models with many parameters for calibration. On the other hand, they represent only the main hydrostratigraphic units (i.e. aquifers and aquitards) leading to so-called "effective models" that have a generalized structure, but are simpler in practical application. In such models, one layer usually represents a larger unit or geologic formation, which is frequently built of several sediment layers. Their hydrogeological parameters are usually input as average or weighted-average values.

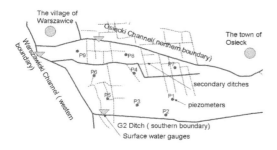

Figure 3. Hydrographic and observation network of the research area.

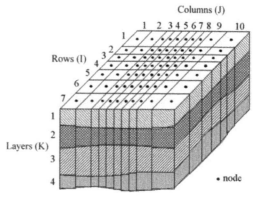

Figure 4. Exemplary model domain (7 rows, 10 columns, 4 layers).

On account of a local scale assumed in the present study, two model layers were used, i.e. layer no. 1 representing top alluvial sand formation and underlying layer no. 2 representing sand and loamy sand. The top sandy layer (no.1) was considered as unconfined (with phreatic water table) while the second was a confined layer (sub-artesian tables). Modflow software also allows the second (underlying) layer to be of convertible type (confined/unconfined). However, here it was resolved that is no such an intensive and deep drainage which would dry the upper layer to the degree where the second one becomes unconfined.

The orthogonal grid, constructed by the modeling software, consisted of 480 columns, 250 rows, and 2 layers, resulting in a total of 117,768 active model cells. All cells outside the research area were inactive, i.e. set as no flow cells. The horizontal dimensions of the grid cells were uniform (10 m by 10 m), and the vertical spacing varied adequately to the depths of the aquifer known from geological logs: layer one—about 3.7 m, layer 2 – depth of 2.2 m, on average.

2.4 Model calibration

After model identification and set up, the calibration process was performed as trial and error procedure, in order to estimate such parameter values for which a reasonable match between modeled and observed ground water levels was achieved. Ground water levels used for calibration were measured in the network of piezometers in 2012, and surface water levels in channels served as boundary conditions. Moreover, they were expressed in meters above the "Vistula River 0 Level" local datum. Hydraulic conductivities and specific yield of layers 1 and 2 were adopted from previous studies, with uncertain values of conductivity for layer 2, which was changed during the calibration process. Since the measured water stages in channels were applied on three boundaries (north, west, south), the ground water inflow (ground water flux—second type boundary condition) was applied on the eastern boundary. It ranged from 0.5 to 2.1 m³/d/m as indicated by the hydrogeological map, and was also subject to calibration.

So called upper boundary condition (recharge flux through the surface into ground water system) could be treated in the Modflow software as P-E, that is precipitation rate minus evapotranspiration. Precipitation totals were achieved by linear interpolation over a wider area, on the basis of the records from three stations: Warszawa (37 km), Siedlce (about 40 km) and Kozienice (about 37 km). However, only meteorological data from Kozienice were used to calculate potential evapotranspitation according to Penman—Monteith method (Brandyk et al., 2016), since that station largely

reflects the conditions outside the city area. It should be stressed that those values are not fully representative for the Całowanie Peatland and should also undergo calibration. Nevertheless, it was omitted in the present study, so as to keep the calibration procedure as simple as possible. Multi-parameter optimization algorithms were not applied here. On the basis of the ground water measurement availability it was decided to run the model with a decade time step; hence, the input fluxes: precipitation and evapotranspiration were calculated as decade totals.

Ground water inflow from the east and hydraulic conductivity of the second layer were changed individually, followed by a visual comparison of observed and modeled heads, correlation coefficient and standard deviation between model output and observations. The values were finally considered as acceptable, while standard deviation reached 0.3 m. and correlation coefficient was equal to 0.77.

Next, the model was validated, on the basis of ground water levels measured throughout 2013. Correlation coefficients for the validation period varied from 0.60 for p7 piezometer to 0.79 for p2 piezometer, which proved a satisfactory model quality.

2.5 Water level criteria

Application of certain hydrological criteria for assessment of wetland restoration has been discussed in scientific literature (Okruszko 2005). Adopting mean water levels and its extremes (threshold values), as well as other hydrological characteristics, e.g. flooding frequency is mainly based on expert knowledge, resulting mostly from observations of reference ecosystems. Within those ecosystems, we find such water level dynamics that the existing habitat conditions (soil properties and plant communities first of all) are preserved well and can be treated as patterns to follow in respect to nature conservation. At present, empirical attempts were made to relate ground water levels with certain habitat quality or type, and to find optimum hydrographs for particular plant groups within an ecosystem (Oświt 1991). This still seems to be subject to discussions, as for hydrologic aspects, the water levels and soil water content are the most important parameters; however, other ecological issues need to be addressed as well, i.e. the seed bank, light availability, nutrient cycling, and others. This may lead to more interdisciplinary eco-hydrological approaches. As yet, the formulation of particular criteria tends to be questionable, and one has to decide, which environmental qualities could be assigned to particular water level or moisture ranges. The soil water management approach we decided to present helps to maintain water levels for particular soil types, which secure 8% oxygen content in the root zone as an effect of capillary rise in the soil. For light, sandy soils in the analysed area, we adopted threshold values, on the basis of the works of Szuniewicz (1979): minimum allowable depth (upper limit): 0.35 m, mean depth: 0.40 m, maximum allowable depth: 0.45 m.

3 RESULTS AND DISCUSSION

Calibration process provided a satisfactory quality of the applied model. During that process, the value of hydraulic conductivity for layer 2 was gradually changed, starting with 35 m/d and finally reaching 25 m/d. Afterwards, the inflow of ground water from east (second type—flux boundary condition) was being adjusted, decreasing from 210 to 156 m^3/d for the first layer and from 120 to 90 m^3/d for the second layer. A joint effect of two decreased parameter values was a closer match of modeled and observed heads. The resulting standard deviation between the model and observation was equal to 0.3 m (variance = 0.055 m—Fig. 5). At a scatter diagram (Fig. 5) a perfect fit means all values lie on the line, which is inclined at the 45 deg angle (linear function y = x). Here, some fraction is scattered around it, which is reflected by the calculated variance and correlation analysis, proving a satisfactory match, but not a perfect one.

In this study, it was decided that only the most uncertain parameters will undergo calibration (hydraulic conductivity of layer 2 and inflow from east to both layers—a boundary flux). The input from precipitation and evaporation could also have been subject to adjustments, as they come from interpolation for a wider area, being influenced by spatial variability of hydrometeorological processes. Also, for further simulations, the calibrated inflow from east remained constant, which most likely influenced model output. However, it was justified by the fact that no water level observations existed from that direction, which would provide more accurate data for applying boundary conditions.

Having completed the calibration, the validation of the model was performed, on the basis of the data from 2013. In the simplest manner, we estimated the quality of the model by calculating correlation coefficients between modeled and observed water tables for all piezometers. The achieved values (0.6–0.79) do not indicate good or very good quality, but rather suggest a satisfactory (or acceptable) one.

After calibration and validation, the model was utilized to calculate ground water heads for the assumed channel water table elevation, so as to estimate potential impact on developing wetland and formulate protection goals. It was done through the estimation of ground water depths within the

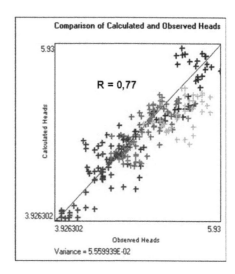

Figure 5. Scatter diagram for modeled and observed ground water heads.

research area, from the point of view of hydrologic unity, i.e. finding out, how big is the surface, on which threshold values have been achieved. On that surface, we claim the existence of similar conditions of wetland habitats development, stressing again that we have only analysed the depth of water as a driving force here, which would secure 8% of oxygen in the root zone, independent of other ecological issues. In this way we can outline the perquisites for water management, in order to maintain wetland meadows in the analysed area.

On the basis of data collected for 2014, simulation of ground water levels for the current status of the system was performed. The input values for that year involved, as it was already mentioned: precipitation and evapotranspiration fluxes, ground water inflow from the eastern direction and measured water levels in the main ditches, which were maintained by the existing hydraulic structures. The physical condition of those structures, as it was stressed before, is a vital issue for the analysed area, with respect to water management and maintenance of environmental status. Their ongoing deterioration needs to be addressed, because it may deteriorate the hydrologic conditions of meadows in the near future. On the basis of the existing assessment of the structures (Pierzgalski et al., 2013) it was found that currently, the maximum allowable rise in channel water table is equal to 0.5 m, which was applied in the second modeling scenario.

The ground water levels, modeled with decade time step, were averaged for 2014 (Fig. 7) and subtracted from terrain elevation (Fig. 8). In this manner, mean ground water depths were determined, which followed the patterns typical for areas

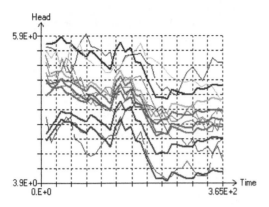

Figure 6. Comparison of the observed (light line) and modeled (thick lines) water tables for 2013.

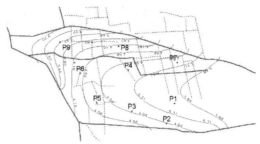

Figure 7. Mean ground water heads for 2014.

Figure 8. Mean ground water depths for 2014.

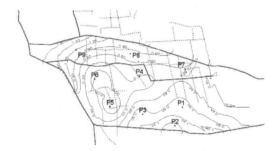

Figure 9. Mean ground water depths in scenario 2—water levels in ditches raised by 0.5 m.

surrounded by drainage—irrigation channels. Maximum depth values were noted close to channels, while in the middle of the area, near the piezometers p4, p5, p6, they approached land surface elevation. An initial guess arises that it would not be feasible to sufficiently increase water table position in the vicinity of ditch zone, rather than in the mid-area. In order to gain an insight, the surfaces with the threshold values of ground water levels were estimated. In the current status of the system, only about 16% of the area (92.6 ha) were found to maintain predefined hydrologic, uniformity", with respect to the applied criteria (water levels within desired range: 0.3–0.45 m). In the second scenario (channel water tables raised by 0.5) that area is extended to 29% (171.6 ha). On these surfaces we can expect changes of soil water conditions to more favourable for the maintenance of wet—low vegetation meadows, underlain by light, sandy soils with shallow water table. The remaining part (71% of the area) would be still subject to lower water table, and insufficient capillary rise in the soil, not reaching the root zone. We wish to emphasize that the above-mentioned observations should be treated as relative, because they are directly dependent on the adopted criteria. Having assumed other allowable limits of hydrologic characteristics, we might achieve a different view on the development of wet meadows versus the obtained water levels.

Attention was paid to rough comparisons between the existing status and the one with channel levels raised by 0.5 m (second scenario, Fig. 9). A future issue is to analyse the effect of channel stages rise "step by step", i.e. by 0.1 m, 0.2 m, 0.3 m and so on, generating "consecutive" scenarios to provide for wetland maintenance. This would enable to choose the restoration "intensity" or "extent" as part of water management planning for Całowanie Peatland area.

4 CONCLUSIONS

Contemporary research works and expert analyses proved the need to find solution for problems related to wetlands restoration and maintenance, with regard to habitat quality, which is strongly dependent on the prevailing hydrologic conditions. In particular, the adopted criteria for analyzing water management goals tend to be crucial, in order to develop proper soil water conditions at the background of the existing surface water-ground water interaction. The case of "developing wetlands" – that is the ones in the initial phase of swamp-forming processes, seems to be especially important here, since we believe that controlled water table management may contribute to maintenance of desired soil water content and finally provide for biodiversity conservation.

For the analysed area, which is underlain by sandy aquifer, bound by drainage- irrigation channels and showing initial stages of wet low-productive meadows development we found following important facts:

- Channel water table clearly exerts influence on ground water levels with a typical pattern—with the highest drawdown close to channels, and the maximum elevation of water table in the middle of the area. We try to describe that influence as conductivity-driven, because its large values for existing sandy soils cause high exchange flows between canals and ground water horizons.
- Depth to water table was analysed according to the adopted criteria which describe soil wetness category for low-productive meadows. The assumed threshold values of ground water table (0.3 to 0.45 m. below land surface) were achieved on maximum 29% of the area if the channel water tables were increased by 0.5 m in the second analysed scenario. The first scenario, which reflected the actual status of the system, guarantees proper water levels only on 16% of the area. We wish to stress, that the restoration target was achieved, when ground water table position was between the adopted extremes. This is of course an assumption. First of all, the criteria seem to be relative, and we also claim that the prescribed water level range is directly assumed as permanent feature of well-preserved habitats (and their plant species composition) and may be used for scenario comparisons in hydrologic, as well as water management analyses.
- Hydrologic "stability" or unity with respect to adopted water level range was possible on a limited surface in the middle of the area. It has not been achieved on its remaining part.
- Adaptive channel water management is an option for the maintenance of wet low-productive meadows within the research area. More detailed scenario studies are indispensable for the decision making on restoration area extent.

REFERENCES

Brandyk, A. 2011. Ground water - fed system restoration on the area of Przemkowsko- Przecławskie Wetlands. *Annals of Warsaw University of Life Sciences-SGGW-Land Reclamation* 43(1): 13–23.

Brandyk, A. & Majewski, G. 2013. Modeling of hydrological conditions for the restoration of Przemkowsko-Przecławskie Wetlands. *Annual Set The Environment Protection* 15(1): 371–392.

Brandyk, A. Kiczko, A. Majewski, G. Kleniewska, M. & Krukowski, M. 2016. Uncertainty of Deardorff's soil moisture model based on continuous TDR measurements for sandy loam soil. *Journal of Hydrology and Hydromechanics* 64: 23–29

Dąbrowski, W., Wiater, J. & Boruszko, D. 2015. Oczyszczanie odcieków z beztlenowej stabilizacji osadów z oczyszczalni ścieków mleczarskich przy zastosowaniu metody hydrofitowej. *Rocznik Ochrona Środowiska/ Annual Set the Environment Protection,* 17(2): 869–879.

Dietrich, O. Schweigert, S. & Steidl, J. 2008. Impact of climate change on the water balance of fen wetlands in the Elbe Lowland. *Proc. of the 13th International Peat Congress.* International Peat Society.

Grootjans, A. & Wołejko, L. 2007. Conservation of wetlands in polish agricultural landscapes. Wydawnictwo Lubuskiego Klubu Przyrodników: Szczecin.

Grygoruk, M. Bańkowska, A. Jabłońska, E. Janauer, G. Kubrak, J. Mirosław-Świątek, D. & Kotowski, W. 2015 Assessing habitat exposure to eutrophication in restored wetlands: Model-supported ex-ante approach to rewetting drained mires. *Journal of Environmental Management* 152: 230–240.

Kasperek, R., Wiatkowski, M. & Rosik-Dulewska, C. 2015. Investigations of Hydrological Regime Changes in an Area Adjacent to a Mine of Rock Raw Materials. *Rocznik Ochrona Środowiska/ Annual Set the Environment Protection,* 17(2): 195–208.

Keizera, F. Schot, P. Okruszko, T. Chormański, J. Kardel, I. & Wassen, M. 2014. A New look at the Flood Pulse Concept: the (ir)relevance of the moving littoral in temperate zone rivers. *Ecological Engineering* 64: 85–99

McDonald, M. & Harbaugh, W. 1988 A modular three—dimensional finite difference groundwater flow model. United States Geological Survey. Openfile Report No. 6.

Okruszko T. 2005. Kryteria hydrologiczne w ochronie mokradeł. Rozpr. hab. Rozprawy Naukowe i Monografie. Wydawnictwo SGGW.

Oświt J. 1991. Roślinność i siedliska zabagnionych dolin rzecznych na tle warunków wodnych. Rocz. Nauk Rol., Ser. D, monografie.

Pierzgalski, E. Pawluśkiewicz, B. Gnatowski, T. & Brandyk, A. 2013. Utrzymanie urządzeń melioracyjnych na obszarach Natura 2000 na przykładzie Bagna Całowanie. In Bogumiła Pawluśkiewicz (ed.), *Gospodarowanie w dolinach rzecznych na obszarach Natura 2000.*

Schot, P. Dekker, S. & Poot, A. 2004. The dynamic form of rainwater lenses in drained fens. *Journal of Hydrology* 293: 74–84.

Stelmaszczyk, S. Okruszko, T. & Meire, P. 2015. Nutrients availability and hydrological conditions of selected wetland ecosystems in the Biebrza river valley. *Annals of Warsaw University of Life Sciences. – SGGW, Land Reclamation* 47 (1): 3–17.

Szuniewicz, J. 1979 Charakterystyka kompleksów wilgotnościowo- glebowych pod kątem parametrów systemu melioracyjnego. Biblioteczka wiadomości IMUZ:58. PWRIL. Warszawa.

Wheeler, B. & Shaw, S.1995. A focus on fens-controls on the composition of fen vegetation in relation to restoration. In: Wheeler, B. Shaw, S. Fojt, W. Robertson, R. (eds.), *Restoration of Temperate Wetlands.* Wiley: Chichester

Environmental Engineering V – Pawłowska & Pawłowski (Eds)
© 2017 Taylor & Francis Group, London, ISBN 978-1-138-03163-0

Detection of potential anomalies in flood embankments

M. Chuchro, M. Lupa, K. Szostek, B. Bukowska-Belniak & A. Leśniak
Department of Geoinformatics and Applied Computer Science, Faculty of Geology,
Geophysics and Environmental Protection, AGH University of Science and Technology, Kraków, Poland

ABSTRACT: Detection of potential anomalies in pore pressure and temperature time series in flood embankments could be used to assess the embankment stability. Temperature and pore pressure sensors were placed in an experimental flood embankment and connected to a database via wireless communication protocols. The analysis was performed on the data stored in a database, using Fast Fourier Transform (FFT). As a result of this spectral analysis, four models for detecting anomalies were created. The occurrence of potential anomalies was examined with determination coefficient R^2 and the Shapiro-Wilk test for model errors. An essential element of the work was the development of a database and applications in C language. The proposed database model reduces the time of processing and changes in structure data sent to the created application for the detection of potential anomalies. During the tests, the presented algorithms detected 17 out of 18 possible events (potential anomalies) in a third of the presented methods.

Keywords: flood embankments, anomaly detection, database, time series, FFT

1 INTRODUCTION

Floods occur with increasing frequency all over the world, leading to large losses and posing a threat to humans and animals. Therefore, it is important to counteract the effects of flooding with emergency management and maintenance of the technical aspects of flood protection; namely, river embankments. Monitoring the status of embankments is a very important issue, especially locating weakened sections that might fail during high water levels. In Europe, several flood risk analysis and crisis management projects have been implemented (Baliś et al. 2011, Krzhizhanovskaya et al. 2011, Pengel et al. 2013, Piórkowski & Leśniak 2014, Mrozik et al. 2015). For example, the international Ijkdijk (Pyayt, 2012) project was implemented in the Netherlands that concerned integration and validation of flood protection and sensor systems in an embankment. Another example is the European UrbanFlood project, which uses real embankments to create an early warning system and crisis management. A simulated flood zone during the interruption of the embankment in a particular place and at a certain water level was created. In both these projects, the physical parameters of the embankment were measured under normal conditions and at higher water level. Observations included both slow natural filtration of water through the embankment, and rapid breakage (www.urbanflood.eu).

The Danube Floodrisk project focused on the most cost-effective measures for flood risk reduction: risk assessment, risk mapping, involvement of stakeholders, risk reduction by adequate spatial planning (www.danube-floodrisk.eu).

FLOODsite project covers the physical, environmental, ecological and socio-economic aspects of floods from rivers, estuaries and the sea. It considers flood risk as a combination of hazard sources, pathways and the consequences of flooding on the "receptors"—people, property and the environment (Różyński et al. 2006, www.floodsite.net).

This work was done as part of the ISMOP Computer System for Monitoring River Embankments Project (in Polish: Informatyczny System Monitorowania Obwałowań Przeciwpowodziowych, www.ismop.edu.pl). The aim of this project is to study processes occurring in soil embankments during floodwater filtration and to create a system for monitoring the state and stability of embankments. In the project, an experimental embankment of normal size was created, consisting of two parallel embankments with 150 m of combined meanders. This creates a reservoir that can be flooded to simulate high water level during flooding. Different types of soil materials within individual sections of the embankment were used and reference sensors were located in each section. This system of reference sensors was installed in profiles along three cross-sections in the embankment (Fig. 1).

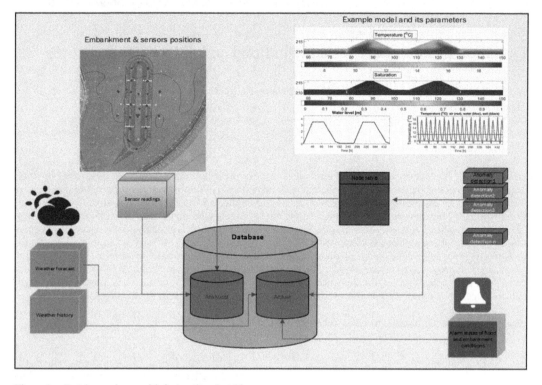

Figure 1. Database schema with featured node table.

In the experimental embankment, various types of sensors were mounted, creating a unique test unit. The measuring system records the physical parameters inside the embankment: temperature, pore pressure, distortion, and weather conditions. The system allows the tracking of the physical parameters of the embankment and its surroundings. In addition to the permanent measurement system, additional measurements are conducted: geodetic (displacements), geophysical, geotechnical, radar and infrared. The suitability of different measurement methods for the rapid location of leaks and areas at risk of failure was tested.

2 DATA BASE

The primary and initially planned functionality of the database (in addition to archiving the generated results), was its role in the conducted analyses (Kasprowski 2012, Bajerski 2009). Therefore, on the basis of the information stored in the database, a number operations are performed, such as time series analysis, inference expert system, or statistical analysis to detect anomalous information.

The database structure was designed with a view of storing information covering both traditional (numeric, text) and spatial data (storing information about objects, their shape, coordinates and coordinate system). Nonetheless, the challenge was to provide a method of storage and effective processing of data resulting from numerical modelling, generated using finite difference methods (Lupa et al. 2014, Pięta & Krawiec 2015).

The authors originally expected that the system would store full numerical models (simulated waveforms of flood wave). These models were used to assess the embankment stability on the basis of comparisons with actual measurements. Additionally, the structure of the models was designed on the basis of the relation of one scenario model to many simulations. Nevertheless, the data magnitude (an example scenario consists of one table containing 75,442,752 records, which gives 16 MB × 924.758 hundred scenarios) meant the initially planned analytical form of database was not possible.

Therefore, the existing database structure was modified by isolating the database dedicated to archiving (full numerical models, as well as storage of historical information, e.g. weather conditions, past alarm states etc.), and creating a second database with an analytical character (light data with efficiency improvement).

The analytical character of the database structure assumed data simplification to form grid-nodes that match the coordinates of the sensors in the flood embankment. In this way, it is possible to compare readings from sensors and time series recorded with an assigned (numerical model) node identifier. Thus, the generalised table structure allows for a significant qualitative leap in the context of searching and processing data.

It is worth emphasising that the node table (Fig. 1), in contrast to the real measurements from the sensors (stored in the database), is not a dynamic table. The structure of sensor tables is dependent on the position (geographical coordinates) of the sensor and its type (optical fibre or independent sensor). The sensor reading data has a certain lifetime (depending on the needs and the required time efficiency calculations) and they are then sent to the archive (the other part of the database).

3 ANOMALY DETECTION

The aim of the ISMOP project is to assess the stability of a flood embankment. The created assessment method should be suitable for implementation in newly established and existing flood embankments. The current stage of the project has been realised on a full size experimental flood embankment, located on the Vistula flood plains in Czernichow near Krakow, Poland. One component of ISMOP was to make models that detect anomalies in temperature and pore pressure measurement time series from the experimental flood embankment.

Additionally, we should present what we consider a potential anomaly. For the ISMOP project, potential anomalies were assumed to be non-recurring or brief events in the analysed time series that changed the standard (average) values recorded for the phenomenon. Permanent changes in the development of the phenomenon were the assumed potential anomalies (Wiatr et al. 2014).

For each temperature and pore pressure, sensor detection of potential anomalies is carried out separately. The results allow for a visualisation of potential anomalies for each sensor and indirect assessment of flood embankment condition (Chuchro et al. 2014).

Potential anomaly detection starts with data collection. Data should have a specific structure and a precisely defined amount of collected data per hour (time stamp resolution). Data needs to be pre-processed before it is read by the anomaly detection models. During the process, incorrect values that relate to sensor malfunctions or data transfer errors are detected and removed. Gaps in the data set are replaced by interpolating results

from nearest neighbour sensors. A new time stamp attribute is inserted in the database which is rounded to 1 minute, based on all data sent from sensors at one time (for each 15-minute period).

In order to evaluate the occurrence of potential anomalies, four methods were selected. Three of them are based on the Fast Fourier Transform (Welch 1967). The last one is related to the standard deviation properties for normal distributed data. The first three methods require a minimum of two sets of time series from a two-day period, measured at an interval of 15 minutes, giving 192 observations for each analysed time series. The first time series set is for the most recent period and is tested for potential anomalies. The second one is from a comparable period, started at the same hour and during similar weather conditions and classified as a period without anomalies for a specific parameter measured by the specified sensor. The easiest method is to observe a period from two to four days before the latest analysed observation. Data prepared in this way limits the errors associated with seasonality of data and the influence of weather conditions.

Both time series are standardised with a Z transformation:

$$Z = (X - \mu) / \sigma \qquad (1)$$

where:

X—particular value of the variable,
μ—mean value of the variable,
σ—standard deviation of the variable.
and then transformed with a Fast Fourier Transform (FFT) (Welch 1967):

$$X_k = \sum_{n=0}^{N-1} x_n e^{-\frac{2\pi i}{N} nk}, \quad k \in \{0, 1, \dots, N-1\} \qquad (2)$$

where:

$x_0, \dots, x_{N-1}$.

3.1 Analysis of amplitudes components for consecutive time series

For the selected frequencies, i.e. 0.005208, 0.010416, 0.015625, 0.020833, 0.026042, 0.03125, 0.036458, and 0.041667, power spectral densities were computed for both analysed time series. Power spectral density describes how the power of a time series is distributed with frequency. Mathematically, it is defined as the Fourier transform of the autocorrelation sequence of the time series. Values of power spectral densities were computed with Welch's method (Cochran et al. 1967). This method determines the power spectral density by averaging over the spectrum signal designated in future time intervals.

Next, we determine the similarities between the computed values of power spectral densities. Three functions were tested as a similarity measure: Mean Square Error (MSE), Pearson coefficient and ANOVA (Wei 2006). The function that was sufficiently accurate and simplest in interpretation was mean square error. For the test, MSE was used as a similarity measure and threshold value equalled 0.75.

If the computed MSE value exceeded the threshold value, we located the potential anomaly in the analysed data. The threshold value is determined during the test and each value for each parameter is measured by a particular sensor.

3.2 Analysis of model fitting

A Fourier frequency model was computed for the comparable time series and the same chosen frequencies as in 3.1, expressed by the formula (Tolstov 1978):

$$f(t) = \frac{a_0}{2} + \sum_{n=1}^{\infty}(a_n cosn\omega t + b_n sinn\omega t) \qquad (3)$$

where:
a_n, b_n – Fourier coefficient,
ω – frequency.

The quality of a model was assessed by determination coefficient (R^2)

$$R^2 = \frac{\sum_{t=1}^{n}(\hat{y}_t - \overline{y})^2}{\sum_{t=1}^{n}(y_t - \overline{y})^2} \qquad (4)$$

where:
y_t—variable Y real data value in time t,
$\hat{y}_t$ —variable Y theoretical value in time t,
$\overline{y}$ —mean value of Y variable.

Then the residuals of the model was calculated and tested for normality the Shapiro-Wilk test. The null hypothesis assumes that the test variable is a normal critical area. In this test we reject the null hypothesis if the computed value of the test statistic is less than or equal to the critical value, i.e. W≤W (α, n). It was assumed the critical value alpha = 0.95 to 0.05.

Shapiro-Wilk statistics are expressed as (Royson 1992):

$$W = \frac{\left[\sum_i a_i(n)(X_{n-i+1} - X_i)\right]^2}{\sum_{j=1}^{n}\left(X_j - \overline{X}\right)^2} \qquad (5)$$

where:
$a_i(n)$ – constant value in an array,
$X_{n-i+1} - X_i$—difference between the ends.

In order to detect potential anomalies, the previously created frequency model is used to compute the determination coefficient for the tested time series (the most recent 192 observations). In addition, the frequency model is subtracted from the tested period to compute the residuals of the model. The new residuals of the model are tested with the Shapiro-Wilk test. The results obtained for the test period are compared with the results data for which the model was created (R^2 and Shapiro-Wilk test). If the value of the coefficient of determination (R2) for the tested period is 10% or more, we assumed that the tested data contained a potential anomaly. Moreover, we assumed a potential anomaly existed if the rest of the model for the tested period did not have normal distribution.

3.3 Analysis of phase shifts for pairs of sensors

This method allows the observation of changes in parameters for each cross section in the flood embankment. Performing analysis with this method requires data from two sensors: one located closest to the riverbed and another most distant from the riverbed for each cross section. As in the case of the first method, for each sensor we need to have two time series of 192 observations first without potential anomalies and second from tested period of time (the newest 192 observations). As in the second method, frequency models are created for previously chosen frequencies for the tested time series. For phase shifting, linear Pearson coefficients were used to find the highest correlation value for pairs of models for each period and each of the two sensors. This kind of correlation is referred to as moving correlation, and during the test moving from plus to minus 20 were tested. The highest correlation value for the pair of models from the two-sensor call as delay is an information we compare with the second computed delay for a period without a potential anomaly. If the delay for the tested period has shortened, it means that a potential anomaly could be developing in a cross section. The delay could be present in hours, in which case we multiple the delay by 0.25 hours.

3.4 Change in the average level of the phenomenon

In this method, threshold data for each parameter and sensor location is assessed. The first method includes the basic statistics of the time series: standard deviation and mean value. Values lower than this mean value minus three times standard deviation or values higher than mean value plus three times standard deviation are only 1% of the normal distributed data. The occurrence of such values in data could be an error or a potential anomaly. Some outlier values occurring in a data set cannot be detected during preprocessing. This is because the preprocessing included the tabular data for one moment of time, but the potential anomaly detection contained a time series of defined length and structure. The second way to assess the threshold value is an expert method.

4 APPLICATIONS

All programs were written using C language and compiled using Microsoft Visual C++ (MSVC) compiler for Windows and GCC 4.8 for Linux. Either the master program or the user should manage these modules or use them individually. Modules were written as batch programs, which mean they can be run from the system command line. This solution is easier to manage and parallelise on multiple processors or virtual machines. All modules perform operations on measurements, which may be only temperature values or temperature with pore pressure, depending on the sensor type.

The input data for the applications is based on the CSV files. These may contain a collection of all measurements for the specified time stamp or the time series of an individual sensor. The files contain sensor location, its type and ID, timestamp of the measurements and measured values, temperature and pore pressure, as presented in Listing 1.

```
x,y,z,r-c,temp,ppress,time_stamp,Sensor_no
50.00,0.00,9.00,0,3.691,6011.687,1425945600, 6
50.00,0.00,9.00,0,3.887,5970.803,1425946800, 6
50.00,0.00,9.00,0,2.381,6028.871,1425948000, 6
50.00,0.00,9.00,0,4.149,5980.794,1425949200, 6
50.00,0.00,9.00,0,2.881,5995.689,1425950400, 6
50.00,0.00,9.00,0,2.924,6008.357,1425951600, 6
50.00,0.00,9.00,0,3.219,5980.416,1425952800, 6
50.00,0.00,9.00,0,2.175,6000.619,1425954000, 6
50.00,0.00,9.00,0,3.436,6015.840,1425955200, 6
50.00,0.00,9.00,0,2.941,6005.010,1425956400, 6
...
```

Listing 1. An example of the input CSV file of one time series. The first line contains a header, which should be omitted during reading.

Before starting, the anomaly detection data is pre-processed. The preprocessing module is used for data adjustment of all measurements at a specified time. Timestamps of the recorded values and the values themselves may contain errors, which must be corrected before further processing. Therefore, the program reads specified data and rounds timestamps: It reads the first record timestamp, rounds it down to 15 minutes and saves that value to all timestamps, because this is the nominal sampling time.

Next, all temperature and pore pressure measurements are checked for correctness and limited according to the parameters passed to the program. What is more, if a value read from sensor is corrupted, the program interpolates it using other sensors.

After program execution, the results are either displayed on the screen or saved to a file for potential use in further processing. A report of changes made to the data can also be displayed.

The anomaly detection module uses four separate methods. The analyses are performed on a time series of 192 observations, sampled at 15-minute intervals. For different sampling rates, interpolation should be used to transform the data into the desired form.

The module is executed as before, with command line parameters. Most of them depend on the selected analysis method or are optional. In results, an anomaly is usually denoted as "1" or "0×1", in contrast to "2", which does not suggest an anomaly.

The list of parameters is presented below:

– a—anomaly detection method (1, 2, 3 or 4),
– if—input CSV file,
– offset—number of rows to skip,
– cf—input CSV file for comparison,
– dA—previous anomaly, required in 1st method,
– dT—previous temperature delay,
– dPP—previous pore pressure delay,
– vsT—comma separated frequencies for Fourier model of temperature, required for 2nd and 3rd method,
– vsPP—as above, but for pore pressure,
– w—moving window width for 2nd method.
– of—optional output file,
– c—time series length, default 192,
– v—verbose output.

4.1 Analysis of amplitudes components for consecutive time series

The first implemented method analyzes time series by searching amplitude component changes. The application compares time series of the last 192 (2 days) with previous data. Therefore, a module should be provided with paths to the file containing the current and previous time series, whose anomaly status should be known (otherwise, no anomaly is assumed).

The loaded time series are transformed into the amplitude-frequency domain using Fourier transform and the amplitude spectrums are compared using mean square error. If the value exceeds the selected threshold, the anomaly is reported in the analysed time series. Results are always written to the output file or to the screen. Anomaly results are coded and can be interpreted by the user or manager application (Listing 2).

```
$ anomalies.exe -if data\pomiary\UT1_X1.csv -cf
data\pomiary\UT1_X2.csv -dA 2
Anomaly detection - method 1
mse Temp: 5.089508
old anomaly: 0x2
new anomaly: 0x1
```

Listing 2.   Example command line and result of anomaly detection. Anomaly was detected.

4.2   *Analysis of model fitting*

This method detects anomalies in times series by comparing them to an anomaly-free time series. First, the program loads two time series: one without anomalies, and another to test. As before, it is necessary to provide frequencies that are used to construct the Fourier model. The program creates two models and calculates the Shapiro-Wilk test (with 0.947 thresholds) and coefficient of determination. If the coefficient changes more than 10% or the Shapiro-Wilk test result changes, the tested time series is assumed to contain an anomaly (Listing 3).

```
$ anomalies.exe -a 3 -if da-
ta\Tout12_bez_anomali01062015h1300.csv -cf da-
ta\Tout12_anomalia2_01072015h1300.csv -vsT
0.005208,0.010416,0.015625,0.020833,0.026041,0.
03125,0.036458,0.041666
Anomaly detection - method 3
Coefficients of determination:
Temp 0x2 (9.080025e-001 -> 9.005312e-001)
SW-W:
Temp 0x1 (0.987137 -> 0.693760)
```

Listing 3.   Second method results. Anomaly was detected because of Shapiro-Wilk test divergence.

4.3   *Analysis of phase shifts for pairs of sensors*

The third method detects anomalies from phase shifts for pairs of sensors. Phase shift can be observed if one of the sensors is closer to the anomaly. Therefore, it is necessary to collect phase shifts from the previous analyses, which can be done by the managing application or user.

The application uses a Fourier model that is built from the frequencies that are selected and provided to the module as parameters. Two time series are loaded from files and Fourier models are constructed. Then, correlations are calculated in a moving window and the maximum value is used to determine the phase shift. The resulting value is compared with the previous shift and the analysis result is printed or saved to a file (Listing 4).

```
$ anomalies.exe -a 2 -if data\UT5.csv -cf da-
ta\UT3.csv -dT 7 -dPP 7 -vsT
0.005208,0.010416,0.015625,0.020833,0.026041,0.
03125,0.036458,0.041666 -vsPP
0.005208,0.010416,0.015625,0.020833,0.026041,0.
03125,0.036458,0.041666
Anomaly detection - method 2
Temp  UT5.csv vs UT3.csv: Delay: 1, Old Delay:
7, Anomaly: 1
PPres UT5.csv vs UT3.csv: Delay: 25, Old Delay:
7, Anomaly: 2
```

Listing 4.   Example of anomaly detection. Required frequencies are given in the command line. Anomaly was detected in time series of pore pressure only.

4.4   *Change in the average level of the phenomenon*

The last method implemented to detect an anomaly is based on the significant parameter change. If the value of temperature or pore pressure changes more than selected threshold, the time series is assumed to have an anomaly. There is only one input file and it should contain a time series to check. Thresholds are supplied to the program in the same manner, as arguments.

5   TEST

For testing methods and application, slowly varying series of sine function times were created with and without anomalies. For one time series without an anomaly (comparing period), three time series were prepared with the most common types of anomalies: single high values, short-term elevated values, new trends. The test data set is presented in Table 1 below. An example of time series with and without anomaly is presented in Table 1 and Figure 2.

5.1   *Analysis of amplitude components for consecutive time series test*

Z standardisations were made for the tested data. Fast Fourier transform and power spectral density for the chosen frequencies presented above were computed. The difference between power spectral densities measured as mean square error is presented in Table 2. In addition, in the same table correlation coefficients between power spectral densities for both time series (with and without anomaly) are shown. As can be seen in Table 2, only 3 anomalies were not recognised for the chosen threshold (MSE = 0.75). If the results of this test are compared with the test data shown in Table 1, it is clear that the chosen algorithm works fine in most cases.

Table 1. Tested dataset.

Sensor	Without anomaly	Anomaly 1	Anomaly 2	Anomaly 3
Tout 12	Sin + Rnd(1) + 6	T = (191–192) + 7.0	T = (183–192) + 2.0	T = (180–192) + 5.2 + n*0.3, n = 1,...,13
Tout 13	Sin(v1) + rnd(2) + 5,5	T = (190–192) + 10.0	T = (186–192) + 6.0	T = (173–192) + 6.3 + n*0.1, n = 1,...,20
UT 6 temp	sin(v1) + rnd(0,75) + 5,5	T = (190–192) + 10.0	T = (183–192) + 1.0	T = (178–192) + 6,5 + n*0,2, n = 1,...,15
UT 6 ppress	sin(v1)*150 + 25000 + rnd(100)	T = (191–192) + 2000	T = (188–192) + 7000	T = (183–192) + 2580 + n*5000, n = 1,...,10
UT 19 temp	sin(v1) + rnd(0,65) + 5,5 + rnd(0,5)	T = (191–192) + 4.5	T = (190–192) + 1.5	T = (185–192) + 5.75 + n*0.15, n = 1,...,8
UT 19 ppress	sin(v1)*170 + 22000 + rnd(120)	T = (188–192) + 22000	T = (180–192) + 6000	T = (179–192) + 25000 + n*400, n = 14

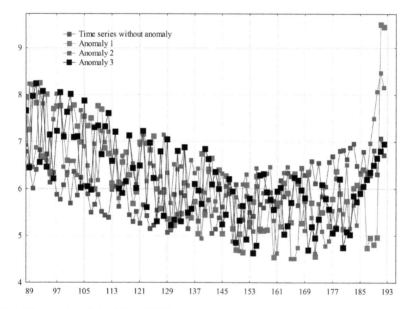

Figure 2. Tested temperature dataset from UT19 sensor.

The method might only fail if an anomaly starts to develop with small changes in value. Table 2 also shows the correlation values between the data with and without anomaly. The similarity between both data sets is high and in most cases above 0.9.

5.2 Analysis of model fitting test

In method 2, a Fourier frequency model was created for a period without anomalies. In next step, the model was compared with data from the tested period (newest 192 observations). Fitting of models for the period without anomalies and the tested period is compared according to the criteria presented above in 3.2. Anomalies detected during the test are marked in bold font. As can be seen in Table 3, only one anomaly was not detected.

5.3 Analysis of phase shifts for pairs of sensors

For this test, data generated before for temperature sensor 'Tout 12' was used, but during the test it was shifted and some Gaussian noise was added. As can been seen in Table 4, the results obtained by this method are weakest for all three methods. This method recognises 3 out of the 5 anomalies in the data set.

Table 2. Results of testing with method 1.

Sensor		Pearson coeff.	MSE
Tout 12			
Temp	Anomaly 1	0.98	2.08
	Anomaly 2	0.99	1.15
	Anomaly 3	0.95	1.58
Tout 13			
Temp	Anomaly 1	0.97	2.38
	Anomaly 2	0.99	0.77
	Anomaly 3	0.99	0.36
UT 6			
Temp	Anomaly 1	0.89	1.34
	Anomaly 2	0.98	0.59
	Anomaly 3	0.98	0.88
Ppress	Anomaly 1	0.45	3.39
	Anomaly 2	0.65	2.81
	Anomaly 3	0.79	2.27
UT 19			
Temp	Anomaly 1	0.98	0.73
	Anomaly 2	0.97	1.06
	Anomaly 3	0.98	0.76
Ppress	Anomaly 1	0.69	1.39
	Anomaly 2	0.93	0.98
	Anomaly 3	0.94	0.95

Table 3. Results of testing with method 2.

Sensor, Parameter	R2		SW-W
Tout 12			
Temp	without anomaly	0.85	0.99
	Anomaly 1	0.46	0.37
	Anomaly 2	0.83	0.73
	Anomaly 3	0.77	0.66
Tout 13			
Temp	without anomaly	0.97	0.97
	Anomaly 1	0.23	0.24
	Anomaly 2	0.57	0.61
	Anomaly 3	0.8	0.91
UT 6			
Temp	without anomaly	0.41	0.98
	Anomaly 1	0.11	0.48
	Anomaly 2	0.36	0.99
	Anomaly 3	0.28	0.97
Ppress	without anomaly	0.85	0.99
	Anomaly 1	0.12	0.08
	Anomaly 2	0.43	0.15
	Anomaly 3	0.56	0.26
UT 19			
Temp	without anomaly	0.48	0.94
	Anomaly 1	0.4	0.94
	Anomaly 2	0.55	0.98
	Anomaly 3	0.53	0.98
Ppress	without anomaly	0.35	0.98
	Anomaly 1	0.08	0.16
	Anomaly 2	0.08	0.34
	Anomaly 3	0.03	0.34

Table 4. Results of testing method 3.

Real delay		Computed delay	
Number	Pearson coeff.	Number	Pearson coeff.
5	0.45	4	**0.53**
6	0.78	6	**0.78**
9	0.82	12	**0.87**
10	0.76	10	**0.75**
11	0.67	11	**0.67**

6 CONCLUSIONS

The anomaly detection methods presented in this paper allow the detection of undesired events occurring in flood embankments. The selected methods and algorithms are easy to implement and interpret, and proved their high efficiency during testing. Only one of the above-mentioned methods was not satisfactorily effective for the test data; therefore, it requires further testing and modification. However, this method could still detect weakening of flood embankment cross-sections.

The first of the presented methods recognised 15 out of 18 anomalies, which is 83% efficient. The second method recognised 17 out of 18 anomalies, which is more than 90% efficient. The third method recognised only 3 out of 5 anomalies, so, at 60%, this is the least effective method. The effectiveness of the last presented method depends on threshold values designated by the person assessing the flood embankment.

Written applications encapsulating the presented anomaly detection methods enable them to be used by a person with no statistic knowledge. Applications are run in batch for data with a specified structure. The results are presented on the screen or saved to a file. The most important information in the output file is that in the analysed time series a potential anomaly was detected. In addition, evaluation results are collected by the root application and can be visualised in order to create maps of flood embankments with marked positions of potential anomalies.

ACKNOWLEDGEMENT

This work was financed by the National Centre of Research and Development (NCBiR), Poland, project PBS1/B9/18/2013 (no. 180535).

This work was partly supported by the AGH University of Science and Technology, Faculty of Geology, Geophysics and Environmental Protection, as a part of statutory project no. 11.11.140.613.

REFERENCES

Bajerski, P. & Kozielski, S. 2009. Computational Model for Efficient Processing of Geofield Queries. *Proceedings of the International Conference on Man-Machine Interactions*: 573–583.

Balis, B., Kasztelnik, M., Bubak, M., Bartynski, T., Gubała, T., Nowakowski, P. & Broekhuijsen, J. 2011. The urbanflood common information space for early warning systems. *Procedia Computer Science* 4: 96–105.

Chochran, W.T., Bell Telephone Laboratories, Inc., Holmdel, N. J, Cooley, J.W., Favin, D.L & Helms, H.D. 1967. What is the fast Fourier transform?, *Proceedings of the IEEE*:.55(10).

Chuchro, M., Lupa, M., Pięta, A., Piórkowski, A. & Leśniak, A. 2014. A concept of time windows length selection in stream databases in the context of sensor networks monitoring, *New trends in database and information systems II*: 173–183.

Habrat, M., Lupa, M., Chuchro, M. & Leśniak, A. 2015. A decision support system for emergency flood embankment stability. *Procedia Computer Science* 51: 2957–2961.

Kasprowski, P. 2012. Choosing a persistent storage for data mining task. *Studia Informatica* 33(2B): 509–520.

Krzhizhanovskaya, V.V., Shirshov, G.S., Melnikova, N.B., Belleman, R.G., Rusadi, F.I., Broekhuijsen, B. J., Gouldby, B.P., Lhomme, J., Balis, B., Bubak, M. et al. 2011. Flood early warning system: design, implementation and computational modules. *Procedia Computer Science* 4: 106–115.

Lupa, M., Chuchro, M., Piórkowski, A., Pięta, A., Leśniak, A. 2014. A proposal of hybrid spatial indexing for addressing the measurement points in monitoring sensor networks. *Beyond Databases, Architectures and Structures, Springer International Publishing*: 437–447.

Mrozik, K., Przybyła, C., Pyszny, K. 2015. Problems of the Integrated Urban Water Management. The Case of the Poznań Metropolitan Area (Poland). *Rocznik Ochrona Środowiska/ Annual Set the Environment Protection,* 17(1): 230–245.

Pengel, B.E., Krzhizhanovskaya, V.V., Melnikova, N.B., Shirshov, G.S., Koelewijn, A.R., Pyayt, A.L, and Mokhov, I.I. 2013. *Flood early warning system: sensors and internet.*

Pięta, A., Krawiec, K. 2015. Random set method application to flood embankment stability modelling, *Procedia Computer Science* 51: 2668–2677.

Piórkowski, A., Leśniak, A. 2014. Using data stream management systems in the design of monitoring system for flood embankments. *Studia Informatica* 35(2): 297–310.

Pyayt, A.L., Mokhov, I.I, Kozionov, A.P., Kusherbaeva, V.T., Lang, B., Krzhizhanovskaya, V.V, Meijer, R.J. 2012. Data-driven modelling for flood defence structure analysis. *Comprehensive Flood Risk Management: Research for Policy and Practice*: 77.

Royston, P. 1992. Approximating the Shapiro-Wilk W-test for non-normality, *Statistic and Computing* 2: 117–119.

Różyński, G., Ostrowski, R, Pruszak, Z., Szmytkiewicz, M., Skaja, M. 2006. Data-driven analysis of joint coastal extremes near a large non-tidal estuary in North Europe. *Estuarine, Coastal and Shelf Science* 62(1–2): 317–327.

Wei, W.W.S. 2006. *Time Series Analysis. Univariate and Multivariate Methods*. Second Edition, Pearson Addison Wesley, New York.

Welch, P.D. 1967. The use of Fast Fourier Transform for the Estimation of Power Spectra: A method based on time averaging over short, modified eriodograms, *Trans. Audio and Electroacoust*. AU-15: 70–73.

Wiatr, K., Kitowski, J., Bubak, M. 2014. An approach to monitoring, data analytics, and decision support for levee supervision, *Proceedings of the Seventh ACC Cyfronet AGH Users' Conference*, ACC CYFRONET AGH, Kraków: 75–76.

http://www.urbanflood.eu - webpage of UrbanFlood project.

http://www.danube-floodrisk.eu—webpage of DANUBE FLOODRISK project.

http://www.floodsite.net—webpage of FLOODsite project.

http://www.ismop.edu.pl—webpage of ISMOP project.

Environmental Engineering V – Pawłowska & Pawłowski (Eds)
© 2017 Taylor & Francis Group, London, ISBN 978-1-138-03163-0

Stochastic model for estimating the annual number of storm overflow discharges

B. Szeląg
Faculty of Environmental, Geomatic and Energy Engineering, Kielce University of Technology, Kielce, Poland

ABSTRACT: Emergency overflow weirs are important components of stormwater drainage systems. During intensive rainfalls significant amounts of stormwater and contaminants contained therein are discharged through these outlets into receiving bodies of water, which may cause the disturbance of the biological-chemical balance. Therefore it is advisable to model the performance of overflow weirs in order to reduce the detrimental impact of the stormwater on the receiver. The article presents a stochastic model for estimating the annual number of storm overflow discharges. Defining the occurrence of an overflow discharge event involved the application of logistic regression, whereas the Monte Carlo method was used for computing annual sequences of rainfall events. 8-year sequences of rainfall measurements were used to estimate the parameters of statistical rainfall distributions. The calculations showed that the knowledge of two parameters i.e. the total and maximum 30-minute rainfall depth is sufficient to estimate the overflow performance.

Keywords: logistic regression, Monte Carlo, SWMM, overflow, stormwater

1 INTRODUCTION

The fundamental function of the stormwater drainage system is draining away precipitation water from the surface of an area to a stormwater treatment plant. When the plant's capacity is exceeded, the excess stormwater is discharged directly into the receiving water via the emergency overflow weir, which leads to an abruptly increased watercourse flow and may contribute to scouring and the disturbance of biological-chemical balance (Martino et al. 2011, Szeląg et al. 2013, Bąk et al. 2012). Therefore, in order to limit a potential threat for the receiving water, it is necessary to know the number and volume of discharges and the quality of the effluent discharged directly into the receiving water bodies. The above considerations resulted in the development of computational methods (Mantegazza et al. 2010, ATV—A 128, Zabel et al. 2001, Dąbrowski 2007) for modelling the storm overflow performance. The methods of determining the number of storm overflow discharges in the combined sewer system can be divided into: empirical ones (Shigorin's equations, equations derived by Fidala-Szope et.al (1999) based on many years of research), deterministic ones (Kuiper's (Dąbrowski 2007, Mrozik et al. 2015) methods of routing drain retention above overflow weirs) and stochastic ones.

The method proposed by Fidala-Szope was based on the results of the measurements performed in a research catchment in Warsaw with an area of 600 hectares and the runoff coefficient of 0.40. The method involved determining the dependencies for defining the number of storm overflow discharges, taking into account rain intensity and the volume of stormwater accumulated in the stormwater drainage system. This method as well as other solutions described by Urcikán and Rusnák (2006) or Dąbrowski (2007) can be applied on a limited scale for catchments of similar physical-geographical and meteorological characteristics. In Kuiper's method calculating the number of overflow discharges is based on multiannual rainfall measurement sequences. When the volume of the rainfall and the stormwater detained in the drainage system are known, it is possible to determine the position of the overflow crest for which there is no stormwater discharge. Since these methods disregard the variability of rainfall intensity distribution, catchment detention (Dayarante & Perera 2008, Gironas et al. 2008), a stochastic nature of rainfalls (Licznar et al. 2011, Rupp et al. 2011) or the dynamics of the inflowing stormwater, they are of merely estimative character and do not make it possible to determine clearly the effect of stormwater on the aquatic environment in the receiving waters.

Alternatively, in order to determine the number of storm overflow discharges one can use numerical programmes (SWMM, PCSWMM, MIKE URBAN MOUSE) as well as simplified mathematical models (Vaes & Berlamont 1999, Dempsey et al., 1997, Wu et al. 2011) based on one-dimensional unsteady flow equations and applied to estimate the performance of systems. Literature review (Cambez et al. 2008, Szeląg et al. 2013, Zawilski & Sakson 2013) shows that the SWMM programme (Storm Water Management Model) is widely used for assessing the performance of a stormwater drainage system. This is mainly due to the possibility of modifying the source code of a program written in the C ++ programming language to one's own needs. Moreover, it should be emphasized that the SWMM has been successfully applied for modelling the performance of separators (Mrowiec 2009), green roofs (Krebs et al. 2013) infiltration trenches or treatment plants (Szeląg et al. 2013). The assessment of the storm overflow performance is based on continuous simulations with the use of multiannual observation sequences. The calculations with regard to the performance of the sewer system network (Carleton et al. 1990, Vaes & Berlamont 1999, Andrés-Domenech et al. 2010) showed that the simulation results can be used in practical discussions, allowing for the evaluation of the performance of the sewer network and objects located therein. Nevertheless, the development of hydrodynamic model is not always viable because of the cost of multiannual rainfall and high resolution flow measurements. In addition, due to the number of parameters that describe the characteristics of the catchment and the stormwater drainage network in the numerical model, there are problems with clear identifation of the values of parameters providing a satisfactory solution (Romanowicz & Beven 2006, Kleidorfer et al. 2009, Kiczko et al. 2013).

Owing to numerous simplifications in empirical and deterministic methods such as frequent problems in the design and calibration of hydrodynamic models, stochastic models were developed to estimate the annual number of storm overflow discharges. Hitherto Thorndahl and Willems (2008) presented a probabilistic model to evaluate the overflow performance in which the FORM (First Order Reliability Model) method was applied to identify the event of an overflow discharge. This method requires the implementation of complex numerical algorithms to determine the model parameters, which greatly limits its application in engineering solutions. In view of the above it is advisable to develop a model for estimating the overflow weir performance, taking into account not only the stochastic nature of rainfall and the variability of rainfall intensity during the various episodes of rainfall, but also the transformation of rainfall into runoff. The suggested solution would not require the development of a hydrodynamic model.

These analyzes show the possibility of using logistic regression to assess the probability of an overflow discharge based on the total and maximum 30-minute rainfall depth in the event. The Monte Carlo method was applied to compute the parameters discussed above, using the statistical distributions determined on the basis of the rainfall measurement results conducted in the years 2008–2015. The identification of the overflow discharge event involved the application of the hydrodynamic simulations results performed in the SWMM programme.

2 MATERIAL AND METHODS

2.1 Object of the study

The paper analyzes the Si9 sewer catchment with the area of $F = 62$ ha located in the city of Kielce, built up with residential quarters, public buildings, main and side streets. A detailed description of the catchment can be found in the works of Dąbkowski et. al. (2010), Szeląg (2013). Stormwater flowing through the storm drain Si9 from the catchment is directed to the Stormwater Treatment Plant (STP) through a division chamber. In the STP the stormwater passes a rectangular settling basin, a coalescence separator and an inspection chamber. When the chamber is filled to a depth of less than 0.42 m, the stormwater flows through four pipes Ø 400 mm to a settling basin, whereas when the division chamber is filled to a depth exceeding 0.42 m, the stormwater is directed to the discharge canal through the stormwater overflow (Fig. 1) and then to the receiver, i.e. the Silnica river. The stormwater from the rectangular settling basin flows simultaneously: to the separator through the Ø 200 mm pipe and to the control chamber through two Ø 500 mm pipelines. Then the stormwater flows from the control chamber to the receiver.

An ultrasonic flow meter Teledyne type 2150 was installed in a Ø 1250 mm—diameter drain, at a distance of 3 meters form the Si9 outlet to the division chamber, which is additionally equipped with sensor for measuring the filling level. The flow meter operates by measuring water pressure and average velocity in the cross—section of the channel. The devices measures the filling and the flow rate of the stormwater at different interval. The measurement is performed every 5 minutes when $h < 0.07$ m, and every 30 seconds when $h < 0.07$ m.

2.2 Precepitation data

The duration of the analysed rainfall events (t_d) was from 20 to 2366 minutes, the period of dry weather (t_{dw}) varied within the range of 0.16 to 60 days, the

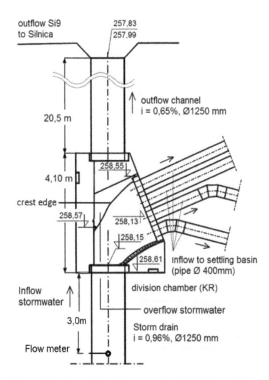

Figure 1. Operational diagram of the Division Chamber (DC).

total rainfall depth (P_c) amounted to 1.0 ÷ 45.2 mm, whereas the maximum 10, 15 and 30 —minute rainfall depths ($P_{t = 10,15,30}$)—reflecting the varied rainfall intensity distribution in the event—were: $P_{t = 10} = 0.2 ÷ 14.4$ mm, $P_{t = 15} = 0.2 ÷ 19.3$ mm and $P_{t = 30} = 0.3 ÷ 21.2$ mm, respectively.

In order to model the performance of the emergency overflow weir this paper uses synthetic rainfall sequences obtained by the Monte Carlo method. In view of the above, the measured rainfall sequences provided the base for the empirical distribution functions of the selected parameters (P_c, $P_{t = 30}$, $P_c(t-1)$, t_d, t_{dw}) depicting the variation in rainfall dynamics during particular events. The following statistical distributions were analyzed: Weibull, chi-square, exponential, GEV, Fisher-Tippet, gamma, log-normal, Pareto and beta in order to obtain the best fit of the theoretical distribution to the empirical one. The goodness of fit between the empirical and theoretical distribution was evaluated on the basis of the results of the Kolmogorov-Smirnov and Chi-square tests.

2.3 Rainfall event and the dry weather period

The evaluation of the storm overflow performance involved the use of 8 – year rainfall sequences (2008–2015) from a raingauge station located app. 2 km from the Si9 storm drain catchment boundary. The evaluation of the emergency overflow weir was based on 8 – years rainfall sequences with a 5 – minute interval recorded by an RG 50 rain gauge produced by SEBA Hydrometrie GmBH, operating in accordance with the requirements of the World Meteorological Organization. The rain gauge is located app. 2 km from the boundary of the Si9 sewer catchment.

On the basis of the reviewed reference sources (Fu & Butler 2014, Fontanazzza et al. 2012, Balistracchi & Biachi 2011, Thornadhl 2008) it can be concluded that "rainfall" is not defined clearly, since a rainfall event is regarded as a rainfall whose depth ranges from 1 to 10 mm. Since in this paper the stochastic model is used to predict the number of overflow discharges in an annual cycle, the adopted minimum rainfall layer depth that provides the basis for the development of the statistical distribution should also allow a case identification when the overflow discharge does not occur. Consequently, the value of the rainfall layer depth specified in the guideline ATV—A 118 (2000) equal to $P_c = 10$ mm, defining an intensive rainfall event, appears to be too high in the analyzed case, whereas the minimum value $P_c = 1$ mm is insufficient, as it may be encumbered with a considerable measurement error resulting from the applied measuring devices (Kotowski et al. 2011). Therefore in the above case for the rainfall to be classified as a rainfall event, its minimum depth must be 2 mm, which, as specified in numerous sources (Fu & Butler 2014, Fontanazza et al 2012, Balistracchi & Biachi 2011), provides the basis for determining the multidimensional distributions of extreme values to evaluate the performance of the stormwater drainage system. It was adopted that the separation time for the successive rainfall events is 4 hours (ATV A—118).

Statistical distributions (empirical distribution functions) were determined for the analysed precipitations and for the parameters describing their variability during rainfall events, and the obtained results were fitted with one of the following statistical distributions: Weibull, chi-square, exponential, GEV, Fisher-Tippet, gamma, log-normal, and beta. The goodness of fit between the empirical and theoretical distribution was evaluated on the basis of the results of the tests Kolmogorov-Smirnov and Chi-square.

2.4 Hydrodynamic model

The hydrodynamic flow model developed in the SWMM programme consisted of 92 subcatchments of 0.12 ÷ 2.10 ha, 200 manholes and 72 pipe segments. The values of the Manning roughness coefficient for various surfaces in the catchment are equal to 0.015 m$^{-1/3}$·s and 0.15 m$^{-1/3}$·s, whereas the roughness coefficient for the sewers' walls has a value of 0.015 m$^{-1/3}$·s. The mathematical model of the Stormwater Treatment Plant (STP) was developed based on the project documentation, the field measurements and the facility project documentation (Butech, 2003). The absolute roughness of the pipe was assumed as k = 1.5 mm (ATV A – 110). Due to a variable width of the overflow crest, a broad-crested hydraulic model of an overflow was applied in the analyses. The calibration of the division chamber model was performed on the basis of the filling level measurements by means of a sampler. The relation of the measured maximum filling level of the Division Chamber (DC) to the simulated one was used to evaluate the conformity of the measured and simulated filling levels of the stormwater division chamber, expressed as:

$$\delta = \frac{h_{DC(mes)}}{h_{DC(sim)}} \quad (1)$$

where: $h_{DC(mes)}$—the maximum measured division chamber filling level (m), $h_{DC(sim)}$—the maximum division chamber filling level calculated using the SWMM program (m).

The calibration of the hydrodynamic model of the division chamber was performed on the basis of the rainfall and flow measurements in the years 2008 ÷ 2012.

2.5 Logistic regression model

Logistic regression model (McFadden 1973), also called the binomial logit model, is primarily used for the analysis of binary data, so it can define the probability of occurrence or non-occurrence of events (McFadden 1973). This model is widely used in economics, social and medical sciences, it is also applied in the river engineering (Pradhan 2009; Sun et al. 2011; Heyer et al. 2012), geotechnics (Ingelmo et al. 2011), geomorphology (Ayalew et al. 2005; Kawagoe et al. 2010) and ecology (Graniero et al. 1999; Horssen et al. 2002), but it has not yet been used to evaluate the performance of an overflow of another structure located in the stormwater drainage system.

The logistic regression model represents a special case of the generalized linear model, which can be written as:

$$g(\mu) = \beta_0 + \beta_1 \cdot X_1 + \beta_2 \cdot X_2 + ... + \beta_j \cdot X_j \quad (2)$$

where: g—the link function expressing the relation between the average value of the dependent variable μ = E(Y|X$_1$ = x$_1$, X$_2$ = x$_2$, X$_j$ = x$_j$) and a linear combination of predictors, β_0 –intercept, β_1, β_2,... β_j—regression coefficients, X$_j$—dependent variables, which in the above study include: rainfall duration time (t$_d$), the dry weather period (t$_{bd}$), rainfall depth P$_i$(t−1) and rainfall duration time during the preceding event t$_d$(t−1).

In the analyzed model p = P(Y|X$_1$ = x$_1$, X$_2$ = x$_2$, X$_j$ = x$_j$) the link function, commonly known as logit, takes the form:

$$g(p) = \log it(p) = \ln\left(\frac{p}{1-p}\right) \quad (3)$$

Summing up, this model can be expressed as:

$$p = P(Y|X_1 = x_1, X_2 = x_2, ...X_j = x_j)$$
$$= \frac{\exp\left(\beta_0 + \sum_{i=1}^{j} \beta_i \cdot x_i\right)}{1 + \exp\left(\beta_0 + \sum_{i=1}^{j} \beta_i \cdot x_i\right)} \quad (4)$$

Model parameters β_0, β_1,..., β_j are determined by the greatest likelihood method maximizing the log-likelihood function with respect to the model parameters using iterative numerical methods.

In the above discussion the fundamental criterion for the overflow discharge occurrence is the probability value P (X) = 0.50 (Graniero and Price 1999) assigned to a set of variables describing the phenomenon, defined by the following relationship:

$$\beta_0 + \beta_1 \cdot t_d + \beta_2 \cdot t_{bd} + \beta_3 \cdot t_d(t-1)$$
$$+ \beta_i \cdot P_i + \beta_j \cdot P_j(t-1) = 0 \quad (5)$$

Equation (5) describes the multidimensional space of precipitation events, during which there is no stormwater discharge through the overflow.

The quality of the model, i.e. the fit of the measurement results with the calculations in the logistic regression model is normally determined using the coefficients: R^2 McFadden's, R^2 Cox–Snell's, R^2 Nagelkerke's (Harrell 2001). Apart from fitting, the predictive capacity (accuracy of estimation) of the logistic regression model is evaluated based on the sensitivity (SENS), specificity (SPEC) and the counting error (R$_z^2$). In order to evaluate the accuracy of estimation, the information criteria are applied as well (Akaike, Bayesian) (Harell 2001). The independent variables in the logistic regression model were selected using the backward stepwise method. The estimation of the β_i coefficients in the logistic regression model was based on the results of numerical calculations performed using SWMM for the years 2008–2015, 10% of which was used to validate the model.

2.6 Estimating the annual number of discharges

Since the estimation of the number of discharges was based on merely 8-year rainfall sequences, in the analysis the average annual number of rainfall events was assumed as $M = N/P = 464/8 = 58$ (where: N—the number of rainfall events in the analysed period of time) (Muhaisen et al. 2009, Osorio et al. 2009). This was followed by two thousand Monte Carlo samplings of rainfall sequences (t_d, t_{bd}, P_i) from statistical distributions taken into account in the logistic regression model in the M-number of rainfall events. The results of simulations from different samples were applied in equation (5) and the probability of storm overflow discharge was calculated, followed by determining whether the discharge event will occur. On this basis an annual number of discharges was defined in the generated sequences and empirical cumulative distribution functions were developed to describe the probability of not exceeding the annual number of storm overflow discharges. In the case when the logistic regression model revealed the possibility of binding independent variables, a linear regression model was applied to reduce the number of variables in the multidimensional probability density distribution.

3 RESULTS AND DISCUSSION

Based on the collected results of the measurements of: rainfall, flows in the outflow storm drain from the catchment, the filling level in the division chamber and the calculations in the SWMM, the parameters of the treatment plant and the overflow were determined through calibration, which allowed for calculating the fitting parameters of the simulated and measured filling levels of the division chamber. The model calibration was performed on the basis of seven rainfall events (Table 1); in the case of the intensive rainfalls of 8th July 2011, 15th Sept. 2010, 30th July 2010 and 8th July 2009 the stormwater was discharged through the outflow, whereas in the remaining cases the discharge did not occur. The obtained values of the δ parameter: $\delta = 0.89 \div 0.97$ (Table 1) indicate the model's satisfactory projection of the hydraulic conditions in the DC because the developed model overestimates the DC filling levels by no more than 11%. Moreover, the performed computations showed that the yield coefficient for the emergency overflow weir is 0.36, whereas the values of local head loss coefficient at the inlet and outlet of pipelines outflowing stormwater to the settling basin varied within the ranges: $0.48 \div 0.53$ and $0.95 \div 1.10$.

That is confirmed by the results of the simulations performed in the SWMM, which showed that in the case of 371 rainfall events registered in the period of $2008 \div 2012$ the developed model correctly predicted the occurrence of the storm overflow discharge in 93% (94 out of 101 events), whereas the lack of discharge was correctly predicted in 95% of the analyzed cases (153 out of 162 events), which indicates a satisfactory fit of the measurement results with calculations.

The calibrated hydrodynamic model of the system was used to define the annual number of overflow discharges in the years 2008–2015; the simulation results are presented in Figure 2. On this basis it can be concluded that the annual number of discharges in the analyzed period ranged from 15 to 29, which indicates a considerable variation of the rainfall intensity distribution in the rainfall events. Next, using the simulation results, a logistic regression model was constructed by backward stepwise method. The defined rainfall parameters P_c, $P_{t=30}$, $P_c(t-1)$, t_d and t_{bd} and the corresponding identified overflow discharge events or lack thereof provided the input data for calculations. The obtained logistic regression model can be expressed as:

$$p = \frac{\exp(\beta_1 \cdot P_{t=30} + \beta_2 \cdot P_c + \beta_0)}{1 + \exp(\beta_1 \cdot P_{t=30} + \beta_2 \cdot P_c + \beta_0)} \qquad (6)$$

Table 1. Calibration results of DC filling levels.

Date	P_c mm	t_d min	δ –
08.07.2011	8.6	60	0.88
15.09.2010	9.2	286	0.95
30.07.2010	12.5	107	0.97
08.07.2009	16.5	270	0.94
03.07.2009	4.2	26	0.96
31.05.2010	5.4	56	0.91
26.04.2010	3.6	92	0.92

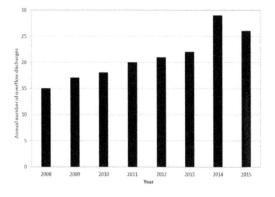

Figure 2. The annual number of overflow discharges in $2008 \div 2015$.

Table 2. Results of the parameters estimation in the logistic regression model.

Variable	Coefficient (β_i)	Standard error	p-value
β_0	−7.269	0.911	0.01
$P_{t=30}$	3.122	0.437	0.0159
P_c	0.135	0.054	0.000001
AIC = 123.03	R^2 McFadde = 0.721		SENS = 0.904
SBC = 133.15	R^2 Cox and Snell = 0.606		SPEC = 0.965
	R^2 Nagelerke = 0.835		R_z^2 = 0.940

where the values of the determined β_i coefficients and parameters defining the fit of measurement results with hydrodynamic calculations are presented in Table 2.

The calculated value of SENS = 0.904 (Table 2) indicates that out of 145 probable cases of overflow discharge in the period of 2008 ÷ 201, the model correctly classified 130 events, whereas based on SPEC = 0.965 it can be stated that out of 273 events of a lack of discharge the logistic regression model properly classified 263 events. The above results show that out of 418 investigated cases the model properly classified 393 events in the overflow performance. The calculations performed for the validation set showed that the model classified correctly 18 out of 20 discharge events, whereas in the case of the no-discharge events 25 out of 26 results were in agreement with the measurements.

Fig. 3 presents a graphic interpretation of the logistic regression model and depicts the impact of both: the total and the maximum 30-minute rainfall depth on the probability of a storm overflow discharge. The performed calculations results (Fig. 3) reveal that it is the maximum 30-minute rainfall depth that has the crucial impact on the occurrence of an overflow discharge, e.g. the increase in the value of $P_{t=30}$ from 1.25 mm to 2.5 mm for P_c = 10 mm leads to the rise of the discharge probability from $p(x_i)$ = 0.12 to $p(x_i)$ = 0.77.

The performance of the storm overflow is also significantly affected by the total rainfall depth in the rainfall event (Fig. 3). As a result of the increase in the total rainfall depth from P_c = 10 mm to P_c = 20 mm, the maximum 30-minute rainfall depth at which the overflow discharge will occur $p(x_i)$ = 0.50 decreases from 1.80 mm to 1.47 mm. In addition, the increase of P_c from 5 mm to 15 mm for $P_{t=30}$ = 1.5 mm causes an increase in the likelihood of an overflow discharge from $p(x_i)$ = 0.12 to $p(x_i)$ = 0.35. Since the probability of an overflow discharge is determined by P_c and $P_{t=30}$, the statistical distributions were defined for the analysed variables, checking whether the variables are independent before proceeding with further analyses.

For this purpose, the value of Spearman's rank correlation coefficient was determined; the calculations showed a moderate relationship between P_c and $P_{t=30}$ and the correlation coefficient was r = 0.60, accordingly, meaning that the variables are not independent. Given the above results and knowing that the development of two-dimensional probability density distributions of the dependent variables involves at least 14-year rainfall sequences (Muhasien et al. 2009, Balistrocchi & Bacchi 2011), the total rainfall depth was determined from the following relation:

$$P_c = 1.46(\pm 0.12) \cdot P_{t=30} + 3.56(\pm 0.46) \, (r = 0.60) \quad (7)$$

Taking into account the obtained results of calculations, a statistical distribution of $P_{t=30}$ based on 8-year rainfall sequences was determined (Table 2), while the value of P_c is calculated from formula (7).

The analyses (Table 3) show that the measurement data of the maximum 30-minute rainfall depth are best fitted with the log-normal distribution (Kolmogorov—Smirnov test p = 0.389 and Chi—square p = 0.056 for α = 5%) with a standard deviation of σ = 0.854 mm, and the mean value μ = 0.591 mm (Fig. 4). The resulting empirical cumulative distribution function for $P_{t=30}$ (Fig. 4) indicates a satisfactory fit of the empirical data with those calculated using the obtained statistical distribution.

Given a specific statistical distribution and regression relations, the annual synthetic sequences of precipitation were determined, sampling 58 rainfall events 5000 times ($P_{t=30}$, P_c) with the Monte Carlo method. The results of the simulation are shown in equation (6) and the fact that overflow discharge occurs when $p(x_i) > 0.5$ allowed for the identification of the events during which the storm

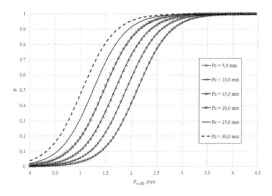

Figure 3. Impact of the total and maximum 30-minute rainfall depth on the probability of the storm overflow discharge.

Table 3. Results of calculating the value of (p) in Kolmogorov-Smirnov and Chi-square tests of the goodness of fit of the empirical distribution to the theoretical distribution obtained using statistical distributions.

Distribution	Test Kolmogorov-Smirnov	Test Chi-square
Beta	0.0001	0.0002
Chi—square	0.0002	0.0003
Weibull	0.0253	0.0312
Expotential	0.0007	0.0005
GEV	0.0002	0.0003
Fisher—Tippet	0.0003	0.0004
Log—normal	0.3130	0.2526

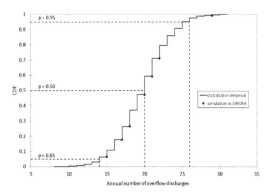

Figure 5. The theoretical Cumulative Distribution Function (CDF) describing the probability of not exceeding the annual number of overflow discharges.

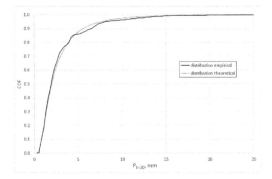

Figure 4. Theoretical and empirical cumulative distribution function of the maximum 30-minute rainfall depth.

overflow discharge took place in the generated sequences.

This allowed for the calculation of an annual number of discharges in the determined sequences, followed by the determination of the theoretical cumulative distribution function describing the likelihood of not exceeding the yearly number of discharges (Fig. 5). In addition, Fig. 5 presents the results of the calculation of an annual number of discharges obtained in the SWMM program for different years. Based on the results of calculations (Fig. 5) it can be stated that the expected value of the annual number of overflow discharges corresponding to the percentile p = 0.50 is equal to 19; while the percentile p = 0.05 and p = 0.95 correspond to 13 and 25 discharges, respectively.

The simulations conducted in the SWMM program show that the annual numbers of discharges for the years 2008 ÷ 2016 are in the range of the obtained probabilistic solution, which confirms the applicational character of the method. The number 15, being the minimum annual number of overflow discharges obtained by hydrodynamic simulation, represents the percentile p = 0.10; while the maximum annual number of discharges obtained in the SWMM program—the number 29 – corresponds to the percentile p = 0.996. Since the total rainfall depth is underestimated, and it has a significant impact on the likelihood of an overflow discharge, it can be concluded that the determined annual number of storm overflow discharges (Fig. 5) in this case is underestimated.

Therefore, it is recommended to conduct further rainfall measurements in order to develop a two-dimensional probability density distribution taking into account the boundary distribution functions of both the total and the maximum 30-minute rainfall depth; as an alternative solution it is acceptable to make a mathematical model $P_c = f(P_{t=30})$ with predictive capabilities determining a higher value of the coefficient of determination than the one obtained using linear regression. Moreover, taking into account the direction of further analyses, it is advisable to investigate the uncertainty of the estimated parameters in the hydrodynamic model in order to determine reliable values of parameters describing the characteristics of different sub-catchments and drains.

4 CONCLUSION

Owing to the fact that the surface runoff is a complex phenomenon, the modelling of the system operation, including the overflow weir performance, involves the application of hydrodynamic models. However, due to the cost of constant rainfall and flow measurements as well as the problems with calibrating the model parameters, it is advisable to develop methods providing high accuracy and simplicity of use.

This article presents the possibility of using logistic regression to determine the probability of a storm overflow discharge. The analyzes showed that with regard to the catchment under discussion two parameters are sufficient for the performance description, i.e. the total and maximum 30-minute rainfall depth.

The results of calculations indicate that the stochastic model developed to assess the annual amounts of discharges can be used in practical discussion but in this case it is useful to improve the predictive capability of the relations used to forecast the total rainfall depth. Nevertheless, it must be emphasized that the annual number of storm overflow discharges, obtained by hydrodynamic simulations in the SWMM model, is in the range of probabilistic solutions, which indicates the applicational character of the method discussed in the paper. Moreover, due to the fact that the logit model presented hereby was developed for a single urban catchment, it is advisable to conduct further research on other urban catchments with various physical-geographic characteristics in order to verify the statistic model analyzed in the paper.

Due to the fact that the determination of empirical parameters (μ, σ) of the log-normal distribution depicting the maximum 30-minute rainfall depth involved the use of 8-year rainfall sequences, it is advisable to perform further measurements of rainfalls in order to verify their values. Moreover, it is advisable to develop a two-dimensional probability density distribution taking into account also the boundary distribution functions of the total and maximum 30-minute rainfall depth, which will allow for the elimination of the regression relation $P_c = f(P_{t=30})$ and the improvement of predictive capabilities of the developed stochastic model, designed to forecast an annual number of overflow discharges. An alternative solution may be the construction of a mathematical model $P_c = f(P_{t=30})$ with higher predictive abilities than the linear regression.

Given the uncertainty in the parameter estimation in hydrodynamic models, statistical distributions and the logistic regression model, further research is recommended in order to determine the impact of the aforementioned factors on the annual number of storm overflow discharges.

REFERENCES

ATV DVWK—A110. 1988. Richtlinien für die hydraulische Dimensionierung und den Leitstungsnachweis von Abwasserkanälen und—Leitungen, GFA, Hennef 1988.

Andrés-Doménech I., Múnera, J. C., Francés F. & Marco J. B. 2010. Coupling urban event-based and catchment continuous modelling for combined sewer overflow river impact assessment. *Hydrology Earth and System Sciences* 14(10): 2057–2072.

Ayalew, L. & Yamagishi, H. 2005. The application of GIS-based logistic regression for landslide susceptibility mapping in the Kakuda-Yahiko Mountains, Central Japan, *Geomorphology* 65: 15–31.

Balistrocchi M. & Bacchi, B. 2011. Modelling the statistical dependence of rainfall event variables through copula functions. *Hydrology Earth and Systems Science* 15(6): 1959–1977.

Butech Sp. z o.o. 2003. Construction of stormwater treatment on the collector Si9, Projekt budowlany. Kielce (in Polish).

Bąk Ł., Górski J., Górska K. & Szeląg B. 2012. Suspended Solids and Heavy Metals Content of Selected Rainwater Waves in Urban Catchment Area: A Case Study, *Ochrona Środowiska* 34(2): 49–52.

Cambez, M.J., Pinho J. & David L. M. 2008. Using SWMM in continuous modeling of stormwater hydraulics and quality. 11th International Conference on Urban Drainage, Edinburgh, Scotland.

Carleton, M.G. 1990. Comparison of Overflows from Separate and Combined Sewers—Quantity and Quality, *Water Science and Technology*, 22(1–10): 31–38.

DWA ATV-A128., 1992. Standards for the Dimensioning and Design of Stormwater Overflows in Combined Wastewater Sewers. German Association for Water, Wastewater and Waste Hennef.

Dayarante S.T. & Perera B.J.C. 2004. Calibration of urban stormwater drainage models using hydrograph modeling, *Urban Water Journal* 1(4): 83–297.

Dempsey P., Eadon A. & Morris G. 1997. SIMPOL a simplified urban pollution modeling tool. *Water Science and Technology* 36(8–9): 83–88.

Dąbkowski, S.L., Górska, K., Górski, J. & Szeląg, B., 2010. Introductory results of examining precipitation in one of Kielce channels. *Gaz Woda Technika Sanitarna*, 10: 20–24 (in Polish).

Dąbrowski, W. 2007. Estimating, computing and measuring multiplication factors for the working of storm overfalls, *Gaz, Woda i Technika Sanitarna* 11: 19–22 (in Polish).

Fidala-Szope M., Sawicka-Siarkiewicz H. & Koczyk A. 1999: *Protection of surface waters from discharges of stormwater from wastewater system. Handbook*, Warszawa, Instytut Ochrony Środowiska (in Polish).

Fontanazza, C.M., Freni, G., La Loggia, G. & Notaro, V., 2011. Definition of synthetic rainfall events for urban flooding estimation: the integration of multivariate statistics and cluster analysis. 12th International Conference on Urban Drainage, Porto Alegre, Brazil.

Fu, G. & Butler D. 2014. Copula-based frequency analysis of overflow and flooding in urban drainage systems, *Journal of Hydrology* 510:49–58.

Gironás, J., Niemann, J. D., Roesner L.A, Rodriguez F. & Andrieu H. 2009. A morpho-climatic instantaneous unit hydrograph model for urban catchments based on the kinematic wave approximation. *Journal of Hydrology* 377(3–4): 317–334.

Graniero, P.A. & Price J.S. 1999. Distribution of bog and heath in a Newfoundland blanket bog complex: Topographic limits on the hydrological processes governing blanket bog development, *Hydrology Earth and System Sciences* 3(2): 223–231.

Gräler, B., Berg, M.J., Vandenberghe, Petroselli, S.A., Grimaldi, S., Baets, B. & Verhoest, N.E.C. 2013.

Multivariate return periods in hydrology: a critical and practical review focusing on synthetic design hydrograph estimation. *Hydrology Earth and System Sciences* 17: 1281–1296.

Harrell, F. 2001. *Regression Modeling Strategies with Application to Linear Models, Logistic Regression, and Survival Analysis*, New York, Springer—Verlag.

Heyer, T. & Stamm, J. 2013. Levee reliability analysis using logistic regression models—abilities, limitations and practical considerations, *Georisk: Assessment and Management of Risk for Engineered Systems and Geohazards* 7(2): 77–87.

Horssen, P.W., Pebesma, E.J. & Schot, P.P. 2002. Uncertainties in spatially aggregated predictions from a logistic regression model. *Ecological Modelling* 154: 93–101.

Ingelmo, F., Molina, M.J., Paz, J.M. & Visconti, F. 2011. Soil saturated hydraulic conductivity assessment from expert evaluation of field characteristics using an ordered logistic regression model, *Soil & Tillage Research* 115–116: 27–38.

Kawagoe, S., Kazama, S. & Sarukkalige, P.R. 2010. Probabilistic modelling of rainfall induced landslide hazard assessment, *Hydrology Earth System Sciences* 14(6): 1047–1061.

Kiczko, A., Romanowicz, R.J., Osuch, M. & Karamuz E. 2013. Maximising the usefulness of flood risk assessment for the River Vistula in Warsaw, *Natural Hazards and Earth System Sciences*, 13: 3443–3455.

Kleidorfer, M., Deletic, A., Fletcher, T.D. & Rauch W. 2009. Impact of input data uncertainties on urban stormwater model parameters, *Water Science and Technology*, 60(6): 1545–1554.

Kotowski, A., Dancewicz, A. & Kaźmierczak, B. 2011. Accuracy of measurements of precipitation amount using standard and tipping bucket pluviographs in comparison to Hellmann rain gauge. *Environment Protection Engineering* 37(2): 23–33.

Krebs G., Kokkonen T., Valtanen M., Koivusalo H. & Setälä H. 2013. A high resolution application of a stormwater management model (SWMM) using genetic parameter optimization *Urban Water Journal* 10(6): 394–410.

Licznar, P., Łomotowski, J. & Rupp, E.D. 2011. Random cascade driven rainfall disaggregation for urban hydrology: An evaluation of six models and a new generator. *Atmospheric Research* 99(3): 563–578.

Mantegazza S. A., Alessandro G., Mambretti S. & Camylyn L. 2010. Designing CSO storage tanks in Italy: A comparison between normative criteria and dynamic modelling methods, *Urban Water Journal* 7(3): 211–216.

Martino, G., Paola, F., Fontana, N., Marini G. & Ranucci A. 2011. Pollution Reduction in Receivers: Storm-Water Tanks. *Journal of Urban Planning Development* 137(1): 29–38.

McFadden, D., 1973. Conditional logit analysis of qualitative choice behavior. In P. Zarembka (ed.): *Frontiers of Econometrics*: 105–142 New York: Academic Press.

Mrowiec M. 2009. *Efficient dimensioning and dynamics regulation of sewage reservoirs*, Częstochowa, Wydawnictwo Politechniki Częstochowskiej (in Polish).

Mrozik, K., Przybyła, C. & Pyszny, K. 2015. Problems of the Integrated Urban Water Management. The Case of the Poznań Metropolitan Area (Poland). *Rocznik Ochrona Środowiska/Annual Set the Environment Protection,* 17(1), 230–245.

Muhaisen, O.S., Osorio, F. & García P.A. 2009. Two-copula based simulation for detention basin design, *Civil Engineering and Environmental Systems*, 26(4): 355–366.

Osorio, F., Muhaisen, O.S. & García, P.A. 2009. Copula-Based Simulation for the Estimation of Optimal Volume for a Detention Basin, *Journal Hydrologic Engineering*, 14(12): 1378–1382.

Pradhan B. 2009. Flood susceptible mapping and risk area delineation using logistic regression, GIS and remote sensing, *Journal of Spatial Hydrology* 9(2): 1–18.

Romanowicz, R. & Beven, K.J. 2006. Comments on generalised likelihood uncertainty estimation, *Reliability Engineering and System Safety* 91(10–11): 1315–1321.

Rup, D.E., Licznar, P., Adamowski, W. & Leśniewski, M. 2012. Multiplicative cascade models for fine spatial downscaling of rainfall: parameterization with rain gauge data. *Hydrology and Earth System Sciences* 16(3): 671–684.

Sun, X.Y., Thompson, C.J. & Croke, B.F.W. 2011. Using a logistic regression model to delineate channel network in southeast Australia. 19th International Congress on Modelling and Simulation, Perth, Australia, 12–16 December 2011.

Szeląg, B., Górski, J., Bąk, Ł. & Górska, K. 2013. Modelling of stormwater quantity and quality on the example of urbanised catchment in Kielce, *Ecological Chemistry and Engineering A*, 20(11): 1305–1316.

Thorndahl S. 2008. Stochastic long term modelling of a drainage system with estimation of return period uncertainty. 11th International Conference on Urban Drainage, Edinburgh, Scotland, UK, 2008.

Thorndahl, S. & Willems, P. 2008. Probabilistic modelling of overflow, surcharge and flooding in urban drainage using the first-order reliability method and parameterization of local rain series, *Water Research*, 42(1–2): 455–466.

Urcikán P. & Rusnák D. 2006. A Complex Method of Calculating Critical Rain Intensity for Needs of Storm Overflow Design, *Ochrona Środowiska* 28(1): 33–38 (in Polish).

Vaes, G. & Berlamont, J. 1999. Emission predictions with a multi—linear reservoir model, *Water Science and Technology* 39(2): 9–16.

Wu, J.Y., Thompson, J.R., Kolka, R.K., Franz, K.J. & Stewart, T.W. 2013. Using the Storm Water Management Model to predict urban headwater stream hydrological response to climate and land cover change. *Hydrology Earth System Sciences* 17(12): 4743–4758.

Zabel T., Milne I. & Mckay G. 2001. Approaches adopted by the European Union and selected Member States for the control of urban pollution, Urban Water Journal 3(1–2): 23–32.

Zawilski M. & Sakson G. 2013. The assessment of the emission of suspended solids transported through the stormwater sewer system from urban areas, Ochrona Środowiska 35(2): 33–40 (in Polish).

Environmental Engineering V – Pawłowska & Pawłowski (Eds)
© 2017 Taylor & Francis Group, London, ISBN 978-1-138-03163-0

Assessment of water supply diversification using the Pielou index

J. Rak & K. Boryczko
Faculty of Civil and Environmental Engineering and Architecture, Rzeszow University of Technology, Rzeszow, Poland

ABSTRACT: The aim of the paper is to present the methodology for determining the diversification degree of water supply in water supply systems. Three parameters which impact on diversification degree of water supply were selected in the presented methodology: maximum water production by water intakes, water tanks capacity, cross sectional area of discharge lines of pumping station stage II. With these data it is possible to calculate the dimensionless Pielou ratio and assessment of water supply diversification of specific water supply system and to compare any water supply systems. The paper presents calculations of diversification of selected water supply systems.

Keywords: diversification, water supply systems, water intakes, water tanks, Pielou index

1 INTRODUCTION

The concept of portfolio diversification, known in economics, is the most popular and described as one of the most effective methods of reducing investment risk. This concept means the division of the portfolio into different types of investments, among others, in terms of the type of market (e.g. raw materials, currency, shares, bonds), trade (in the case of shares) or geographical coverage of given entities (e.g. shares, shares funds of enterprises from a specific region). In technical application, for the purpose of comparing different systems of collective water supply, the concept of diversification can be used to describe differences in water resources of Collective Water Supply Systems (CWSS).

The basics of reliability simply show that in order to improve the reliability of the given element it should be doubled in the parallel reliability structure as a hot reserve. In the case of water resources in collective water supply systems we usually have several water intakes with different capacities. One intake makes the given system of collective water supply significantly exposed to the lack of water supply due to its failure or inadequate water quality (Boryczko & Tchórzewska-Cieślak 2013, Boryczko et al. 2014, Boryczko & Tchórzewska-Cieślak 2014, Kaleta et al. 2009, 2016; Kowalski & Miszta-Kruk 2013, Kwietniewski 2006, Li et al. 2016, Lindhe et al. 2009, Papciak et al. 2013, Petkovic et al. 2011, Pietila 2016, Tchórzewska-Cieślak 2011, Tchórzewska-Cieślak & Szpak 2015, Zimoch & Łobos 2012). Emergency shutdown of basic and only water intake poses a threat to health and lives of the residents of the given settlement. A similar situation appears in the case of network water tanks, (Aini et al. 2001,

Bajer 2007, Bajer 2013, Pietrucha-Urbanik 2014, Rende 2007, Studziński 2014) but not in every water supply system there is a tank. In such case, same as when there is one tank in the water supply network, we deal with a lack of water volume diversification (Rak 2015, Dąbek 2015). Diversity is also topic of many other research in finance, sports, biology, chemistry, social science (Ball & Crawford 2006, Blanch et al. 2001, Corradini & De Propris 2015, Humphrey et al. 2015, Hong & Herk 1996, Kashcheeva & Tsui 2015, Maj 2015, Mariano et al. 2016, Sterkowicz-Przybycien 2010).

The subject of the article is the methodology for calculating the degree of water supply diversification in water supply systems. The presented methodology proposed three parameters affecting the degree of water supply diversification: the maximum production capacity of water intakes, the volume of water tanks, cross-sectional area of discharge pipes in the second stage pumping stations. On the basis of these data it is possible to calculate the dimensionless Pielou ratio and assess the degree of diversification of the given collective water supply system, as well as compare even very different water supply systems. The paper presents the calculation of diversification indexes for selected CWSS in Poland.

2 METHOD FOR DIVERSIFICATION ANALYSIS

2.1 *The Shannon-Weaver index*

Claude Elwood Shannon was received diplomas in mathematics and electrical engineering in the University of Michigan. Warren Weaver was graduated from the University of Wisconsin-Madison in civil engineering. Weaver was fascinated by the work of Shannon on cryptography and theory of information developed during the war. Together with C. Shannon he created the work, in which they discussed the mathematical theory in communication. In that work Shannon focused more on the engineering aspects of the mathematical model, while Weaver approached the matter in a more philosophical way.

From the joint work of Shannon and Weaver (Shannon 1948, Shannon & Weaver 1962) the diversification index was taken:

$$d_{SW} = -\sum_{i=1}^{n} (u_i) \cdot (\ln(u_i)) \qquad (1)$$

where u_i = share of the i-th element in the entirety, n = number of elements.

2.2 The pielou index

Evelyn Crystal Pielou was an ecologist by education. She significantly contributed to the development of mathematical ecology, mathematical modelling of natural systems. She is the author of six scientific books on this subject (Pielou 1966, Pielou 1969, Pielou 1975, Pielou 1977, Pielou 1979, Pielou 1984).

From her work on the assessment of the biodiversity degree of biocoenosis, a measure to assess the degree of diversification of water supply to the city was adopted, determined by the formula:

$$d_P = \frac{d_{SW}}{d_{SW\,max}} \qquad (2)$$

where d_{SW} is the Shannon-Weaver index, calculated according to the formula (1) and value d_{SWmax} is calculated as:

$$d_{SWmax} = \ln(n) \qquad (3)$$

Analysing the formulas (1) and (3) it can be concluded that: for n = 1 the index d_P is undefined, for n = 2 the index d_P = 1.0. Tables 1÷3 present the values of d_P.

The analysis of the indexes d_P from the Table 1 to 3 shows that:

- with the lack or little unevenness of shares of the index d_P obtain values near or equal to 1.0,
- with the lack of significant balance of shares, the rule that the bigger number of n the higher index d_P is not applicable.

In Table 4 the numerical values of the Pielou indexes for equal shares were summarized.

While analysing Table 4 it should be noted that for any number n, with the same shares the Pielou index obtains the maximum value of an even diversification of 1.0.

3 METHOD FOR ASSESSMENT OF WATER SUPPLY DIVERSIFICATION

For the assessment of diversification the following parameters were proposed: Q—parameter associated with the resource of water in the Water Supply Subsystems (WSS) (Rak & Włoch 2015), V—parameter associated with the volume of water in the network water tanks (Rak 2015), F—parameter associated with the cross-sectional area of collective discharge pipelines of the second stage pumping stations.

Water resources in the individual WSS mean the maximum daily production capacity. The parameter V includes the volume of the network tanks, i.e.

Table 1. The numerical values of the indexes d_P for n = 2.

u_1	u_2	d_P
0.5	0.5	1.0
0.6	0.4	0.97
0.7	0.3	0.88
0.8	0.2	0.72
0.9	0.1	0.47
0.95	0.05	0.286
0.99	0.01	0.081

Table 2. The numerical values of the indexes d_P for n = 3.

u_1	u_2	u_3	d_P
0.33	0.33	0.33	1.0
0.4	0.3	0.3	0.991
0.5	0.3	0.2	0.938
0.6	0.3	0.1	0.817
0.6	0.2	0.2	0.865
0.7	0.2	0.1	0.730
0.8	0.1	0.1	0.582

Table 3. The numerical values of the indexes d_P for n = 4

u_1	u_2	u_3	u_4	d_P
0.25	0.25	0.25	0.25	1.0
0.3	0.3	0.2	0.2	0.985
0.4	0.3	0.15	0.15	0.936
0.5	0.3	0.1	0.1	0.843
0.6	0.2	0.1	0.1	0.786
0.7	0.1	0.1	0.1	0.582

Table 4. The numerical values of the Pielou indexes for equal shares.

n	u_i	d_P
2	0.5	1.0
3	0.33	1.0
4	0.25	1.0
5	0.2	1.0
6	0.167	1.0
8	0.125	1.0
10	0.1	1.0
20	0.05	1.0

all the tanks located in the technological line after the second stage pumping stations. The parameter F includes all the collective discharge pipelines of the second stage pumping stations and their cross-sectional areas. Diversification degree for parameter Q, V, F were determined according to:

$$d_Q = \frac{-\sum_{i=1}^{m}(u_i \cdot lnu_i)}{lnm} \quad (4)$$

$$d_V = \frac{-\sum_{j=1}^{n}(u_j \cdot lnu_j)}{lnn} \quad (5)$$

$$d_F = \frac{-\sum_{k=1}^{p}(u_k \cdot lnu_k)}{lnp} \quad (6)$$

where d_Q = the diversification index of water resources (intakes) in the CWSS; d_V = the diversification index of the volume of water in the network water tanks; d_F = the diversification index of cross-sectional area of collective discharge pipelines of the second stage pumping stations; u_i = share of maximum daily production capacity of the i-th CWSS (m³/d) in the total maximum daily capacity of water supply to CWSS; u_j = share of the volume of the j-th tank (m³) in a total volume of water in the network water tanks; u_k = share of the cross-sectional area of the k-th collective discharge pipeline of the second stage pumping stations in a sum of cross-sectional area of collective discharge pipelines of the second stage pumping stations; m = number of WSS; n = number of network tanks; p = number of collective discharge pipelines of the second stage pumping stations.

Global diversification degree was calculated according to:

$$d = d_Q + d_V + d_F \quad (7)$$

In the calculations it was assumed that if $u_i = 1.0$, $u_j = 1.0$ or $u_k = 1.0$, the value of the index d_Q, d_V or d_F taken to calculate the index d from the equation (7) is 0.0.

The following standards for the ratio d were adopted:

- lack of diversification d = 0
- low diversification 0 < d ≤ 0.6
- average diversification 0.6 < d ≤ 1.8
- sufficient diversification 1.8 < d ≤ 2.4
- very satisfactory diversification 2.4 < d ≤ 3.0

4 RESULTS AND DISCUSSION

Calculations of the assessment of the diversification degree for selected CWSS are presented below.
Rzeszów

- Q—two water intakes for which shares of capacity are as follows:

  $u_1 = 0.43$
  $u_2 = 0.57$
  $d_Q = -(0.43 \cdot ln\,0.43 + 0.57 \cdot ln\,0.57)/(ln\,2) = 0.986$
- V—eight network tanks for which shares are as follows:

  $u_1 = 0.018$
  $u_2 = u_3 = 0.054$
  $u_4 = u_5 = u_6 = u_7 = 0.088$
  $u_8 = 0.522$
  $d_V = -(0.018 \cdot ln\,0.018 + 2 \cdot 0.054 \cdot ln\,0.054 + 4 \cdot 0.088 \cdot ln\,0.088 + 0.522 \cdot ln\,0.522)/(ln\,8) = 0.762$

- F—four discharge pipelines of the second stage pumping stations for which shares of cross-sectional area are:

  $u_1 = 0.05$
  $u_2 = u_3 = 0.086$
  $u_4 = 0.778$
  $d_F = -0.05 \cdot ln\,0.05 + 2 \cdot 0.086 \cdot ln\,0.086 + 0.778 \cdot ln\,0.778)/(ln\,4) = 0.553$
  According to (7):
  d = 0.986 + 0.762 + 0.553 = 2.281

The corresponding calculations for CWSS in Poland were shown in Table 5 and Figure 1.

Four out of the analysed CWSS, were classified in the category of sufficient diversification. Highly satisfactory diversification was found for five CWSS, which is due to balanced water resources, volumes and a high value of the index d_F. As for the proposed in the method three parameters, there are always two of them in the given CWSS, i.e. water intakes and discharge pipelines. Analysing the results of the research it was found that the lack of water tanks has an impact on the low value of the index d (Bydgoszcz, $d_V = 0$). In addition, the

Table 5. The numerical values of the Pielou indexes analysed CWSS.

Miasto	m	n	p	d_Q	d_V	d_F	d
Wadowice	2	5	2	0.957	0.778	1.0	2.735
Poznań	3	2	7	0.914	0.918	0.902	2.734
Krosno	3	2	3	0.835	1.0	0.877	2.712
Racibórz	2	3	2	0.811	1.0	0.857	2.668
Częstochowa	3	11	6	0.755	0.943	0.732	2.43
Tarnów	3	14	6	0.679	0.718	0.913	2.31
Kraków	4	10	8	0.752	0.57	0.902	2.224
Przemyśl	1	3	2	0	0.93	0.977	1.907
Bydgoszcz	2	0	6	0.876	0	0.884	1.76
Brzesko	2	2	2	0.286	0.439	1.0	1.725
Jasło	2	2	3	0.141	0.827	0.698	1.666

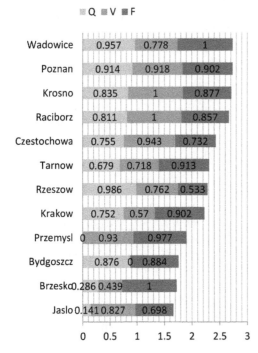

Figure 1. Collective water supply systems chart ranked according to the index d.

lack of diversification of water resources in the CWSS Przemysl (one intake, $d_Q = 0$) is the cause of the low value of the global index of diversification, despite the highest, after Czestochowa, value of the index d_V.

With the increase in the number of water intakes, water tanks or collective discharge pipelines of the second stage pumping stations the value of the indexes d_Q, d_V, d_F, grows at balanced shares (except the CWSS in Krakow where despite 10 water tanks the value of the index d_V is lower than for the CWSS in Rzeszow which has 8 tanks).

4 CONCLUSIONS

The degree of diversification of water supply in the CWSS is greatly affected by:

– number of the WSS, network tanks, collective discharge pipelines of the second stage pumping stations, evenness of distribution of WSS productivity, volume of network tanks, specific cross-sectional area of collective discharge pipelines of the second stage pumping stations
– share of the volume of accumulated water in the size of its collection from the water supply system (the development of criteria and standards is in the process of research).

The CWSS safety is closely related to the diversification of water supply to consumers. It is a desirable situation when during the failure of one of the CWSS elements the recipient of water does not feel any inconvenience related to the operation of the water supply network. The presented method for determining the degree of diversification is a simple tool in the hands of the CWSS operators, showing how the CWSS is prepared to a statutory obligation to supply water.

Dimensionless values of the global diversification index predispose it to analyse the degree of diversification of water supply in different CWSS. They make it possible to compare even very different systems, with one or several intakes, equipped or not equipped with water supply tanks, with different numbers and different specific cross-sectional area of collective discharge pipelines of the second stage pumping stations.

The authors are aware that the analyses of the diversification performed in particular cities are result of the technical conditions, the size of water resources in each source, the need for expansion of the CWSS elements. The designers of these systems certainly were not guided by the ability to calculate and assess the degree of diversification because they did not know such concept. We realize that a large number of cities are supplied from a single source (e.g. for Montreal it is a big St. Lawrence river) and the CWSS operate in those cities with satisfactory reliability. The content of the work is a new contribution to the possibility of conscious design of the CWSS expansion, taking into account the diversification as the basis of preliminary reliability analysis.

REFERENCES

Aini, M. S., Fakhru'l-Razi, A. & Suan K. S. 2001. Water crisis management: Satisfaction level, effect and coping of the consumers. *Water Resources Management* 15(1): 31–39.

Bajer, J. 2007. Reliability analysis of variant solutions for water pumping stations, In M. R Dudzińska, L. Pawłowski & A. Pawłowski (eds), *Environmental Engineering*: 253–261. New York, Singapore: Taylor & Francis Group.

Bajer, J. 2013. Ecomical and reliability criterion for the optymization of the water supply pumping stations designs, In M. R Dudzińska, L. Pawłowski & A. Pawłowski (eds), *Environmental Engineering IV*: 21–28. London: Taylor & Francis Group.

Ball, C. L. & Crawford, R. L. 2006. Bacterial diversity within the planktonic community of an artesian water supply. *Canadian Journal of Microbiology* 52: 246–259.

Blanch, A. R., Cerda-Cuellar, M. & Hispano, C. 2001. Diversity of Vibrio spp. populations in several exhibition aquaria with a shared water supply. *Letters in Applied Microbiology* 33: 137–143

Boryczko, K. & Tchórzewska-Cieślak, A. 2013. Analysis and assessment of the risk of lack of water supply using the EPANET program, In M. R Dudzińska, L. Pawłowski & A. Pawłowski (eds), *Environmental Engineering IV*: 63–68. London: Taylor & Francis Group.

Boryczko, K. & Tchórzewska-Cieślak, B. 2014, Analysis of risk of failure in water main pipe network and of developing poor quality water. *Environment Protection Engineering* 40(4): 77–92.

Boryczko, K., Piegdoń, I. & Eid, M. 2014. Collective water supply systems risk analysis model by means of RENO software, In P.H.A.J.M. Van Gelder, R.D.J.M. Steenbergen, S., Miraglia & A.C.W.M. Vrouwenvelder (eds), *Safety, Reliability and Risk Analysis: Beyond the Horizon*: 1987–1992. London: Taylor & Francis Group.

Corradini, C. & De Propris, L. 2015. Technological diversification and new innovators in European regions: evidence from patent data. *Environment and Planning A* 47: 2170–2186.

Dąbek, L. 2015. Zastosowanie sorpcji i zaawansowanego utleniania do usuwania fenoli i ich pochodnych z roztworów wodnych. *Rocznik Ochrona Środowiska/Annual Set the Environment Protection*, 17(1), 616–645.

Hong, C. S. & Herk, L. F. 1996. Incremental risk aversion and diversification preference. *Journal of Economic Theory* 70: 180–200.

Humphrey, J. E., Benson, K. L., Low, R. K. Y. & Lee, W. L. 2015. Is diversification always optimal? *Pacific-Basin Finance Journal* 35: 521–532.

Kaleta, J., Papciak D. & Puszkarewicz A. 2009. Natural and modified minerals in remediation of groundwaters. *Gospodarka Surowcami Mineralnymi-Mineral Resources Management* 25(1): 51–63.

Kashcheeva, M. & Tsui, K. K. 2015. Political oil import diversification by financial and commercial traders. *Energy Policy* 82: 289–297.

Ke, W., Leia, Y., Shaa, J., Zhangb, G., Yana, J., Lind, X. & Pana X. 2016. Dynamic simulation of water resource management focused on water allocation and water reclamation in Chinese mining cities. *Water Policy* 18(1): 1–18.

Kowalski, D. & Miszta-Kruk, K. 2013. Failure of water supply networks in selected Polish towns based on the field reliability tests. *Engineering Failure Analysis* 12(35): 736–742.

Kwietniewski, M. 2006, Field reliability tests of water distribution system from the point if view of consumer's needs. *Civil Engineering and Environmental Systems* 23(4): 287–294.

Li, F., Wang, W. & Ramírez, L. H. G. 2016, The determinants of two-dimensional service quality in the drinking water sector—evidence from Colombia. *Water Policy* 18(1).

Lindhe, A., Rosen L., Norberg, T., & Bergstedt O. 2009, Fault tree analysis for integrated and probabilistic risk analysis of drinking water systems. *Water Research* 43(6): 1641–1653.

Maj, G. 2015. Diversification and Environmental Impact Assessment of Plant Biomass Energy Use. *Polish Journal of Environmental Studies* 24: 2055–2061.

Mariano, M., Hartmann, R. W. & Engel, M. (2016). Systematic diversification of benzylidene heterocycles yields novel inhibitor scaffolds selective for Dyrk1 A, Clk1 and CK2. *European Journal of Medicinal Chemistry* 112: 209–216.

Papciak, D., Kaleta, J. & Puszkarewicz, A. 2013. Removal of Ammonia Nitrogen from Groundwater on Chalcedony Deposits in Two-stage Biofiltration Process. *Rocznik Ochrona Srodowiska* 15(2): 1352–1366.

Petkovic, S., Gregoric E., Slepcevic, V., Blagojevic, S., Gajic, B., Kljujev, I., Žarković, B., Djurovic, N. & Draskovic, R. 2011, Contamination of local water supply systems in suburban Belgrade. *Urban Water Journal* 8(2): 79–92.

Pielou, E. C. 1966. The measurement of diversity in different types of biological collections. *Journal of Theoretical Biology* 13: 131–144.

Pielou, E. C. 1969. *Introduction to Mathematical Ecology*. New York: Wiley-Interscience.

Pielou, E. C. 1975. *Ecological diversity*. New York: Wiley.

Pielou, E. C. 1977. *Mathematical ecology*. New York: Wiley.

Pielou, E. C. 1979. *Biogeography*. New York: Wiley.

Pielou, E. C. 1984. *The interpretation of ecological data: a primer on classification and ordination*. New York: Wiley.

Pietila, P. E. 2013. Diversity of the water supply and sanitation sector: roles of municipalities in Europe. *Water Services Management and Governance: Lessons for a Sustainable Future* 1: 99–111.

Pietrucha-Urbanik, K. 2014. Assessment model application of water supply system management in crisis situations. *Global Nest Journal* 16(5): 893–900.

Rak, J. & Włoch, A. 2015. Models of level diversification assessment of Water Supply Subsystems, In A. Kolonko, C. Madryas, B. Nienartowicz & A. Szot (eds), *Underground Infrastructure of Urban Areas 3*: 237–244. London: Taylor & Francis Group.

Rak, J. 2015. Propozycja oceny dywersyfikacji objętości wody w sieciowych zbiornikach wodociągowych. *Journal of Civil Engineering, Environment and Architecture* 32(62): 339–349.

Rende, M. 2007, Water transfer from Turkey to water-stressed countries in the Middle East. *Water Resources in the Middle East* 2: 165–173.

Shannon, C. & Weaver, W. 1962. *The Matematical Theory of Communication*. Urbana: University of Illionois Press.

Shannon, C. 1948. A mathematical theory of communication. *Bell System Technical Journal* 27: 379–423, 623–656.

Sterkowicz-Przybycien, K. 2010. Technical diversification, body composition and somatotype of both heavy and light Polish ju-jitsukas of high level. *Science & Sports* 25: 194–200.

Studziński, A. 2014. Amount of labour of water conduit repair, In P.H.A.J.M. Van Gelder, R.D.J.M. Steenbergen, S. Miraglia, & A.C.W.M. Vrouwenvelder (eds), *Safety, Reliability and Risk Analysis: Beyond the Horizon*: 2081–2084. London: Taylor & Francis Group.

Tchórzewska-Cieślak, B. & Szpak, D. 2015. A Proposal of a Method for Water Supply Safety Analysis and Assessment. *Ochrona Środowiska* 37(3): 43–47.

Tchórzewska-Cieślak, B. 2011. Fuzzy Model for Failure Risk in Water-pipe Networks Analysis. *Ochrona Środowiska* 33(1): 35–40.

Zimoch, I. & Łobos E. 2012. Comprehensive interpretation of safety of wide water supply systems. *Environment Protection Engineering* 38(3): 107–117.

Environmental Engineering V – Pawłowska & Pawłowski (Eds)
© 2017 Taylor & Francis Group, London, ISBN 978-1-138-03163-0

Reduction of water losses through metering of water supply network districts

T. Cichoń
Municipal Water and Sewerage Company MPWiK S.A, Cracow, Poland

J. Królikowska
Faculty of Environmental Engineering, Technical University of Cracow, Cracow, Poland

ABSTRACT: The article describes the studies on water losses of water supply system in Kraków and the method of reducing these losses by creating metered districts of the water supply network. One of the most efficient methods of active leakage control involves balancing the water flow rates in metered districts of water supply network (District Metering Areas—DMA). It is necessary to determine the sets of water meters of all recipients supplied in a given district. An assessment of leakage level is performed in these districts through a cyclic calculation of water balance between the district supply and the sum of recipients' water meters. The best way of applying DMA districts method is with the use of automatic readouts of all water meters in a district, performed on a daily basis. The article also discusses the functional assumptions of software used for an assessment of water loss levels in districts.

Keywords: water supply system, failure, reliability, water loss

1 INTRODUCTION

The level of water losses is a parameter of paramount importance for every water company. The losses are defined as the difference between the amount of water pumped into the water supply network and the amount of water sold. The level of losses is reported by every water company, in addition to other key figures, such as the sales volume and revenues. Water losses are usually reported as a percentage in relation to the sales volume. Such presentation of data is a far-fetched simplification, because it does not relate to the length of the water supply network, pressure, or the number of recipients. However, most of all, it does not indicate what possible actions could be taken in order to reduce these losses. The actual aim of curbing the water loss levels is reducing financial losses borne by a company. Therefore, in line with International Water Association standards, economic level of leakage is used more frequently than the aforementioned percentage index. On the other hand, the main component of economic level of leakage is the volume of so called unavoidable real loses, calculated for per kilometre of the network and for each water supply connection.

In accordance with IWA, the actions aimed at reducing water losses consist in a quick reaction to any failures and their elimination, successive rehabilitation of pipes, active leakage control and pressure control.

This article describes the use of metered water supply districts as a tool for an active leakage and—both real and apparent—water loss control.

2 MATERIAL AND METHODS

2.1 *Characteristics of research area*

Water supply system in Kraków is one of the largest in Poland. It is made up of 2166 km of water supply network. Metering of water supply is performed by means of more than 56 thousand main water meters. Water meters installed at recipients are readout every 30 or 60 days (Cichoń 2015). Metering of the water amount flowing in the supply system is conducted at the outflow of water treatment stations, as well as in numerous metering points installed on the municipal network. Water supply system in Kraków, due to the diversified terrain level and its sheer size, is divided into pressure districts, separated from individual supply districts. Pressure districts are of varying size and the largest ones supply ten-odd thousand recipients, which is reflected in the failure rate of the network (Zimoch 2010). The main districts boast several water supply connections, as well as parts where water flows in different directions, depending on its reception.

In order to calculate the water balance in the districts, it is necessary to meter the water flowing into each zone by means of a readout system of water meters installed at recipients. The readout data needs to be transmitted and archived.

On the basis of tests involving various readout systems it is assumed that the desired frequency of gathering all readout data should amount to 24h. This assumption necessitates, among others, conducting readouts of all recipients' water meters once a day.

At present, it is not possible to calculate the water balance in Kraków water supply system on a daily basis, as water meters are readout only once per accounting period. A solution to this problem involves the technology of remote readouts which allows quick and accurate readouts at any time, without the recipients' participation and the necessity of entering their property (manholes) or rooms (cellars, flats). This technology allows for a full monitoring of metering devices and the obtained data from any accounting period may be processed and analyzed, both at the level of building administrator, as well as utility provider (Cichoń 2015, Tuz 2006a, Tuz 2006b, Usidus & Drozdowicz 2010, Żuchowicki & Gawin 2013).

The application of a remote readout system has many benefits, e.g. it improves the sense of security and privacy of residents, lowers the costs related to the readouts of utilities meters, eliminates the risk of readout error due to a human mistake, and enables to calculate water use balance for the entire district (comparison of readouts with the main water meter), thus becoming a tool for an early detection of failures of both the water supply network, as well as individual meters or installations. This greatly contributes to the limitation of water use, which positively influences not only the budget of recipients, but also the water resources in the environment (Cichoń & Królikowska 2015, Zydroń & Szoszkiewicz 2013).

Modern technological solutions which allow for a remote, credible, and accurate readout of water meters replace the collector going from door to door or the transfer of readouts via telephone, and are slowly becoming the standard for the Municipal Water and Sewerage Company in Kraków. However, in order to utilize the system in a daily balancing of the district metered areas, it must be improved or the readout devices must be replaced by more advanced ones.

Vastly more difficult and costly is metering the water supply connections of individual network districts. Many of them have no metering device whatsoever. Moreover, there are no chambers in the network, which would enable the installation of metering devices. Construction of such chambers in densely populated city centres is extremely difficult or even impossible.

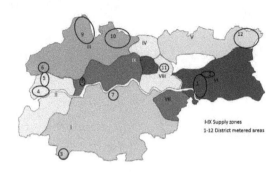

Figure 1. District metered areas created on the basis of existing objects.

In order to perform effective water balancing, it is necessary to separate zones with several hundred main water meters at most. Looking at the map of water supply network in Kraków, it is obvious that smaller zones must be separated on the basis of the existing districts, which would constitute a proper basis for the active control of water losses.

In the future, such a structure will be created for an entire network; however, a detailed analysis shows that many parts of the network may constitute metering districts at present. This is because there are small hydrophore or reduction zones which supply water to the areas inhabitant by several dozens or hundreds of recipients. Figure 1 presents the district metered areas created in the first stage.

3 RESULTS AND DISCUSSION

Employing district metered areas in the water supply network for the detection of water losses was shown on the example of one of the created districts. A district with block of flats, which is supplied by a pipe branching out from the water main (Cichoń 2015) was chosen for the purpose of the research. Next to the branching there is an underground chamber for the installation of pressure regulator and a water meter measuring the amount of water flowing into this district. The district is inhabited by about 2 thousand people. The GPS map of the district in question, with the water supply network marked on it, is presented in Fig. 2.

Water metering of the district is performed by means of a single-jet R315 water meter with the diameter of 100 mm, equipped with a radio readout add-on. There are 15 water meters in the studied district, installed on water supply connections of the blocks of flats (main water meters). All water meters in the district metering area were replaced with new ones, which could operate jointly with radio add-ons enabling remote readout. The utilized water meters were advanced R160 and—in the

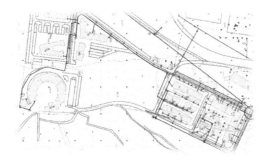

Figure 2. GIS map view of the considered district with water supply network and branching water main marked (Cichoń 2015).

case of industrial meters—R315 models. Three different diameters were selected: 20 [mm]—6 pieces, 40 [mm]—4 pieces, and 50 [mm]—5 pieces. The radio add-ons operate in one-way communication mode, which enables a rapid readout regardless of the transmission sequence. Such setup enabled a simultaneous readout of all water meters in the district area.

The analysis of readout data enables to identify failures and their potential causes. Register of the flow rate in the pipeline supplying water to the district was presented in Fig. 3.

The peak flow rate reaches [m³/h]. While the water meter does not stop a night, the flow rate drops to approximately 1 [m³/h], which constitutes roughly 4% of the peak value. Such distribution of flow rates in the district water supply network allows to draw the conclusion that the analyzed network is operational and watertight. Simultaneously, regular readouts of all water meters began. Throughout the studies, readouts were carried out by means of a device installed in a car which drove through the district, each time in the same direction. A single readout of water meters in the district always started with the supplying water meter and lasted no more than three minutes; thus, the obtained readouts could be considered as simultaneous. Readouts were performed over a week in order to observe trends of changes for individual water meters and the entire district.

The first readouts showed that an average daily water inflow to the entire zone amounts to 240 m³, whereas after balancing the main water meters, the difference in relation to the supply water meter equals 3.68%. Such a difference is the result of standard measurement errors of all water meters used in district metering. The balance difference is within the maximum permissible errors. Subsequent measurements confirmed the obtained values, both in regard to the daily flow rates, as well as the balance difference for the zone. Following several readouts, an increase in the balance difference of the zone was observed. After

a few consecutive measurement series it exceeded 10% of the amount of water supplied to the zone, and after another ones, it already amounted to almost 40 m³ per day. The graph showing the balance difference of water in the zone was presented in Fig. 4.

While analyzing the values of average daily readouts of individual main water meters at recipients it was observed that none of them showed significant deviations from the previous readouts, which could signify a failure of one of them (Fig. 5). The calculated standard deviation of the zone supply water meter amounted to 13.76, whereas the deviation of summed up recipients' water meters readouts equalled 7.48. Such values of standard deviation and the graph of average daily readouts of the supply water meters and the sum of recipients' water meters readouts suggested that the difference in water balance of the studied district could result from real losses stemming from a leakage in the water supply network. The network diagnostic team confirmed this conclusion. A leakage from the water supply network was detected, where water infiltrated into the sewerage network. On the basis of measurements, the size of leakage was determined as greater than 1.5 m³/h.

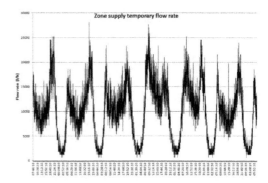

Figure 3. Graph of the flow rate in the pipeline supplying the district zone (Cichoń 2015).

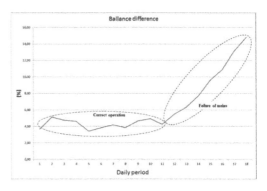

Figure 4. Graph of water supply balance difference in the supply zone.

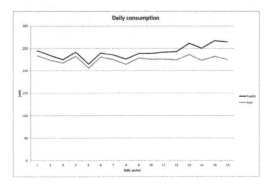

Figure 5. Graph of average daily readouts of the supply water meter and the sum of recipients' water meters.

After the failure was eliminated, the value of balance difference in the considered zone returned to the one from the initial stage of the conducted research.

4 CONCLUSIONS

Such results of flow rate measurements prompted to conduct the analysis of the tested material which aimed to pinpoint the causes of balance difference, and to draw conclusions regarding the procedure of dealing with similar situations.

Implementing the system of the automatic fixed area readouts may ensure the creation of a database containing water meter readouts from each day. Such database would serve not only to carry out a systematic settlement of charges with recipients, but also to analyze the water losses in the distribution network and apparent losses at the recipients' water meters (Podwójci et al. 2011).

Proper permanent service of supply zones requires the implementation of special software with its own database, which will contain all the readout data. Constant data exchange of this application with billing systems and GIS system is necessary. The readout software requires mapping of the supply tree structure within the metered network zones. The implemented system should include a mechanism of calculating average daily values on the basis of available readouts. Thus, the value of each received readout is compared with the projected one. If there is a deviation which exceeds a certain set threshold, it is immediately signalled to the operator.

Another, more advanced tool for controlling water losses is a module of billing software called the supply tree. This module contains a mechanism of marking a water meter as supplying a given zone, and others as receiving meters belonging to this zone. On this basis, a balance difference is calculated following each readout series. Further analysis of the obtained data which allows drawing conclusions pertaining to the location of occurring losses in line with the presented algorithm, is also possible. This tool will be especially valuable and useful after the implementation of automatic readouts, when the basic period between readouts will amount to 24h.

The conducted studies show a method of automatic control of water losses. It pertains both to the apparent losses and the possibility of indicating real leakages from the water supply network. Implementation of modern technologies of water meter readouts and settlement of charges is a huge step towards mitigating the results of meter malfunctions and network failures. The value of the presented method mainly consists in shortening the period of cyclic readouts of metering devices, as well as the possibility of analyzing real daily readouts of all water meters, including the one on the pipeline supplying a given zone. In the system with 60 or 90 day-long accounting period, such solutions offers a vastly improved quality of data gathered from readouts. This will also allow for a detailed analysis of the applied water meters in regard to optimizing the nominal diameter or class. For network users, it may constitute an important tool for determining possible ways of modernization or development, not only with respect to metering the supply, but resource management and modernization requirements as well.

REFERENCES

Cichoń, T. 2015. Multi-element evaluation of reliability of a metering system for water consumption and water losses based on operaticing data, *PhD. Dissertation, Cracow University of Technology*.

Cichoń, T & Królikowska, J. 2015. Analysis of water meters failures and its impact on a reliability of water meter management, *Gas, Water and Sanitation* No. 3; 92–95.

Measuring Instrument Directive of European Parliament and Council of 31March 2004 on measuring instruments, No 2004/22/WE.

Miłaszewski, R. 2009. Methods of Assessment of Environmental and Resource Costs Generated by Water Use. *Annual Set. The Environment Protection*, 11: 339–353.

Podwójci, P & Kozłowski, M & Krysiuk, M. 2011. Water Consumption in the Multi-family Housing-Selected Issues. *Annual Set. The Environment Protection*, 13: 1653–1665.

Tuz, P. 2006a. Why do we monitor water connections, *Instal* 4–5: 42–47.

Tuz, P. 2006b. Monitoring of water connections, *Magazine of the Instalator* 7–8: 95–96.

Regulation of the Minister of Economy of 23 October 2007 on the requirements to be met by water meters and the detailed scope of tests performed during legal metrological control of these measuring instruments (Dz. U. No. 209, item 1513) Based on Art. 9a, Act of 11 May 2001. – Measuring Instrument Directive (Dz. U. 2004. No. 243, item. 2441, as amended).

Usidus, D. & Drozdowicz, A. 2010. Analysis of Water Use on the Selected Area of Central Pomerania—in Sianów Municipality. *Annual Set. The Environment Protection.* 15: 553–558.

Zimoch, I. 2010. Reliability analysis of water-pipe networks in Cracow, Environmental Engineering III—Pawłowski, Dudzińska & Pawłowski, Taylor & Feancis Group, London, 561–565.

Zydroń, A & Szoszkiewicz, K. 2013. Value of the Natural Environment and Willingness of the Society to Pay for this Good, *Annual Set. The Environment Protection*, 15: 2874–2886.

Żuchowicki, A W & Gawin, R. 2013. Structure of Water Consumption in Single-family Buildings, *Annual Set. The Environment Protection*, 15: 924–929.

Environmental Engineering V – Pawłowska & Pawłowski (Eds)
© 2017 Taylor & Francis Group, London, ISBN 978-1-138-03163-0

Potential DBD-jet applications for preservation of nutritive compounds on the example of vitamin C in water solutions

D. Bozkurt
Plasma Aided Bioengineering and Biotechnology Research Group, Hacettepe University, Ankara, Turkey

M. Kwiatkowski, P. Terebun, J. Diatczyk & J. Pawłat
Institute of Electrical Engineering and Electrotechnologies, Lublin, Poland

ABSTRACT: Among the nutrients in fruit and vegetables, vitamin C (L-ascorbic acid) is being given attention due to the health benefits it provides consumers. Unfortunately, pasteurization processes aimed at protection of food from rotting, often result in the reduction of the concentration of the vitamin. In the present study, the DBD (Dielectric Barrier Discharge) non-equilibrium plasma jet source has been applied for checking the applicability of the plasma sterilization method without causing appreciable changes on vitamin C in food. The measurements were performed for the main factors that may affect the amount of generated active particles: time of treatment, composition of working gas, the flow rate value and the distance from the end of discharge. The obtained results allow to conclude that DBD plasma application causes negligible changes in temperature and concentration of vitamin, and can be used in food application

Keywords: DBD plasma jet, vitamin C, non thermal plasma

1 INTRODUCTION

For the food industry, it is important to produce reliable products that are free of hazards and preserve them for long periods. Many microorganisms can grow in food without causing any changes in physical properties like odor, color, texture of food, while simultaneously producing metabolites that threaten human health causing illness and even deaths, damaging the national and global economy (Pawłowski 2015, Piecuch & Hewelt 2013, Piecuch & Piecuch 2013). Sterilization is a physical or chemical process that eliminates or kills all forms of life, especially microorganisms present on a surface or contained in a fluid such as biological culture media (Moisan et al. 2002). It is a key process for food processing. Harmful microorganisms can cause detrimental effects such as rotting or disease, consequently leading to economic losses. Inactivation of these microorganisms can be accomplished by conventional techniques such as heat, steam, chemical solutions or gases, radiation, as well as the recently developed techniques. However, most of these sterilization methods can cause damages to the material or limit complete sterilization (Park et al. 2007a). This opens new research areas for development of alternative sterilization methods (Lee et al. 2006). Plasma sterilization is considered to be a promising alternative to the conventional sterilization methods. It is a versatile, fast and efficient method that protects and conserves food-contacting materials (Lee et al. 2006). The use of plasma is a relatively new concept in food safety. Bacteria inactivation was achieved in apple, melon and lettuce samples (Critzer et al. 2007); in apple juices (Montenegro et al. 2002); in almond samples (Deng et al. 2007) with using atmospheric air plasma and in cheese and ham samples with atmospheric helium plasma. (Song et al. 2009). Selçuk et al. (2008) investigated the inhibition effect of low pressure SF6 plasma treatment on molds in the cereal grains and legumes. In another study, it was stated that Aspergillus parasiticus inoculated nut surfaces were decontaminated with using low-pressure SF6 and air plasmas and also a reduction was provided in the amount of aflatoxin (Basaran 2008). Park et al. succeeded to degrade 3 different mycotoxins: Aflatoxin B1 (AFB1), Deoxynivalenol (DON) and Nivelenol (NIV) in 5 seconds via plasma system (Park et al. 2007b). These results show that plasma could be an effective system for degrading mycotoxins. However, there are some concerns related to applications of plasmas on food materials and few published studies are present about effects of plasmas on necessary components of foods. Huge fraction of the export materials in Turkey and Poland is composed of foodstuff. For this reason, sterilization of food and materials coming into contact with it is important from economic and public health point of view.

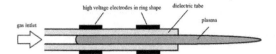

Figure 1. Geometry of the DBD reactor with ring electrodes.

Food materials contain various nutritive compounds such as vitamins, polyphenols etc. These compounds are very sensitive especially to heat. Plasma contains a wide variety of active particles, such as electrons, ions, radicals, metastable excited species, and ultraviolet radiation that has sufficient energy to break covalent bonds and initiate certain reactions and form volatile compounds (Sen et al. 2012). All these active species and heat lead to the death of cells. In addition to this, they can affect the nutritive compounds found in the food products. In the present study, the DBD (Dielectric Barrier Discharge) non-equilibrium plasma jet source has been applied to check the applicability of the plasma sterilization method without causing appreciable changes on vitamin C (L-ascorbic acid) content of food. In a non-equilibrium plasma the electron energies are much larger than the energy of other particles (ions and neutral particles), whereby it is possible to carry out a biochemical and physical reaction at a relatively low temperature of the working gas. Due to the selectivity of energy and the possibility of adjusting parameters of the plasma over a wide range, it has found many applications in environmental engineering, food industry and medicine (Weltmann et al. 2010, Pawłat et al. 2011, Stryczewska et al. 2013, Pawłat 2013, Raniszewski 2013, Hensel et al. 2015, Brisset & Pawłat 2016). Nozzle-shape plasma jet reactors can be used for these applications (Weltmann et al. 2010, Pawłat 2013, Pawłat et al. 2013, Kwiatkowski et al. 2014, Pawłat et al. 2015, Zaplotnik et al. 2015). The plasma generated inside such reactor is transported outside of the nozzle under forced gas flow. In the present study, the DBD reactor with two ring electrodes was used (Fig. 1).

The measurements were performed for the main factors that may affect the amount of generated active particles: time of treatment, composition of working gas, the flow rate value and the distance from the end of discharge.

2 MATERIAL AND METHODS

2.1 Preparing the stock solution

The method of Selimovic et al. was used for analysis of L-ascorbic acid (Selimovic et al 2011). Buffer solution (pH = 5.4), a mixture of potassium dihydrogenphosphate (0.03 mol/dm^3) and disodium hydrogenphosphate (8.99 × 10–4 mol/dm^3) was prepared by dissolving 4.08 g of KH$_2$PO$_4$ (Merck) and 0.16 g of Na$_2$HPO$_4$ •2H$_2$O (Merck) in 1000 cm^3 of distilled water. Sodium oxalate solution (0.0056 mol/dm^3) was prepared by dissolving 0.75 g of sodium oxalate (Merck) in 1000 cm^3 of the buffer solution. Stock L-Ascorbic acid solution (1.13 × 10^{-3} mol/dm^3) was prepared by dissolving of 0.05 g of L-ascorbic acid (Merck) in 250 cm^3 of the sodium oxalate solution.

2.2 Preparing the sample solution used in experiments

For the experiments, 37.5 mL stock solution was transferred to 500 mL standard volumetric flask, then diluted to the mark with 0.0056 mol/dm^3 sodium oxalate solution. This concentration was used for analysis of each parameter.

2.3 Plasma application

The system for DBD included power supply, flow meter, gas source and plasma pen. Flow rate was controlled by an electronic flow meter (Fig. 2).

Following parameters applied to 5 mL sample solution:

- Time: 4, 8, 12, 16, 20, 60 seconds.
- Gas composition: Helium and helium-oxygen mixture.
- Flow rates for pure helium and in mixture: 0.112; 0.140; 0.168 L/min.
- Flow rates for oxygen in mixture: 0.063; 0.100; 0.141 L/min.
- Distance (between discharge and top of the solution): 2, 3, 5 cm.

Similar experiments were performed for both helium and helium-oxygen gas mixtures. All experiments were carried out at room temperature and the highest temperature of sample was 24.5°C.

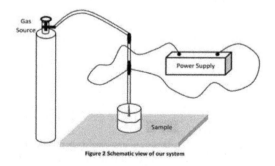

Figure 2. Schematic view of the treatment system.

2.4 Spectrophotometric measurements

Absorbance of standard solutions prepared from stock solution, initial absorbance of sample solution and final absorbance of every sample after each treatment were measured at 243 nm.

3 RESULTS AND DISCUSSION

Firstly, the absorbance of standard solutions was measured. Then calibration curves were graphed as presented in Figures 3 and 4. Next, absorbance of samples was measured. According to the calibration curve, the concentration of samples was calculated. Following graphs show the loss of vitamin (%) between the initial and final concentrations of samples.

The results of plasma treatment with pure helium of different flow rates as a substrate gas are shown in Figs. 5–7. At 2 cm distance, only the highest flow rate was performed to compare the effect of the distance.

Generally, reactive oxygen and nitrogen species can cause degradation of bioactive compounds. Low dose of oxygen was added to the helium in order to investigate its effect on the vitamin solutions. Results depicted in Figs. 8 and 9 indicated degradation of vitamin C. After a small admixture of oxygen, change in the vitamin C loss decreased. It can be explained by reducing volume of discharge for this mixture and less amount of reactive particles, related to the more difficult electrical_breakdown.

As expected, the highest loss of vitamin C amounting to 30% in both substrate gases was observed after the longest treatment time: 60 s and shortest distance between sample and reactor

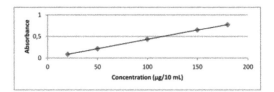

Figure 3. Calibration curve for helium.

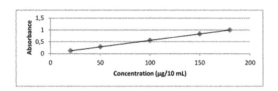

Figure 4. Calibration curve for helium-oxygen gas mixture.

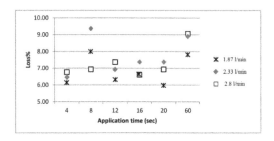

Figure 5. Absorbance after plasma application with helium at 5 cm distance.

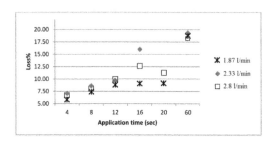

Figure 6. Absorbance after plasma application with helium at 3 cm distance.

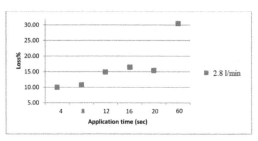

Figure 7. Absorbance after plasma application with helium at 2 cm distance.

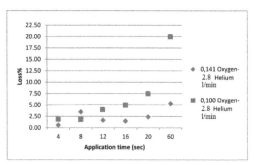

Figure 8. Loss% after plasma application with helium-oxygen gas mixture at 3 cm distance.

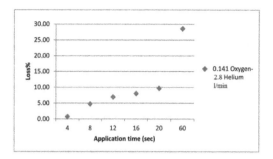

Figure 9. Loss% after plasma application with helium-oxygen gas at 2 cm distance.

nozzle: 2 cm. Loss in such a value range is quite acceptable, comparing to the traditional food preservation and decontamination methods, where losses of vitamin C may reach 90%.

4 CONCLUSION

Experimental results indicated that increase of vitamin C loss depends on time and distance. The loss reaches the highest value in 60 seconds for both selected substrate gases and for all investigated distances. The increase of sample temperature was not observed during the conducted experiments and deteriorating influence of DBD post-glow gas temperature on processed sample was negligible.

Due to amount of generated active particles and cost reduction it seems that the most optimal working conditions of plasma treatment is distance of 2 cm at time ranging from 8 to 20 seconds with pure helium as the working gas. After further development of proposed system, DBD plasma can become a good alternative for food conditioning.

ACKNOWLEDGEMENTS

We would like to thank prof. Jacek Czerwiński from LUT for his great help in all spectrophotometric experiments, prof. Henryka Stryczewska and prof. Mehmet Mutlu for their valuable advices and assistance during the research.

Conducted research was supported by COST Actions TD 1208 and MP1101, KORANET and by Lublin University of Technology.

REFERENCES

Basaran, P., Basaran-Akgul, N. & Oksuz, L. 2008, Limination of Aspergillus parasiticus from Nut Surface with Low Pressure Cold Plasma (LPCP) Treatment, *Food Microbiology*, 25, 626–32.

Brisset J.-L., Pawłat J. 2016, Chemical effects of air plasma species on aqueous solutes in direct and delayed exposure modes: discharge, post-discharge and plasma activated water, Plasma Chemistry and Plasma Processing, 36(2),355–381

Critzer, F.J., Kelly-Wintenberg, K., South, S., Roth, J.R. & Golden, D.A. 2007, Atmospheric Plasma Inactivation of Foodborne Pathogens on Fresh Produce Surfaces, *J. Food Prot.*, 70, 2290–96.

Deng, S., Ruan, R., Mok, C.K., Huang, G., Lin, X., Chen, P. 2007, Inactivation of Escherichia coli on Almonds Using Nonthermal Plasma, *J. Food Sci.*, 72(2): 62–66.

Hensel K., Kučerová K., Tarabová B., Janda M., Machala Z., Sano K., Mihai C. T., Gorgan L. D., Jijie R., Pohoata V. & Topala I. 2015, Effects of air transient spark discharge and helium plasma jet on water, bacteria, cells and biomolecules, *Biointerphases* 10 (2), 029515.

Kwiatkowski M., Terebun P., Krupski P., Samoń R., Diatczyk J., Pawłat J. & Stryczewska H., 2014, Właściwości i zastosowania reaktorów plazmowych typu dysza plazmowa, *IAPGOŚ*, nr 3, 31–35

Lee, K., Paek, K., Ju, W.T., & Lee, Y. 2006, Sterilization of bacteria, yeast, and bacterial endospores by atmospheric-pressure cold plasma using helium and oxygen. *The Journal of Microbiology*, 44: 269–275.

Moisan, M., Barbeau, J., Crevier, M.-C., Pelletier, J., Philip, N. & Saoudi, B., 2002, Plasma sterilization. Methods and mechanisms, *Pure and Applied Chemistry* 74 (3): 349–358.

Montenegro, J., Ruan, R., Ma, H. & Chen, P. 2002„ Inactivation of Escherichia coli O157:H7 Using a Pulsed Non-thermal Plasma System, *J.Food Sci.*, 67(2): 646–48.

Park, B.J., Takatori, K., Lee, M.H., Han, D.W., Woo, Y.I., Son, H.J., et al. 2007a. Escherichia coli sterilization and lipopolysaccharide inactivation using microwave-induced argon plasma at atmospheric pressure. *Surface and Coatings Technology*, 201: 5738–741.

Park, B.J., Takatori, K., Konishi, Y.S., Kim, I., Lee, M., Han, D., Chung, K., Hyun, S.O., Park, J. 2007b, Degradation of Mycotoxins Using Microwave-Induced Argon Plasma at Atmospheric Pressure, *Surface and Coating Technology,* 201(9–11): 5733–37.

Pawłat J. 2013, Atmospheric pressure plasma jet for decontamination purposes, *European Physical Journal—Applied Physics*, 61: 11.

Pawłat J., Diatczyk J. & Stryczewska H. 2011 Low-temperature plasma for exhaust gas purification from paint shop—a case study, *Przegląd Elektrotechniczny*, 1(87), 245–248.

Pawłat J., Kwiatkowski M., Terebun P. & Murakami T. 2015, Rf-powered atmospheric-pressure plasma jet in surface treatment of high-impact polystyrene *IEEE Transactions on Plasma Science*, no 99: 1–7.

Pawłat J., Samoń R., Stryczewska H., Diatczyk J. & Giżewski T. 2013, RF-powered atmospheric pressure plasma jet for surface treatment, *European Physical Journal—Applied Physics*, 61: 6.

Pawłowski L. 2015, Where Is the World Heading? Social Crisis Created by Promotion of Biofuels and Nowadays Liberal Capitalism, *Rocznik Ochrona Środowiska/ Annual Set the Environment Protection,* 17(1): 26–39.

Piecuch I. & Hewelt G. 2013, Environmental Education—first knowledge and then the habit of environment protection, *Rocznik Ochrona Środowiska/Annual Set the Environment Protection,* 15: 136–150.

Piecuch I. & Piecuch T. 2013 Environmental education and its social effects, *Rocznik Ochrona Środowiska/Annual Set the Environment Protection,* 15: 192–212.

Raniszewski G. 2013 Temperature measurements in arc-discharge synthesis of nanomaterials dedicated for medical applications *EPJAP* 61: 24311.

Selçuk, M., Öksüz, L. & Basaran, P. 2008, Decontamination of Grains and Legumes Infected with Aspergillus spp. and Penicillum spp. by Cold Plasma, *Bioresource Technology,* 99(11): 5104–09.

Selimovic, A., Salkic, M. & Selimovic, A. 2011, Direct Spectrophotometric Determination of L-Ascorbic Acid in Pharmaceutical Preparations Using Sodium Oxalate as a Stabilizer, *International Journal of Basic & Applied Sciences,* 11(2).

Sen, Y., Bağcı, U., Guleç, H.A. & Mutlu, M. 2012, Modification of Food Contacting Surfaces By Plasma Polymerisation Technique: Reducing The Biofouling of Microorganisms on Stainless Steel Surface, *J Food and Bioprocess Technology,* 5(1): 166–175.

Song, H.P., Kim, B., Choe, J.H., Jung, S., Moon, S.Y., Choe, W. & Jo, C. 2009, Evaluation of Atmospheric Pressure Plasma to Improve The Safety of Sliced Cheese and Ham Inoculated by 3-Strain Cocktail Listeria monocytogenes, *Food Microbiol.,* 26: 432–6.

Stryczewska H., Pawłat J. & Ebihara K. 2013, Non-thermal plasma aided soil decontamination, *JAOTs* 1(16): 23–30

Weltmann K.D., Kindel E., von Woedtke T., Hahnel M., Stieber M., Brandenburg R. 2010, Atmospheric-pressure plasma sources: Prospective tools for plasma medicine, *Pure and Applied Chemistry,* 82: 1223–1237.

Zaplotnik R., Bišćan M., Kregar Z., Cvelbar U., Mozetič M. & Milošević S., 2015, Influence of a sample surface on single electrode atmospheric plasma jet parameters, *Spectrochimica Acta B: Atomic Spectroscopy,* 103: 124–130.

Environmental Engineering V – Pawłowska & Pawłowski (Eds)
© 2017 Taylor & Francis Group, London, ISBN 978-1-138-03163-0

Modeling and predicting the concentration of volatile organic chlorination by-products in Krakow drinking water

A. Włodyka-Bergier, T. Bergier, Z. Kowalewski & S. Gruszczyński
Department of Environmental Protection and Management, Faculty of Mining Surveying and Environmental Engineering, AGH University of Science and Technology, Kraków, Poland

ABSTRACT: The goal of this article is to develop the models predicting the occurrence of volatile organic chlorination by-products in two Krakow (Poland) water distribution systems. The research results were used to develop the models predicting the concentration of five groups of chlorination by-products (trihalomethanes, haloacetonitriles, haloketones, chloral hydrate and chloropicrin), on the basis of several parameters of water quality, chlorination and distribution (organic carbon and nitrogen, bromide ions, UV_{254}, temperature, pH, retention time). The multiple linear regression analysis was used to develop the base models, in which all analyzed parameters were included. The backward stepwise regression method was used to optimize these models. The Pearson correlation analysis was used to assess the ability of the models to predict the by-products concentration in the analyzed water distribution systems. The sensitivity analysis was also performed to identify the influence of individual parameters on the formation of each by-products group.

Keywords: disinfection by-products, trihalomethanes, haloacetonitriles, haloketones, chloral hydrate, chloropicrin, predictive models

1 INTRODUCTION

The application of chemical disinfectants (including chlorine) to water prior to its distribution is necessary, due to the health risks associated with the presence of microorganisms and the possibility of their growth during the water transport. However, the use of chemical compounds is inherently associated with their reactivity with other compounds present in the water and the formation of Disinfection By-Products (DBPs). Due to their physical properties, many authors (Nikolaou et al. 2005, Ates et al. 2007, Kim et al. 2002) distinguish a group of Volatile Chlorination By-Products (VCBPs) from a wide range of halogenated organic water chlorination by-products. Among these by-products, the most frequently studied and described group is Trihalomethanes (THM): Trichloromethane (TCM), Bromodichloromethane (BDCM), Dibromochloromethane (DBCM) and Tribomomethane (TBM). The other VCBPs are Haloketones (HK): 1,1-dichloropropanone (1,1-DCP) and 1,1,1-trichloropropanone (1,1,1-TCP) and by-products containing nitrogen, including Halogenoacetonitriles (HAN): Trichloroacetonitrile (TCAN), Dichloroacetonitrile (DCAN), Bromochloroacetonitrile (BCAN) and Dibromoacetonitrile (DBAN); and Chloropicrin

(CP), which belongs to the group of halogenonitomethanes. Volatile halogenated organic by-products may penetrate into the human body through the inhalation, ingestion of chlorinated water and absorption trough skin during bathing (Chowdhury et al. 2010).

There are two main approaches to predicting DBPs content in water distribution networks. The first one employs the simple linear statistical correlation between the concentrations of individual by-products and their groups, or between them and the water quality parameters, including the chlorine dose (Fabbricino & Korshin 2009, Francis et al. 2009, Lee et al. 2001, Włodyka-Bergier & Bergier 2011).

The second approach is to develop the mathematical equations predicting DBPs concentration, on the basis of parameters characterizing: water quality, its chlorination and transport conditions (Chowdhury et al. 2010, Sadiq & Rodriguez 2004). The parameters usually included in such mathematical models are: Total Organic Carbon (TOC), Dissolved Organic Carbon (DOC), UV absorbance at 254 nm (UV_{254}), SUVA (UV_{254}/DOC), pH, temperature (T), bromide ions concentration (Br⁻), chlorine dose (Cl_2), reaction time (t) (Lee et al. 2001, Sadiq & Rodriguez 2004, Chowdhury et al. 2009, Toroz & Uyak 2005, Abdullah et al. 2003,

Table 1. The selected models predicting the concentration of VCBPs in water distribution systems (Chen & Westerhoff 2010, Toroz & Uyak 2005, Abdullah et al. 2003, Sohn et al. 2001).

DBP	$VCBPs = a{\cdot}DOC^b{\cdot}TOC^c{\cdot}UV_{254}^d{\cdot}pH^e{\cdot}Cl_2^f{\cdot}t^g{\cdot}T^h{\cdot}(Br^-)^i{\cdot}(Br+1)^j{\cdot}(DON+1)^k$											
	a	b	c	d	e	f	g	h	i	j	k	R^2
TCM	1805	0.11		1.22							−2.19		0.88
BDCM	137	0.16		0.94							3.66		0.69
ΣTHM	1147	0.00		0.83							0.27		0.87
	11.967		0.398				0.702	0.158					0.83
	$10^{-6.77}$					4.469	1.765						0.60
	$10^{-6.24}$		1.342			3.952	2.902						0.66
	$4.12{\cdot}10^{-2}$	1.1				1.6	0.152	0.26	0.61	0.068			–
DCAN	0.39	1.53		0.00							−0.87		0.62
ΣHAN	1.65	0.87		0.00							3.22		0.63
	17.05	0.15		0.72							3.78	0.67	0.64

Uyak et al. 2005, Uyak et al. 2007, Hong et al. 2007, Rodriguez & Sérodes 2001, Chen & Westerhoff 2010, Kulkarni & Chellam 2010).

Among the publications on mathematical models predicting DBPs concentrations in water distribution networks, almost 90% articles focus on THM (Chowdhury et al. 2009). In recent years, few articles have been published on the predicting the concentration of DBPs containing nitrogen, mainly HAN. Models presented in these articles usually include Dissolved Organic Nitrogen (DON) in addition to the parameters listed above (Chen & Westerhoff 2010). So far, models describing the formation of dynamics HK, CH and CP have not been developed. Table 1 presents the selection of models predicting VCBPs content in water distribution systems, which are based on the multiple linear regression method with the log-transformation of the variables (Toroz & Uyak 2005, Abdullah et al. 2003, Uyak et al. 2007, Chen & Westerhoff 2010, Zimoch & Stolarczyk 2010, Sohn et al. 2001). Apart from the models presented in Table 1, there are many other models employing the variety of mathematical and statistical methods (Sadiq & Rodriguez 2004, Chowdhury et al. 2009, Uyak et al. 2005, Hong et al. 2007, Rodriguez & Sérodes 2001, Czapczuk et al. 2015).

The main purpose of the article is to develop the models predicting the concentration of the comprehensive set of VCBPs, including THM, HAN, HK, CH and CP, in Krakow water distribution system. On the other hand, the authors' ambition is to base the developed models on the broad set of parameters affecting the concentration of studied VCBPs. Thus, the considered controlling (independent) variables include the parameters of water before its chlorination (DOC, UV_{254}, Br⁻, DON), and those characterizing the conditions of water chlorination and transport (Cl_2, pH, T, t).

Those parameters were measured in the water samples taken from two Krakow drinking water distribution system, supplied by the water treatment plants Raba and Bielany. Several statistical methods were employed to develop the predictive models, based on the results of the research on water distribution systems. In the first step, multiple linear regression analysis was used to form the base models, in which all independent variables were included. In the second step, the number of parameters included in the models was reduced by the elimination of variables of no statistical significance. The backward stepwise regression method was used to realize this step. The third step was the correlation analysis, used to compare values predicted by the models with those measured in the water distribution systems. The last step was the sensitivity analysis, which was performed to assess the influence of individual parameters on the VCBPs formation.

2 MATERIALS AND METHODS

2.1 Study area and sampling procedure

The research was conducted over the period of 14 months, from February 2011 to March 2012. The water samples were taken from Water Treatment Plant (WTP) Raba and WTP Bielany, and also from the distribution systems supplied by these plants. Both analyzed plants apply chlorination to disinfect water, but they differ in water sources, treatment processes, their size and production capacity. The annual production of WTP Raba is $58,522,000\ m^3$ of water, which covers 59% of the total Krakow water consumption, while in a case of WTP Bielany it is $4,991,000\ m^3$ (9%). The source of water for WTP Raba is the Dobczyce Reservoir, and the treatment process includes: Ozonation, coagulation with polyaluminum chlorides, sand rapid filtration and chlorine disinfection. Due to the big size of the distribution system supplied by WTP Raba, water is additionally chlorinated during its transport to Krakow. In a case of WTP Bielany, water is taken from the Sanka River, and then treated in the sedimentation and slow filtration processes, after which it is disinfected with chlorine.

Every month five water samples were taken from the distribution system supplied by WTP Bielany, and six from WTP Raba. Thus, totally 154 samples were taken and analyzed within the research period (84 from WTP Raba system and 70 from WTP Bielany). In the samples taken from these networks, the concentration of 12 by-products were measured (TCM; BDCM; DBCM; TBM; TCAN; DCAN; BCAN; DBAN; 1,1-DCP; 1,1,1-TCP; CH; CP). Their individual concentrations were used to calculate the totals for VCBPs groups, which are statistically analyzed in the article (ΣTHM, ΣHAN, ΣHK, CH, CP). During taking the water samples from the networks, temperature and pH were also measured. Simultaneously, water taken directly from both water treatment plants was also tested. For this purpose the samples of water after all treatment processes, but prior to chlorination, were taken. Four water quality parameters were measured in these samples: DOC, UV_{254}, Br⁻, DON. The exact description of sampling and analytical methods can be found in the publication (Włodyka-Bergier et al. 2014).

2.2 Statistical analyses

The goal of statistical analyses, performed within the article, was to determine the set of mathematical equations predicting the concentration of VCBPs, on the basis of values of the parameters of water quality, chlorination and transport.

The first stage of the statistical assessment was the Multiple Linear Regression analysis (MLR) for each studied group of VCBPs (analogously to Uyak et al. 2005, Chen & Westerhoff 2010), on the basis of full set of measured parameters: DOC, Br⁻, UV_{254}, DON, pH, T, Cl_2, t. The last parameter, time of water flow in the network, was determined with EPANET software, on the basis of a distance between a water sampling point and the water treatment plant.

Due to the fact that VCBPs formation dynamics have no linear character, it was necessary to logarithm all variables prior the main analysis, thus the classic linear MLR equation has the form (1).

$$\log(VCBPs) = a + b \cdot \log(DOC) + c \cdot \log(Br^-)$$
$$+ d \cdot \log(UV_{254}) + e \cdot \log(DON)$$
$$+ f \cdot \log(Cl_2) + g \cdot \log(pH)$$
$$+ h \cdot \log(T) + i \cdot \log(t) \qquad (1)$$

This equation (1) has been transformed into the universal equation (2), which allows to predict the concentration of each VCBPs, based on all parameters of water quality, chlorination and transport, studied in the article.

$$VCBPs = 10^a \cdot DOC^b \cdot (Br^-)^c \cdot (UV_{254})^d \cdot (DON)^e \cdot$$
$$\cdot (Cl_2)^f \cdot pH^g \cdot T^h \cdot t^i \qquad (2)$$

Prior to the calibration of the above-mentioned equation for each individual VCBP, it was assessed if its distribution is normal and if there are no outliers (i.e. outside the range: a mean ± 3 standard deviations). The concentration of all analyzed VCBPs met these conditions. As a result of MLR analysis, two independent sets of base equations were created, one for each WTP (Model 1).

The second statistical stage was the optimization of equations, obtained in the previous stage. The backward stepwise regression was conducted for each VCBP. In each iteration, the variable (parameter) with the lowest statistical significance for the given VCBP was removed, and the reduced equation was analyzed in the next iteration. Such an operation was repeated (i.e. the next variable was removed) until an equation consists only statistically important variables. In each iteration the linear regression determination coefficient (R^2) for an equation was also evaluated. As a result of the backward stepwise regression, the set of optimized equations was created for each WTP (Model 2).

The third stage of statistical assessment was to calculate Pearson coefficients (R) of the correlation between the values predicted by the model and the real VCBP concentrations measured in the analyzed distribution systems. It was conducted for both base and optimized model, for both water distribution systems.

As the last stage of statistical assessment, the sensitivity analysis was performed for the base sets of equations (analogously to EPA 2009), in which the influence of each parameter on the given VCBP was evaluated. The variance-based approach was applied, on the basis of test F results from the backward stepwise regression, to calculate the percentage share of each parameter in controlling the given VCBP concentration.

All statistical analyses, described above, were conducted with Statistica software package (ver. 10.0) by StatSoft. They were done separately for the distribution system of WTP Raba and WTP Bielany. While a measured substance was not detected in a sample, the half of its detection limit was taken for statistical analysis.

3 RESULTS AND DISCUSSION

Table 2 contains the characteristics of water from the studied water distribution systems, supplied by WTP Raba and WTP Bielany. The parameters of water quality (DOC, UV_{254}, Br^-, DON), water chlorination and transport (Cl_2, pH, T, t) are presented in Table 2. Mean, minimum and maximum values are given for each of them.

Table 3 presents the base models, which are the results of MLR analysis. The upper part contains the set of equations, creating Model 1 for WTP Raba, and the bottom part—for WTP Bielany. The presented equations include all parameters analyzed in studied water distribution systems, but those with the statistically important relationship to the predicted VCBP concentration were marked. Table 4 presents the optimized models, which are the results of the backward stepwise regression. The similar manner of presenting data was used as in Table 3. However, due to the fact that most of the variables are statistically important, only one which is not was marked.

The results presented in Tables 2–4, as well as the results of other statistical calculations described in the previous section, are discussed in the below part of the paper. This discussion is divided into sections, each of which is dedicated to the individual group of considered VCBPs.

Table 2. The parameters of water quality and conditions of water chlorination and transport (WTP Raba and WTP Bielany).

		WTP Raba			WTP Bielany		
		Mean	Min.	Max.	Mean	Min.	Max.
DOC	mg/L	2.1	1.1	3.7	3.2	1.5	6.0
UV_{254}	cm^{-1}	0.096	0.048	0.135	0.245	0.180	0.365
Br^-	mg/L	0.3	0.2	0.4	0.6	0.4	0.8
Cl_2	mg/L	0.8	0.7	1.1	1.8	1.6	2.3
DON	mg/L	0.11	0.02	0.31	0.18	0.02	0.53
pH	–	7.97	7.60	8.23	7.56	7.40	7.79
T	°C	13.5	0.5	22.0	13.0	0.5	24.0
ΣTHM	µg/L	21.0	6.53	55.23	17.64	3.54	36.73
ΣHAN	µg/L	2.29	0.40	8.86	2.40	nd	13.05
ΣHK	µg/L	3.72	nd	13.88	3.18	nd	9.21
CH	µg/L	2.71	0.34	6.74	2.08	nd	8.55
CP	µg/L	0.08	nd	0.43	0.10	nd	0.54

nd—not detected (0.01 µg/L).

Table 3. Base models (Model 1) predicting the concentration of VCBPs in water from WTP Raba and WTP Bielany.

$$VCBPs = 10^a \cdot DOC^b \cdot (Br^-)^c \cdot UV_{254}^{\ d} \cdot DON^e \cdot Cl_2^{\ f} \cdot pH^g \cdot T^h \cdot t^i$$

	a	b	c	d	e	f	g	h	i	R^2
WTP Raba										
ΣTHM	−2.16	0.23	0.04	−0.35[1]	−0.01	1.44[1]	2.86	0.26[1]	0.18	0.647
ΣHAN	8.08	1.20[1]	0.27	−0.41	0.05	0.25	−8.90	0.00	−0.26	0.370
ΣHK	7.15	0.21	1.60[1]	1.62[1]	0.43[1]	−0.71	−4.65	0.01	0.21	0.402
CH	1.36	0.27	−0.19	−0.69[1]	0.23[1]	−0.38	−0.80	0.17	−0.77	0.153
CP	−9.34	0.58	0.36	0.76	0.01	1.82	9.09	−0.01	0.77	0.074
WTP Bielany										
ΣTHM	−7.45	−0.14	0.26	0.15	0.10	1.93[1]	9.45	0.09	0.17[1]	0.288
ΣHAN	−30.00[1]	0.35	−4.40[1]	−1.52[1]	−0.64[1]	3.84[1]	30.31[1]	−0.20	0.15	0.629
ΣHK	−60.41[1]	−0.73	−3.62[1]	−1.28	−0.43	2.71	66.80[1]	−0.49	0.30	0.309
CH	−53.41[1]	−0.82	−0.35	−1.88[1]	0.19	0.89	59.38[1]	0.30	0.06	0.470
CP	−56.69[1]	−1.99[1]	0.93	0.76	0.69[1]	2.06	65.13[1]	−0.30	0.23	0.504

[1]significant with $p < 0.05$.

Table 4. Optimized models (Model 2) predicting the concentration of VCBPs in water from WTP Raba and WTP Bielany.

VCBPs=$10^a \cdot DOC^b \cdot (Br^-)^c \cdot UV_{254}^d \cdot DON^e \cdot Cl_2^f \cdot pH^g \cdot T^h \cdot t^i$										
	a	b	c	d	e	f	g	h	i	R^2
WTP Raba										
ΣTHM	0.59	0.29	–	–0.39	–	1.29	–	0.28	–	0.639
ΣHAN	–0.16	1.22	–		–	–	–	–	–	0.312
ΣHK	3.70	–	1.83	1.78	0.43	–	–	–	–	0.396
CH	–0.221	–	–	–0.78	0.22	–	–	–	–	0.101
CP^2										
WTP Bielany										
ΣTHM	–7.45	–0.14	0.26	0.15	0.10	1.93[1]	9.45	0.09	0.17[1]	0.288
ΣHAN	–30.00[1]	0.35	–4.40[1]	–1.52[1]	–0.64[1]	3.84[1]	30.31[1]	–0.20	0.15	0.629
ΣHK	–60.41[1]	–0.73	–3.62[1]	–1.28	–0.43	2.71	66.80[1]	–0.49	0.30	0.309
CH	–53.41[1]	–0.82	–0.35	–1.88[1]	0.19	0.89	59.38[1]	0.30	0.06	0.470
CP	–56.69[1]	–1.99[1]	0.93	0.76	0.69[1]	2.06	65.13[1]	–0.30	0.23	0.504

[1] no significant with $p < 0.05$.
[2] optimized model was not created.

3.1 Trihalomethanes

The concentration of compounds from THM group was higher in the network of WTP Raba than in one of WTP Bielany (Table 2), the average values of ΣTHM were 21.00 µg/L and 17.64 µg/L respectively. For WTP Raba the minimum value was 6.53 µg/L and maximum 55.23 µg/L, for WTP Bielany they were 3.54 µg/L and 36.73 µg/L, respectively. Considering the seasonal THM variation, which was studied in the other research (Włodyka-Bergier & Bergier 2013), the high levels of ΣTHM in WTP Raba network were observed in summer and autumn, while in WTP Bielany network it was impossible to identify strong seasonal patterns, however the slightly lower THM levels were observed in winter. In both systems, TCM was observed in the highest concentration among all THM. However, due to the higher concentration of bromides, the higher share of brominated species was observed in WTP Bielany water distribution system in comparison to WTP Raba (Włodyka-Bergier et al. 2014).

MLR was used to obtain the equation to predict the concentration of ΣTHM in WTP Raba distribution network:

$$\Sigma THM = 10^{-2.16} \cdot DOC^{0.23} \cdot (Br^-)^{0.04} \cdot UV_{254}^{-0.35} \cdot$$
$$\cdot DON^{-0.01} \cdot Cl_2^{1.44} \cdot pH^{2.86} \cdot T^{0.26} \cdot t^{0.18} \qquad (3)$$

with relatively good estimation ($R^2 = 0.647$). After the reduction to the statistically important parameters, the equation was optimized to the following form: $\Sigma THM = 10^{0.59} \cdot DOC^{0.29} \cdot UV_{254}^{-0.39} \cdot Cl_2^{1.29} \cdot T^{0.2}$, with similar, good estimation ($R^2 = 0.639$). In a case of WTP Bielany network, the base equation was:

$$\Sigma THM = 10^{7.45} \cdot DOC^{0.14} \cdot (Br)^{0.26} \cdot UV_{254}^{0.15} \cdot$$
$$\cdot DON^{0.10} \cdot Cl_2^{1.93} \cdot pH^{9.45} \cdot T^{0.09} \cdot t^{0.1} \qquad (4)$$

with significantly lower coefficient R^2 (0.288). After the optimization process, the equation predicting the concentration of ΣTHM in Bielany network includes three parameters (Cl_2, pH, t):

$$\Sigma THM = 10^{-10.53} \cdot Cl_2^{1.74} \cdot pH^{12.77} \cdot t^{0.16} \qquad (5)$$

and the determination coefficient R^2 slightly decreased to 0.267.

Figure 1 shows the relations between the values of ΣTHM predicted by both models and measured in a distribution system (Figure 1 A for WTP Raba, Figure 1B for WTP Bielany). Model 1 for WTP Raba predicted the concentration of ΣTHM in its distribution system very well, with Pearson correlation coefficient R = 0.839. On the other hand, Model 1 created for WTP Bielany, was less statistically accurate (R = 0.544). The elimination of parameters with no statistical importance allowed to create the equations (Model 2), which predicted ΣTHM concentration in both networks even better (R = 0.843 for WTP Raba and R = 0.570 for WTP Bielany).

The sensitivity analysis of the base models was used to rank factors controlling THM formation in analyzed water distribution systems. Both in network of WTP Raba and WTP Bielany the chlorine dose had the strongest influence on THM formation (51% in Model 1 for WTP Bielany and 43% for WTP Raba). The order of other parameters were: T (33%) > UV_{254} (12%) > DOC (8%) > pH = t (2%) for WTP Raba system, and t (27%) > pH (9%) > DON (5%) > T (3%) > other parameters (1–2%) for WTP Bielany. The strong relation between the chlorine dose and THM formation is broadly known and well explained, and was observed by several authors (Toroz & Uyak 2005, WHO 2000, Abdullah & Hussona 2013). Conversely, the influence of temperature is not so definite, and the authors studying the seasonal changes of THMs concentration in water distribution systems reported different patterns (Lee et al. 2001, Toroz & Uyak 2005, Uyak et al. 2007, Sérodes et al. 2003). Some of these studies showed the strong dependence of THM concentration on water temperature, but in others such relation was not observed at all. Thus, the temperature of water in the distribution system may or may not be the factor strongly affecting THM formation. Besides the chlorine dose, the most important factors affecting THM formation in WTP Bielany distribution system are retention time (the distance from WTP) and pH. The similar relationships are reported in the literature, e.g. Nikolaou et al. (2004) observed the increase in THM concentration for the longer time of reaction of water with chlorine, and in other research reports (WHO 2000, Abdullah & Hussona 2013, Singer 1994) the similar influence of the higher values of pH was noticed. The positive value of coefficients for the correlation between UV_{254} and ΣTHM is usually reported in the literature (e.g. Toroz & Uyak 2005, Abdullah & Hussona 2013). The statistically important negative value of this coefficient obtained for WTP Raba system is rather surprising. It could be caused by the relatively low levels of UV_{254} in water from this WTP, and its low variability.

3.2 Haloacetonitriles

HAN levels observed in water distribution systems of WTP Raba and WTP Bielany were similar—their average concentrations were 2.29 µg/L and 2.40 µg/L respectively. The maximum HAN concentration (13.05 µg/L) was observed in autumn for WTP Bielany, and in spring for WTP Raba (8.86 µg/L). Among all examined HAN, the highest share of DCAN was observed for both distribution systems (48% for WTP Raba and 38% for WTP Bielany). DBAN accounted for 17% of all HAN in water from WTP Raba and 31% from WTP Bielany, TCAN for 5% and 4% respectively (Włodyka-Bergier et al. 2014, Włodyka-Bergier & Bergier 2013).

The base equation, predicting HAN concentration in water distribution system of WTP Raba, was calibrated into the following form ($R^2 = 0.370$):

$$\Sigma HAN = 10^{8.08} \cdot DOC^{1.20} \cdot (Br^-)^{0.27} \cdot UV_{254}^{-0.41} \\ \cdot DON^{0.05} \cdot Cl_2^{0.25} \cdot pH^{-8.90} \cdot T^{0.00} \cdot t^{-0.26} \quad (6)$$

After the elimination process only DOC remained, and the optimized equation was ($R^2 = 0.370$):

$$\Sigma HAN = 10^{-0.16} \cdot DOC^{1.22} \quad (7)$$

As it can be observed in Figure 2A, Model 1 very well predicted HAN concentration in WTP Raba network very well (R = 0.743); on the other hand, after the reduction to only one variable (Model 2) the statistical accuracy did not decrease drastically (R = 0.615).

In the case of WTP Bielany network the base equation was ($R^2 = 0.629$):

$$\Sigma HAN = 10^{-3.00} \cdot DOC^{0.35} \cdot (Br^-)^{-4.40} \\ \cdot UV_{254}^{-1.52} \cdot DON^{-0.64} \cdot Cl_2^{3.84} \\ \cdot pH^{30.31} \cdot T^{-0.20} \cdot t^{0.15} \quad (8)$$

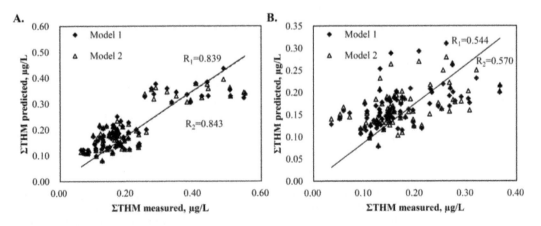

Figure 1. Predicted vs. measured values of trihalomethanes in distribution systems: A. WTP Raba; B. WTP Bielany.

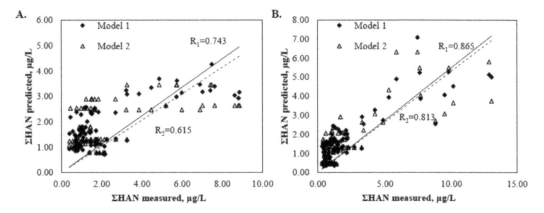

Figure 2. Predicted vs. measured values of haloacetonitriles in distribution systems: A. WTP Raba; B. WTP Bielany.

and after the optimization ($R^2 = 0.596$):

$$\Sigma HAN = 10^{-22.05} \cdot (Br^-)^{-4.46} \cdot UV_{254}^{-1.18} \cdot DON^{-0.54} \cdot Cl_2^{4.16} \cdot pH^{21.53} \quad (9)$$

Both base and optimized models predicted HAN concentration in water distribution system of WTP Bielany very well, with Pearson correlation coefficients 0.865 and 0.813 respectively (Figure 2B).

The effectiveness order of parameters in predicting HAN concentration, obtained as a result of the sensitivity analysis, was following for WTP Raba: DOC (80%) > pH (10%) > UV_{254} (6%) > other parameters (0–2%); and for WTP Bielany: Br⁻ (45%) > DON (18%) > Cl_2 (17%) > pH (8%) > UV_{254} (7%) > other parameters (1–2%). The positive correlations between DOC and ΣHAN, similar to those observed in WTP Raba system, were reported by other authors (Table 1, Wei et al. 2010). Other researchers also observed similar values of the coefficient R^2 for the equations to predict ΣHAN in water distribution systems, on the basis of DOC, among the other parameters (Table 1). In WTP Bielany network, the concentration of bromide ions had the highest, but negative, influence on HAN formation in water. Bromides react quickly with organic compounds, thus their presence in water favors the formation of brominated chlorination by-products, usually THM (WHO 2000); thus more brominated derivatives of THM can be formed, instead of HAN. The presence of bromides in water can also influence the HAN formation dynamics, i.e. BCAN and DBAN are formed instead of DCAN (Bond et al. 2011), which was confirmed in the speciation of HAN formed in WTP Bielany network (Włodyka-Bergier et al. 2014). HAN precursors are organic compounds consisting nitrogen (Lee et al. 2007, Templeton et al. 2010, Reckhow et al. 2001, Mitch et al. 2009, Dotson et al. 2009), thus their formation is influenced by the concentration of DON in chlorinated water (Chen & Westerhoff 2010). The negative influence of DON and UV_{254} on HAN formation, obtained for WTP Bielany system can be explained only by the seasonal changes in the reactivity and quality of organic matter. The positive correlation of HAN concentration with chlorine dose is also reported by other researchers (WHO 2005). The higher values of pH can cause the more intensive HAN formation in time up to 8 h, while after 96 h they rather cause the decrease in the concentration of HAN (WHO 2005). The WTP Bielany system is a small network, where the maximum water transport time is circa 10 h. It explains the positive influence and statistical importance of pH, although the range of pH values observed in this network was relatively narrow (7.40–7.79).

3.3 Haloketones

The minimum HK concentration was under the detection limit in both systems, while the maximum 13.88 µg/L in WTP Raba network and 9.21 µg/L in WTP Bielany network. The highest levels were observed in studied systems in autumn and winter, the lowest in summer (Włodyka-Bergier & Bergier 2013). The distribution of HK was also similar in both systems. The dominant compound from this group was 1,1,1-TCP, which share was 52% in the system of WTP Raba, and 55% in WTP Bielany. While the share of 1,1-DCP was 48% and 45% respectively (Włodyka-Bergier et al. 2014).

For WTP Raba, the base model to predict ΣHK was ($R^2 = 0.402$):

$$\Sigma HK = 10^{7.15} \cdot DOC^{0.21} \cdot (Br^-)^{1.60} \cdot UV_{254}^{-1.62} \cdot DON^{0.43} \cdot Cl_2^{-0.71} \cdot pH^{-4.65} \cdot T^{0.01} \cdot t^{0.21} \quad (10)$$

while after the optimization the equation had the form ($R^2 = 0.396$):

$$\Sigma HK = 10^{3.70} \cdot (Br^-)^{1.83} \cdot UV_{254}^{1.78} \cdot DON^{0.43} \qquad (11)$$

As it can be seen in Figure 3A, both of these models predicting the concentration ΣHK in WTP Raba network with the same Pearson correlation coefficient ($R = 0.571$). For WTP Bielany, the base model had the form

$$\Sigma HK = 10^{-60.41} \cdot DOC^{-0.73} \cdot (Br^-)^{-3.62} \cdot UV_{254}^{-1.28}$$
$$\cdot DON^{-0.43} \cdot Cl_2^{2.71} \cdot pH^{66.80} \cdot T^{-0.49} \cdot t^{0.30} \qquad (12)$$

with the relatively low determination coefficient ($R^2 = 0.309$). The model reduced to only statistically important variants had the form

$$\Sigma HK = 10^{-39.20} \cdot (Br^-)^{-2.13} \cdot pH^{44.33} \qquad (13)$$

and even lower coefficient ($R^2 = 0.141$). Figure 3B also shows that both models had the low ability to predict the real values of ΣHK concentration in WTP Bielany ($R = 0.146$ for Model 1 and 0.103 for Model 2).

The sensitivity analysis for Model 1 indicated that UV_{254} was the parameter with most important effect on the concentration of HK in WTP Raba (41%). HK precursors are not well-studied and there is a limited number of publications on their formation dynamics. However in the other research (Włodyka-Bergier & Bergier 2011) the highest HK formation potential was observed for chlorination of organic matter consisting hydrophobic acids, which are the compounds with the highest value of SUVA (UV_{254}/DOC). The importance of other parameters: DON (29%) and Br^- (26%) is very difficult to explain, because compounds from HK group do not contain bromine nor nitrogen. These positive correlations are probably caused by the transformations of HK and other chlorination by-products.

3.4 Chloral hydrate

The average concentration of CH in the system of WTP Raba was 2.71 µg/L, its minimum 0.34 µg/L and maximum 6.74 µg/L. While in water from WTP Bielany system, the average concentration of this compound was 2.08 µg/L. Its maximum was noticeably higher than in water from WTP Raba, and was 8.55 µg/L. The minimum was under the detection limit. In both analyzed systems, the highest levels of CH were observed in autumn, and the lowest in spring (Włodyka-Bergier et al. 2014, Włodyka-Bergier & Bergier, 2013).

For WTP Raba network, the base model, resulted from MLR analysis, had the form:

$$\Sigma CH = 10^{1.36} \cdot DOC^{0.27} \cdot (Br^-)^{-0.19} \cdot UV_{254}^{-0.69}$$
$$\cdot DON^{0.23} \cdot Cl_2^{-0.38} \cdot pH^{-0.80} \cdot T^{0.17} \cdot t^{-0.7} \qquad (14)$$

However its statistical importance was very low, with the linear determination coefficient R^2 0.153. After the backward stepwise regression, the equation was reduced to the form:

$$\Sigma CH = 10^{-0.22} \cdot UV_{254}^{-0.69} \cdot DON^{0.23} \qquad (15)$$

and the coefficient R^2 furtherly decreased to 0.101. As it can be seen in Figure 4A, these models predict CH concentration with Pearson correlation coefficient 0.361 and 0.240 receptively for Model 1 and 2. The equations for WTP Bielany had noticeably better goodness-of-fit, their determination coefficient R^2 was 0.470 and 0.449 respectively for the base model:

$$\Sigma CH = 10^{-53.41} \cdot DOC^{-0.82} \cdot (Br^-)^{-0.35} \cdot UV_{254}^{-1.88}$$
$$\cdot DON^{0.19} \cdot Cl_2^{0.89} \cdot pH^{59.38} \cdot T^{0.30} \cdot t^{0.06} \qquad (16)$$

and the optimized one:

$$\Sigma CH = 10^{-58.57} \cdot UV_{254}^{-2.72} \cdot pH^{64.75} \qquad (17)$$

The capacity of the base model to predict CH concentration in the distribution water system of WTP Bielany was moderate ($R = 0.558$), and minimally better ($R = 0.566$) for the optimized model.

As the sensitivity analysis showed, pH is the parameter the mostly effects CH concentration in WTP Bielany network (59%). It is consistent with the reports by other authors, which also noticed the strong positive correlation between the concentration of chloral hydrate and pH of water (Wei et al. 2010, Dąbrowska & Nawrocki 2009). The negative correlation with the parameter UV_{254} is possible connected with the fact, that the precursors of CH are aldehydes and acetaldehydes, generally hydrophilic compounds characterized by low SUVA values (Koudjonou et al. 2008).

3.5 Chloropicrin

The levels of CP concentration observed in WTP Raba network was from 0.00 (under the detection limit) to 0.43 µg/L, with the average value 0.08 µg/L. While in WTP Bielany network these concentrations were slightly higher—from 0.00 (under the detection limit) to 0.54 µg/L (average 0.10 µg/L). The highest concentration of CP in water from WTP Raba was observed in autumn, while in water from WTP Bielany in winter. The lowest CP levels were observed in spring for WTP Bielany and in summer for WTP Raba (Włodyka-Bergier et al. 2014, Włodyka-Bergier & Bergier, 2013).

MLR did not result in the satisfactory model to predict the concentration of CP in water distribution system of WTP Raba. The base model:

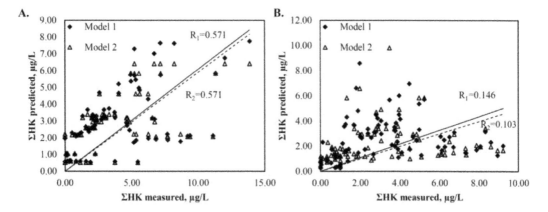

Figure 3. Predicted vs. measured values of haloketones in distribution systems: A. WTP Raba; B. WTP Bielany.

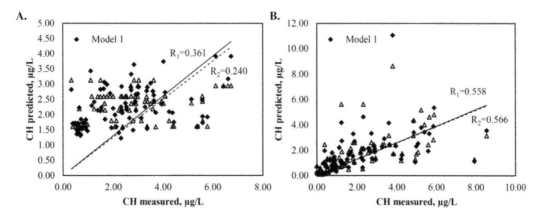

Figure 4. Predicted vs. measured values of chloral hydrate in distribution systems: A. WTP Raba; B. WTP Bielany.

$$\Sigma CP = 10^{-9.34} \cdot DOC^{0.58} \cdot (Br^-)^{-0.36} \cdot UV_{254}^{0.76}$$
$$\cdot DON^{0.01} \cdot Cl_2^{1.82} \cdot pH^{9.09} \cdot T^{-0.01} \cdot t^{0.77} \quad (18)$$

had a very low correlation coefficient ($R^2 = 0.074$). With such low level there was no sense in optimizing this model, thus the backward stepwise regression was not performed and the optimized model was not formed. For the same reason, the sensitivity analysis was not performed for this model. Figure 5A shows how the base model predicted the concentration of CP in WTP Raba network. Its Pearson correlation coefficient was also very low ($R = -0.036$). The model for WTP Bielany network had better statistical parameters and goodness-of-fit. The base model, which had the form:

$$\Sigma CP = 10^{-56.69} \cdot DOC^{-1.99} \cdot (Br^-)^{0.93} \cdot UV_{254}^{0.76}$$
$$\cdot DON^{0.69} \cdot Cl_2^{2.06} \cdot pH^{65.13} \cdot T^{-0.30} \cdot t^{0.23} \quad (19)$$

was characterized by the correlation coefficient $R^2 = 0.504$. The reduction of this model to statistically important parameters (DOC, DON and pH) resulted in the optimized model:

$$\Sigma CP = 10^{-42.78} \cdot DOC^{-1.66} \cdot DON^{0.63} \cdot pH^{48.63} \quad (20)$$

with the correlation coefficient $R^2 = 0.449$. These models predicted CP concentration in WTP Bielany network with Pearson correlation coefficient 0.522 and 0.499 respectively (Figure 5B).

The sensitivity analysis, conducted for WTP Bielany system, identified pH as the parameter mostly affecting CP concentration (59%). The results of the research conducted by other authors did not prove the certain relation between pH and the formation of HNM. In the experiments of Mitch et al. (2009), in which the concentration of CP was analyzed after 72 h reaction, in pH 5, 7 and 9, the highest value was measured in the sample with pH 9, and lowest—pH 7. The further most important parameters controlling CP formation are DOC (28%) and DON (20%). As the several research reports (Koudjonou et al.

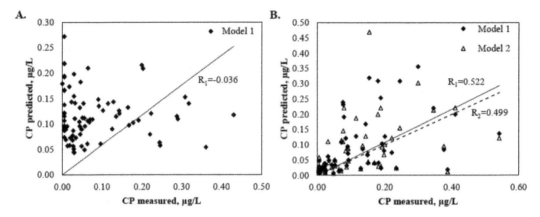

Figure 5. Predicted vs. measured values of chloropicrin in distribution systems: A. WTP Raba; B. WTP Bielany.

2008, Song et al. 2010) showed, the formation of CP is strongly dependent on the sources of precursors consisting nitrogen, including sewage released to natural waters and algae blooms.

4 CONCLUSIONS

An attempt to develop the predictive models to estimate the volatile disinfection by-products occurrence in two water distribution systems in Krakow has been presented in the article. The assessment of models, resulted from the multiple linear regression for several parameters of water quality, chlorination and transport, allowed to form the following conclusions:

- THM concentration in the water distribution network of WTP Raba, which size is relatively large, was accurately predicted by the developed model (R = 0.839). In the case of WTP Bielany system, which is a relatively small network, it was impossible to build the good model basing on the analyzed factors (R^2 = 0.288); however, the correlation between measured values and those calculated by this model was moderate (R = 0.544).
- HAN occurrence was more accurately predicted by the model developed for WTP Bielany system (R = 0.865), but the prediction by the model for WTP Raba was also fairly good (R = 0.743); however, its determination coefficient was relatively low (R^2 = 0.370).
- HK concentration in WTP Raba system was moderately predicted by the developed model (R = 0.571), while it was impossible to develop the satisfactory predictive model for WTP system (R = 0.143).
- For WTP Raba, it was impossible to develop satisfactory models to predict the occurrence of CH and CP in the network (R was respectively 0.361 and −0.036). The situation was better in the case of WTP Bielany, the developed models for this system were able to predict CH and CP concentration on the moderate level (R respectively 0.558 and 0.522).
- The optimization of developed models, which was conducted with backward stepwise regression, allowed to reduce the set of independent variables. The ability of reduced models to predict VCBPs concentration had not been compromised significantly. In some cases, the optimized models had even better correlations with measured values than original ones (i.e. THM in both systems, HAN in WTP Raba, CH in WTP Bielany).

However, the most important conclusion from the conducted analyses is the fact that it is possible to use the multiple linear regression to develop the predictive models not only for THM, but also for other volatile chlorination by-products. Such models, predicting the concentration of HK, CH or CP in real water distribution systems, have not been published yet. Despite the fact, that some of the developed models did not predict accurately the content of these compounds for both studied networks, it is the important first step to study this issue. Hopefully, further studies will help to develop more accurate models, resulting in better understanding of VCBPs formation dynamics, and thus in better protection of consumers against harmful effects of these compounds.

ACKNOWLEDGEMENTS

The work was completed under AGH University statutory research No. 11.11.150.008 for the Department of Environmental Management and

Protection. Sampling of water from water treatment plants Raba and Bielany was possible due to the courtesy of Krakow Municipal Waterworks and Sewer Enterprise.

REFERENCES

Abdullah, A. & Hussona, S. 2013. Predictive model for disinfection by-product in Alexandria drinking water, northern west of Egypt. *Environ Sci Pollut Res* 20: 7152–7166.

Abdullah. M., Yew. C. & bim Ramli. M. 2003. Formation, modeling and validation of Trihalomethanes (THM) in Malaysian drinking water: a case study in the districts of Tampin, Negeri Sembilan and Sabak Bernam, Selangor, Malaysia. *Water Res* 37: 4637–4644.

Ates, N., Kitis, M. & Yetis, U. 2007. Formation of chlorination by-products in waters with low SUVA-correlations with SUVA and differential UV spectroscopy. *Water Res* 41:4139–4148.

Bond, T., Huang, J., Templeton, M. & Graham, N. 2011. Occurrence and control of nitrogenous disinfection by-products in drinking water—A review. *Water Res* 45: 4341–4354.

Chen, B. & Westerhoff, P. 2010. Predicting disinfection by-product formation potential in water. *Water Res* 44: 3755–3762.

Chowdhury, S., Champagne, P. & McLellan, P. 2009. Models for predicting disinfection byproducts (DBP) formation in drinking waters: A chronological review. *Sci Total Environ* 407: 4189–4206.

Chowdhury, S., Rodriguez, M. & Serodes, J. 2010. Model development for predicting changes in DBP exposure concentrations during indoor handling of tap water. *Sci Total Environ* 408: 4733–4743.

Czapczuk, A., Dawidowicz, J. & Piekarski,. J 2015. Artificial Intelligence Methods in the Design and Operation of Water Supply Systems, Rocznik Ochrona Środowiska/Annual Set the Environment Protection, 17(2): 1527–1544.

Dąbrowska, A. & Nawrocki, J. 2009. Controversies about the occurrence of chloral hydrate in drinking water. *Water Res* 43: 2201–2208.

Dotson, A., Westerhoff, P. & Krasner, S. 2009. Nitrogen enriched Dissolved Organic Matter (DOM) isolates and their affinity to form emerging disinfection by-products. *Wat Sci Technol* 60: 135–143.

EPA. 2009. *Guidance on the development, evaluation, and application of environmental models*; EPA/100/K-09/003; Washington: US Environmental Protection Agency, Council for Regulatory Environmental Modeling.

Fabbricino, M. & Korshin, G. 2009. Modelling disinfection by-products formation in bromide-containing waters. *J Hazard Mater* 168: 782–786.

Francis, R., Small, M. & VanBriesen, J. 2009. Multivariate distributions of disinfection by-products in chlorinated drinking water. *Water Res* 43: 3453–3468.

Hong, H., Liang, Y., Han, B., Mazumder, A. & Wong, M. 2007. Modeling of trihalomethane (THM) formation via chlorination of the water from Dongjiang River (source water for Hong Kong's drinking water). *Sci Total Environ*, 385: 48–54.

Kim, J., Chung, Y., Shin, D., Kim, M., Lee, Y., Lim, Y. & Lee, D. 2002. Chlorination by-products in surface water treatment process. *Desalination* 151: 1–9.

Koudjonou, B., LeBel, G. & Dabeka, L. 2008. Formation of halogenated acetaldehydes, and occurrence in Canadian drinking water. *Chemosphere* 72: 875–881.

Kulkarni, P. & Chellam, S. 2010. Disinfection by-product formation following chlorination of drinking water: Artificial neural network models and changes in speciation with treatment. *Sci Total Environ* 408: 4202–4210.

Lee, K., Kim, B., Hong, J., Pyo, H., Park, S. & Lee, D. 2001. A Study on the distribution of chlorination by-products (CBPs) in treated water in Korea. *Water Res* 35: 2861–2872.

Lee, W., Westerhoff, P. & Croué, J.-P. 2007. Dissolved organic nitrogen as a precursor for chloroform, dichloroacetonitrile, N-Nitrosodimethylamine, and trichloronitromethane. *Environ Sci Technol* 41: 5485–5490.

Mitch, A., Krasner, S., Westerhoff, P. & Dotson, A. 2009. *Occurrence and formation of nitrogenous disinfection by-products*. Water Research Foundation: Denver.

Nikolaou, A., Golfinopoulos, S., Lekkas, T. & Arhonditsis, G. 2004. Factors affecting the formation of organic by-products during water chlorination: a bench-scale study. *Water Air Soil Poll* 159: 357–371.

Nikolaou, A., Golfinopoulos, S., Rizzo, L., Lofrano, G., Lekkas, T. & Belgiorno V. 2005. Optimization of analytical methods for the determination of DBPs: Application to drinking waters from Greece and Italy. *Desalination* 176: 25–36.

Reckhow, D., Platt, T., MacNeill, A. & McClellan, J. 2001. Formation and degradation of DCAN in drinking waters. *J Water Supply Res T* 50: 1–13.

Rodriguez, M. & Sérodes, J. 2001. Spatial and temporal evolution of trihalomethanes in three water distribution systems. *Water Res* 35: 1572–1586.

Sadiq, R. & Rodriguez, M. 2004. Disinfection by-products (DBPs) in drinking water and predictive models for their occurrence: a review. *Sci Total Environ* 321: 21–46.

Sérodes, J., Rodriguez, M., Li, H. & Bouchard, C. 2003. Occurrence of THMs and HAAs in experimental chlorinated waters of the Quebec City area (Canada). *Chemosphere* 51: 253–263.

Singer, P. 1994. Control of disinfection by-products in drinking water. *J Environ Eng* 120: 727–744.

Sohn, J., Gatel, D. & Amy, G. 2001. Monitoring and modeling of disinfectionby-products (DBPs). *Environ Monit Assess* 70: 211–222.

Song, H., Addison, J., Hu, J. & Karanfil, T. 2010. Halonitromethanes formation in wastewater treatment plant effluents. *Chemosphere* 79: 174–179.

Templeton, M., Nieuwenhuijsen, M., Graham, N., Bond, T., Huang, L. & Chen, Z. 2010. *Review of the current toxicological and occurrence information available on nitrogen-containing disinfection by-products*. Imperial Consultants: London.

Toroz, I. & Uyak, V. 2005. Seasonal variations of trihalomethanes (THMs) in water distribution networks of Istanbul City. *Desalination* 176: 127–141.

Uyak, V., Ozdemir, K. & Toroz, I. 2007. Multiple linear regression modeling of disinfection by-products formation in Istambul drinking water recervoirs. *Sci Total Environ* 378: 269–280.

Uyak, V., Toroz, I. & Meriç, S. 2005. Monitoring and modeling of trihalomethanes (THMs) for a water treatment plant in Istambul. *Desalination* 176: 91–101.

Wei, J., Ye, B., Wang, W., Yang, L., Tao, J. & Hang, Z. Spatial and temporal evaluations of disinfection by-products in drinking water distribution systems in Beijing, China. *Sci Total Environ* 2010, 408: 4600–4606.

WHO, 2000. *Environmental Health Criteria 216. Disinfectants and disinfectant by-products.* World Health Organization: Geneva.

Włodyka-Bergier, A. & Bergier, T. 2011. The influence of organic matter quality on the potential of volatile organic water chlorination products formation. *Arch Environ Prot* 37: 25–35.

Włodyka-Bergier, A., Bergier, T. 2011. The occurrence of haloacetic acids in Krakow water distribution system. *Arch Environ Prot* 37, 21–29.

Włodyka-Bergier, A. & Bergier, T. 2013. Seasonal variations in volatile organic halogen water chlorination by-product content in Krakow City water distribution system. *Ochr Sr* 35: 23–27 (in Polish).

Włodyka-Bergier, A., Bergier, T. & Kot, M. 2014. Occurrence of volatile organic chlorination by-products in water distribution system in Krakow (Poland). *Desalin Water Treat* 52: 3898–3907.

Zimoch, I., Stolarczyk, A. 2010. Raman spectroscopy in estimating THM formation potential in water pipe network. *Environ Prot Eng* 36, 55–64.

Environmental Engineering V – Pawłowska & Pawłowski (Eds)
© 2017 Taylor & Francis Group, London, ISBN 978-1-138-03163-0

The carbon and nitrogen stable isotopes content in sediments as an indicator of the trophic status of artificial water reservoirs

L. Bartoszek, P. Koszelnik & R. Gruca-Rokosz
Department of Chemistry and Environmental Engineering, Faculty of Civil and Environmental Engineering, Rzeszów University of Technology, Rzeszów, Poland

ABSTRACT: The paper presents the results of studies conducted in the years 2009–2014 within the ecosystems of ten water reservoirs of south-eastern Poland. Most of the reservoirs are rheolimnic, located in the catchment areas with anthropogenic characteristic and classified as small retention. Their trophic state changes from meso-trophy to hypertrophy. The purpose of the study was to analyse the influence of the progressive eutrophication on the effect of isotopic fractionation of carbon and nitrogen during the process. Moreover, the possibilities of using the isotope ratios of carbon ($\delta^{13}C$) and nitrogen ($\delta^{15}N$) as markers of trophic status and performance of the accumulation of autochthonous organic matter in sediments were considered. There has been a statistically significant difference between the mean values of both $\delta^{13}C$ and $\delta^{15}N$. Significant correlation between the isotopic ratios and concentrations of chlorophyll a, which initially confirms the thesis regarding the impact of internally originating organic matter in bottom sediments of the small bodies of water, was identified only in a few the studied bodies of water.

Keywords: reservoirs, eutrophication, stable isotope

1 INTRODUCTION

Cultural eutrophication is one of the important causes leading to deterioration of water ecosystems quality (Chislock et al. 2013, Neverova-Dziopak & Kowalewski 2014). This problem applies also to small retention reservoirs on streams or rivers due to the localisation of strongly anthropogenic areas and supplied by external organic and mineral matter loads (Wiatkowski et al. 2015). In that case, the primary problem is silting and consequential decrease of the reservoir volume (Michalec & Tarnawski 2008). The main reason for that decrease is external inflow of debris, but the organic matter derived from the production within the ecosystem, resulting from the excessive availability of nutrients can have an important contribution to the amount of sediment deposited in reservoirs of stagnant water. Various studies indicate that up to 35% of organic matter produced in the euphotic layer of eutrophic water reservoirs enriches the surface layer of bottom sediments (de Junet et al. 2005, Vreča & Muri 2010). The formation of the sediment, associated with the mineralisation leads on a global scale to the burial in the sediment of only 0.1% of net primary production (Lehmann et al. 2002), although paleo-limnological studies indicate that the process of accumulation of sediment is directly proportional to the increase of primary production (Tyson 2001, Zilius et al. 2015).

So far, the most common method of identifying the origin of organic matter in bottom sediments has been the analysis of the changes in organic carbon content in relation to nitrogen (the ratio C:N). It is characteristic, that the values C:N in the organic matter, derived from sedimentation of terrigenous substances are higher, such as in cellulose, which is a component of terrestrial plants. Autochthonous—plankton matter—is characterised by lower values of the ratio, approximately 6:1, as the remains of algae contain more nitrogen than the terrestrial matter (Gu et al. 1999, de Junet et al. 2005). In recent years, the increasingly used method for assessing the origin of bottom sediments of surface waters is the content analysis of stable isotopes of carbon and nitrogen (Lehmann et al. 2002, de Junet et al. 2005, Brenner et al. 2006, Koszelnik 2009, Gąsiorowski & Sienkiewicz 2013, Oczkowski et al. 2014). These methodologies draw on the fact that the deposition of organic matter in sediments is the final step in a series of transformations of carbon and nitrogen, whose nature is different in terrestrial ecosystems than in aquatic ecosystems. The result is a variety of different isotope fractionations in the process of this transformation, which results in different values of the isotope ratios of 15N:14N and 13C:12C in the bottom sediments of various origins. The chemical composition of the organic matter contained in the bottom sediments of reservoirs and lakes is an important indicator of the nature and intensity of the aforesaid processes ongoing in the water column. The observed variability to the occurrence of different indicators at different depths of sediment corresponds to the trophic state of the reservoir at the moment the material in question was deposited (Gu et al. 1999, Brenner et al. 2006, Oczkowski et al. 2014). The supply of biogenic elements from external sources is usually transformed into an increase in the production of organic matter in the euphotic layer of a reservoir. Some of this material may be destroyed once present in oxygenated lake waters, but, where production is intensive and/ or an oxygen deficit arises, the importance of sedimentation increases (Brenner et al. 2006).

Another method confirming the origin of matter entails isotopic determination of TOC and TN in bottom sediments. The complex transformations that both nitrogen and carbon undergo in terrestrial and aquatic ecosystems ensure that sediments of differing origin finally possess differentiated isotopic compositions manifested as the corresponding $\delta^{15}N$ and $\delta^{13}C$ contents of the elements. Carbon from a catchment may have $\delta^{13}C$ values in the range −10 to −30‰. The isotopic signature of C3 plants shows a higher degree of 13C depletion than that of C4 plants (White 2015). In turn, the matter of planktonic origin may be characterised by more negative values for $\delta^{13}C$, potentially reaching −44‰ (or less than −100 in extreme cases); the latter situations typify methane-rich or tropical wetland ecosystems (Jędrysek 2006; de Junet et al. 2005).

The content of stable isotopes within TN is also dependent on the element origins (Gu et al. 1999). The isotopic compositions of autochthonous origin and particulate nitrogen and non-N-fixing plankton range from −3 to +12‰ δ^{15} N. Non-N-fixing allochthonous plants unaffected by fertiliser use generally have a narrower range of +6 to +13‰, but are isotopically lighter. Values for algae range from −2 to +4‰, with majority in the range +2 to +4‰. Thus, most nitrogen-fixing terrestrial plants are typically heavier than non-N-fixing plants (White 2015).

The objective of the study work was to examine the utility of analysis of bottom sediment N and C stable isotope composition as a part of a management strategy—Assessment of primarily the share of eutrophic source in sediment organic matter.

2 MATERIALS AND METHODS

The samples of the superficial water and bottom sediment were taken from 23 sampling sites within 10 artificial reservoirs in South-Eastern Poland (Fig. 1). Samples from the Chańcza, Nielisz and Maziarnia reservoirs were collected eight times

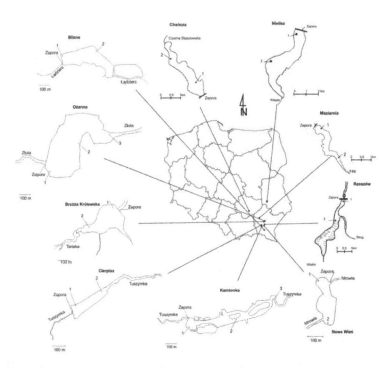

Figure 1. Studied reservoirs.

during the spring-autumn period from 2009 to 2011, but others were studied nine times in 2013–2014. The basic morphological parameters of the examined reservoirs are shown in Table 1.

A gravity sediment corer (KC Kajak of Denmark) was used for sediment collection. The top layer (0–5 cm) was used for chemical analysis. Each cored sample was portioned and the first subsample was dried and further measured for Loss-on-Ignition (LOI) at 550°C for four hours (4h), construed as organic matter content (OM). Another subsample was analysed for determining total content of organic carbon (C_{org}) and total nitrogen (N_{Tot}) using a CN analyser (CN Flash EA 1112, ThermoQuest) and for stable carbon ($\delta^{13}C$) and nitrogen ($\delta^{15}N$) an isotopic composition using IRMS DELTAPlus Finnigan coupled with CN analyser. C_{org} was measured after the removal of inorganic carbon with the use of 1M HCl. The stable isotopic compositions were expressed as "per mil":

$$\delta R = \left[\frac{R_a/R_b \, (sample)}{R_a/R_b \, (standard)} - 1 \right] \cdot 10^3 \qquad (1)$$

where R_a/R_b are the $^{13}C/^{12}C$ or $^{15}N/^{14}N$ ratios relative to the PDB and AIR standards, respectively.

Concentrations of the chosen compounds of carbon, nitrogen and phosphorus and also content of chlorophyll a were determined in water samples. A Shimadzu Total Organic Carbon (TOC) analyser with an adapter for Total Nitrogen (TN) calculation was used in laboratory assays. Spectrophotometry was used (reaction with ammonium molybdate) to determine the total amount of phosphorus content (TP) (after prior mineralisation in the presence of H_2SO_4 and peroxodisulphate) (in accordance with PN-EN ISO 6878:2006). Chlorophyll a was also determined spectrophotometrically (in accordance with PN-ISO 10260:2002). In all the performed analyses an Aquamate Thermo Scientific spectrophotometer was used.

The progress of eutrophication was evaluated by calculating the Trophic State Index values (TSI) by Carlson (1977) based on the annual average content of TP and chlorophyll a in the growing season. Due to the small depth of the researched sites, the index based on the measurement of water transparency was omitted. The Secchi disc measurement is often overlooked when assessing the trophic level of shallow water bodies, as the reduc-

Table 1. Parameters of the studied reservoirs. The last column presents the trophic state on the base of Carlson indexes (E—eutrophy; H—hypertrophy).

Studied reservoir	Year of construction	Coordinate	Volume $10^3 \, m^3$	Mean depth (max.) m	Area ha	Hydraulic ret. time day	Catchment area km^2	Studied Site	TSI— Chla	TSI— TP
Nowa Wieś	1977	50°06'N	75	1(3)	3	1.3	208.1	NW1	H	E
		22°03'E						NW2	H	E
Brzóza Królewska	1978	50°14'N	50	0.7(1.5)	7.05	2.5	30.4	BK3	H	E
		22°19'E						BK4	H	E
Ożanna	1978	50°17'N	275	1.4(3.7)	20	3.5	136.3	OZ5	H	E
		22°31'E						OZ6	H	E
								OZ7	H	E
Rzeszów	1974	50°00'N	1800	0.6(4.9)	68.2	0.8	2025	RZ8	H	E
		21°59'E						RZ9	H	E
								RZ10	H	E
Blizne	2002	49°44'N	137	1.6(3.9)	8.66	18.0	12.0	BL11	E	E
		22°00'E						BL12	E	E
Kamionka	1957	50°08'N	105	1.5(3)	7	4.8	84.8	KA13	E	E
		21°40'E						KA14	E	E
								KA15	E	E
Cierpisz	1956	50°09'N	22	0.9(1.5)	2.3	1.2	54.5	CI16	H	E
		21°43'E						CI17	H	E
Maziarnia	1988	51°20'N	4,2	2.6(8)	160	36	233	MA18	H	E
		22°55'E						MA19	H	E
Nielisz	2008	50°48'N	19,5	2(4)	890	–	–	NI20	H	E
		23°02'E						NI21	H	E
Chańcza	1984	50°38'N	40	11	340	218	475.0	CH22	E	E
		21°03'E								

tion in the transparency of the water often is not a result of the development of algae. TSI were calculated from the equations:

$$TSI_{Chla} = 9.81 \ln(Chl\ a) + 30.6 \quad (2)$$
$$TSI_{TP} = 14.43 \ln(TP) + 4.15 \quad (3)$$

and trophic status was estimated within the ranges:
oligotrophy (O): TSI < 40; mezotrophy (M) 40 < TSI < 50; eutrophy: 50 < TSI < 70; hypertrophy: TSI > 70.

The whole statistical analysis was performed in Excel's Analysis ToolPack.

3 RESULTS

The studied waters had relatively poor levels of TN. Mean concentration ranged from 0.56 (BL11) to 4.33 mgN dm^{-3} (NW2), but in most cases was below 2 mgN dm^{-3} (Fig 2). Those were the values corresponding to the mesotrophic and eutrophic conditions (Nurnberg 2001). Mean concentrations of TP in waters ranged from 0.054 mgP dm^{-3} to 0.557 mgP dm^{-3} noted at NI21 and NW2 respectively. The Nowa Wieś and Maziarnia reservoirs were most enriched with both elements (Fig. 2). The molarity TN:TP ratio calculated on that basis indicates an excess of nitrogen in most of the waters analysed, due to the ratio usually exceeding the stoichiometric 16:1. Lower values were observed only three times but values higher than 20:1 were observed nine times of twenty three sites (Fig. 3). Mean concentrations of TOC in waters changed widely in the range 8.32–23.20 mgC dm^{-3} (RZ9 and MA18 respectively), while chlorophyll a occurred between 11.2 µg dm^{-3} (NW2) and 59.9 µg dm^{-3} (NI20).

The trophic status for each of the studied sites calculated from equations (2) and (3) is presented in Table 1. On the basis of these values it may be concluded that the studied waters were hypertrophic more often due to biomass concentration (TSI$_{Chla}$) while TSI$_{TP}$ classifies all the studied water as eutrophic. Nevertheless, the obtained results indicate low water quality of the considered reservoirs because of intensive eutrophication. The discrepancy noted between TSI$_{Chla}$ and TSI$_{TP}$ and especially the higher trophic status calculated from TSI$_{Chla}$ than TSI$_{TP}$ is often observed in reolimnetic bodies of water. In that case more of the phosphorus is exported via outflow and is not used in primary production.

The organic matter content in the bottom sediment surface layer was not high, in the range 0.82–14.00% of their dry weight (% d.w.) (Fig. 5). Values above 9% d.w. were noted only in three stations, while in nine instances OM were below 3% d.w. Significant correlation between OM and C$_{org}$ was identified ($r^2 = 0.8503$, p = 0.0000). C$_{org}$ in the studied reservoirs ranged from 0.08 to 4.35% d.w. The most enriched with OM as well as C$_{org}$ were Nowa Wieś and station MA19, which differentiated from Kamionka, Cierpisz and Chańcza where bottom sediment was mostly sandy and clay (Fig. 5). N$_{Tot}$ in studied sediment was in the range 0.01–0.39% d.w. and remained in close correlation with C$_{org}$ ($r^2 = 0.9561$, p = 0.0000). Relationships noted suggest the same source of OM, C and N stored in bottom sediment of the reservoirs.

The studied bottom sediments were characterised by C$_{org}$:N$_{Tot}$ values in the range of 5.96–17.91. The average values for this indicator were of approximately 12–13 (Fig. 3). Such results indicate that the organic matter deposited in the studied sediments is of mixed origin. The reference values obtained from

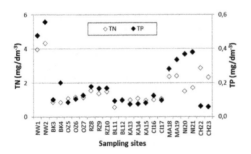

Figure 2. Mean values of Total Nitrogen (TN) and total phosphorus content in waters from studied sites.

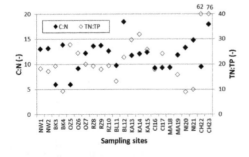

Figure 3. Mean values of nitrogen, phosphorus and organic carbon total forms ratios in waters (TN:TP) and bottom sediment (C:N) from studied sites.

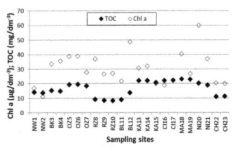

Figure 4. Mean values of Total Organic Carbon (TOC) and chlorophyll a content in waters from studied sites.

the literature for planktonic and terrigenous matter indicator were of 6.8 and 17.1 respectively (Gulia et al. 2004). The obtained values for the $\delta^{15}N$ differ from −1.51 to 5.54, but did not display any clear trends nor did the prevent drawing any definite conclusions (Fig. 6). The $\delta^{13}C$ values suggest that the top surface layer of the sediment from a few reservoirs (Rzeszow, Nielisz, Blizne, Chańcza) was poorer in ^{12}C, this in turn indicates a greater role of autochthonic organic matter (values of approx −25‰ and more, Fig. 6). This would be consistent with the research results obtained by others (Lehman et. al. 2002, Tomaszek et al. 2009, Machiwa 2010, Bartoszek et. al. 2015). Most of the studied sediment had mean $\delta^{13}C$ value next −30‰, that seems to be characteristic for C3 plant e.g. macrophytes. There was no correlation between $\delta^{13}C$ and $\delta^{15}N$ values that suggests mixed source of sediment matter.

Analysis of the stable isotope composition indicates two groups of studied ecosystems. Despite the trophic status deposition of autotrophic matter is at a different rate. That was confirmed by statistical analysis of variance (Table 2). The data show a significant difference between mean values of OM, $\delta^{13}C$ and $\delta^{15}N$ between two groups of studied reservoirs ($p < \alpha = 0.05$). Especially in the case of $\delta^{13}C$ mean values from Rzeszów, Nielisz,

Table 2. Mean values of parameters studied for two groups of reservoirs. Parameter p lower than 0.05 suggests that values differ statically.

	G1	G2	p
OM	5.30	5.22	0.0499
N_{Tot}	0.16	0.23	0.1226
C_{org}	1.72	1.72	0.2060
$\delta^{13}C$	−28.72	−25.59	0.0000
$\delta^{15}N$	1.81	2.59	0.0016
TOC	19.31	12.21	0.2016
Chl a	28.10	33.14	0.0900
C:N	10.66	13.74	0.2550
TP	0.19	0.17	0.2272
TN	1.66	1.59	0.0634

Blizne, Nowa Wieś, Cierpisz and Chańcza (group G1) was significantly lower (−28.82 vc. −25.59 ‰) than from the Brzóza Królewska, Ożanna, Kamionka and Maziarnia (group G2).

4 DISCUSSION

Diversification of OM sources can also affect the value of the $\delta^{13}C$ and $\delta^{15}N$ ratios in bottom sediments of reservoirs. Research conducted in many aquatic ecosystems indicates the depletion of the organic carbon isotope ^{13}C due to the participation of matter of phytoplanktonic origin. This effect is accompanied by higher $\delta^{15}N$ (Gu et al. 1999). In this study variability was found and confirmed only in the case of reservoirs of the G1 group where a strong direct relationship between the concentration of chl a and $\delta^{13}C$ value was observed (Fig. 7). In the case of reservoirs of the G2 group, despite the high value of the chlorophyll index the effect of depletion with ^{13}C isotope was less important.

On the basis of the literature it may be supposed that the eutrophic effect of OM, being the enrichment with the ^{15}N and depletion with ^{13}C isotope, has been obscured by matter of terrestrial origin. The dominance of a single source of organic matter in the sediments is confirmed by significant correlation between $\delta^{13}C$ and $\delta^{15}N$ (Gu et al. 1999, Vreča & Muri 2010). In the studied ecosystems, there was no such correlation, suggesting a parallel share of both sources.

The isotopic ratios values for the analysed cases did not confirm the impact of internal production on the loss of water depth of reservoirs. Probably, this is due to the characteristics of the small bodies of water that are susceptible to short-term external influences. For larger objects and located in the same geographic zone with a large retention time these indices enabled estimation of this impact (Tomaszek et al. 2009).

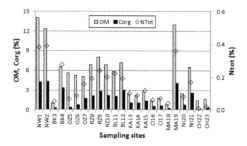

Figure 5. Mean values of Organic Matter (OM), carbon (C_{org}) and total nitrogen (N_{tot}) in the bottom sediment from studied sites.

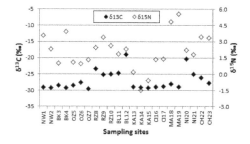

Figure 6. Carbon ($\delta^{13}C$) and nitrogen ($\delta^{15}N$) isotopic signature of the bottom sediment organic matter from studied sites.

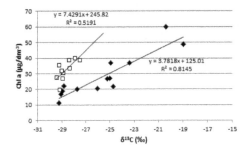

Figure 7. Relationship between chlorophyll and carbon stable isotope composition for groups G1 (white) and G2 (black) of the reservoirs.

5 SUMMARY

The analysis of the results indicates that the values of the isotopic ratios of N and C in the sediment do not always correlate with the indicators determining their trophic status. Thus, it is difficult to confirm the participation of the planktonic primary production in the OM deposited in bottom sediment of small reservoirs only on the basis of these indicators. In the next step, statistical analyses and modelling that account for a number of other variables should be initiated which should enable a quantitative estimation of the impact of eutrophication on the depositions of organic matter onto the bottom sediment of small reservoirs.

ACKNOWLEDGMENTS

The study gained financial support from Polish Ministry of Science via grant no. N N305 077836 and Polish National Science Centre via grant no. 2011/03/B/ST10/04998.

REFERENCES

Bartoszek, L., Koszelnik, P., Gruca-Rokosz, R. & Kida, M. 2015. Assessment of agricultural use of the bottom sediments from eutrophic Rzeszów reservoir. *Rocznik Ochrona Środowiska* 17: 396–409.

Brenner, M., Hodell, D.A., Leyden, B.W., Curtis, J.H., Kenney, W.F., Gu, B., & Newman J.M. 2006. Mechanisms for organic matter and phosphorus burial in sediments of a shallow, subtropical, macrophyte-dominated lake. *Journal of Paleolimnology* 35: 129–148.

Carlson, R.E. 1977. A trophic state index for lakes. *Limnology and Oceanography* 22(2): 361–369.

Chislock, M.F., Doter, E., Zitomer, R.A. & Wison, A.E. 2013. Eutrophication: Causes, Consequences, and Controls in Aquatic Ecosystems. *Nature Education Knowledge* 4(4):10.

Curtis, C.J., Flower, R., Rose, N., Shilland, J., Simpson, G.L., Turner, S., Yang, H. & Pla, S. 2010. Palaeolimnological assessment of lake acidification and environmental change in the Athabasca Oil Sand Region, Alberta. *Journal of Limnology* 69: 92–104.

De Junet, A., Abril, G., Guerin, F., Billy, I. & De Wit, R. 2005. Sources and transfers of particulate organic matter in tropical reservoir (Petit Saut, French Guiana): a multi-traces analysis using δ13C, C/N ratio and pigments. Biogeosciences Discussions 2: 1159–1196.

Gąsiorowski, M. & Sienkiewicz, E. 2013. The sources of carbon and nitrogen in mountain lakes and the role of human activity in their modification determined by tracking stable isotope composition. *Water, Air, & Soil Pollution* 224(4): 1–9.

Gulia, L., Guidi, M., Bonadonna, F. & Macera, P. 2004. A preliminary geochemical study of two cores from Massaciuccoli eutrophic lake, Northern Tuscany and paleoclimatic implications. Atti Della Società Toscana Di Scienze Naturali Residente In Pisa. Memorie. Serie A 109: 97–102.

Jędrysek, O.M., Kurasiewicz, M., Trojanowska, A., Lewicka, D., Omilanowska, A., Kałużny, A., Izydorczyk, K., Drzewicki W., & Zalewski, M. 2006. Diurnal variations in carbon isotope composition of dissolved inorganic carbon (DIC) in a freshwater dam reservoir. *Ecohydrology & Hydrobiology* 6(1): 53–59.

Lehmann, M.F., Bernasconi, S.M. Barbieri, A. & McKenzie J.A. 2002. Preservation of organic matter and alteration of its carbon and nitrogen isotope composition during simulated and in situ early sedimentary diagenesis. *Geochimica et Cosmochimica Acta* 66(20): 3573–3584.

Machiwa, J.F. 2010. Stable carbon and nitrogen isotopic signatures of organic matter sources in near-shore areas of Lake Victoria, East Africa. J Great Lakes Res 36: 1–8.

Michalec, B. & Tarnawski, M. 2008. The influence of small water reservoir operational changes on capacity reduction. *Environment Protection Engineering* 34(3): 117–124.

Neverova-Dziopak, E. & Kowalewski, Z. 2014. Towards Methodological Problems of Trophic State Assessment of Running Waters. *Ecological Chemistry and Engineering S* 21(4): 637–650.

Nürnberg, G. 2001. Eutrophication and trophic state. *LakeLine* 29(1): 29–33.

Oczkowski, A., Markham, E., Hanson, A., & Wigand, C. 2014. Carbon stable isotopes as indicators of coastal eutrophication. *Ecological Applications* 24(3): 457–466.

Tomaszek, J.A., Koszelnik, P. & Gruca-Rokosz, R. 2009. The distribution and isotopic composition of carbon and nitrogen as indicators of organic-matter fluxes in the Solina Reservoir (south-east Poland). *Marine and Freshwater Research* 60: 647–652.

Tyson, R.V. 2001. Sedimentation rate, dilution, preservation and total organic carbon: some results of a modelling study. *Organic Geochemistry* 32(2): 333–339.

Vreča, P. & Muri G. 2010. Sediment organic matter in mountain lakes of north-western Slovenia and its stable isotopic signatures: records of natural and anthropogenic impacts. *Hydrobiologia* 648:b35–49.

White, W.M. 2015. Isotope Geochemistry. Oxford: Wiley-Blackwell

Wiatkowski, M., Rosik-Dulewska, C. & Kasperek, R. 2015. Inflow of Pollutants to the Bukówka Drinking Water Reservoir from the Transboundary Bóbr River Basin. *Rocznik Ochrona Środowiska* 17: 316–336.

Zilius, M., de Wit, R., & Bartoli, M. 2015. Response of sedimentary processes to cyanobacteria loading. *Journal of Limnology*. http://dx.doi.org/10.4081/jlimnol.2015.1296.

Environmental Engineering V – Pawłowska & Pawłowski (Eds)
© 2017 Taylor & Francis Group, London, ISBN 978-1-138-03163-0

Design errors in working water and sludge installations in industrial plants

T. Piecuch, J. Piekarski & A. Kowalczyk
Division of Water-Sludge Technology and Waste Utilization, Koszalin University of Technology, Koszalin, Poland

ABSTRACT: Two existing examples of incorrectly designed and, unfortunately, working water and sludge installations are discussed in the paper. The first one, marked as A, is water and sludge installation in coal mechanical processing plant of coal mine operating coking coal. Here the first design and then implementation error had been introduction of vacuum filters into installation, and then after a few years, next error was replacing vacuum filters with decantation-filtration centrifuges. Design and following it implementation error of application of fixed vertical screen at inlet to a hydrocyclone and use of a hydrocyclone with specific custom parameters, without its testing in water and sludge installation in gravel pit, marked as B, is discussed. Both cases were subjected to descriptive technical analysis based on operational block diagrams.

Keywords: design errors, water and sludge installations

1 INTRODUCTION

The water and sludge installations working in industrial plants are most often (though not always) small wastewater pre-treatment plants, which mainly aim to separate the solid and liquid phases (Battaglia 1961, Nowak 1970, Turek, Włodarczyk-Makuła & Nowacka, 2014, Uliasz-Bocheńczyk, Mazurkiewicz, & Mokrzycki 2015). Most often, the liquid phase is still polluted with soluble compounds and only conducting further processes, e.g. physicochemical or chemical ones, may achieve sufficient treatment level in order to enable the re-use this wastewater for production or discharge to sewerage or open water bodies, etc. (Piecuch 1980). On the other hand, the other product obtained in such water and sludge installations (which constitutes waste) is usually sludge with certain humidity and transportability. If the need for additional drying of sludge arises, it significantly increases the costs of the entire technology (Piecuch 1980).

The notion of water and sludge installations can be understood as a certain system formed by a chain of co-operating devices, in which the processes involving separation of solid and liquid phase (fine grains dispersed in a liquid) occur. The separation, depending on the characteristics and properties of wastewater, is usually gradual, and rarely ultimate, although it might as well be achieved in the first device of water and sludge installation.

Water and sludge installations, depending on the type of industry, may be characterized by varying degrees of complexity. For instance, in the case of small processing plants, they may comprise two consecutive junctions. The most complex water and sludge installations are found in mining, steel, and chemical industries. Smaller installations (lower amount of post-production wastewater, fewer devices, etc.) are found in food, wood, tannery industries, etc.

Designing and implementing water and sludge installations in industrial plants requires knowledge and experience of the designer, obtained through education, industrial practice, and before he becomes an independent designer—designing practice as well. Such path towards obtaining independence in designing by a given person used to be respected (around 50, 40, or 30 years ago), but later the situation deteriorated. The changes in political system that took place after 1989 allowed ambitious and brave people who looked for employment to take initiative and become independent. Unfortunately, this trend resulted in the fact that designing is performed by poorly-educated people, who not only lack experience, but knowledge and imagination as well. Hence, elementary designing errors are made.

As a result, significant problems arise when such defective project is implemented, and the water and sludge installations do not operate properly.

Finally, the deteriorating quality of education of students at technical universities needs to be mentioned (due to the lack of appropriate enrolment, the requirements put before students during examinations are very low), which also negatively impacts the above-mentioned status quo.

In view of this situation, the authors, based on their own designing and research experience, as well as the related supervision and implementation of own technologies, would like to present the examples of typical designing errors, which have been repeated for many years, and the society suffers from their consequences, related both to investments and exploitation (manpower, energy, materials, renovation, etc.). Due to the obvious reasons, the authors do not disclose the names of companies and location of faulty water and sludge installations, aiming only to present what should not be done and what is the possible outcome.

2 EXAMPLE OF DESIGN ERRORS IN WATER AND SLUDGE INSTALLATION IN MECHANICAL COAL PROCESSING PLANT

Polish mining industry, disregarding the discussion about the purposefulness of its existence and functioning, is still a key industry in Poland, being the basis of Polish power industry (or, more precisely, the Polish power industry is adjusted to it) (Blaschke & Nycz 2003, Blaschke 2005, Blaschke, 2004, Blaschke, 2001, Blaschke, 2001, Grudziński, 2013, Poros & Sobczyk, 2013, Stachowski, Oliskiewicz—Krzywicka & Kozaczyk 2013, Uliasz-Boheńczyk, Mazurkiewicz & Mokrzycki 2015).

In regard to designing water and sludge installations of pre-treatment plants, where treatment is usually performed to achieve the level of production waters, i.e. with parameters enabling their reuse in mechanical coal processing plants, there are certain standards which determine when, where, and how one should conduct phase separation, as well as what devices should be used.

Of course, the industry should be open to innovations, new technologies and new inventions, but the industry representatives should have appropriate knowledge, constantly educate themselves in order to avoid the implementation of dubious innovations, which only worsen the existing operation of water and sludge installations.

In the mechanical coal processing plant (coking coal), a system was designed, in which the flotation concentrate was transported directly from the floatation machine to vacuum filters and then redirected for production, whereas the filtered sludge, so called filter cake had to be dried (which generated extra costs) prior to being sold—which is presented in Fig. 1.

Most often, a mixture of water and fine grains of solids (in this case—coal) with the concentration ranging from 100 up to 200 g/dm^3 is directed to the flotation process and in this range, depending on other process parameters, optimal density is adjusted for the size analysis of the feed, aeration of slurry and dispersion of air in flotation pulp. The size analysis of flotation pulp shows than coal water slurry comprises grains with smaller than 0.3 mm, usually with granulation smaller than 0.1 mm (100 microns), and—very often—in major part even smaller than, for instance, 50 microns.

According to the studies conducted by the Division of Water-Sludge Technology and Waste Utilization of Koszalin University of Technology, the size grade in the considered water and sludge installation was smaller than 0.04 mm in 99.29% of size analysis. Therefore, in such situation, the application of vacuum filters for dewatering of such finely-grained flotation concentrates (Fig. 1) was a blatant design error (Piecuch 1985, Piecuch 1980, Piecuch 2010). Despite the publication of numerous papers by prof. Tadeusz Piecuch, pertaining to the selection of filtering and decanting devices for mixtures (Piecuch 1985, Piecuch 1992, Piecuch & Anielak 1982, Piecuch 1980, Piecuch 2010), this error was still made.

Hence, it is important that the designers read up-to-date literature, because major research units conduct experiments, and on the basis of the obtained results, guidelines and formulas are created (Nowak 1970, Piecuch 1985, Piecuch & Anielak 1982, Piecuch 1980, Piecuch 2010, Piecuch, Andiyewska, Dąbrowski, Dąbrowski, Juraszka & Kowalczyk 2015).

Such guidelines were devised by prof. Tadeusz Piecuch for coal mixtures, based on Dahlstrom filtration index and published in the form of monograph already in 1975 (habilitation dissertation entitled Analytical-empirical model of coal sludge vacuum filtration process).

Implementation of such botched design resulted in a search for better solutions, and after many

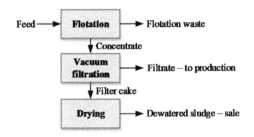

Figure 1. Operational diagram showing fragments of poorly designed water and sludge installation.

years, on the basis of institute's (name undisclosed) research team evaluation, the vacuum filters (Fig. 1) were swapped with decantation-filtration centrifuges, which were placed in water and sludge installation as shown in Fig. 2.

In the first part of these centrifuges, the mixture undergoes centrifugal decantation and drainage from this part should be clarified as much as possible, so that it can be redirected to production (Piecuch 1992, Piecuch & Anielak 1982, Piecuch 1980, Piecuch 2010). The operation of such centrifuges is often supplemented with flocculant, in that particular case it was MAGNAFLOK. Well-operating decantation centrifuges, which reach up to 3000 revolutions per min^{-1}, produce drainage with density of a few to ten-odd g/dm^3 (Piecuch & Anielak 1982, Piecuch 1980, Piecuch 2010). Meanwhile, in the considered mechanical coal processing plant, the installed USA-made centrifuges operated too slowly, with approximately 1000 min^{-1} (Piecuch 1985, Piecuch 1992, Piecuch 2010). In this case, the obtained mixture, i.e. flotation concentrate fed to the centrifuge, was characterized by the density of 332.89 g/dm^3, while the drainage from the first part of the centrifuge (decantation part) had density as high as 318.50 g/dm^3. Simultaneously, the filtrate from the second part of the centrifuge (filtration) was characterized with very high density of about 309.81 g/dm^3.

In view of the above, the designed modernization of water and sludge installation (Fig. 2), involving the swapping vacuum filters with decantation-filtration centrifuges not only did not improve the situation, but even made it worse. Attention should be drawn to the costs of the design, purchase of centrifuges, and operation, because centrifuges are the most energy-consuming devices in water and sludge technology, which in this case operate 24 hours a day (Piecuch 1992). Moreover, it should be added that the centrifuges were purchased and imported from the USA; hence, significant shipping costs had to be added, while excellent German—e.g. WESTFALIA SEPARATOR or WEDAG—or Swedish centrifuges such as NOXON are available. The decent Polish centrifuges manufactured in Wronki (Greater Poland Voivodeship) are also worth mentioning.

Therefore, a question arises—what was expertise of the mine and mechanical coal processing plant managerial staff? Who and why wanted to import such expensive devices from USA and how was the tender conducted? Finally—who is liable and suffers (if at all) the consequences of implementing such a faulty design?

According to the media, mining industry is in a difficult situation; and here we have just mentioned an example where the financial losses are counted in millions of zlotys. The centrifuges operate, consume power, use MAGNAFLOK flocculant, bearings oil, etc., virtually without performing any mechanical separation of solid and liquid phases. In that particular situation, another improvement was suggested, with the third modification of water and sludge installation. The managerial staff of the mine and mechanical coal processing plant suggested introduction of highly-concentrated drainage from decanting centrifuges to a tank (Dorr-Oliver separator), where gravitational decantation process occurs. From this tank, the concentrated product, i.e. lower outflow was planned to be introduced to the pressure filtration process on industrial filter presses—see Figure 3.

Filter presses are a new element in this part of water and sludge installation. These presses were not a new purchase, since they were operating in this mine and were used for dewatering the flotation waste.

This means that in the first version of water and sludge installations, a surplus of filter presses was purchased and thus unnecessary costs were incurred, but this is not all. The third redesign

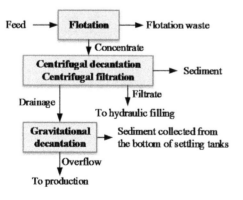

Figure 2. Operational diagram showing fragments of water and sludge installation which was incorrectly modified according to the suggestion of research team from Institute X.

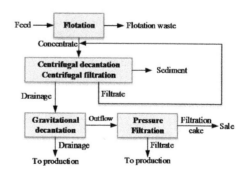

Figure 3. Operational diagram showing fragments of water and sludges installation after the implementation of improvements.

(Fig. 3) issued by the managerial staff of the mine and mechanical coal processing plant, involves improvement of the second modernization, which was implemented a few years earlier. The authors of this improvement suggestion demand payment of huge sums of money from the coal company for having conducted the practical implementation of their proposal (Fig. 3), which was done when they were in charge of that particular mine.

To sum up, we draw attention to the losses suffered by the Polish economy and taxpayers, which were incurred by implementing the first and then second faulty design, and finally—a correct one but for huge amount of money, which will be transferred to private accounts of managerial staff. These people should diligently fulfil their duties, not necessarily for extra money (which they earn apart from their salary and bonuses). Losses amounting to many millions, incurred by the Polish economy due to the irresponsibility of managerial staff and indifference or lack of knowledge, can be easily estimated.

3 EXAMPLE OF DESIGN ERRORS IN WATER AND SLUDGE INSTALLATION IN A GRAVEL PIT

The most important elements of water and sludge installation in a gravel pit comprise two devices, i.e. fixed screen (placed vertically in a special housing) and hydrocyclone, specially manufactured for the needs of the gravel pit.

The first device in water and sludge installation is a non-propelled trailing suction dredger, which is used for aggregate excavation in closed water bodies or in unnavigated rivers. The dredger in question can pump the excavated material with the range of approximately 600 m and discharge it on land. Obviously, it can be also used to dredge water bodies or rivers. The dredger is composed of two side pontoons, the central pontoon with a pump, and the head pontoon which supports the boom of the suction pipe. The total length, including the pontoons, is 30.75 m, while the total width is 6.18 m. The height at the dredger side is 1.4, and the total height from the table of water is about 4.0 m, with the side pontoons submerged to about 0.9 m. The length overall is 13.0 m, whereas the width is about 1.6 m. The pontoon which supports the boom is U-shaped. The dredger in question can be transported on land and its dimensions are smaller then, amounting to 13.0 m, 3.1 m, and 3.0 m for length, width, and height, respectively. The most important technical parameter, i.e. the hydromixture output ranges from about 1450 to 1650 m³/h. The nominal output of the excavated material is about 240 m³/h, assuming transportation over the distance of 600 m. The dredger can dig up to the depth ranging from about 9.5 to 15 m (counting from the table of water). The concentration of the hydromixture reaches roughly 14 to 18%.

The diameter of suction pipeline equals 400 m, while the diameter of delivery pipeline amounts to 350 mm. The dredger supplies the hydromixture onto a special screen made of wire with the diameter of 3 mm, forming 5×5 mm mesh. The height of the screen equals 980 mm. The screen is installed in a set of bars, forming a housing with the inflow nozzle going inside it and cylindrical outflow nozzle fixed from the bottom. The dredger supplied the hydromixture through the inlet (cylindrical-conical) to the housing with grain size distribution ranging from 0 to several of even ten-odd centimetres. The mixture hits the vertically fixed screen. Grains with size greater than about 5 mm are stopped on the screen, fall dawn gravitationally and are subsequently discharged through the outlet. These grains, after their dewatering and drying, constitute the first commercial product, i.e. coarse-grained gravel. Obviously, this separation is not very precise with turbulent hydromaterial stream occurring inside the screen housing.

Part of hydromixture drains, passes through a narrowing conical reducer and screen mesh, where it subsequently goes through the inlet with the diameter of 400 mm, eventually reaching the cylindrical part of hydrocyclone.

The stream in hydrocyclone splits into two parts, one of them enters the overflow and passes is discharged outside through an overflow nozzle with the diameter of 508, while the other one enters the outflow nozzle and is discharged outside as the outflow, which should be concentrated as much as possible, i.e. with the highest concentration of grains in the total volume of the obtained outflow. It turns out that the considered hydrocyclone with the diameter of the cylindrical part equalling about 1500 mm and the height of about 980 mm operates poorly and does not fulfil the basic function of a clarifying and thickening device, i.e. clarifying the overflow and maximally thickening the outflow. Out of 1600 m³ of the fed hydromixture, approximately 400 Mg of solid phase should be obtained, i.e. sand grains and fine gravel with the size of 0÷5 mm. Meanwhile, in the considered case, the hydrocyclone overflow, which should be clarified as much as possible, about 200 Mg/h of sand and gravel escapes, and the rest—approximately 200 Mg—is discharged through the outflow at the bottom. Therefore, the hydrocyclone does not clarify (overflow) or thicken (outflow) at all.

One might suppose that the gravel pit owner intended to design and then construct a water and sludge installation with the lowest possible cost, which would allow him to recover coarse- and finely grained gravel (smaller than 5 mm) by means of a hydrocyclone, i.e. the device that is the cheapest in purchase, but most difficult to operate (within water-sludge technology).

Unfortunately, while looking for the project contractor the investor found a private design studio, which offered a broad range of services, not only designing, but also execution; however, it did not mean that they had the necessary experience in the exploitation of hydrocyclones. The designed and constructed hydrocyclone does not, and probably never will, meet the expectations of the investor.

Hydrocyclone is a device, which requires constant, stable feeding of hydromixture. The dredger excavating gravel from the bottom of the water body does not guarantee the concentration stability of this solid phase in the entire volume of the mixture transported to the hydrocyclone. Therefore, the application of hydrocyclone requires stabilization ensured through an appropriately constructed gravitational tank. Only the outflow from such tank can be fed to the hydrocyclone by means of a sludge pump.

Placing a vertical screen in the housing before the hydrocyclone inlet is a major design error. The pressurized stream, which is characterized by mass and velocity (i.e. momentum) that degrades the screen, expands in the housing. The recovery of coarse fraction with the size of 5 mm, in such screen is imprecise and some fine grains smaller than 5 mm will not pass through the screen and will fall dawn gravitationally along with the coarse fraction. This means that a significant amount of sub-grains will be present in the assortment, which will lower the commercial value of this material. The screen with such housing constitutes an unnecessary obstacle for the optimal and stable operation of the hydrocyclone, as it significantly slows down the stream fed into the device. Therefore, the screen and hydrocyclone should not operate jointly in such system.

Hydrocyclone, which is a device extremely sensitive to any changes in variable independent technical, structural and technological parameters, should be tested beforehand at a test stand for a given stream, which is a mixture of a liquid and solids. Extensive research on the operation of a hydrocyclone was conducted in the 50 s by prof.

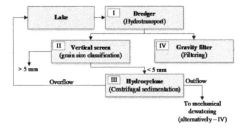

Figure 4. Simplified diagram of the current water and sludges installation in the gravel pit.

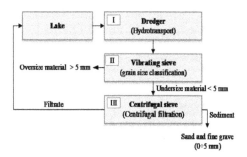

Figure 5. Block diagram of water and sludges installation according to the authors' suggestion.

Andrzej Battaglia at AGH University in Kraków (Battaglia 1963, Battaglia 1961, Battagia 1962, Battaglia 1967, Battaglia 1969, Battaglia 1962), in the 60 s by prof. Stanisław Bednarski at Cracov University of Technology (Bednarski 1964, Bednarski 1980), and in the 70 s by prof. Zygfryd Nowak (Nowak 1965, Nowak 1970, Nowak 1982). The papers of above-mentioned professors can be consulted when selecting standard hydrocyclones. However, it the case of that particular gravel pit, a custom, non-standard hydrocyclone was used without conducting prior tests on the utilized stream.

Obviously, there are a few variants for such water and sludge installation operating in a gravel pit. It seems that the optimal solution would be the one presented in the operational block diagram in Fig. 5. In such case, the gravel and sand which was excavated with the dredger should be directed in the form of a stream onto the dewatering vibrating sieve with 5 × 5 mm mesh. The dewatered oversize material from this sieve should fall onto the conveyor belt and eventually onto the gravel pile. On the other hand, the undersize material, which constitutes a mixture of liquid and grains smaller than 5 mm, should fall into a buffer tank with a tilted bottom protected from clogging. Only then the outflow material can be fed into the centrifuge with a sludge pump. Such centrifuge is an expensive piece of equipment and should be made of abrasion-resistant steel.

The presented variant enables to obtain transportable sediment of sand and gravel with the size of 0÷5 mm, which can be received and directed with a conveyor belt onto a storage pile.

It should be noted that the investor unnecessarily spent huge amounts of money, which unfortunately in most part constitute losses.

4 SUMMARY

The two presented examples of design errors are not isolated cases. The authors know many examples from their scholars' practice. The information given in papers by prof. Tadeusz Piecuch, published in the 10th and 12th issue of the *Rudy i Metale Niezależne* journal already in 1978, describe a special large-scale case.

The most important conclusions which can be drawn from the above-mentioned publication are as follows:

An investor, who incurs significant costs connected with the design, purchase and installation of devices should also devote a certain sum of money for the research. The savings made as a result of disregarding specialized studies often cause additional losses suffered by the investor.

An investor who establishes co-operation with a design office should pay attention to their references, especially to the relevant design experience in the considered field.

Prior to the final implementation, the design should be subjected to an opinion issued by an experienced specialist. Therefore, one should not be too stringent with spending money in such situation, as design errors can generate much greater losses.

REFERENCES

Battaglia, A. 1963. *Odwadnianie produktów wzbogacania i obiegi wodne płuczek*. Wydawnictwo Górniczo-Hutnicze, Katowice, wyd. II, 1963.

Battaglia, A. 1961. Badania nad działaniem hydrocyklonu zagęszczająco-klarującego D 282. *Prace Głównego Instytutu Górnictwa, 270*.

Battaglia, A. 1969. Wzór na wydajność objętościową hydrocyklonu typu konwencjonalnego. *Archiwum Górnictwa, 2*.

Battaglia, A. et al. 1969. Nowe badania nad natężeniem przepływu przez hydrocyklony. *Przegląd Górniczy* 10.

Battaglia, A. 1967. *Hydrocyklony klarująco-zagęszczające. Katalog urządzeń przeróbczych*. Wydawnictwo Biura Projektów SEPARATOR, Katowice, 1967.

Battaglia, A. 1962, Formula for calculation of the rate of flow through cyclones of conventional type. *Archiwum Górnictwa*, VII.

Bednarski, S. 1964. Hydrozyklone in der Starkeindustrie. *Die Starke,* 16.

Bednarski, S. 1980. Osadniki z wypełnieniem. Hydrocyklony. XIV Krakowska Konferencja Naukowo-Techniczna Przeróbki Kopalin, Kraków-Jaszowiec, Zeszyt Naukowy AGH Kraków.

Blaschke, W. & Nycz, R. 2003. Problemy produkcji czystych energetycznych węgli kamiennych. Zeszyty Naukowe Wydziału Budownictwa i Inżynierii Środowiska, Politechnika Koszalińska, nr 21, seria: Inżynieria Środowiska, Wydawnictwo Uczelniane Politechniki Koszalińskiej, Koszalin, 2003.

Blaschke, W. 2005. Krytycznie o planach dotyczących przyszłości polskiego górnictwa przesłanych do Komisji UE. *Wieści* (dodatek do *Przeglądu Górniczego*) SITG, 2005.

Blaschke, W. 2004. Problemy produkcji czystych węgli jako źródło wytwarzania czystej energii. Mat. Międzynarodowej Konf.: Przyszłość węgla w gospodarce świata i Polski, PK ŚRE-GIPH, Katowice.

Blaschke, W. 2001. Rola węgla w polityce energetycznej państwa. Mat. Konf.: Reforma Pol-skiego Górnictwa Węgla Kamiennego, Politechnika Śląska, PAN, PARG, Szczyrk.

Blaschke, W. 2001. Rozwiązanie problemu poziomu cen węgla warunkiem harmonijnego roz-woju kompleksu paliwowo-energetycznego. Studia, Rozprawy, Monografie, nr 91, Wydawnictwo IGSMiE PAN, Kraków.

Grudziński, Z. 2013. Koszty środowiskowe wynikające z użytkowania węgla kamiennego w energetyce zawodowej. *Rocznik Ochrona Środowiska (Annual Set The Environment Protection)*, 15(1).

Nowak, Z. & Olszowski, J. 1965. Zastosowanie multihydrocyklonów w obiegu wodno-mułowym płuczki. Dokumentacja prac Głównego Instytutu Górnictwa, Katowice, 1965.

Nowak, Z. 1970. Hydrocyklony w Przeróbce Mechanicznej Kopalin. Wydawnictwo ŚLĄSK, Katowice.

Nowak, Z. 1982. Gospodarka Wodno-Mułowa w Zakładach Przeróbki Mechanicznej Węgla. Wydawnictwo ŚLĄSK, Katowice.

Palica, M. & Grotek, A. Opis odwadniania zawiesiny zrzutowej po wirówce filtracyjno-sedymentacyjnej BIRDa modelem SORENSENA. *Rocznik Ochrona Środowiska (Annual Set The Environment Protection)*, 9.

Piecuch, T. 1985. Równanie czasu przepływu rotacyjnego ścieku przez wirówkę filtracyjną. *Archiwum Ochrony Środowiska*, 3–4.

Piecuch, T. 1992. Analiza Studialna Procesu Rozdziału w Wirówce Sedymentacyjnej. Monografia, nr 39, Wydawnictwo Politechniki Koszalińskiej, Koszalin.

Piecuch, T. & Anielak, A.M. 1982. Sedymentacja rotacyjna poflotacyjnych odpadów miedzi. Zeszyty Naukowe Politechniki Częstochowskiej, Seria Nauki Podstawowe, nr 22.

Piecuch, T. 1980. Podstawy sedymentacyjnej teorii procesu filtracji. Zeszyty Naukowe Politechniki Częstochowskiej, Seria Nauki Podstawowe, nr 21.

Piecuch, T. 2010. Technika Wodno-Mułowa. Urządzenia i Procesy. Wydawnictwo NaukowoTechniczne, Warszawa.

Piecuch, T., Andriyevska, L., Dąbrowski, J., Dąbrowski, T., Juraszka, B. & Kowalczyk, A. 2015. Oczyszczanie ścieków ze stacji naprawy samochodów. *Rocznik Ochrona Środowiska (Annual Set The Environment Protection)*, 17(1).

Poros, M. & Sobczyk, W. 2013. Rewitalizacja terenu pogórniczego po kopalni surowców skalnych na przykładzie kamieniołomu Wietrznia w Kielcach. *Rocznik Ochrona Środowiska (Annual Set The Environment Protection)*, 15(1).

Stachowski, P. Oliskiewicz-Krzywicka A. Kozaczyk P. 2013. Ocena warunków meteorologicznych na terenach pogórniczych Konińskiego Zagłębia Węgla Brunatnego. *Rocznik Ochrona Środowiska (Annual Set The Environment Protection)*, 15(1).

Turek, A., Włodarczyk-Makuła, M., Nowacka, A. 2014. Usuwanie związków organicznych i wybranych węglowodorów aromatycznych ze ścieków koksowniczych. *Rocznik Ochrona Środowiska (Annual Set The Environment Protection)*, 16(1).

Uliasz-Bocheńczyk, A., Mazurkiewicz, M., Mokrzycki, E. 2015. Fly ash from energy production—a waste, byproduct and raw material. *Gospodarka Surowcami Mineralnymi (Mineral Resources Management)*, 31(4).

Environmental Engineering V – Pawłowska & Pawłowski (Eds)
© 2017 Taylor & Francis Group, London, ISBN 978-1-138-03163-0

Impact of UV disinfection on microbial growth in drinking water

A. Włodyka-Bergier & T. Bergier
Department of Environmental Protection and Management, Faculty of Mining Surveying
and Environmental Engineering, AGH University of Science and Technology, Kraków, Poland

ABSTRACT: The application of UV radiation to disinfect water becomes more and more popular in Polish treatment plants producing drinking water. UV disinfection is often sequenced with chemical disinfectants to prevent the growth of bacteria in water distribution systems. However, there are also cases where water is pumped to a water distribution system directly after the UV disinfection (without additional chemical protection). In the article, the results of research on the influence of UV disinfection on microbial stability of water have been presented. The water samples were taken from four water treatment plants, in which UV lamps are used as an element of water disinfection process. The water microbial stability was determined by the observation of heterotrophic bacteria growth rate in time and microbial growth potential. The observations were conducted for time up to 21 days. The analysis of obtained results showed that UV radiation can negatively influence the water microbial stability in distribution systems.

Keywords: water disinfection, microbial stability, UV irradiation

1 INTRODUCTION

The process of treatment of the water which is intended for human consumption aims to provide water of good quality. The combination of various treatment methods reduces the concentration of pollutants to the allowable limit values and ensures such a quality of water that it can be safely transported in distribution systems. Secondary pollution, reactions with the materials which the water is in contact with, biological processes, and reactions between the water components are the main threats to its quality. Biological processes in distribution systems may cause hygienic (the growth of microorganisms), organoleptic (taste and unpleasant smell) and technical (corrosion) threats (van der Kooij 2000, Sun et al. 2014). These processes, however, can be controlled by the dosage of chemical disinfectants in amounts allowing to protect the water against secondary microbial growth or the distribution of biologically stable water (van der Kooij 2000). Biostability can be defined as the inability of water or materials which are in contact with water, to promote the growth of microorganisms in the absence of a disinfectant (van der Kooij 2000, Polańska et al. 2005). Organic matter is one of the main factors determining the microbial stability of water. Treatment of the water fed into the distribution system in order to reduce the content of organic compounds, which can be a culture medium for microorganisms during its transportation, is a big challenge. However, not every organic compound may be assimilated by the bacteria, but only those with low molecular weight which, for the purpose of assessing the water for its microbial stability, is classified as biodegradable organic carbon and assimilable carbon (van der Kooij 2000, Polańska et al. 2005). The single processes removing organic matter with low molecular weight include coagulation, slow filtration, adsorption on activated carbon (Wen et al. 2014, Yang et al. 2011). Some processes used at water treatment plants may cause degradation of organic compounds of higher molecular weight, which are not available for microorganisms, to biodegradable compounds. These processes include chemical oxidation processes, such as ozonation (Yang et al. 2011), and (as proved by some research studies) the use of UV radiation. After irradiating water with UV rays, the amount of biodegradable and hydrophilic compounds increases, and the molecular weight of organic compounds decreases (Liu et al. 2012, Choi & Choi 2010, Miksch et al. 2015).

UV radiation is a very good disinfectant, which results in an effective inactivation of microorganisms, even of those resistant to chlorine, such as Cryptosporidium and Giardia (Lyon et al. 2012, Koivunen & Heinonen-Tanski 2005, Hijnen et al. 2006). The use of UV radiation along with chlorine disinfectants is more effective in removing microorganisms than the use of UV radiation or chlorination alone (Rand et al. 2007). However, the studies conducted by some authors show that the use of UV radiation for water disinfection may promote secondary microbial contamination of water. The studies carried out on a model distribution system proved that using UV radiation for water disinfection resulted in the biofilm being more sensitive (compared to the test samples which were not irradiated) to the supply of organic compounds, which indicates the formation of secondary microbial cells (Pozos et al. 2004). Other studies, conducted on the samples disinfected with the sequence of UV irradiation—chlorine disinfectant, demonstrated that at low amounts of chlorine, the bacteria of Escherichia coli survived, which was not the case when chlorination was applied alone (Murphy et al. 2008).

There are numerous methods to assess the microbial stability of water; however, the most commonly used one is determining the value of the biodegradable or assimilable organic carbon (van der Kooij 2000, Polańska et al. 2005, Wang et al., 2014). The studies presented in this article use the method of direct assessment of water biostability, involving observation of the microbial growth potential in the water inoculated with heterotrophic bacteria (Lehtola et al. 2001, Lehtola et al. 2003). The water collected from four treatment plants which used UV light for disinfection—before and after the UV treatment—was the subject of this research study. The collected samples were assessed for the growth potential of heterotrophic bacteria and the build-up of these bacteria in time.

2 MATERIALS AND METHODS

2.1 Samples

The water samples were collected from four water treatment plants, in which UV radiation lamps were used as part of the water disinfection process before feeding the water into the distribution system. The water was collected after all the treatment processes, but before applying the UV irradiation and immediately after the UV irradiation. Names of the objects were coded and denoted as WTP1, WTP2, WTP3 and WTP4. The plants were selected so as to differ in both the water treatment process and the kind of a UV lamp (medium-pressure and low-pressure):

- WTP-1: coagulation → rapid filtration → irradiation with low-pressure UV lamp;
- WTP-2: coagulation → rapid filtration → irradiation with medium-pressure UV lamp;
- WTP-3: coagulation → rapid filtration → ozonation → adsorption on activated carbon → irradiation with low-pressure UV lamp;
- WTP-4: ozonation → coagulation → rapid filtration → irradiation with medium-pressure UV lamp;

Table 1 presents the measured parameters of the water collected before and after the irradiation with a UV lamp, according to the methodology described in (Włodyka-Bergier et al. 2014).

2.2 Biostability assessment procedure

The samples before and after the UV irradiation were subjected to the assessment procedure for microbial stability, according to the method recommended by Lehtola et al. (2001, 2003). Microbial stability of the water was assessed directly by observing the growth of heterotrophic bacteria (HGR—*heterotrophic bacteria growth*). The HGR observations were carried out for 21 days on water samples inoculated with the microorganisms from the river (the Vistula River) and incubated in glass bottles at 15°C. The water samples were collected every few days and they were inoculated by the method of culture on a nutrient agar R2A. The inoculated plates with agar were incubated at $22 \pm 2°C$ for 7 days, after which the grown colonies of heterotrophic bacteria were counted. A diagram presenting the experiment procedure has been illustrated in Figure 1.

Table 1. Quality of water before and after UV treatment.

		Concentration of water parameter				
	Sampling point	DOC, mg/L	UV_{254}, cm^{-1}	$N-NO_3$, mg/L	Conductivity, mS/cm	Br^-, mg/L
WTP-1	before UV	1.8	0.050	1.3	0.355	0.24
	after UV	1.7	0.048	1.4	0.378	0.20
WTP-2	before UV	1.5	0.028	0.9	0.336	0.38
	after UV	1.3	0.025	1.0	0.341	0.45
WTP-3	before UV	1.6	0.023	1.4	0.398	0.50
	after UV	1.1	0.022	1.5	0.413	0.53
WTP-4	before UV	3.2	0.029	0.3	0.312	0.43
	after UV	1.8	0.034	0.3	0.322	0.35

*DOC—dissolved organic carbon.

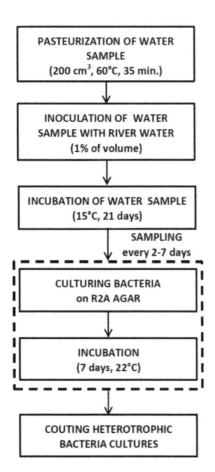

Figure 1. Biostability test scheme.

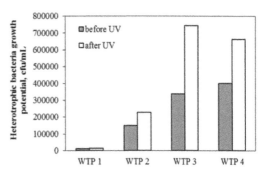

Figure 2. Heterotrophic bacteria growth potential.

3 RESULTS AND DISCUSSION

3.1 *Microbial growth potential*

In order to assess the effect of UV radiation on the microbial stability of water, the maximum HGR (HGR_{max}) was applied in water samples in the analyzed incubation time (Lehtola et al. 2003). The number HGR_{max} demonstrates the bacterial growth potential in a given water sample, thus showing its susceptibility to the growth of microorganisms. Figure 2 illustrates the potential of HGR in the water samples collected before and after the UV treatment.

In the case of all the analyzed plants, the water samples collected after the UV irradiation exhibited poorer microbial stability. In the case of WTP-1 and WTP-2, the plants in which the water was treated using coagulation and rapid filtration, in the water samples before the UV irradiation, HGR_{max} was relatively low and amounted to 12,112 colony forming units (cfu), and 148,500 cfu in the samples collected from WTP-1 and WTP-2, respectively. Irradiating water with low-pressure UV lamp (WTP-1) resulted in an increase of HGR_{max} by 18% (up to 14,256 cfu), whereas the medium-pressure UV lamp used for water disinfection (WTP-2) increased the potential for the build-up of bacteria by 53% (up to 227,700 cfu).

The water in plants WTP-3 and WTP-4, prior to the UV process, in addition to coagulation and rapid filtration, was also treated via organic matter oxidation as a result of ozonation. The potential build-up of bacteria in the samples collected before the UV treatment was higher than in the case of the plants WTP-1 and WTP-2, and it amounted to 336,363 cfu for WTP-3 and 401,200 cfu for WTP-4. Irradiation of the water collected from WTP-3 with the low-pressure UV lamp resulted in the increased HGR_{max} by 122% (up to 745,455 cfu), while using the medium-pressure lamp in WTP-4 resulted in the increased potential for the buildup of heterotrophic bacteria up to the value of 663,000 cfu (the increase by 65%). The presented results demonstrated negative effects of UV radiation on the microbial stability of water in the water distribution system. The study also proved that UV radiation may increase the buildup of bacteria in the water which was chemically pre-oxidized. The tests were carried out on the water samples collected from various water treatment plants, and so the quality of the water before the process of UV irradiation differed, but the results obtained from the plants WTP-1 and WTP-2 demonstrated that the use of a medium-pressure lamp results in a greater build-up of bacteria than in the case of a low-pressure lamp. The results obtained in the water samples from WTP-3 and WTP-4 did not confirm this assumption. The use of the ozonation process before the UV treatment results in the degradation of organic matter to the particles of lower molecular weight, which is more easily decomposed. The results summarized in Table 1 illustrate that organic carbon in the water sample from WTP-4, after irradiation with the medium-pressure UV lamp, is decomposed by about 44%, whereas the low-pressure UV lamp installed in WTP-3 results in a reduction in organic carbon by about 30%. The increased decomposition of organic matter by using the medium-pressure UV lamp in the sample collected from WTP-4 could have affected the microbial stability by making it better than in the sample irradiated with low-pressure lamp (WTP-3).

3.2 *Microbial growth in time*

In order to assess the microbial stability of water, the maximum value of the build-up of bacteria over a long period of observation is used. However, considering the time needed for the transportation of water in distribution systems, the observation of the buildup of microorganisms in time, especially in the first days of the observation, is essential for the waterworks. Figure 3 presents the number of heterotrophic bacteria over the time period of up to 21 days in the water samples collected from the analyzed water treatment plants before and after the UV irradiation process. The obtained result is the number of heterotrophic bacteria counted from the formula according to the standard PN-ISO 728: 1998/A1:2004:

$$N = \frac{\sum c}{V \cdot [n_1 + (0,1 + n_2)] \cdot d}$$

where: c—sum of the colonies on plates from two subsequent dilutions, of which at least one contains 15 colonies; V—volume of the material applied to each plate; n_1—number of plates from the first counted dilution; n_2—number of plates from the second counted dilution; d—dilution factor corresponding to the first counted dilution (d = 1 when undiluted sample is cultured).

Taking into account the influence of incubation time of the sample in the water collected from WTP-1 (Fig. 3A), the highest average number of heterotrophic bacteria in both the water collected before and after the UV treatment was observed on the last day of incubation. The water collected from this treatment plant was characterized by the best microbial stability. It was obvious that microorganisms needed a long time to get to the absorbable forms of dietary substrates. In the water sample collected from WTP-2 (Fig. 3B), the UV radiation resulted in a relatively high HGR (126,000 cfu) on the fourth day of incubation, followed by a reduction in the amount of bacteria on the seventh day of incubation, and then the number of bacteria grew in time to reach its maximum (227,700 cfu) on the eleventh day of the incubation. The water which was not irradiated was characterized by a slow build-up of bacteria rate in time, with a maximum of 148,500 cfu also on the eleventh day of the observation. In the early days of the observation, the buildup of bacteria was low and amounted to 87 cfu on the first day of the observation, and 390 cfu on the fourth.

The samples which were taken before the irradiation with ultraviolet rays from WTP-3 and WTP-4 were characterized by a relatively high HGR on the 3–5 day of the observation. In the case of the sample from WTP-3, on the fifth day, the maximum number

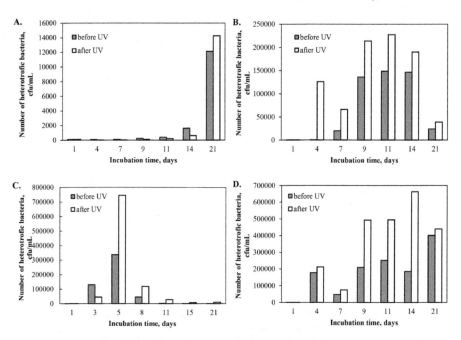

Figure 3. The number of heterotrophic bacteria in time (A) WTP-1, (B) WTP-2, (C) WTP-3, (D) WTP-4.

of bacteria (336,363 cfu) was observed, whereas the water sample from WTP-4, after a slight increase in the number of bacteria on the fourth day of the observation (up to 177,480 cfu), on the seventh day of the observation a decrease in HGR was observed, and then an increase in the number of bacteria until day 21 of the observation to the value of 401,200 cfu. Irradiation with UV rays resulted in increased numbers of bacteria in individual samples, but the growth of the buildup of bacteria was similar. In the sample collected from WTP-3 after the UV treatment, HGR_{max} was observed on the fifth day of the observation, whereas in the water collected from WTP-4, after a slight increase in HGR on the fourth day (up to 211,980 cfu), there was a decrease in the amount of bacteria on the seventh day, and then a large increase in the number of bacteria, with a maximum reached on the 14th day of the observation (663,000 cfu).

4 CONCLUSIONS

In order to improve water quality, many water treatment plants before chemical disinfection of water started to use irradiation with UV rays. This is a very effective process for the inactivation of microorganisms. However, properly conducted process of disinfection not only means the process which was successfully carried out at the plant, but also providing a microbiologically pure water to the consumer in the water distribution network. As it was proven by the research studies presented in this article, the use of UV radiation for disinfection of water can adversely affect the microbial stability of water in the distribution system. The conducted studies of the water collected before and after the UV treatment from four water treatment plants demonstrated that the microbial stability after the UV irradiation in the water samples which had been chemically oxidized before, decreased more than in the samples treated conventionally with coagulation and rapid filtration. The type of the lamp which is used can also affect the microbial stability of water.

ACKNOWLEDGEMENTS

The work was completed under AGH University statutory research No. 11.11.150.008 for the Department of Management and Protection of Environment.

REFERENCES

Choi, Y. & Choi, Y.-J. 2010. The effects of UV disinfection on drinking water quality in distribution systems. *Water Research* 44: 115–122.
Hijnen, W., Beerendonk, E. & Medema, G. 2006. Inactivation credit of UV radiation for viruses, bacteria

and protozoan (oo)cysts in water: A review. *Water Research* 40: 3–22.
Koivunen, J. & Heinonen-Tanski, H. 2005. Inactivation of enteric microorganizms with chemical disinfectants, UV irradiation and combined chemical/UV treatments. *Water Research* 39: 1519–1526.
Lehtola, M., Miettinen, I., Vartiainen, T., Myllykangas, T. & Martikainen, P. 2001. Microbially available organic carbon, and microbial growth in ozonated drinking water. *Water Research* 35(7): 1635–1640.
Lehtola, M., Miettinen, I., Vartiainen, T., Rantakokko, P., Hirvonen, A. & Martikainen, P. 2003. Impact of UV disinfection on microbially available phosphorus, organic carbon and microbial growth in drinking water. *Water Research* 37: 1065–1070.
Liu, W., Zhang, Z., Yang, X., Xu, Y. & Liang, Y. 2012. Effects of UV irradiation and UV/chlorine co-exposure on natural organic matter in water. *Science of the Total Environment* 414: 576–584.
Lyon, B., Dotson, A., Linden, K. & Weinberg, H. 2012. The effect of inorganic precursors on disinfection by product formation during UV-chlorine/chloramine drinking water treatment. *Water Research* 46: 4653–4664.
Miksch, K, Cema, G., Felis, E., & Sochacki, A. 2015. Nowoczesne techniki i technologie inżynierii środowiska. *Rocznik Ochrona Środowiska/Annual Set the Environment Protection,* 17(1), 245–253.
Murphy, H., Payne, S. & Gagnon, G. 2008. Sequential UV- and chlorine-based disinfection to mitigate Escherichia coli in drinking water biofilms. *Water Research* 42: 2083–2092.
Polanska, M., Huysman, K. & Keer, C. 2005. Investigation of assimilable organic carbon (AOC) in flemish drinking water. *Water Research* 39: 2259–2266.
Pozos, N., Scow, K., Wuertz, S. & Darby, J. 2004. UV disinfection in a model distribution system: biofilm growth and microbial community. *Water Research* 38: 3083–3091.
Rand, J.L., Hofmann, R., Alam, M.Z.B., Chauret, C., Cantwell, R., Andrews, R.C. & Gagnond, G.A. 2007. A field study evaluation for mitigating biofouling with chlorine dioxide or chlorine integrated with UV disinfection. *Water Research* 41: 1939–1948.
Sun, H., Shi, B., Bai, Y. & Wang, D. 2014. Bacterial community of biofilms developed under different water supply conditions in a distribution system. *Science of the Total Environment* 472: 99–107.
Van Der Kooij, D. 2000. Biological stability: a multidimensional quality aspect of treated water. *Water, Air, and Soil Pollution* 123: 25–34.
Wang, Q., Tao, T. & Xin, K. 2014. Experimental study using the dilution incubation method to assess water biostability. *Journal of Environmental Sciences* 26: 1994–2000.
Wen, G., Ma, J., Huang, T. & Egli, T. 2014. Using coagulation to restrict microbial re-growth in tap water by phosphate limitation in water treatment. *Journal of Hazardous Materials* 280: 348–355.
Włodyka-Bergier, A., Rajca, M. & Bergier, T.: Removal of halogenated by-products precursors in photocatalysis process enhanced with membrane filtration. *Desalination and Water Treatment* 52 (19–21): 3698–3707.
Yang, B.M., Liu, J.K., Chien, C.C., Surampalli, R.Y. & Kao, C.M. 2011. Variations in AOC and microbial diversity in an advanced water treatment plant. *Journal of Hydrology* 409: 225–235.

Environmental Engineering V – Pawłowska & Pawłowski (Eds)
© 2017 Taylor & Francis Group, London, ISBN 978-1-138-03163-0

Electron microscopy assessment of the chemical composition of sediments from water supply pipes

J. Bąk, J. Królikowska & A. Wassilkowska
Faculty of Environmental Engineering, Cracow University of Technology, Kraków, Poland

T. Żaba
Municipal Water and Sewerage Company MPWiK S.A., Kraków, Poland

ABSTRACT: Qualitative studies of the chemical composition of sediments from water supply pipes were performed by means of Scanning Electron Microscope (SEM). Sediment samples were taken from different areas of one water supply system. The study was carried out using a variable pressure SEM equipped with an Energy-Dispersive X-ray spectrometer (EDS). The EDS method employed for analysing the chemical composition of sediment surface is non-invasive and standardless, since the theoretical spectra of the elements are embedded in a software. The analyses made it possible to compare the elemental composition of sediments from a system supplied by different water sources. The composition of the sediments which had formed on the materials, namely steel and cement-lined cast iron pipes, was also compared.

Keywords: water distribution system, tubercle deposits, microscopic analysis, EDS, SEM, contamination

1 INTRODUCTION

Environmental engineering encompasses issues related to the various types of residues. These included among others, sewage sludge, sludge from water treatment processes and sediments from reservoirs. Numerous studies and publications both in Polish language and English-language technical literature are devoted to them. As examples of works merely from 2015 year, the following publications may be mentioned—in the field of sewage sludge (Czechowska-Kosacka et al. 2015, Wolski et al. 2015), on research related to the problem of sludge from water treatment – (Totczyk et al. 2015), about the bottom sediments (Bartoszek et al. 2015, Karwacka et al. 2015). In addition to listed types of deposits, they are those of water supply systems.

Numerous publications and studies are currently devoted to the phenomenon of contamination, which is to say, the secondary water pollution in a water supply network and changes in the quality of the water in the distribution system. Sediments in water supply pipes have a significant impact on the quality of water being transported. The publications in the Polish technical literature pertaining to research into these sediments are few and appear intermittently. Papers such as (Zerbe & Zawadzka 1973, Rudnicka & Świderska-Bróż

1995, Świderska-Bróż & Wolska 2006, Łomotowski 2007, Wassilkowska & Dąbrowski 2012, Grabarczyk 2015, Zimoch 2016) are worth to be mentioned. Only publications on biofilm are more numerous in Poland. Among the far more numerous foreign publications on testing various types of deposits found in water supply systems may be mentioned, among others, such studies as (Chawla et al. 2012, Echeverría et al. 2009, Gauthier et al. 2001, Lin et al. 2001, Sly et al. 1990, Zhang et al. 2015).

The chemical composition of sediments depends on several factors, including the material of the pipe on which they form, the quality of the water coming into contact with the interior of the pipeline and the consecutive deposition layers accumulating on the pipe's internal walls. The quality of water, particularly as far as its chemical stability or lack is concerned, governs the mechanism of sediment creation. The process of sediment deposition is also influenced by the fact that water from different sources is mixed in some areas of a distribution system. For instance, one study (Bray et al. 2011) determined that combining water from the Cretaceous Period with water from the Quaternary Period may result in the loss of its chemical and biological stability and the formation of sludge, even though quality requirements are met.

The sediments which form in water supply systems can be divided into four groups (Grabarczyk 2015):

– attached sediments, primarily found in water supply networks;
– tubercle sediments, primarily found in water supply networks;
– bottom sediments, primarily found in raw water pipelines;
– biofilm, primarily found in raw water pipelines.

Water treatment technology, particularly coagulation and water additives such as phosphate corrosion inhibitors, may also have an impact on the quality of the sediments. In the studies described by (Świderska-Bróż & Wolska 2006), the highest quantities of aluminium compounds were found in the sediments from pipes coming into contact with surface water treated with aluminium coagulants. On the other hand, significant amounts of manganese compounds were found in the sediments from areas supplied with ground- and infiltrated water. Phosphate corrosion inhibitors, primarily orthophosphates, polyphosphates, and mixtures thereof, form a thin, protective layer of poorly soluble phosphates on pipe walls. However, when sediments are already present, the inhibitor additives remove the accumulated material by its slowly washing out, creating a protective layer at the same time (Biłozor 2016).

The chemical composition of sediments is also relevant to the phenomenon of tap water contamination, since it governs which types of pollutants are released into the water (Rudnicka & Świderska-Bróż 1995). The aim of the present study was to assess the chemical composition of sediments from water supply pipes.

2 RESEARCH MATERIAL

The research material comprised sediments taken from a number of pipes within a large water supply system. The water supply network in question is supplied from four surface water sources treated in four water treatment plants and from one groundwater source.

The analysis covered sediment samples taken from three pipes of one water supply system. The samples were acquired from two steel pipes and

Table 1. Research material data.

Sample	Pipe diameter [mm]	Pipe material	Year of commissioning	Number of sources supplying water to the pipe
Pipe 1	DN 800	steel	1975	2 (mixing zone)
Pipe 2	DN 500	steel	1973	3 (mixing zone)
Pipe 3	DN 800	cement-lined cast iron	1955 (cement lining from the 90 s)	1

Table 2. Macroscopic examination results.

Sample	Sediment colour	Sediment characteristics	Photo of sediment surface structure
Pipe no.1	blackened brown with rusty spots	Tubercle of sediment with maximum size of 9-13 mm, which are difficult to separate from the surface of pipe;	
Pipe no.2	bright brown, rusty	Tubercle of sediment with maximum size of 13-18 mm, which are easily separated from the surface of pipe;	
Pipe no.3	brown	Very thin layer of sediment, high roughness of the wall	

one cement-lined, cast iron pipe dating from the nineteen nineties. They were gathered from different water supply areas. Two of the sediment samples originated from mixing zones, which is to say, a water mix from as many as three different sources in the case of one sample.

Table 1 presents the basic characteristics of each sample. Throughout this paper, the number of the pipe from which a sediment sample was taken corresponds to that sample.

3 METHODS

The research was divided into three stages. The first stage involved macroscopic examination, includ-

Table 3. Elemental composition of sediment surface from pipe no.1 (steel).

Measurement series	Percentage by weight of elements, %											
	C	O	Na	Mg	Al	Si	P	K	Ca	Mn	Fe	Total
1.1	20.0	43.6	0.2	0.2	3.5	2.2	0.3	0.1	1.1	0.4	28.4	100
1.2	14.1	58.5	–	–	–	–	–	–	–	–	27.4	100
1.3	19.9	42.6	0.1	0.1	2.6	1.5	0.3	0.1	0.6	0.2	32.0	100

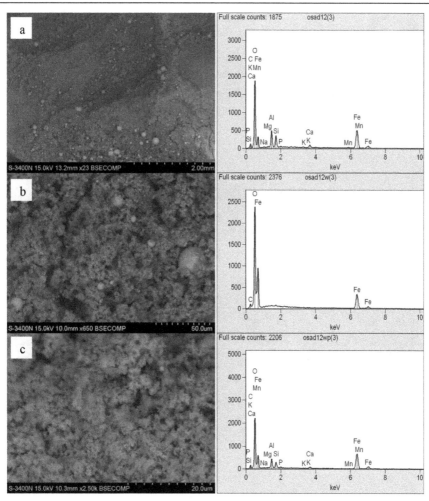

Figure 1. Structure of sediment surface from pipe 1 a) image at a magnification of 23× and the corresponding EDS spectrum of this sample; b) image at a magnification of 650× and point spectrum from a single spherical particle; c) image at a magnification of 2500× and EDS spectrum of the entire sample.

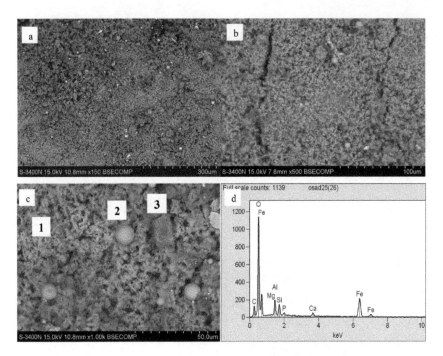

Figure 2. Structure of sediment surface from pipe 1: a) image at a magnification of 150×; b) image at a magnification 500×; c) The observed morphology of sediment 1 – flocs; 2 – spherical particles; 3 – lumps, image at a magnification of 1000×; d) exemplary EDS spectrum from a surface with flocs present (200 × 200 μm sample).

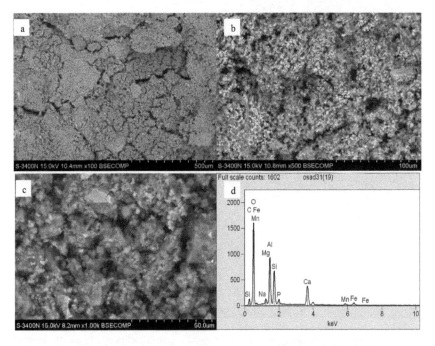

Figure 3. Structure of sediment surface from pipe 3: a) image at a magnification of 100×; b) image at a magnification of 500×; c) image at a magnification of 1000×; d) exemplary EDS spectrum from 200 × 200 μm sample.

ing the visual analysis of sediment color and structure and the height measurement of larger tubercles using a digital caliper. The second and third stages were conducted using a HITACHI S-3400 N Scanning Electron Microscope (SEM) equipped with an Energy Dispersive X-ray Spectrometer (EDS) manufactured by ThermoScientific. In the second stage the fresh sediment surface has been examined using backscattered mode in SEM. The final stage employed the non-invasive EDS method to analyze the chemical composition in micro-areas of the sediment surface.

The advanced qualitative EDS study of the chemical composition is non-invasive and standardless, since the theoretical spectra of the elements are embedded in a software of the equipment. The technique for the spectral analysis of micro-samples was described in details in (Wassilkowska et al. 2014).

The SEM used in the study can operate in variable pressure mode, making it possible to analyze the morphology of samples which do not conduct electricity in their natural state, in other words, without the need to vacuum coating of a sample with a conductive layer first. The imaging of the structure was performed by means of a Backscattered Electron detector (BSE). The BSE image of structural constituents shows a contrast of different atomic number where the brighter areas are characterized by a higher mean atomic number for the chemical elements (Wassilkowska et al. 2014). The imaging was carried out with an electron beam accelerating voltage of 15 kV.

4 RESULTS

The results of the macroscopic examinations are presented in Table 2. The results of the microscopic examinations with imaging of the sediment structure are given in Figures 1, 2 and 3, which correspond to pipes 1, 2 and 3 and also provide exemplary EDS spectra corresponding to the samples in question. Tables 4 to 7 show the elemental composition of the sediment surfaces along with the percentage of those elements obtained, by weight.

5 DISCUSSION

The microscopic examinations of the sediment surface which formed on the steel pipes, namely pipes 1 and 2, showed that the structure in both samples is predominantly composed of loose flocs (Figs. 1–2). In a BSE image contrast, some bright spherical particles and some dark thickened sludge constituents appear on the tubercle surfaces investigated.

As Table 3 demonstrates, the sediment from pipe 1 contains mainly iron and oxygen. Elements with the atomic number $Z < 5$, from hydrogen to boron, are not detected by EDS method (Wassilkowska et al. 2014). The carbon peak is always visible in the EDS spectrum; however, it is not determined quantitatively as a result of the significant error in the analysis of the light elements ($Z < 11$). The floc structure of the sediment collected from pipe 1 also contained Al, Si, Ca, Mn, and P, as well as trace amounts of Mg, Na, and K.

Table 4. Elemental composition of sediment surface from pipe no.2 in the area with flocs present.

| Measurement series | Percentage by weight of elements, % | | | | | | | | |
	C	O	Mg	Al	Si	P	Ca	Fe	Total
2.1	28.2	48.2	0.1	2.0	1.5	0.5	0.8	18.7	100
2.2	19.6	45.1	0.1	1.4	1.2	1.0	1.6	30.0	100
2.3	17.5	45.3	0.1	1.5	1.3	1.1	1.2	32.0	100

Table 5. Elemental composition of sediment surface from pipe no. 2 in the area with single lumps.

| Measurement series of lumps | Percentage by weight of elements, % | | | | | | |
	C	O	Al	Si	P	Ca	Fe
1	17.8	55.4	24.6	0.2	0.1	0.0	1.8
2	29.5	49.3	3.4	3.3	0.5	0.7	13.3
3	28.1	46.0	2.2	7.1	0.6	0.8	15.1
4	25.0	33.0	0.5	27.9	0.1	0.4	13.0

Table 6. Elemental composition of sediment surface from pipe no. 2 in the area with single spherical particles.

| Measurement series of spherical particles | Percentage by weight of elements, % | | |
	C	O	Fe
1	15.6	49.2	35.2
2	14.4	45.6	40.0
3	10.6	53.0	36.4
4	9.8	48.1	42.1

Table 7. Elemental composition of sediment surface from pipe no.3 (cement lining).

Measurement series	Percentage by weight of elements, %										
	C	O	Na	Mg	Al	Si	P	S	Ca	Mn	Fe
3.1	22.5	49.9	0.3	0.5	5.5	5.3	0.7	0.2	3.3	7.5	4.3
3.2	23.4	51.7	0.2	0.3	7.3	5.1	0.9	-	3.1	5.3	2.8
3.3	25.1	51.4	0.1	0.8	6.5	4.9	0.9	0.1	6.0	1.6	2.5

Compounds and minerals containing calcium, manganese and magnesium may precipitate from water; however, the main element of the sediment from pipe 1 includes compounds of iron and oxygen. Hydrogen is not analyzed using the EDS method.

In comparison with the fresh surface of the sediment from pipe 1, a significantly lower number of bright spherical particles was observed in BSE contrast on the surface of the sediment from pipe 2. However, compact, dark-colored structures, which is to say lumps, were more apparent. The predominant morphological forms of the deposition constituents in the samples of water pipe sediments are shown in Figure 2c.

The elemental composition of the dark-colored lump structures differs from the composition of the flocs (Tables 4 and 5), with a significantly higher percentage, by weight, of aluminum or silicon being observed for the lumps. The study showed that lumps may be formed not only from aluminum and silicon oxides, but also from iron oxides. While EDS point analysis confirmed only iron and oxygen constituents in the bright-colored spherical particles.

Figure 3 shows the structure and elemental analysis of thin brownish sediment deposits from pipe 3, in other words, on the cement mortar lining from an old cast iron pipe. Table 7 demonstrates that the main components of the sediment formed on the cement mortar lining include Al, Si, Ca, Mn, and Fe, as well as smaller quantities of Mg and P. In comparison with the sediment, which is to say, floc, samples from the steel pipes, a higher content of such elements as Al, Si, Ca, and Mn was observed. Moreover, sulphur, which was not found in any other sludge materials under study, was here evidently detected in two of the EDS measurement (Table 7). Certain similarities can be seen in the studies cited in (Świderska-Bróż & Wolska 2006), which were carried out on sediment taken from an asbestos-cement pipeline which had remained in service for twenty-six years. In the case of that sediment, low amounts of iron compounds, high amounts of magnesium compounds and relatively abundant silicates were found. The analysis of sediment from the cement-lined pipe no.3 showed a significantly lower percentage, by weight, of iron, as compared to the samples of sediment taken from the steel pipes no.1 and 2. Additionally,

enhanced percentage of silicon and magnesium were detected. Another study (Echeverría et al. 2009) revealed the contents of such elements as C, O, Fe and Si in outer shell of tubercle deposits. In conducted study, the same elements had the largest shares in the percentage composition, by weight, for the sample from the pipe no.1.

To sum up, the main components of the sediments from the steel pipes include compounds of iron and oxygen, with the exception of the lump structures. For the sediments deposited on the cement mortar lining of cast iron pipe, the percentage, by weight, of iron was lower than that one determined in the sediment samples from steel pipes.

In the previously mentioned studies, described by (Świderska-Bróż & Wolska 2006), the highest amounts of aluminum were found in the sediment from pipelines supplied with surface water treated with aluminum coagulants, which may explain the presence of aluminum in the sediment taken from all three pipelines. The presence of phosphorus might be connected with the addition of phosphate corrosion inhibitors to the water. According to (Grabarczyk 2015), in the case of corroded pipes, calcium carbonate sediment may precipitate from water even when it is not supersaturated, which could explain the presence of calcium and large amounts of iron in the sludge analyzed in this study. The presence of manganese and iron may also be related to the participation of manganese and iron bacteria.

The elemental composition of the sediments under study here is similar to the qualitative composition of other sediments from water supply systems in Poland (Rudnicka & Świderska-Bróż 1995, Świderska-Bróż & Wolska 2006, Łomotowski 2007).

6 CONCLUSIONS

Samples of sediments acquired from pipes from one water supply system were examined using the scanning electron microscope method. The structural analysis of the surface of sediments formed on steel pipes transporting water in mixing zones, in other words, from two or more sources, confirms that the differences in the sediment color are connected with the varying chemical composition of samples' surfaces. Both the qualitative composition and by weight percentage of the individual elements forming the sediment layer on the rugged surface of the cement lining revealed significant differences in comparison with the sediments from steel pipes.

The research results obtained in respect of the chemical composition of sediments from water supply pipes in the system under analysis closely resemble the results of studies on the sediment composition from other water supply networks in Poland.

The elemental composition of sediments may prove useful in determining the mixing zones of water from different sources. The area of such zones is very important, because they may be prone to sediment formation and the phenomenon of secondary water contamination. Further studies are required for better understanding of the influence of water mixing on the sediment deposition process and secondary water contamination. The studies will be conducted into that direction.

REFERENCES

Bartoszek, L., Koszelnik, P., Gruca-Rokosz, R. & Kida, M. 2015. Assessment of Agricultural Use of the Bottom Sediments from Eutrophic Rzeszów Reservoir, *Annual Set the Environmental Protection*, 17: 396–409.

Biłozor, S. 2016. Zmiany parametrów wody w sieci wodociągowej w *Magazyn Instalatora* 5.02.2016r. http://www.instalator.pl/2016/02/zmiany-parametrow-wody-w-sieci-wodociagowej/dostęp 9.04.2016 r.

Bray, R., Sokołowska, A., Jankowska, K. & Olańczuk-Neyman K. 2011. Wpływ mieszania wody podziemnej z różnych pięter wodonośnych na jej stabilność chemiczną i biologiczną w sieci wodociągowej, *Ochrona Środowiska* 3: 19–23.

Chawla, V., Gurbuxani, G.P. & Bhagure, G.R. Corrosion of water pipe: a comprehensive study of deposits 2012. *Journal of Minerals & Materials Characterization & Engineering*,11(5): 479–492.

Czechowska-Kosacka, A., Cao, Y. & Pawłowski, A. 2015. Criteria for Sustainable Disposal of Sewage Sludge, *Annual Set the Environmental Protection*, 17: 337–350.

Echeverría, F., Castaño, J.G., Arroyave, C., Penuela, G., Ramírez, A. & Morató, J. Characterization of deposits formed in a water distribution system 2009. *Ingeniare. Revistachilena de ingeniería*, 17(2): 275–281.

Gauthier, V., Portal, J.M., Yvon, Y., Rosin, C., Block, J.C., Lahoussine, V., Benabdallah, S., Cavard, J., Gatel, D. & Fass, S. 2001. Characterization of suspended particles and deposits in drinking water reservoirs, *Water Science and Technology: Water Supply*, 1(4): 89–94.

Grabarczyk, C. 2015. *Hydraulika urządzeń wodociągowych. Tom 1,* Warszawa, Envirotech Wydawnictwo WNT.

Karwacka, A., Niedzielski, P. & Staniszewski, R. 2015. Ocena stanu osadów dennych wybranych jezior powiatu poznańskiego, *Rocznik Ochrona Środowiska*, 17: 1684–1698.

Lin, J., Ellaway, M. Adrein, R. 2001. Study of corrosion material accumulated on the inner wall of steel water pipe, *Corrosion Science*, 43(11): 2065–2081.

Łomotowski, J. 2007. *Przyczyny zmian jakości wody w systemach wodociągowych*, Warszawa, Polska Akademia Nauk Instytut Badań Systemowych.

Rudnicka, Ł. & Świderska-Bróż, M. 1995. Skład chemiczny osadów z wrocławskiej sieci wodociągowej, *Ochrona Środowiska* 3(58): 63–65.

Sly, L.I., Hodgkinson, M.C., Arunpairojana, V. 1990. Deposition of manganese in a drinking water distribution system, *Applied and Environmental Microbiology*, 56(3): 628–639.

Świderska-Bróż, M. & Wolska, M. 2006. Główne przyczyny wtórnego zanieczyszczenia wody w systemie dystrybucji, *Ochrona Środowiska* 4: 29–34.

Totczyk, G., Klugiewicz, I., Pasela, R. & Górski, Ł. 2015. Usuwanie fosforanów z wykorzystaniem osadów potechnologicznych pochodzących ze stacji uzdatniania wody, *Rocznik Ochrona Środowiska*, 17: 1660–1673.

Wassilkowska, A., Czaplicka-Kotas, A., Zielina, M. & Bielski A. 2014.An analysis of the elemental composition of micro-samples using EDS technique, *Technical Transactions*, 1: 133–148.

Wassilkowska, A. & Dąbrowski, W. 2012. Badania osadów korozyjnych pobranych z rurociągu przy użyciu mikroskopu elektronowego w *Nowe Technologie w sieciach i instalacjach wodociągowych i kanalizacyjnych*, 105–115.

Wolski, P., Wolny, L., Zawieja, I. 2015. Ultrasonic Processors and Drainage of Sewage Sludge, *Annual Set The Environmental Protection*, 17: 450–460.

Zerbe, J. & Zawadzka, H. 1973. *Przyczyny powstawania i metody zapobiegania tworzeniu się osadów w przewodach wodociągowych. Praca przeglądowa*, Warszawa, Zakład Informacji Naukowo—Technicznej i Ekonomicznej Instytut Gospodarki Komunalnej.

Zhang, H., Tian, Y., Wan, J. & Zhao, P. 2015. Study of biofilm influenced corrosion on cast iron pipes in reclaimed water, *Applied Surface Science*, 357(A): 236–247.

Zimoch, I. 2015. Ocena toksyczności osadów wodociągowych, *Instal* nr 1 (369): 58–59.

Environmental Engineering V – Pawłowska & Pawłowski (Eds)
© 2017 Taylor & Francis Group, London, ISBN 978-1-138-03163-0

Effect of chemical coagulants on the sedimentation properties of activated sludge

A. Masłoń & J.A. Tomaszek

Faculty of Civil and Environmental Engineering and Architecture, Rzeszow University of Technology, Rzeszów, Poland

ABSTRACT: Bulking sludge and foam formation on the surface of activated sludge reactors constitute significant problems in biological wastewater treatment systems. The improvement of the sedimentation properties of the sludge can be achieved using chemical oxidants or coagulants. In this study, coagulants—polyaluminum chloride, polyaluminum chloride + iron (II) chloride, sodium aluminate, iron (II) chloride, iron (III) chloride and iron (III) sulfate—were used to improve the sedimentation performance of activated sludge. The experimental assays were conducted in batch systems. The results showed varied effects of improvement of the activated sludge sedimentation properties, depending on the type and amount of the applied coagulant. The efficient use of coagulants aims to improve settleability, to decrease the sludge volume index, and to intensify the sedimentation velocity and to decrease the concentration of total suspended solids in treated wastewater. The most effective coagulant in the advancement of sedimentation-activated sludge was sodium aluminate.

Keywords: activated sludge, bulking sludge, sedimentation properties, coagulants

1 INTRODUCTION

Review of the literature shows that the activated sludge technology has experienced operational problems since its inception. Bulking sludge and foam formation on the surface of activated sludge reactors is a significant problem in biological wastewater treatment systems. Excessive and uncontrolled development of certain filamentous bacteria is an immediate factor in the bulking sludge phenomenon. Massive growth of filamentous microorganisms leads to disturbances mainly in the sedimentation process of activated sludge, and consequently reduces the efficiency of wastewater treatment. In addition, there are numerous operational difficulties such as problems with the recirculation of sludge, reducing excess sludge properties in the process of thickening and dewatering, as well as difficulties with scrapers in secondary clarifiers etc. The Sludge Volume Index (SVI) is an indicator of the deterioration of the activated sludge sedimentation properties. Bulking sludge, termed as poorly sedimenting sludge, is characterized by the SVI greater than 150 mL·g^{-1}. It is assumed that bulking sludge does not occur at low values of SVI (Gerardi 2002, Jenkins et al. 2004, Przybyła et al. 2009, Naidoo et al. 2011, Guo et al. 2014, Kida et al. 2015).

Technological operations such as the organic loading rate of sludge and oxygen concentration,

the use of a high degree of recirculation, or addition of digested sludge should be applied at the first instance to prevent and counteract bulking and foaming sludge (Jenkins et al. 2004, Martins et al. 2004). The improvement of the sedimentation properties of the sludge can be achieved using chemical oxidants (ozone, sodium hypochlorite, hydrogen peroxide) (Walczak & Cywińska 2007), inorganic coagulants, i.e.—aluminum salts (Drzewicki 2009), iron salts (Mamais et al. 2011), and organic synthetic polymers (Juang 2005). It is possible to use hybrid reagents such as polyaluminum chloride and a cationic polymer.

The aim of this study is to determine the effect of chemical coagulants on the sedimentation properties of activated sludge.

2 MATERIAL AND METHODS

Industrial commercial coagulants used in this study included: polyaluminum chloride (PAX 16), polyalu-minum chloride + iron (II) chloride (PAX 25), sodium aluminate (SAX 18), iron (II) chloride (PIX 100), iron (III) chloride (PIX 111) and iron (III) sulfate (PIX 113). The characteristics of the applied coagulants are shown in Tables 1–2. Studies were conducted using coagulants in each of the five measurement series. The scope of research

Table 1. Characteristic of Al-coagulants (Kemipol 2013).

Characteristics	PAX 16	PAX 25	SAX 18
	polyaluminum chloride	polyaluminum chloride + iron (II) chloride	sodium aluminate
Chemical formula	$Al(OH)_xCl_y+H_2O$ $(x + y = 3\ 1.05 < x < 2)$	$Al(OH)_xCl_y + H_2O$ $FeCl_3$	$NaAlO_2$
Form	light yellow solution	light brown solution	straw colored liquid
Chemical composition			
Al_2O_3 (%)	15.5 ± 0.4	11.9 ± 0.4	18.0 ± 1.0
Al^{+3} (%)	8.2 ± 0.2	6.3 ± 0.2	9.5 ± 0.5
Fe^{+2} (%)	–	2.2 ± 0.1	–
Density (20°C) (g·L⁻¹)	1330 ± 20	1360 ± 30	1450 ± 50
Viscosity (mPa·s)	25.0	10.0 ± 5.0	105.0 ± 10.0

Table 2. Characteristic of Fe-coagulants (Kemipol 2013).

Characteristics	PIX 100	PIX 111	PIX 113
	iron (II) chloride	iron (III) chloride	iron (III) sulfate
Chemical formula	$FeCl_2$	$FeCl_3$	$Fe_2(SO_4)_3$
Form	dark brown solution	dark brown solution	dark brown solution
Chemical composition			
Total iron (Fe) (%)	10.3 ± 0.7	13.4 ± 0.6	11.8 ± 0.4
Fe^{+2}/Fe^{+3} (%)	10.3 ± 0.7	< 0.3	0.4 ± 0.3
Density (g·L⁻¹)	$1250 \div 1280$	$1380 \div 1500$	$1500 \div 1570$
Viscosity (mPa·s)	–	–	60.0

included (depending on the content of the coagulant in the activated sludge) analysis of settleability after 30 (V_{30}) and 60 (V_{60}) minutes, designation of the sludge volume index, velocity sedimentation sludge and measurements of pH and redox potential (by means of a flexi HQ30d HACH). All of the obtained data was analyzed statistically.

Activated sludge in the volume of 1200 mL was introduced to each of the five beakers with a capacity of 2 L and relevant quantities of coagulants were added afterwards, so that the final concentrations of 30.0 (1); 60.0 (2); 90.0 (3) and (4) 120.0 mg·L⁻¹ were attained. The reference level (0) represented activated sludge without the addition of coagulant. The quality of activated sludge used in this study showed a strong sludge bulking and was characterized by a wide range of sludge volume index (Masłoń 2013). The coagulant dose was determined on the basis of literature data (Walczak & Cywińska 2007).

Beakers were placed in a mechanical stirrer with rapid (1 minute) and slow mixing (10 minutes) programmes. The concentrations of the activated sludge were quantified as dry weight of the suspended solid (X), and were determined for each sample. Then, cylinders with the capacity of 1 L were supplemented with a sample of sludge. Sedimentation observation was carried out for the period t = 90 minutes. After 90 minutes, 50 mL samples of wastewater were collected to determine the total suspended solids (Masłoń 2013). The sedimentation velocity (v_s) of activated sludge was calculated from the formula (1) (Giokas et al. 2003, Mines et al. 2001). The height of activated sludge layer h_i was determined on the basis of volume of sedimenting activated sludge flocs (V_i) at time t_i (2).

$$v_s = 7.27 \cdot e^{-(0.0281 + 0.00229 \cdot svl) \cdot x} \left[m \cdot h^{-1} \right] \quad (1)$$

$$h_i = v_i \cdot 0.34 \left[mm \right] \quad (2)$$

3 RESULTS AND DISCUSSION

This study showed different efficiencies in the improvement of activated sludge sedimentation performance, depending on the type and amount of coagulants. It was observed that the settleability of activated sludge with coagulants proceeded with a comparable intensity for 10–15 minutes, and after that time, the progress of the settleability was recorded in a differentiated manner, as evidenced by the more diffused sludge sedimentation curves (Fig. 1).

Sedimentation tests showed that SAX 18 and PIX 113 coagulants have the greatest influence on the improvement of V_{30} and V_{60} settleability over the range of doses used. In turn, the operations of other chemicals depended on their concentration. It was also observed that in some cases, the addition of coagulants, especially PAX 25 and PIX 100, caused a considerable deterioration of sedimentation and outflow of the activated sludge (Fig. 2).

In terms of SVI reduction, the addition of higher concentrations of aluminium and iron salts causes more beneficial changes than small doses. The average initial sludge value index for the control samples was: 230.6 mL·g^{-1} (PAX 16); 238.9 mL·g^{-1} (PAX 25); 225.0 mL·g^{-1} (SAX 18); 237.4 mL·g^{-1} (PIX 100); 232.1 mL·g^{-1} (PIX 111) and 236.6 mL·g^{-1} (PIX 113). Decreases of SVI values are shown in Figure 3. Studies indicate that the SVI decreased significantly as a result of SAX 18 addition. More than a 30% reduction of SVI was noted at a dose of 120 mg SAX 18·L^{-1}. On the other hand, negligible or trace effects were identified in relation to the use of PIX 111 and PAX 25, a 120 mg dose of PAX 25·L^{-1} coagulant caused only a 14% reduction of SVI. A relatively small decline in the SVI was found in the case of the PIX 113, PIX 100 and PAX 16. The maximum reduction of 24.0% was achieved in the sludge volume index when 120 mg of PIX 100·L^{-1} was applied. Walczak & Cywińska (2007) obtained similar results. Application of iron (II) chloride at a concentration of 100 mg·L^{-1} enabled a reduction of about 20% SVI. In turn, while using higher doses of coagulant, i.e. 200 mg·L^{-1} the authors achieved over 50% SVI decrease. However, such significant effect coagulant addition was influenced by the condition of the test sludge (SVI = 124 mL·g^{-1}) (Walczak & Cywińska 2007).

The dosage of coagulants allowed a slight intensification of the activated sludge sedimentation velocity. The average value of v_s was 0.827 m·h^{-1}; 0.849 m·h^{-1}; 0.866 m·h^{-1} and 0.898 m·h^{-1} for a coagulant dose (1), (2), (3), (4), respectively and 0.801 m·h^{-1} in the case of control sample (0). The best results of sedimentation velocity intensification were obtained in the case of SAX 18 and PIX 113, while the weakest—with PIX 100 (Fig. 4). Deterioration of v_s in the case of PIX 100 and PAX 25 application—in relation to the control sample—has also been noted.

The analysis of results of the coagulants effect on the improvement of activated sludge sedimentation properties indicates that a significant role is played by

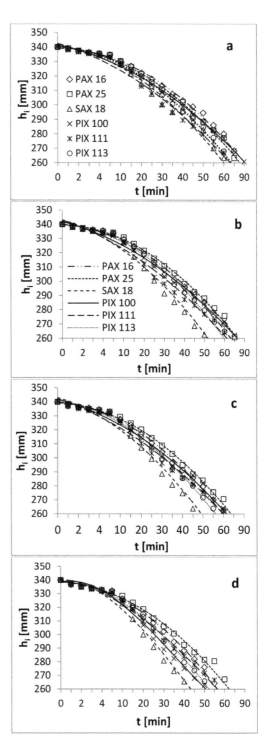

Figure 1. Comparison of sedimentation of activated sludge with the coagulants addition; (a) 30 mg·L^{-1}; (b) 60 mg·L^{-1}; (c) 90 mg·L^{-1}; (d) 120 mg·L^{-1}; (average values).

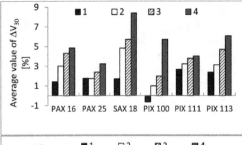

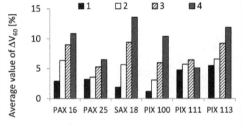

Figure 2. Effect of coagulants on the improvement of activated sludge settleability; 30 mg·L^{-1} (1), 60 mg·L^{-1} (2), 90 mg·L^{-1} (3), 120 mg·L^{-1} (4).

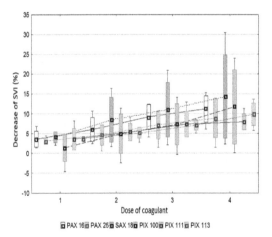

Figure 3. Influence of the type and dose of coagulant to decrease of sludge volume index; 30 mg·L^{-1} (1), 60 mg·L^{-1} (2), 90 mg·L^{-1} (3), 120 mg·L^{-1} (4).

the content of aluminum or iron in samples. Depending on the type and dose of coagulant, the content of aluminum and iron in the activated sludge amounted to 1.89÷11.4 mg Al^{3+}·L^{-1} and 0.66 ÷ 16.08 mg Fe^{3+}·L^{-1} (Tab. 3). Therefore, the content of aluminum and iron in activated sludge in sedimentation tests was: 0.5 ÷ 3.08 mg Al^{3+}·g^{-1} (PAX 16); 0.59 ÷ 3.83 mg Al^{3+}·g^{-1} (SAX 18); 0.44 ÷ 2.45 mg Al^{3+}·g^{-1} and 0.15 ÷ 0.86 mg Al^{3+}·g^{-1} (PAX 25); 0.69 ÷ 4.43 mg Fe^{2+}·g^{-1} (PIX 100); 0.92 ÷ 5.25 mg Fe^{3+}·g^{-1} (PIX 111) and 0.77 ÷ 4.57 mg Fe^{3+}·g^{-1} (PIX 113).

Studies showed that the application of high amounts of iron and aluminum leads to a significant improvement of the activated sludge sedimentation parameters. An important correlation between aluminum content and the SVI change was obtained with respect to the Al-coagulants. For the Fe-coagulants, the correlation between iron content and the SVI decrease was definitely lower (Fig. 5). Despite the technological justification in terms of bulking sludge limitation, using high amount of coagulants or synthetic polymer induces toxic effects of aluminium and iron on activated sludge biocenosis (Kucharski et al. 2000, Juang & Chiou 2007, Naumczyk et al. 2013, Budzińska et al. 2015).

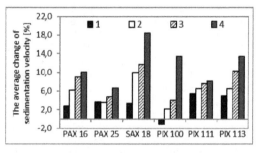

Figure 4. Influence of the coagulant dose on change of sedimentation velocity (average values); 30 mg·L^{-1} (1), 60 mg·L^{-1} (2), 90 mg·L^{-1} (3), 120 mg·L^{-1} (4).

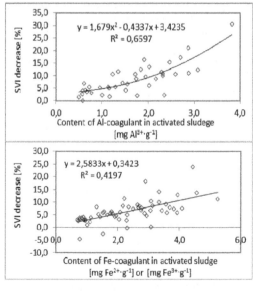

Figure 5. Relationship between aluminium/iron content in activated sludge and decrease of sludge volume index.

Table 3. Dose of aluminum and iron in this study.

Type	Unit	\multicolumn{5}{c}{Dose of coagulant (mg·L⁻¹)}
		0	30	60	90	120
PAX 16	mg Al^{3+}·L⁻¹	0	2.46	4.92	7.38	9.84
PAX 25	mg Al^{3+}·L⁻¹	0	1.89	3.78	5.67	7.56
	mg Fe^{3+}·L⁻¹	0	0.66	1.32	1.98	2.64
SAX 18	mg Al^{3+}·L⁻¹	0	2.85	5.70	8.55	11.40
PIX 100	mg Fe^{2+}·L⁻¹	0	3.09	6.18	9.27	12.36
PIX 111	mg Fe^{3+}·L⁻¹	0	4.02	8.04	12.06	16.08
PIX 113	mg Fe^{3+}·L⁻¹	0	3.54	7.08	10.62	14.16

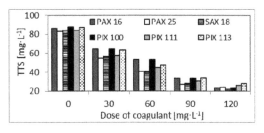

Figure 6. The quality of treated wastewater (average values).

Table 4. Statistical model describing improvement of sedimentation properties.

Type of coagulant	Statistical model
PAX 16	$Y = 3.72x_1 - 0.03x_2 + 32.02$
PAX 25	$Y = 4.68x_1 + 0.01x_2 - 11.29$
SAX 18	$Y = 4.10x_1 - 0.05x_2 + 48.65$
PIX 100	$Y = 3.52x_1 - 0.04x_2 + 37.84$
PIX 111	$Y = 1.44x_1 - 0.02x_2 + 21.01$
PIX 113	$Y = 1.64x_1 - 0.03x_2 + 37.02$

Y—decrease of sludge volume index (%) x_1 – Al^{3+} or Fe^{2+}/Fe^{3+} content in activated sludge (mg Al^{3+}·g⁻¹), (mg Fe^{2+}·g⁻¹) or (mg Fe^{3+}·g⁻¹) x_2 – V$_{30}$ settleability (mL)

For comparative purposes, in a study by Drzewicki (2009) polyaluminum chloride (PAX 18) was used in an amount of 0.63 ÷ 2.13 mg Al^{3+}·g⁻¹ and caused a small improvement in the efficiency of activated sludge sedimentation because the low values of SVI were not achieved. In turn, Geneja (2008) proved that prolonged use of PAX 16 in the amount of 1.4 ÷ 3.2 mg Al^{3+}·g⁻¹ contributes to a significant reduction in the degree of bulking sludge. This revealed significantly lower maximum values of SVI in relation to the extreme level (247 ÷ 317 mL·g⁻¹) observed prior to the application of aluminum salts. The author noted the SVI values ranging from 135 to 155 mL·g⁻¹ during the period of application with polyaluminum chloride (Geneja 2008).

Statistical analysis of the results showed that the degree of improvement of sedimentation properties determined the level of SVI decrease, depending on the content of Al^{3+} or Fe^{2+}/Fe^{3+} in the activated sludge. On the other hand, settleability describes the linear regression model (Tab. 4). On the basis of a statistical model, the highest improvement of sedimentation was found for the PAX 25 and SAX 18.

The tested chemical coagulants also allowed a significant improvement in the quality of treated wastewater (Fig. 6). Studies showed that increasing the dose of a reagent leads to a reduction of total suspended solids in the treated wastewater.

The degree of wastewater turbidity reduction was varied and depended on the coagulants used. The content of total suspended solids in wastewater in the control samples ranged between 54.0 ÷ 105.0 mg·L⁻¹. Applying reagents in an amount of 30, 60 and 90 mg·L⁻¹ made it possible to decrease the Total Suspended Solids (TSS) concentration to a level of 32.0 ÷ 88.0; 24.0 ÷ 65.0 and 20.0 ÷ 44.0 mg TSS·L⁻¹, respectively. In turn, the TSS concentration in the effluent at 120 mg·L⁻¹ coagulant applications remained at the level of <38 mg TSS·L⁻¹. Studies showed that using high doses of SAX 18 allows to maintain the concentration of total suspended solids at 14 mg TSS·L⁻¹ (Fig. 6).

4 CONCLUSIONS

The obtained results showed varied effects of improvement of the activated sludge sedimentation properties, depending on the type and amount of the applied coagulant. Application of coagulants aims to improve settleability, decrease the sludge volume index, intensify the sedimentation velocity and decrease the concentration of TSS in treated wastewater.

Finally, the results of this study indicated that the most effective chemical coagulants in the advancement of sedimentation activated sludge was SAX 18, for which a dose of 120 mg·L⁻¹ resulted in highest improvement of V$_{30}$ and V$_{60}$ settleability at the level exceeding 8% and 13%, respectively. Up to a 30% reduction in SVI was found for this coagulant, and over 18% intensification of sedimentation velocity and a significant improvement in wastewater quality. For other chemical reagents, improvements of parameters of activated sludge were smaller.

The analysis of the obtained results allowed for the designation of effectiveness of the use of chemical coagulants by following a series of:

PIX 111 < PIX 100 < PIX 113 < PAX 16 < PAX 25 < SAX 18

ACKNOWLEDGEMENTS

The research was funded by the project "Subcarpathian scholarship fund for graduate students" in 2011–2013, co-financed by the European Union through the European Social Fund.

REFERENCES

Budzińska, K., Bochenek, M., Traczykowski, A., Szejniuk, B., Pasela, R. & Jurek, A. 2015. Eliminacja bakterii nitkowatych w osadzie czynnym pod wpływem wybranych koagulantów i związków utleniających. *Rocznik Ochrona Środowiska (Annual Set the Environmental Protection)* 17(2): 1569–1582.

Drzewicki, A. 2009. Effect of application of polyaluminium chloride on reducing exploitation problems as the wastewater treatment plant in Olsztyn. *Polish Journal of Natural Sciences* 24: 158–168.

Geneja, M. 2008. Use of aluminum for controlling the filamentous bacteria growth in the activated sludge systems. *Przemysł Chemiczny* 87: 452–455 (in Polish).

Gerardi, M. H. 2002. *Settleability problems and loss solids in the activated sludge process.* John Wiley & Sons, Inc., New Jersey.

Giokas, D., Daigger, G. T., von Sperling, M., Kim, Y. & Paraskevas, P. A. 2003. Comparison and evaluation of empirical zone settling velocity parameters based on sludge volume index using a unified settling characteristics database. *Water Research* 37: 3821–3836.

Guo, J., Peng, Y., Wang, S., Yang, X. & Yuan, Z. 2014. Filamentous and non-filamentous bulking of activated sludge encountered under nutrients limitation or deficiency conditions. *Chemical Engineering Journal* 255: 453–461.

Jenkins, D., Richard, M. G. & Daigger, G. T. 2004. *Manual on the causes and control of activated sludge bulking, foaming, and other solids separation problems.* 3rd Edition, IWA Publishing, London.

Juang, D.F. 2005. Effects of synthetic polymer on the filamentous bacteria in activated sludge. *Bioresource Technology* 96: 31–40.

Juang, D.F. & Chiou, L.J. 2007. Microbial population structures in activated sludge before and after the application of synthetic polymer. *International Journal of Environmental Science and Technology* 4(1): 119–125.

Kemipol Sp. z o.o. 2013. The manufacturer of chemical coagulants. Materials.

Kida, M., Masłoń, A., Tomaszek, J. & Koszelnik, P. 2015. The possibilities of limitation and elimination of activated sludge bulking. 35–49. In J. A. Tomaszek & P. Koszelnik (eds.), *Progress in Environmental Engineering.* CRC Press, Balkema Taylor & Francis Group.

Kucharski, B., Kowalski, J. & Uberna, A. 2000. Effects of sodium aluminate NaAlO$_2$ in wastewater treatment. *Conference proceedings of IX Ogólnopolskie Seminarium PZITS O/Kielce.* Warsaw University of Technology (in Polish)

Mamais, D., Kalaitzi, E. & Andreadakis, A. 2011. Foaming control in activated sludge treatment plants by coagulants addition. *Global NEST J.* 13: 237–245.

Martins, A. M. P., Pagilla, K., Heijnen, J. J. & van Loosdrecht, M. C. M. 2004. Filamentous bulking sludge—a critical review. *Water Research* 38: 793–817.

Masłoń, A. 2013. *The improvement of activated sludge technology by use of powdered keramsite in sequencing batch reactor.* PhD Thesis. Unpublished manuscript.

Mines, R. O., Vilagos, J. L., Echelberger, W. F. & Murphy, R. J. 2001. Conventional and AWT mixed-liquor settling characteristics. *Journal of Environmental Engineering* 127(3): 249–258.

Naidoo, D., Kumari, S. & Bux, F. 2011. Characterization of *Nocardia farnica*, a filamentous bacterium isolated from foaming activated sludge samples. *Water Environment Research* 83(6): 527–531.

Numczyk, J., Marcinowski, P., Bogacki, J. & Wiliński, P. 2013. Oczyszczanie ścieków z przemysłu kosmetycznego za pomocą koagulacji. *Rocznik Ochrona Środowiska (Annual Set the Environmental Protection)* 15: 873–891.

Przybyła, C., Bykowski, J. & Filipiak, J. 2009. Efektywność funkcjonowania gminnych oczyszczalni ścieków. *Rocznik Ochrona Środowiska (Annual Set the Environmental Protection)* 11: 231–239.

Walczak, M. & Cywińska, A. 2007. Application of selected chemical compounds to limit the growth of filamentous bacteria in activated sludge. *Environment Protection Engineering* 33: 221–230.

Environmental Engineering V – Pawłowska & Pawłowski (Eds)
© 2017 Taylor & Francis Group, London, ISBN 978-1-138-03163-0

Bacteria from on-site wastewater treatment facilities as enzymes producers for applications in environmental technologies

Ł. Jałowiecki, J. Chojniak & G. Płaza
Institute for Ecology of Industrial Areas, Unit of Environmental Microbiology, Katowice, Poland

E. Dorgeloh & B. Hegedusova
Development and Assessment Institute in Waste Water Technology at RWTH Aachen University, Aachen, Germany

H. Ejhed
IVL Swedish Environmental Research Institute, Natural Resources and Environmental Effects Stockholm, Sweden

ABSTRACT: The Effluent Treatment Plants (ETPs) represent microbial communities which could contain many bacteria with technological/biotechnological properties. Industries are looking for alternative technologies using the bio-based materials and nature's production processes known as enzymatic processing which could be used to either replace or supplemented conventional technologies in moving toward cleaner production processes. Advancements in technologies have resulted in the need for searching of microorganisms with various enzymatic properties. Up to now, industrially produced enzymes are used in a broad variety of production processes such as pulp and paper production, remediation technologies, wastewater treatment technologies, chemical production, biomasses pretreatment, biofuel production. The need new isolates with better properties is growing. The present work deals with screening of enzymatic properties of bacterial strains isolated from the effluents of onsite wastewater treatment facilities. Most of the investigated bacteria belong to the multi-enzyme producers and constitute a possible source of *potentially technological* important *enzymes use*.

Keywords: wastewater, onsite biological wastewater treatment plants, enzymes, BIOLOG™ system

1 INTRODUCTION

Nowadays, the microbial enzymes are widely applied in multiple industries like food, feed, detergent, tanning, textiles, laundry, cosmetics, chemical, biotechnological and pharmaceutical industries, as well as environmental engineering. These applications account for over 80% of the global market of enzymes (Kumar et al. 2014). Over 500 industrial products are being produced using enzymes (Adrio & Demain 2014). The total market for industrial enzymes reached $3.3 billion in 2010. Food and beverage enzymes comprise the largest segment of the industrial enzymes with revenues of nearly $1.2 billion, while the market for enzymes for technical applications alone making up $1.1 billion. Proteases currently constitute the largest market, although carbohydrases and lipases are the faster growing segments. Estimates of future demand vary, with future markets of $6 billion–$8 billion is expected in 2016 (BBC Research 2015). Enzymes form a crucial pillar of industrial biotechnology and can be used in a wide range of bioindustrial sectors for example in the production of biofuels, washing detergents, food and feed production and in the preparation of bio-based chemicals, enhanced of biological treatment of biomass. The importance of enzymes in the development of the bio-based economy is undisputed. About 150 different bioprocesses use enzymes or whole microbial cell catalysts (Adrio & Demain 2014, Jegannathan & Nielsen 2013, Kumar et al. 2014). The broad variability of enzyme applications are presented in 28 technological processes that savings on raw material energy and/or chemicals obtained by implementation of enzymatic processing.

Many authors support the hypothesis that enzyme technology is a promising means of moving toward cleaner industrial production and offers a great potential for bioeconomy development. Parawira (2011) gives an updated review of the biotechnological advances to improve biogas production by microbial enzymatic hydrolysis of different complex organic matter for converting them into fermentable structures, for example: the improvement of biogas production from lignocellulolytic materials, one of the largest and renewable sources of energy, after pretreatment with cellulases and cellulase-producing microorganisms, lipid-rich wastewaters pretreatment with lipases and lipase-producing microorganisms, and the enzymatic treatment of mixed sludge by added enzymes prior to anaerobic/Aerobic Digestion (AD). Aerobic microbial pretreatment can be carried out with naturally occurring mixed cultures. The concept behind this pretreatment is that some aerobic organisms produce cellulose, hemicellulose and/or lignin degrading enzymes rapidly and in large amounts, and these solubilise the substrate. The arguments in favor of enzymes to pretreat complex biomass are compelling. The high cost of commercial enzyme production still limits application of enzymatic hydrolysis in full-scale biogas production plants, although production of low-cost enzymes from microorganisms or using of microorganisms with specific enzyme activities are addressing this issue.

The paper described by Das et al. (2015) reviews pretreatment techniques including physical, physicochemical, chemical, biological methods respectively. The various effects of pretreatment on organic solid wastes are discussed separately and pretreatment methods have been compared on the basis of cost, efficiency and suitability to substrate. The advantages of biological pretreatment over chemical or thermal pretreatment is that biological pretreatment can take place at low temperature without using chemicals. One disadvantage is that it can be slower than non-biological methods. Developing new methods for the pretreatment of lignocellulosic biomass that increase cellulose conversion and require less energy inputs would make lignocellulosic biofuels more attractive for investors. Pretreatment has the potential to reduce costs and improve biofuels yields. In general, cellulose-degrading, hemicellulose-degrading and starch-degrading enzymes work best between pH 4 and 6 at temperatures from 30 to 50°C, so the pre-acidification step increases the degradation rate by creating an optimal environment for these enzymes. Several batch Anaerobic Digestion (AD) studies have indicated that the addition of enzymes to the first stage of a two-stage anaerobic digestion process leads to slightly higher substrate solubilisation (leading to higher biogas yield), such as with cellulases on grass (Romano et al. 2009) or with cellulosic enzyme cocktails on wheat straw (Quéméneur et al. 2012).

Some studies showed that enzymatic pretreatment in a dedicated vessel leads to higher substrate solubilisation or biogas yields in batch AD tests, for example with pectinase on hemp (Pakarinen et al. 2012), pectinase on switchgrass (Frigon et al. 2012), or various agricultural residues with a cellulolytic enzyme cocktail (Suárez-Quiñones et al. 2012). A study by Warthmann et al. (2012) looked at the effect of 25 different commercially available enzyme preparations including enzyme mixtures marketed to biogas plants as well as pure enzymes normally marketed to other industries. Industries are looking for new microbial strains in order to produce different enzymes to fulfill the current bio-based industries requirements. The discovery of new microbial enzymes through extensive and persistent screening is an open and simple route for biosynthetic processes and, consequently, new ways of solving environmental problems are developed (Ogawa & Shimizu 1999). The need for discovering new isolates with better enzymatic properties is growing. In view of the above, the present work deals with the evaluation of various enzymatic activities of bacterial strains isolated from the effluents of on-site wastewater treatment facilities.

2 METHODS

2.1 Isolation and identification of the strains

The wastewater samples were collected from three different biological facilities of on-site wastewater treatment, named A, B and C. Plants A and B are based on the biofilm technology where microorganisms degrade organic contaminants in the wastewater while being attached to different carrier materials and forming a biofilm. On the other hand, plant C uses a combination of the activated sludge technology and the biofilm technology. Detailed description of the technologies is presented by Jałowiecki et al. (2016). A 1000 ml volume was chosen for every sample. All the samples collected were stored in the sterile Polypropylene (PP) bottles at 4°C for microbiological analysis within 24 h from the sampling. The procedure of the bacteria isolation is described by Jałowiecki et al. (2016). The identification of isolated bacteria was performed by new GEN III MicroPlate™ test panel of the Biolog system. The GEN III Micro-Plates™ enable testing of gram negative and gram positive bacteria in the same test panel. The test panel contains 71 carbon sources and 23 chemical sensitivity assays. GEN III dissects and analyzes the ability of the cell to metabolize all major classes of compounds, in addition to determining other important physiological properties, such as pH, salt and lactic acid tolerance, reducing power, and chemical sensitivity. All the reagents applied were from Biolog, Inc. (Hayward, CA, USA). Fresh overnight cultures of the isolates were tested, as recommended by the manufacturer. The bacterial suspensions were prepared by removing bacterial colonies from the plate surface with a sterile cotton swab and agitating it in 5 ml of 0.85% saline. The bacterial suspension was adjusted in IF-0a to achieve a 90–98% Transmittance (T90) using a Biolog turbidimeter. The amount of suspension equal to 150 μL was dispensed into each well of a Biolog GEN III microplate. The plates were incubated at 26°C in an Omnilog Reader/Incubater (Biolog). After incubation, the phenotypic fingerprint of purple wells is compared to Biolog's extensive species library. If a match was found, a species level identification of the isolates could be made. Out of the isolated bacteria, 30 were chosen for evaluation of enzyme activities. The bacteria were selected on the basis of their different morphological characteristics and identification results.

2.2 API ZYM test

API ZYM (API bioMerieux Ltd.) is the semiquantitative micromethod. Nineteen tests for the presence of the following enzymes were carried out: Alkaline phosphatase (Bph), Esterase (Est), Esterase lipase (Esl), Lipase (Lip), leucine arylamidase (Leu), Valine arylamidase (Val), Cysteine arylamidase (Cys), Trypsin (Try), Chymotrypsin (Chy), Acid phosphatase (Aph), Naphtol-AS-Bi-phosphopydrase (Nap), α-Galactosidase (αGa), β-Galactosidase (βGa), β-Glucuronidase (βGl), α-Glucosidase (αGs), β-Glucosidase (βGs), N-acetyl-β-glucosaminidase (Nac), α-Mannosidase (αMa), and α-Fucosidase (αFu). Bacteria were cultured on SMA (Standard Methods Agar) for 48 h at 30°C. Afterwards, the bacterial colonies were picked from the plates and rinsed in 5 cm^3 of salinity solution, and adjusted to the turbidity of 4 MacFarland standard corresponding to 10^9 bacterial cells per 1 cm^3. In accordance with the manufacturer's instructions, 65 μl of this suspension was inoculated on a plastic strip to cupules containing different substrates. All strips were incubated at 30°C for 24 h; afterwards, API reagents ZYM 1 and ZYM 2 were added to the wells. The obtained results were compared with the colour chart provided by the kit manufacturer.

2.3 Plate assays for screening of enzymes activities

Amylase activity. In order to detect the amylase activity, bacteria were streaked on starch (1%) supplemented nutrient agar plates. The plates were then incubated at 28 ± 1°C for 48 h. After the appearance of the colonies on the starch-agar medium, the culture plates were flooded with 1% Lugol's iodine solution to identify amylase activity.

Cellulase activity. For the detection of cellulose activities, isolates were grown Carboxymethyl Cellulose agar (CMC) containing plates and flooded with 1% Lugol's iodine solution to identify enzyme activity. The presence of cellulolytic activity was confirmed by the yellow color around the colonies.

Inulinase activity. The bacteria were inoculated on MHI agar plates which contained inulin as the carbon source and incubated at 28 ± 1°C for 3 to 5 days. The composition of the MHI medium (modified MH medium) was as described by Li et al. (2011). Inulin was sterilized separately and added before pouring the medium into the plates, with the final concentration of 2%. Following incubation, plates were flooded with Lugol's iodine solution for 3 to 5 min. Subsequently, the staining solution was poured off and the plates were washed twice with a suitable amount of distilled water. A clear halo appeared around the colonies, where the inulin has been degraded.

Laccase activity. Bacteria were screened for the production of lacasses enzyme on Nutrient agar plates supplemented with 0.01% guiacol. The bacterial cultures were streak inoculated on the plate and incubated for 4–7 days at 28°C. An uninoculated plate was used to check contamination. After incubation, the appearance of red-brown color was observed around the laccase positive colonies.

Lipase activity. Prior to the lipolytic activity assay, the bacterial strains were streaked onto blood agar plates and later incubated at 28°C for 24–48 hours. After autoclaving, the Blood agar (BioMerieux) was cooled, then 5% (v/v) defibrinated sheep blood was added. The detected lipolityc enzyme bacterial samples were plated on the Tween-80 agar plates with phenol red indicator and incubated at 28°C for 4 days. The lipase activity will bring about change in coloration from pink to lemon yellow.

Protease activity. The extra-cellular protease production was evaluated with two methods. In the first method the isolates were streaked on enriched Nutrient agar (2% starch) plates and incubated at 28 ± 1°C for 48 h. The appearance of a clear zone around the colony after flooding the plate with 1% Lugol's iodine solution confirmed the presence of proteolytic activity. In the second method, Skimmed Milk Agar (SMA) was used as a screening medium for the protease production. The composition of SMA medium was as follows (g/l): skimmed milk powder – 28; casein enzyme hydrolysate – 5.0; yeast extract – 2.5; dextrose – 1.0; agar – 20.0. The plates were incubated at 28°C for 48 hrs. Development of clear zones around the bacterial strains confirmed the proteolytic activity of the strains.

Urease activity. In order to evaluate the urease activity, the bacteria were streaked on the *Urea Agar Base (UAB) of the following composition (g/l): peptone – 1; glucose – 1; NaCl – 5;* Na$_2$PO$_4$ – 1.2;

NaH$_2$PO$_4$ – 0.8; phenol red – 0.012; agar – 20. Then, bacteria were incubated at 28°C for 48 hrs. The urease activity was changed in coloration to purple.

3 RESULTS AND DISCUSSION

Many enzymes from microbial sources are already being used in various industrial processes, and many industries are currently pursuing enzymatic approaches for developing green cleaner production processes. Enzymes or whole microbial cell catalysts are used in about 150 different industrial processes (Adrio & Demain 2014). Selected microorganisms, including bacteria, fungi and yeasts, have been studied for the biosynthesis of economically viable preparations of various enzymes for commercial applications. In this study, 30 bacterial isolates has been screened for 7 enzymatic activities in the plate assays and 19 enzymes in API ZYM test that were considered to have potential for biotechnological

development. The isolates were examined for their ability to produce industrially important enzymes. Table 1 highlights the enzyme activities: amylase, celullase, inulinase, laccase, lipase, protease and urease of the isolated bacteria. Among the bacteria, 22 were found to be protease producing bacteria, 20 – urease producing bacteria, 17 – inulinase and celullase producing bacteria. Figure 1 presents the examples of enzymatic activities.

Laccase producing *Streptococcus australis and Streptococcus criceti, Bacillus horti and Bacillus alcalophilus, Variovorax paradoxus, Carnobacterium dovergens, Pseudomonas fulva, Flavobacterium tirrenicum and resinovorum* have been reported. Laccase (p-diphenol-dioxygen oxidoreductases, EC 1.10.3.2) is found in a wide range of higher plants and fungi, and recently some bacterial laccases have been also characterized from *Azospirillum lipoferum, Bacillus subtilis, Streptomyces lavendulae, S. cyaneus and Marinomonas mediterranea* (Kunamneni et al. 2007). Laccase belongs to

Table 1. Summary of enzyme activities evaluated by the conventional plate assays.

Strain	Laccase	Inulinase	Cellulase	Urease	Protease	Lipase	Amylase
Streptococcus australis	+	–	+	+	+	–	+
Pseudomonas fluorescens	–	+	+	+	+	–	+
Stenotrophomonas maltophilia	+	–	+	+	+	–	+
Pseudomonas fragi	–	+	–	+	+	–	+
Stenotrophomonas rhizophila	–	–	–	+	+	–	–
Microbacterium flavescens	–	+	+	+	+	–	+
Lactobacillus coryniformis	–	–	+	–	+	–	+
Microbacterium maritypicum	–	+	+	+	+	–	+
Alcaligenes faecalis ss faecalis	–	–	+	+	–	+	–
CDC group II-E A	–	–	–	–	–	+	–
Pseudomonas chlororaphis	–	–	–	+	+	–	–
CDC group II-H	–	+	+	–	+	–	–
Flavobacterium hydatis	–	+	+	+	+	+	+
Flavobacterium resinovorum	+	+	+	–	+	+	+
Mycobacterium brumae	–	+	+	+	+	–	+
Flavobacterium hydatis	–	–	–	+	–	–	–
Bacillus horti	+	–	+	–	–	+	–
Variovorax paradoxus	+	+	+	+	+	+	–
Bacillus alcalophilus	+	–	–	+	+	+	–
Acinetobacter johnsonii	–	+	–	–	–	+	–
Chryseobacterium balustinum	–	+	+	+	+	–	–
Aeromonas bestiarum	–	+	–	+	+	–	+
Enterococcus haemoperoxidus	–	+	–	–	–	–	–
Paenibacillus azoreducens	–	–	–	–	–	–	–
Carnobacterium dovergens	+	–	–	–	–	+	–
Streptococcus criceti	+	+	–	+	+	+	–
Pseudomonas fulva	+	+	+	–	+	+	+
Flavobacterium tirrenicum	+	+	+	+	+	+	+
Sphingobacterium multivorum	–	+	+	+	+	–	–
Serratia marcescens	–	–	–	+	+	–	+

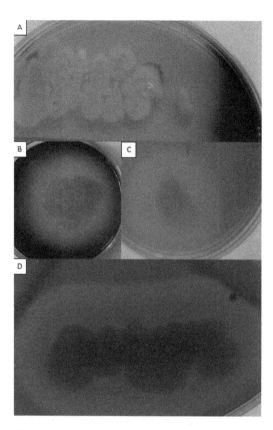

Figure 1. Qualitative plate assay for: protease (a), inulinase (b), cellulase (c), lipase (d).

the complex enzyme system consists of oxidative ligninolytic enzymes: Lignin Peroxidase (LiP) and Manganese Peroxidase (MnP) which are applicable in the hydrolysis of lignocellulosic agricultural residues, particularly for the degradation of the complex and recalcitrant constituent lignin (biomasses). The enzyme application is established in bioremediation, pollution control and in the treatment of industrial effluents containing recalcitrant and hazardous chemicals such as textile dyes, phenols and other xenobiotics. Laccases have mostly been isolated and characterized from plants and fungi. In contrast, little is known about bacterial laccases (Muthukumarasany & Murugan 2014; Pannu & Kapoor 2014). The majority of laccases have been characterized from fungi, particularly from basidiomycetes, white-rot fungi which are efficient lignin degraders. In the last years, laccases have received much attention due to their ability to oxidize both phenolic and non-phenolic compounds. Bacterial enzymes are expected to serve as useful tools for the conversion of lignin into intermediate metabolites. Muthukumarasany & Murugan (2014) describe the distribution of bacterial laccases and their overview in industrial application in various sectors such as paper processing, discoloration of wine, environmental pollutants detoxification, pretreatment technologies, and chemical production from lignin. Lipolitic and amylolitic activities were observed in 12 and 14 strains, respectively. Lipases (triacylglycerol acylhydrolase; EC 3.1.1.3) are important hydrolytic enzymes that catalyze the hydrolysis of Triacylglycerols (TAGs), esterification and transesterification reactions such as acidolysis, alcoholysis and interesterification. Another important enzyme detected in the 17 bacteria was inulinase (2, 1-β-D-fructan fructanohydrolase, EC3.2.1.7) catalyses the hydrolysis of inulin, producing inulo-oligosaccharides, fructose and glucose as main products. Inulinase is one of the key industrial food enzymes which have gained much attention recently. It could simplify the fructose syrup production process, resulting in cost reduction and increasing the amount of products. The traditional production requires at least two enzymes to complete all processes and provide only 75% yield in of a product. However, if inulinase is used in this process, it can reduce the number of necessary steps and achieve a 95% yield (Vijayaraghavan et al. 2009; Chi et al. 2011). In addition, inulinase has been used in the production of fuel ethanol and inulooligosacharides from inulin. A large number of bacteria, such as: *Bacillus subtilis 430A*, *Bacillus polymyxa*, *B. polymyxa* 722, *B. subtilis* 68, *Bacillus polymyxa* MGL21, *Bacillus sp.* LCB41, *B. stearothermophilus* KP1289, *Bacillus cereus* MU-31, *Marinimicrobium sp.* LS-A18, *Pseudomonas mucidolens*, *Arthrobacter ureafaciens*, *Streptococcus salivarius*, *Staphylococcus sp.*, *Clostridium acetobutylicum*, *Clostridium thermoautotrophicum*, *Streptomyces sp.*, yeast (*Kluyveromyces sp.*) and fungal strains (*Penicillium*, *Fusarium* and *Aspergillus*) were described in the literature as inulinase producers (Neagu & Bahtim 2011; Aruna & Hati 2014). Most of the tested bacteria could produce more than one enzyme. Moreover, the production of more than 3 enzymes was detected in 16 bacterial strains. *Pseudomonas fulva* and *Flavobacterium tirrenicum* produced all the tested enzymes. Most of the selected bacteria belong to the multi-enzyme producers and constitute an interesting source of *potentially biotechnological* important *enzymes. It was peculiar that* no tested enzymes were produced by *Paenibacillus azoreducens*.

Enzymatic activities were also evaluated by the API ZYM test. The reactions obtained were divided into negative (grade 0 to 1), and positive reactions (grade 2 to 5). The results obtained with the test are summarized in Table 2. α-galactosidase, β-glucuronidase, α-fucosidase and α-mannosidase (except for *Aeromonas bestiarum*) were absent in all

Table 2. Enzymatic activities of the isolates detected with the API ZYM system.

Enzymes	*Pseudomonas fluorescens*	*Stenotrophomonas maltophilia*	*Stenotrophomonas rhizophila*	*Microbacterium flavescens*	*Alcaligenes faecalis ss faecalis*	*Flavobacterium hydatis*	*Variovorax paradoxus*	*Acinetobacter johnsonii*	*Aeromonas bestiarum*	*Serratia marcescens*
Control	–	–	–	–	–	–	–	–	–	–
Alkaline phosphatase (EC3.1.3.1)	–	+	+	+	+	+	+	+	+	+
Esterase C4 (EC3.1.1)	–	–	+	–	–	+	–	+	–	–
Esterasa lipasa C8 (EC3.1.1)	–	+	+	+	–	+	+	+	–	+
Lipasa C14 (EC3.1.1)	–	+	+	–	–	+	+	+	–	–
Leucine arylamidase (EC3.4.11.1)	+	–	+	+	+	+	+	+	+	–
Valine arylamidase (EC3.4.11.2)	–	–	+	–	–	+	–	–	–	–
Cystine arylamidase (EC3.4.11.3)	–	+	–	–	–	+	–	–	–	–
Trypsin (EC3.4.21.4)	–	+	–	–	–	+	–	–	+	–
α chymotrypsin (EC3.4.21.1)	–	+	–	–	–	+	–	–	–	–
Acid phosphatase (EC3.1.3.2)	+	+	+	–	+	+	+	–	+	+
Naphthol-AS-BI-phosphohydrolase (EC3.2.1.50)	+	+	+	+	–	+	–	–	–	+
α-galactosidase (EC3.2.1.22)	–	–	–	–	–	–	–	–	–	–
β- galactosidase (EC3.2.1.23)	–	–	–	–	–	+	–	–	+	–
β-glucuronidase (EC3.2.1.31)	–	–	–	–	–	–	–	–	–	–
α-glucosidase (EC3.2.1.20)	–	–	–	+	–	+	–	–	–	–
β- glucosidase (EC3.2.1.21)	–	–	+	+	–	+	–	–	+	–
N-acetyl-β-glucosaminidase (EC3.2.1.96)	–	–	–	–	–	+	–	–	+	–
α-mannosidase (EC3.2.1.24)	–	–	–	–	–	–	–	–	+	–
α-fucosidase (EC3.2.1.51)	–	–	–	–	–	–	–	–	–	–

the tested strains. The most frequently produced enzymes by the bacteria were: alkaline and acid phosphatases, esterase lipase C8, leucine arylamidase. Additionally, 15 out of 19 enzyme activities were found in *Flavobacterium hydatis*.

The practical value of the API-ZYM system for identification and characterization of enzymatic profiles has been examined to a limited extent (Hofstad 1980; Muytjens et al. 1984; Boiron & Provost 1990). Recently, this system has been used for the detection of enzymatic activity in environmental samples like soil, compost, water (Tiquia 2002; Mudryk & Podgórska 2006; Wasilkowski et al. 2012; Boluda et al. 2014; Martínez et al. 2016, Stańczyk-Mazanek et al. 2015).

Nigam (2013) presents an overview of the enzymes produced by microorganisms, which have been extensively studied worldwide for their special characteristics of technological importance. Some enzymes (protease, keratinase, amylase, xylanase, ligninase, cellulase, lipase) which possess special characteristics useful in various bio-processes, are characterized by the author, and microorganisms produced the enzymes are presented. In addition, Bhat (2000) included an overview of the biotechnological state-of-the-art for cellulases and related enzymes. Currently, cellulases, hemicellulases and pectinases are widely used in food, brewery and wine, animal feed, textile and laundry, paper and pulp industries, pretreatment technologies as well as in research and development (Parawira 2011; Frigon et al. 2012; Jegannathan & Nielson 2013; Das et al. 2015). In some of these applications, one or two selected components of cellulase, hemicellulase or pectinase are preferred, while other require mixtures of cellulases, hemicellulases and pectinases for maximum benefit (Dąbrowska 2015, Białowiec et al. 2015).

4 CONCLUSION

The ongoing progress and interest in enzymes constitute further success in areas of industrial biocatalysis (bioindustry). The unique properties of enzymes such as high specificity, fast action and biodegradability allow enzyme-assisted processes improve yields and reduced waste generation and gas emission. The enzymatic processes are favorable to the environment compared with the traditional ones.

Bacteria isolated from the onsite wastewater treatment plant are strong candidates for biodiscovery research. Enzymatic activities with good potential for biotechnology were widely distributed among major bacterial strains included in this study. Our findings suggest a potential of these isolates for biotechnological applications. Among the tested bacteria, some appear to be more promising owing to their enzyme producing abilities. Thus, our work signifies the possible applications of these isolates in the biological pretreatment methods.

ACKNOWLEDGEMENTS

This work was done under the project entitled: "Optimization of small wastewater treatment facilities (OPTITREAT)" no. 2112932-1 financed by BONUS EEIG.

REFERENCES

Adrio, J.L. & Demain, A.L. 2014. Microbial enzymes: Tools for biotechnological processes. *Biomolecules* 4(1): 117–139.

Aruna, K. & Hati, A. 2014. Optimization of inulinase production by *Bacillus sp.* B51f isolated from rhizosphere soil of *Agave sisalana. International Journal of Pure & Applied Biosciences* 2(2): 161–176.

BBC Research. Report BIO030. 2015. F: Enzymes in Industrial Applications: Global markets. BBC Research. Wellesley, MA, USA.

Bhat, M.K. 2000. Cellulases and related enzymes in biotechnology. *Biotechnology Advances* 18(2): 355–383.

Białowiec, A., Wiśniewski, D., Pulka, J., Wiśniewski, A. 2015. Wpływ obciążenia hydraulicznego Okresowego Bioreaktora Beztlenowego odciekami na warunki termiczne fermentacji oraz skład wytwarzanego biogazu. *Rocznik Ochrona Środowiska/Annual Set the Environment Protection,* 17(2): 1259–1273.

Boiron, P. & Provost, F. 1990. Enzymatic characterization of *Nocardia* spp. and related bacteria by API ZYM profile. *Mycopathology* 110(1): 51–56.

Boluda, R., Roca-Perez, L., Iranzo, M., Gil C. & Mormeneo, S. 2014. Determination of enzymatic activities using a miniaturized system as a rapid method to assess soil quality. *European Journal of Soil Science* 62(2): 286–294.

Chi, Z.M., Zhang, T., Cao, T.S., Liu, X.Y., Cui, W. & Zhao, C.H. 2011. Biotechnological potential of inulin for bioprocesses. *Bioresource Technology* 102(9): 4295–4303.

Dąbrowska, L. 2015. Wpływ sposobu prowadzenia fermentacji osadów ściekowych na produkcję biogazu. *Rocznik Ochrona Środowiska/Annual Set the Environment Protection,* 17(1): 943–957.

Das, A., Mondal, C. & Roy, S. 2015. Pretreatment methods of ligno-cellulosic biomass: A Review. *Journal of Engineering Science and Technology Review* 8(5): 141–165.

Frigon, J.-C., Mehta, P. & Guiot, S.R. 2012. Impact of mechanical, chemical and enzymatic pre-treatments on the methane yield from the anaerobic digestion of switch grass. *Biomass and Bioenergy* 36(1): 1–11.

Hofstad, T. 1980. Evaluation of the API ZYM system for identification of *Bacteroides* and *Fusobacterium* species. *Medical Microbiology & Immunology* 168(1): 173–177.

Jałowiecki, Ł., Chojniak, J.M., Dorgeloh, E., Hegedusova, B., Ejhed, H., Magnér, J. & Płaza, G.A. 2016. Microbial community profiles in wastewaters from onsite wastewater treatment systems technology. *PlosOne.* doi:10.1371/journal.pone.0147725.

Jegannathan, K.R. & Nielson, P.H. 2013. Environmental assessment of enzyme use in industrial production—a literature review. Journal of Cleaner Production 42 (1): 228–240.

Kumar, V., Singh, D., Sangwan, P. & Gill, P.K. 2014. Global market scenario of industrial enzymes. In Beniwal V. & Shrama A.K. (eds), *Industrial Enzymes, Trends, Scope and Relevance*: 173–196. Nova Science Publishers, Inc. New York.

Kunamneni, A., Ballesteros, A., Plou, F.J & Alcalde, M. 2007. Fungal laccase—a versatile enzyme for biotechnological applications. In Andrea Méndez-Vilas (ed.), *Communicating current research and educational topics and trends in applied microbiology*: 233–245, Formatex.

Li, A.X., Guo, L.Z., Fu, Q. & Lu, W.D. 2011. A simple and rapid plate assay for screening of inulin-degrading microorganisms using Lugol's iodine solution. *African Journal of Biotechnology* 10(12): 9518–9521.

Martínez, D., Molina, M.J., Sánchez, J., Moscatelli, M.C. & Marinari S. 2016. API ZYM assay to evaluate enzyme fingerprinting and microbial functional diversity in relation to soil processes. *Biology & Fertility of Soils* 52(1): 77–89.

Mudryk, Z.J. & Podgórska, B. 2006. Enzymatic activity of bacterial strains isolated from marine beach sediments. *Polish Journal of Environmental Studies* 15(3): 441–448.

Muthukumarasamy, N.P. & Murugan, S. 2014. Production, purification and application of bacterial laccase: A review. *Biotechnology* 13(2): 196–205.

Muytjens, H.L., van der Ros-van deRepe, J. & van Druten, H.A.M. 1984. Enzymatic profiles of *Enterobacter sakazakii* and related species with special reference to the α-glucosidase reaction and reproducility of the test system. *Journal of Clinical Microbiology* 20(4): 684–686.

Neagu, C. & Bahrim, G. 2011. Inulinases–A versatile tool for biotechnology. *Innovative Romanian Food Biotechnology*. 9(1): 1–11.

Nigam, P.S. 2013. Microbial enzymes with special characteristics for biotechnological applications. *Biomolecules* 3(6): 597–611.

Ogawa, J. & Shimizu, S. 1999. Microbial enzymes: new industrial applications from traditional screening methods. *TIBTECH* 17(1): 13–23.

Pakarinen, A., Zhang, J., Brock, T., Maijala, P., Viikari, L. 2012. Enzymatic accessibility of fiber hempis enhanced by enzymatic or chemical removal of pectin. *Bioresource Technology* 107(1): 275–281.

Pannu, J.S. & Kapoor, R.K. 2014. Microbial laccases: A mini-review on their production, purification and applications. *International Journal of Pharmaceutical Archives* 3(12): 528–536.

Parawira, W. 2011. Enzyme research and applications in biotechnological intensification of biogas production. *Critical Reviews in Biotechnology* 4(1) 1–15.

Quemeneur, M., Bittel, M., Trably, E., Dumas, C., Fourage, L., Ravot, G., Steyer, J.-P. & Carrere, H. 2012. Effect of enzyme addition on fermentative hydrogen production from wheat straw. *International Journal of Hydrogen Energy* 37(11): 10639–10647.

Romano, R.T., Zhang, R., Teter, S. & McGarvey, J.A. 2009. The effect of enzyme addition on anaerobic digestion of Jose Tall Wheat Grass. *Bioresource Technology* 100(5): 4564–4571.

Stańczyk-Mazanek, E., Kępa, U. & Stępniak, L., 2015. Drug-Resistant Bacteria in Soils Fertilized with Sewage Sludge. *Rocznik Ochrona Środowiska/Annual Set the Environment Protection,* 17(1): 125–142.

Suarez Quiñones, T., Plöchl, M., Budde, J. & Heiermann, M. 2012. Results of batch anaerobic digestion test–effect of enzyme addition. *Agricultural Engineering International: CIGR Journal* 14(1): 38–50.

Tiquia, S.M. 2002. Evolution of extracellular enzyme activities during manure composting. *Journal of Applied Microbiology* 92(4): 764–775.

Vijayaraghavan, K. et al. 2009. Trends in inulinase production—a review. *Critical Review of Biotechnology* 29(1): 67–77.

Warthmann, R., Baum, S. & Baier, U. 2012. Massnahmen zur Optimier ungder Vergärung durch Vorbehandlung, Prozess-undVerfahrenstechnol und Hilfsstoffe. End of project report, Bundesamt für Energie BFE, Switzerland No. 103312/154366.

Wasilkowski, D., Krzyżak, J., Płaza, G. & Mrozik, A. 2012. Ocena aktywności enzymatycznej w glebie skażonej metalami ciężkimi i poddanej rekultywacji techniką fitostabilizacji wspomaganej przy zastosowaniu testu API®ZYM. W: T.M. Traczewska (red.), *Interdyscyplinarne Zagadnienia w Inżynierii i Ochronie Środowiska*: 509–514, Oficyna Wydawnicza Politechniki Wrocławskiej, Wrocław.

Environmental Engineering V – Pawłowska & Pawłowski (Eds)
© 2017 Taylor & Francis Group, London, ISBN 978-1-138-03163-0

Evaluation of the possibilities of water and sewage sludge disposal

J. Górka & M. Cimochowicz-Rybicka
Environmental Engineering Department, Institute of Water Supply and Environmental Protection,
Cracow University of Technology, Cracow, Poland

ABSTRACT: The study focuses on a possible utilization of water sludge during anaerobic digestion of sewage sludge. The water sludge for the study was produced during coagulation, ozonation, and backwashing of rapid filters and collected at a water treatment plant with a capacity of about $100\,000$ m$^3 \cdot$ d^{-1}. In addition, the analysis of the sludge structure was conducted, which showed the presence of organisms (algae) that could have an impact on the efficiency of the digestion process. The proposed methodology of the research, as well as its preliminary results showed that co-digestion of water and sewage sludge increases the biogas production rate. The biogas production ranged from 0.16 to 0.31 m$^3 \cdot$ kg^{-1} VSS. Apart from the increased gas production, it was been found that water sludge has a negative effect on sewage sludge dewatering characteristic. It is hoped that the results of this research study will help develop a future methodology related to the sludge control and disposal.

Keywords: water sludge, sewage sludge, anaerobic co-digestion, disintegration

1 INTRODUCTION

During water treatment a large amount of sludge is generated; it constitutes 2 to 5% of the volume of treated water (Kyncl et al. 2012). Its reuse is an essential element of planning and management of water resources in order to reduce and limit the amount of waste generated during the water treatment process. Modern understanding of the idea of sludge management at water treatment plants is seen as all activities that allow for its (Krajewski et al. 2010, Szerzyna 2013):

- placement in the natural environment or disposal in a safe way, according to the local regulations,
- landfilling,
- industrial or economic use,
- reuse at the wastewater treatment plants (recovery of chemicals or recycle of supernatant),
- H$_2$S binding,
- disposal with sewage sludge or solid waste (joined composting at a fixed mass ratio).

The ultimate methods of sewage sludge disposal result mainly from the physical and chemical properties of sludge (Tables 1 and 2). These are determined mostly by the quality and type of treated water. It should also be noted that changes in a surface waters composition result in formation of precipitates of different quantity and quality. Clay minerals, clay and sand particles, colloidal and dissolved organic matter, as well as plant and animal residues are the main compounds removed from the surface water and found in the sludge (Verrelli et al. 2009). The chemical composition of sludge depends primarily on (Leszczyńska et al. 2009, Balcerzak & Rybicki 2011):

- type and concentration of chemical compounds present in raw water,
- types and doses of reagents used in the water treatment process,
- treatment efficiency, determined by the removal of particular compounds,
- type and concentration of intermediate and by-products generated during the treatment process.

Ahmad & others (2016) discussed the methods of water sludge disposal—see Table 3. The interesting solution is the use of water sludge for conditioning and dewatering of sewage sludge. The studies have confirmed that the use of water sludge with aluminum improves the sludge dewatering characteristic and sludge density. Aluminum in sewage sludge acts as a "skeleton" and makes sludge less compressible, which results in its more efficient dewatering. Additionally, it was proven that the use of water sludge improves the removal of phosphorus from the supernatant (Lai et al. 2004). The mixture of sewage sludge and water sludge (at a volumetric ratio 2:1) can reduce the phosphorus load in the supernatant by 99% (Yang et al. 2007). In turn, Płonka & Barbusiński (2010) used water sludge from the water treatment plant in the co-digestion process of sewage sludge.

Table 1. Physicochemical composition of alum and ferric sludge.

Para-meter	Units	Alum sludge	Ferric sludge	References
pH	–	6.5 ± 0.3	7.0 ± 1.3	(Ahmad et al. 2016)
Total solids	$mg \cdot dm^{-3}$	2500.0 ± 52.3	2132.0 ± 50.7	
Al	$mg \cdot kg^{-1}$	118.7 ± 24.3	61.4 ± 35.9	
Fe	$mg \cdot kg^{-1}$	37.0 ± 19.7	220.9 ± 32.2	
Ca	$mg \cdot kg^{-1}$	10.36 ± 4.30	nd	
Mg	$mg \cdot kg^{-1}$	2407 ± 572	nd	
Na	$mg \cdot kg^{-1}$	355 ± 142	nd	
K	$mg \cdot kg^{-1}$	3547 ± 582	nd	
S	$mg \cdot kg^{-1}$	6763 ± 2955	nd	
Fe	$mg \cdot kg^{-1}$	37.0 ± 19.7	220.9 ± 32.2	
Mn	$mg \cdot kg^{-1}$	2998 ± 1122	1088 ± 178	
Zn	$mg \cdot kg^{-1}$	98 ± 31	36 ± 4	
Cu	$mg \cdot kg^{-1}$	624 ± 581	46 ± 12	
Ni	$mg \cdot kg^{-1}$	28 ± 10	64 ± 14	
Pb	$mg \cdot kg^{-1}$	22 ± 12	47 ± 1	
Cr	$mg \cdot kg^{-1}$	20 ± 7	38 ± 4	
Cd	$mg \cdot kg^{-1}$	0.1 ± 0.02	nd	(Babatunde et al. 2007)
Hg	$mg \cdot kg^{-1}$	0.5	nd	
As	$mg \cdot kg^{-1}$ of total solids	9.2 ± 32.0	nd	(Płonka et al. 2012)
Ba	$mg \cdot kg^{-1}$	19.0 ± 323.0	nd	(Leszczyńska et al. 2009)
Sr	$mg \cdot kg^{-1}$	20.2 ± 160	nd	
Cl	$mg \cdot kg^{-1}$	16000 ± 16200	nd	(Yang et al. 2006)

Table 2. The impact of the treatment process and type of additives on the water sludge properties (Kowal et al. 2007).

Type of sludge	Dewaterability	Value of hydration at which the sludge loses the liquidity [%]
Sludge after coagulation process containing silica, low organic matter, low Fe and Al.	Good	50–60
Sludge after coagulation process rich in humus, a large amount of Fe and Al. Depending on the type of coagulant.	Very low	70–80
Sludge after coagulation process containing a large amount of Fe and Al, low organic matter and silica.	Low	60–65
Sludge after coagulation process containing dusty carbon.	Good	60–65
Sludge from iron and manganese removal of groundwater.	Moderate low if $[Fe] > [Mn]$	65–70
	Good if $[Fe] < [Mn]$	50–60
Sludge from iron and manganese removal of water containing humus and organic matter.	Low	65–80
Sludge after carbon removal with high content of $CaCO_3$	More than good	30–40
Sludge after lime coagulation process with high content of $Mg(OH)_2$	Low	60–75

They proved that increasing participation of water sludge in the sludge mixture causes the decrease of the biogas production, which is the result of presence aluminum cations that delay the growth of anaerobic bacteria. However, this research confirms the beneficial role of water sludge in the sewage sludge dewatering.

The authors tried to assess the possibility of co-digestion of water and sewage sludge. A special attention was paid to the impact of the water sludge on sludge dewatering characteristic, after digestion. Methane fermentation is a very important process at sewage treatment plants. It comprises a series of biochemical reactions, involving microorganisms and results in a biogas production. It is considered one of the best methods used in waste management, because it utilizes valuable (in terms of energy) organic materials present in waste (Heidrich 1999, Bień 2008, Rybicki 2014).

2 MATERIALS AND METHODS

Studies were conducted at the Environmental Engineering Department of the Cracow University of Technology. The experiments were carried out on the sewage sludge from wastewater treatment plant and water sludge from water treatment plant, serving urban area in Southern Poland. Samples have been taken directly from the process treatment lines at the facility. The basis of energy research included the respirometric tests, which allow an analysis of the quantity and quality of the fermentation gas (Cimochowicz-Rybicka 2013). In the respirometer (Fig. 1) used in anaerobic process, the

Table 3. The reuseability of water sludge (Ahmad et al. 2016).

Reuse options	Advantages	Disadvantages
Coagulant recovery and reuse	Cheap	Complicated and laborious recovery process
	Reduction in sludge volume and disposal cost	Limited purity and possibility of contamination
As coagulant in wastewater treatment	Excellent removal efficiency	Restricted use
	Enhanced removal of turbidity, COD and TSS	Haulage cost
As adsorbent for contaminants and heavy metals wastewater	Adsorbs heavy metals such as Cd, Cr, As etc.	Extent of adsorption is under research
	Suitable for P removal	Possible release of substances
As substrate in constructed wetlands	Substitute to wetland media	Haulage distances
	Improved P and N removal in constructed wetlands	Still under investigation
In sewage sludge dewatering	Improved settling of sludge	Haulage cost
	Improves sewage sludge dewaterability	Disposal of mixed sludge
	P removal from the reject water	
In cement production	Chemical component similar to cement clay	Retardation of setting time
	Possess cementitious property	Inclusion of deleterious components
		Affect the mechanical properties of final product
In manufacturing lightweight aggregates	Comparable to commercial lightweight aggregates	Requires high sintering temperature
	Best suitable for non-structural concrete	
In brick and ceramic production	Raw material for clay bricks	Reduction in mechanical and tensile strength with increase in sludge production
	Reduced shrinkage	
	Rich colour of the final product	
	Organic content may present some energy savings	
As raw material for concrete and mortar	Substitute to sand and cementitious material	Greater water absorption
	Safe for non-load bearing members	Reduction in compressive strength at higher percentage
In agricultural practice and other land based uses	Improved soil aggregation	Risk of metal accumulation
	Increased moisture holding capacity and water permeability	Little fertilizer value
	Effective P reduction at low cost	Potential fixation of available P

Figure 1. Respirometer.

gas production is measured by designating of fermentation gas volume. An example of writing the data read by the device is shown in the Figure 3.

The second stage of research was to determine the sludge dewaterability by Capillary Suction Time measurements (CST). Measurements have been made by means of CST gauge of Envolab company (Fig. 2). CST defines the speed of liquid donation by the tested sludge. When the CST is smaller, the tested sludge easier (faster) releases liquid. The main advantage of the CST test is its short time and a relatively simple device used to

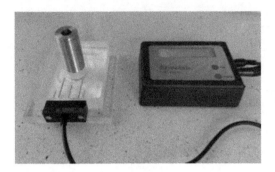

Figure 2. CST device.

Figure 3. Data chart for respirometric tests AER-208.

conduct measurements. It should be noted that the CST to a certain extent depends on sludge concentration and the equipment used.

Measurements were carried out over 30 days at mesophilic (35°C) conditions in three separate runs. Configurations of samples were selected based on the dry organic solids content and included excess, digested and water sludge. Co-substrate concentrations ranged from 4.5 to 8.5 gVSS · dm^{-3}. In order to ensure the appropriate process conditions, the pH of samples was adjusted to pH = 7.0 with NaOH. Before measurements, the sample were purged with technical nitrogen. The laboratory analysis included:

- dry solids, volatile dry solids,
- COD,
- alkalinity,
- pH,
- ammonia nitrogen,
- total phosphorus.

The impact of the water sludge on methane fermentation of sewage sludge was evaluated on the basis of:

- digestion and removal of organic compounds,
- process stability (based on pH and alkalinity),
- biogas yield and energy efficiency, as a function of a water sludge dose,
- dewatering characteristic.

3 RESULT AND DISCUSSION

During the tests, the mixtures of following types of sludge were analyzed: digested sludge (P), excess sludge (N) and water sludge (W). The samples were collected directly from sludge drying beds, in November 2015. Water sludge was produced during coagulation (PAX 19, XL 10, PAX 16), ozonation, dosing of powdered activated carbon and sand filters backwashing. The plant flow (water treatment plant capacity) is approx. 100 000 m^3 · d^{-1}, so a proper handling of by-products (e.g. sludge) is a serious task to solve, especially in such a large object.

It is worth noting that the structure of the sludge stored in the sludge drying beds changes over time (see Fig. 4) due to the humification process. Additionally, some algae species may be observed growing on sludge surfaces (Fig. 4c). Algae are commonly used in the co-digestion of sewage sludge, because these organisms besides their energetic properties can also absorb nutrients (Umble 2010). Therefore, there has been a concept of using these organisms in both digestion and sewage treatment. The sewage sludge, due to its high carbon content and various active microorganisms, produces in combination with the algal biomass a larger volume of fermentation gas during digestion (Werle 2014). The analyzed water

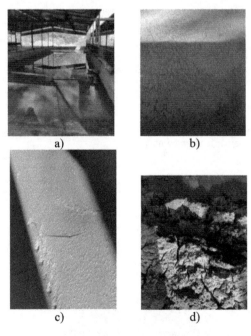

Figure 4. a—Sludge plot, b—the 'fresh' sludge structure, c—the algal structure, d—the structure of dehydrated sludge.

sludge was also examined with a microscope. During the observations, numerous species of micro and macro algae were identified, which may affect the digestion process (Fig. 5).

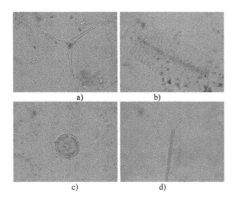

Figure 5. Microscopic photos of algae in water sludge: a—*Staurastum*, b—*Fragilaria*, c—*Cyclotella*, d—*Scytonema*.

The impact of the water sludge on methane fermentation of sewage sludge (digested) was examined for two mixtures configurations, i.e. 1:1 and 1:0.5. The second series of measurements has been further expanded to include excess sludge. The control samples were carried out on digested sludge. Introduction of sewage sludge as a substrate resulted in changes in the concentrations of all the parameters in the reactor input (Table 4). In series D, the concentrations of dissolved organic matter (sCOD) amounted to 249 mgO$_2$ · dm^{-3} (D2) and 489 mg O$_2$ · dm^{-3} (D3), respectively, and were 40% and 174% higher than for the digested sludge. On the other hand, in series P the values were: 3562 mgO$_2$ · dm^{-3} (P2), 55 mgO$_2$ · dm^{-3} and 120 mgO$_2$ · dm^{-3} (P3 and P4), i.e. samples with addition of excess sludge had lower dissolved organic content than the digested sludge (60–80%).

After 500 hours of incubation, the concentration of SCOD was greatly reduced (Table 5); the observed reduction was 50–70%, when compared to the mixtures tested in series D and P. The highest

Table 4. Characteristics of the sludge samples before fermentation process.

		Series D			Series P			
		D1	D2	D3	P1	P2	P3	P4
		P	P:W	P:W	P	P:W	P:N:W	P:N:W
Parameter		–	1:1	1:0,5	–	1:1	1:1:1	1:1:0,5
pH	–	7.47	7.75	7.74	7.27	7.30	7.24	7.13
Alkalinity	[mgCaCO$_3$ · dm^{-3}]	575	540	580	580	620	680	635
TS	[gTS · dm^{-3}]	9.02	8.78	9.09	6.42	16.03	15.76	10.07
VSS	[gVSS · dm^{-3}]	5.54	5.04	5.07	4.43	6.92	8.26	5.83
tCOD	[mgO$_2$ · dm^{-3}]	3111	4444	4800	6359	4977	3041	6544
sCOD	[mgO$_2$ · dm^{-3}]	178	249	489	290	356	55	120
CST	[s]	676	521	457	536	453	186	256

Table 5. Characteristics of the sludge samples after fermentation process.

		Series D			Series P			
		D1	D2	D3	P1	P2	P3	P4
		P	P:W	P:W	P	P:W	P:N:W	P:N:W
Parameter		–	1:1	1:0,5	–	1:1	1:1:1	1:1:0,5
pH	–	7.40	7,45	7,42	6,95	6,98	6,76	6,79
Alkalinity	[mgCaCO$_3$ · dm^{-3}]	925	1025	888	835	970	970	960
TS	[gTS · dm^{-3}]	7.04	8.34	8.69	5.55	14.91	11.14	7.64
VSS	[gVSS · dm^{-3}]	4.11	4.31	4.32	3.44	6.18	5.33	3.97
tCOD	[mgO$_2$ · dm^{-3}]	1067	1333	1422	2022	1450	1551	3556
sCOD	[mgO$_2$ · dm^{-3}]	54	80	356	36	71	18	107
CST	[s]	463	606	545	323	509	496	417

drop was observed in the samples of the digested sewage sludge and water sludge, in both series. Despite the COD reduction, co-digestion of sewage and water sludge did not result in lower concentrations of dry volatile solids, probably due to a low content of easily biodegradable organic fraction in the sludge. The highest loss of dry volatile solids was observed in the mixtures of water, digested and excess sludge—approx. 30–35% – while for the digested sludge it was only 22–26%.

Figures 6a and 6b illustrate the fermentation gas production in each series. Biogas production during co-digestion with the water sludge was higher than for digested sludge alone. In series D, the gas production was: 0.24 m^3 · kg^{-1}VSS and 0.31 m^3 · kg^{-1}VSS for samples D2 and D3, respectively; these figures were 84% and 138% higher, respectively, than the ones obtained for the digested sludge (0.13 m^3 · kg^{-1}VSS). Similar results were reported in series P. Comparing the digestion results, an increase in efficiency was observed: for the sample P2 by 50% (0.21 m^3 · kg^{-1}VSS) and for the samples P3 and P4 by 14% (0.16 m^3 · kg^{-1}VSS) and 71% (0.24 m^3 · kg^{-1}VSS), when compared to the digested sludge (0.14 m^3 · kg^{-1}VSS). It should be noted that the samples with higher water sludge content digested poorly comparing to the samples mixed at a lower ratios.

Unfortunately, it was also found that co-digestion of water and sewage sludge had a negative impact on the sludge dewatering characteristic. The analysis showed that the capillary suction time for the mixed digested samples was higher than before digestion (except for the digested sludge). This is probably caused by the coagulant in water sludge, which takes a gel form and has a negative effect on a concentration of sludge particles. Therefore, the breakdown of these particles through various methods of disintegration becomes an important issue.

The preliminary studies showed some changes in the CST and COD of a water sludge sample after its thermal and ultrasonic disintegration, before digestion—see Table 6. As can be seen, disintegration had a positive impact on the sludge COD. Especially during a hybrid disintegration a more efficient utilization of organic matter during fermentation was noted. However, the more complicated method of disintegration, the greater CST. While studying the CST of sludge after ultrasonic disintegration, the highest value was found for 64 kW/m^2 and time equal to 3 minutes—resulting in an approximately 40% increase. In turn, the smallest value was obtained for the intensity 64 kW/m^2, time equaling 7 minutes. The lower CST values were observed only

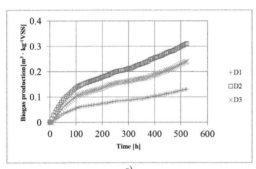

a)

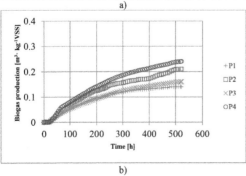

b)

Figure 6. The biogas production: a—series D, b—series P.

Table 6. The value of sCOD and Capillary Suction Time (CST) for water sludge after disintegration.

Series 1 – ultrasonic disintegration

Parameters of process		sCOD	CST
24 kW/m^2	3 min	88.00 mgO$_2$/dm^3	135.93 s
	7 min	240.53 mgO$_2$/dm^3	139.66 s
64 kW/m^2	3 min	111.47 mgO$_2$/dm^3	171.22 s
	7 min	205.33 mgO$_2$/dm^3	124.69 s
Sludge without disintegration	–	99.73 mgO$_2$/dm^3	122.99 s

Series 2—thermal disintegration

Parameters of process		sCOD	CST
70°C	30 min	130.47 mg O$_2$/dm^3	116.40 s
	45 min	204.23 mg O$_2$/dm^3	119.01 s
	60 min	204.23 mg O$_2$/dm^3	115.59 s
Sludge without disintegration	–	97.85 mg O$_2$/dm^3	125.23 s

Series 3—hybrid disintegration

Parameters of process	sCOD	CST
70°C (60 min)	88.00 mgO$_2$/dm^3	115.60 s
24 kW/m^2 (7 min)	105.60 mgO$_2$/dm^3	143.27 s
24 kW/m^2 (7 min) + 70°C (60 min)	162.80 mgO$_2$/dm^3	145.38 s
Sludge without disintegration	44.00 mgO$_2$/dm^3	123.31 s

during the thermal disintegration –5–8% reduction. Even the combination of these two methods (pasteurization + ultrasound) increases this parameter by 18%. These measurements will serve as an introduction to further research on intensification of co-digestion and on possible improvements of dewatering characteristic of the sludge mixture.

4 CONCLUSIONS

1. The need for proper management of water sludge initiated the search for rational methods of its disposal. Reuse is an essential element of planning and management of water resources in order to reduce and limit the amount of waste generated during the water treatment process.
2. The analyzed water sludge contains algal organisms, which have good energy properties. Their number depends primarily on the water sludge retention time at the sludge drying beds. The organisms can affect the efficiency of co-digestion of sewage and water sludge.
3. The data analysis showed that co-digestion of water and sewage sludge increases the production of biogas, and thereby increases the energetic potential of thise poorly biodegradable waste. Biogas production in the co-digestion process ranges from 0.16 to 0.31 $m^3 \cdot kg^{-1}$ VSS, while for the digested sludge it was 0.13 $m^3 kg^{-1} \cdot$ VSS. At the same time, utilization of water sludge as a substrate has a negative impact on sludge dewatering characteristic, and therefore on the economy of the plant. Sometime, co-digestion may not be profitable.
4. The longer capillary suction time is related to the structure of the sludge fractions. The use of disintegration can change the particle structure making the sludge it more prone to dewatering processes. Thermal disintegration reduces the capillary suction time.

REFERENCES

Ahmad, T., Ahmad, K. & Alam, M. 2016. Sustainable management of water treatment sludge through 3R concept. *Journal of Cleaner Production*. 1–13.

Babatunde, A.O. & Zhao, Y.Q. 2007. Constructive approaches towards water treatment works sludge management: a review of beneficial reuses. *Crit. Rev. Environ. Sci. Technol.* 37: 129–164.

Balcerzak, W. & Rybicki, S.M. 2011. Assessment of Water Eutrophication Risk Exemplified by the Swinna Poreba Dam Reservoir *Ochrona Środowiska*, 33(4): 67–69.

Bień, J. 2008. *Osady ściekowe. Teoria i praktyka.* Częstochowa: brak nazwiska, 2008.

Chiang, K.Y., Chou, P.H., Hua, C.R. & Chien, K.L., Cheesman C. 2009. Lightweight bricks manufactured from water treatment sludge and rice husks. *J. Hazard. Mater.* 171: 76–82.

Cimochowicz-Rybicka, M. 2013. *Aktywność metanogenna osadów ściekowych poddanych beztlenowej stabilizacji z zastosowaniem dezintegracji ultradźwiękowej.* Kraków: Poliechnika Krakowska, 2013.

Heidrich, Z. 1999. Stabilizacja beztlenowa osadów ściekowych. *Seria: Wodociągi i Kanalizacja.* Warszawa: brak nazwiska, 1999.

Kowal, A. & Świderska-Bróż, M. 2007. *Oczyszczanie wody.* Warszawa: PWN, 2007.

Krajewski, P. & Sozański, M. 2010. Możliwości i metody wykorzystania osadów z uzdatniania wody. *Technologia Wody.* 5: 30–36.

Kyncl, M., Cihalova, S., Jurkova, M. & Langarova, S. 2012. Unieszkodliwianie i zagospodarowanie osadów z uzdatniania wody. *Inżynieria Mineralna.* 2012, pp. 11–20.

Lai, J.Y. & Liu, J.C. 2004. Co-conditioning and dewatering of alum sludge and waste activated sludge. *Water Science and Technology.* 50(9): 41–48.

Leszczyńska, M. & Sozański, M. 2009. Szkodliwość i toksyczność osadów i popłuczyn z procesu uzdatniania wody. *Ochrona Środowiska i Zasobów Naturalnych.* 40: 575–585.

Płonka, I. & Barbusiński, K. 2010. Preliminary research into the digestion of post-coagulation sludge. *Environment Protection Engineering.* 36(3): 59–67.

Płonka, I., Pieczykolan, B., Amalio-Kosel, M. & Loska, K. 2012. Metale ciężkie w osadach powstających przy uzdatnianiu wody. *Proceedings of ECOpole.* 1(6): 337–342.

Rybicki, S.M. 2014. Role of Primary Sludge Hydrolysis in Energy Recovery from Municipal Wastewater Sludge, *Pol. J. Env. Studies*, 23(3): 1033–1037.

Szerzyna S. 2013. Możliwości wykorzystania osadów powstających podczas oczyszczania wody. *Eko-Dok.* 609–617.

Umble, D. 2010. How to utilize nutrients to your adventage. *Water Online The Magazine.* 44–45.

Verrelli, D.I., Dixo, n D.R. & Scales, P.J. 2009. Effect of coagulation conditions on the dewatering properties of sludges produced in drinking water sludge. *Colloids and Surfaces A: Physicochemical and Engineering Aspects.* 2009, 348: 14–23.

Werle, S. *Pozyskiwania paliwa gazowego z biomasy niekonwencjonalnej w procesie zgazowania*

Yang, Y., Tomlinson, D., Kennedy, S. & Zhao, Y.Q. 2006. Dewatered alum sludge: a potential adsorbent for phosphorus removal. *Water Science & Technology.* 54(5): 207–213.

Yang, Y., Zhao, Y.Q., Babatunde, A.O. & Kearney, P. 2007. Co-conditioning of the anaerobic digested sludge of municipal wastewater treatment plant with alum sludge: benefit of phosphoous reduction in reject water. *Water Environmental.* 79(13): 2468–2476.

Environmental Engineering V – Pawłowska & Pawłowski (Eds)
© 2017 Taylor & Francis Group, London, ISBN 978-1-138-03163-0

Application of iron sludge from water purification plant for pretreatment of reject water from sewage sludge treatment

K. Piaskowski & R. Świderska-Dąbrowska

Faculty of Civil Engineering, Environmental and Geodetic Sciences, Koszalin University of Technology, Koszalin, Poland

ABSTRACT: In the domestic wastewater treatment, the processes aiming to control and reduce the pollution of recirculated reject water generated during treatment of sewage sludge are more and more frequently considered. The increasing costs of applying chemical methods necessitate using various types of waste containing aluminum and iron ions in wastewater-sludge treatment. Their application for the removal of contaminants from wastewater or reject water helps to reduce the operating costs of sewage treatment plants, and thus makes it a practical method of disposal of certain waste. The aim of the study presented in the paper was to determine the effectiveness of using Drinking-Water Treatment Sludge (DWTS) from two underground water purification plants for pretreatment of reject water from anaerobic sludge digestion. Conducted examinations indicate that it is possible to use DWTS for reduction of contaminants concentration in reject water, especially for the removal of orthophosphates.

Keywords: DWTS, ferric sludge, iron hydroxides, reject water

1 INTRODUCTION

Technologies used in processing and disposal of sewage sludge, generate highly contaminated reject water. Increase in the efficiency of sewage sludge dewatering and the use of effective methods of its disintegration, prior to methane digestion, additionally increases the contaminants load. It can reach the level of 20–30% of the total nitrogen and phosphorus load flowing into wastewater treatment plants along with raw sewage. Ammonia nitrogen concentration in the reject water may be as high as 500–1500 mgN/l, which, at unfavourable low value of C/N ratio, causes problems during nitrification and denitrification. High phosphorus concentration (60–160 mgP/l) is also the result of its release by Polyphosphate-Accumulating Organisms (PAOs) during fermentation (Penget et al. 2012; Battistoniet et al. 2000).

Treatment of reject water may have an important impact on the efficiency of wastewater treatment plants, due to the substantial quantitative and qualitative variability and many fold higher concentrations of contaminants than analysed in raw sewage. Recirculation of the reject water into raw sewage adversely affects the operation of biological reactors. Energy demand for oxygenation of aeration chambers increases and expensive chemical aiding of the process is required. Current possibilities of treatment or at least pre-treatment of the reject water include many methods and processes, which are more and more often used prior to their recirculation into the main wastewater treatment.

The main types of methods currently used for treatment of reject water are based on biological and physico-chemical (usually chemical precipitation with industrial coagulants) processes. Increasingly popular short nitrification processes: ANAMMOX, CANON, OLAND, SHARON, DEAMOX are used in biological methods (Hill & Khan 2008; Masłoń & Tomaszek 2009). Membrane separation is also increasingly proposed. However, application of membrane method alone has a substantial disadvantage—the problem involving disposal of obtained concentrate is still present.

The increasing costs utilizing chemical methods justify research on the application of various waste containing aluminium and iron ions in sewage and sludge treatment. The studies were conducted on such waste as: fly ash, blast furnace slag, spent alum sludge, waste aluminium and iron deposits from hydrometallurgical processes, waste from mineral processing industry (Zeng et al. 2004). Recovery of coagulants used in water treatment is increasingly common (Yang et. al. 2014). Researchers look for new applications of iron deposits produced during underground water treatment as well. Common methods of their disposal are: transportation to wastewater treatment plants through the sewer system and storage, e.g. in lagoons. Therefore it is necessary directional using of ferric sludge among others in wastewater treatment plants (Piaskowski 2009; Klinksieg et al. 2008). The use of drinking water treatment sludge (iron sludge/backwash sludge/DWTS) for the removal of contaminants from wastewater or reject water reduces the operational costs of wastewater treatment plants, and also constitutes a practical and sustainable method of their disposal. Several studies indicate the possibility of using aluminium and iron deposits in wastewater treatment processes. They may be used: for the removal of phosphorus from wastewater during adsorption or a chemical precipitation, for the removal of organic substances and suspended solids from wastewater, as well as to control of swelling of activated sludge and sludge during fermentation and to enhance its dewatering (Razali et al. 2007; Song et al. 2011; Guan et al. 2005). Practical application of iron sludge requires specifying the process conditions in which the best results are obtained. Many studies, conducted in this field, omit an important aspect: the form of during their use.

The aim of the study, presented in the paper, was to determine the possibility and efficiency of pretreatment of reject water with the use of DWTS containing substantial amount of iron ions from underground water treatment.

2 MATERIALS AND METHODS

2.1 Reject water characterization

The research was conducted using mixture of reject water from gravitational thickening of primary and excess sludge and effluent from dewatering centrifuges of digested sludge. That mixture is recirculated to wastewater treatment. The quality of reject water during research period was highly variable and this is a constant element of the operation of sewage sludge treatment. The results of analyses of 17 reject water samples, collected within period of 5 months, statistically elaborated, are presented in Table 1. The observed high concentration of carbon, nitrogen and phosphorus compounds in reject water result from the processes occurring during the mesophilic sludge digestion, i.e. – hydrolysis of polyphosphates accumulated in the cells of PAOs bacteria and intensive process of ammonification of proteins. Due to the low values of $BOD_5/N_{tot} = 2.1$ and $BOD_5/P = 17.6$ in reject water, its treatment in biological processes will not yield the required efficiency.

2.2 DWTS characterization

The iron sludge (DWTS) used during experiments was taken from settlers of filter backwash water of

Table 2. Results of chemical analysis of DWTS (FAAS method).

	Metal [mg/g of DWTS]	
	DWTS I	DWTS II
Fe	291.0	331.6
Mn	72.6	25.3
Ca	17.4	18.8
Mg	0.5	0.7

two underground water treatment plants. Those plants use the following system of treatment: aeration—gravity filtration on of quartz sand deposits. Two different sludges were used in two forms: primarily hydrated and dried at room temperature.

The basic chemical analysis of sediments from two underground water treatment plants (I and II) were conducted using material, dried at room temperature, of grain size <0.75 mm. Selected metals (Fe, Mn, Ca, Mg) were determined in the mineralized iron sludge samples using FAAS method and Philips PU 9100X apparatus (Table 2).

Diversified chemical characteristics of studied DWTS is mainly a result of the quality of raw water and the operation of the plant associated with it (double-layer deposits in plant II and the single-layer in plant I). Plant I is characterized by a high concentration of iron and manganese – 2.01 mgFe/l and 0.25 mgMn/l in the raw water. Station II – 0.25 mgFe/l and 0.10 mgMn/l. About 5 970 kg of Fe in plant I and 1 173 kg of Fe in plant II is produced yearly form backwash water.

The chemical composition of DWTS affects its structure. The iron sludge I easily deagglomerated and sifted after drying, it had a lighter dark orange colour. DWTS II (higher Fe/Mn ratio) was characterized by a darker, browner colour. The iron sludge II also created a much stronger structure after drying.

2.3 Experimental setup

The experiments were conducted under static conditions (jar test). 100 ml of reject water was put into 250 ml Erlenmeyer flasks and then a dose of DWTS was added to obtain its relevant concentration. The samples were mixed on a laboratory shaker table at 130 RPM. After a certain time the samples were centrifuged at 10 000 RPM for 15 min. Following parameters were determined in the obtained effluent: pH, specific conductivity, orthophosphates, total nitrogen and TOC.

Following issues were examined during experiments conducted using described methodology: (i) impact of reaction time (from 0 to 24 h) on pretreatment of reject water with use of DWTS

(5 g/l); (ii) impact of DWTS dose (from 0 to 10 g/l) and the initial concentration of pollutants on quality of reject water for 2 h of reaction time; (iii) impact of pH of reject water on the efficiency of orthophosphates removal by DWTS. The pH adjustment within the range from 3.0 to 9.4 was performed using solutions of HCl and NaOH, prior to addition of DWTS (dose 3 g/l). The studies also determined the process effectiveness of sludge depending on its origin (DWTS I and DWTS II) and its form (primarily hydrated and dried).

3 RESULTS AND DISCUSSION

3.1 Impact of reaction time on the efficiency of pre-treatment of reject water with the use of iron sludge DWTS

The use of alternative reagents in wastewater or reject water treatment depends on their effectiveness and reaction time. Research conducted by authors on the use of sludge from drinking-water treatment plants for removal of contaminants from prepared water containing only one component has proven that the best results were obtained for removal of orthophosphate ions (Piaskowski 2013). Apart from ammonium ions, they are, quantitatively, the most significant pollutant of reject water. Therefore kinetic studies were conducted in order to determine the impact of time on reaction of orthophosphates removal from real reject water (the dose of DWTS I and DWTS II was 5 g/l).

The obtained results show the most dynamic changes of orthophosphates concentration during the first hour of reaction for primarily hydrated DWTS I and II (Fig. 1). Hydrated DTWS also show higher efficiency of orthophosphates removal as compared to dried ones. During drying the iron and manganese hydroxides contained in DWTS turn into oxides. Oxides are less soluble and therefore less chemically active even after rehydration.

After 1 hour of reaction the final concentration of the analyzed ions amounted to 12–13 mgP/l for the dried DWTS and 2–3 mgP/l for the hydrated DWTS. Differences between the results of orthophosphates removal obtained for DWTS I and II for each different forms of iron sludge were not significant.

The obtained results indicate that the use of iron sludge from an underground water treatment plant is in practice possible without its additional preparation or drying processes, which facilitates dosing or transportation. Since the first minutes of contact the hydrated sludge had caused more than 90% efficiency of orthophosphate ions removal and that value did not significantly increase over the next hours (up to 24 hours). In the case of dried sludge dynamics of the process was much slower.

3.2 Impact of the dose of DWTS on reject water quality

The use of DWTS from both plants (I and II) and in both forms indicated a similar property of buffering of the variability of reject water. Dosage of iron sludge regulated pH of reject water into neutral values. DWTS also contain a certain amount of calcium ions, apart from iron and manganese. Calcium ions, along with the decreasing pH, migrate to the solution, thus buffering the pH of reject water. Changes of the absolute value of the pH parameter depending on the initial pH and the dose of sludge are presented in Figure 2.

Dosage of iron sludge also had an impact on the conductivity of reject water. The tendency and value of changes this parameter depended not only on the form of DWTS, but also on its origin (DWTS I or DWTS II). Experiments carried out for three different values of the initial reject

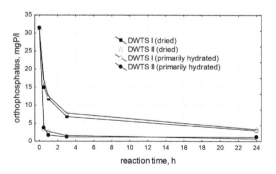

Figure 1. Impact of time on the changes in reject water parameters during the reaction with backwash sludge. DWTS dose 5 g/l, initial concentration 31.37 mgP/l.

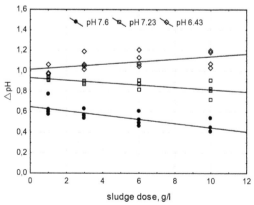

Figure 2. Impact of DWTS on the changes of pH of reject water at variable values of initial pH (block chart for sludge I and II and their both forms), contact time 2 hours.

water conductivity (1.34; 1.72; 2.26 mS/cm) indicated the same trend of alignment. An example of results for the initial specific conductivity 2.26 mS/cm is shown in Figure 3.

Primarily hydrated sludge had a stronger impact on decrease of value of parameter than the dried sludge. Moreover, the origin of DWTS was important as well because iron sludge II (the higher Fe content), in both forms, caused a higher decrease of conductivity than sludge I. Such results may be caused by the formation of hydroxo complexes of iron ions, which provide a surface for the adsorption of colloids and other ions. Calcium ions could also affect the process. The quantity of calcium ions migrating from the sludge to the reject water increased along with pH decrease. Studies on sludge washout had proven the migration of constituents from sludge to a solution, causing an increase of conductivity (Piaskowski 2013). An opposite trend in the real reject water indicates reactions between contaminants from reject water and components washed out from iron sludge, causing a decrease of conductivity.

The use DWTS for pre-treatment of reject water caused a relatively insignificant removal of organic substances, expressed by TOC (Fig. 4a) and total nitrogen (Fig. 4b). During experiments it was noted that along with the increase of the initial concentration, the efficiency of TOC removal decreased, while in case of total nitrogen it increased. The use of primarily hydrated iron sludge was more effective than in case of dry sludge. On the other hand, differences between the effectiveness of DWTS I and DWTS II were insignificant.

The highest efficiency of DWTS was obtained in case of removal of orthophosphates. Taking into account the analysis based on the dose of iron sludge, given in g/l, the effectiveness of removal of orthophosphates per gram of iron sludge increased along with increasing of initial concentration of PO_4^{-3} ions in the reject water (within range 19–74 mgP/l). The obtained results indicate that the specific orthophosphates removal efficiency of used DWTS I and II was very similar but only in specified form of iron sludge, e.g. for DWTS I–II in dried form.

However, the form of DWTS, also affected the efficiency of the process. Primarily hydrated form of iron sludge showed a higher specific removal efficiency of orthophosphates (Fig. 5) than the dried one.

For example: for the primarily hydrated DWTS (Initial concentration = 74 mgP/l, dose 1g/l) removal of 35 mg of P per gram of iron sludge was obtained, for the dried DWTS, under the same conditions, a removal of 15 mg of P per gram of DWTS was obtained.

Due to the different content of iron ions, the main component of DWTS, the dose of iron sludge was also converted to unit amount of Fe. The obtained results for the highest initial concentration of orthophosphates in reject water shown in Figure 6.

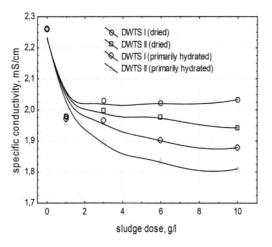

Figure 3. Impact of DWTS on changes of specific conductivity of reject water—contact time 2 hours, the value of initial conductivity—2.26 mS/cm.

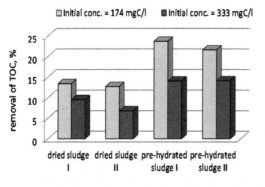 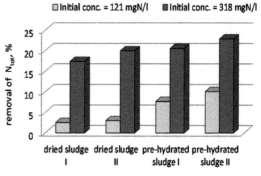

Figure 4a–b. Removal of TOC and N_{tot} for selected two initial concentrations in reject water and dose of iron sludge 10 g_{dm}/l—centrifuged samples, contact time 2 hours.

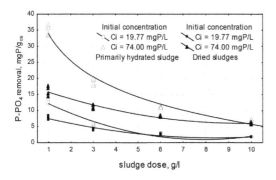

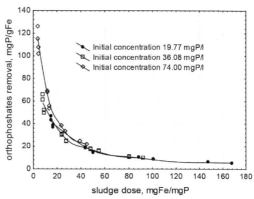

Figure 5. Impact of DWTS dose sludge and the initial concentration of orthophosphates in the reject water on their specific removal efficiency by the dried and primarily hydrated DWTS (combined data for DWTS I and II), contact time 2 hours.

Figure 7. Impact of dose of DWTS on specific removal of orthophosphate ions (combined data for the primarily hydrated DWTS I and II).

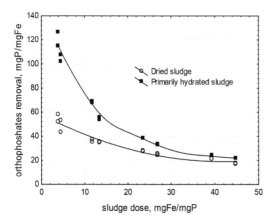

Figure 6. Impact of the DWTS dose and its form on the specific removal efficiency of orthophosphate ions. Initial concentration 74 mgP/l (combined data for DWTS I and II), contact time 2 hours.

Despite the differences of iron content in DWTS I and II, their specific removal efficiency was very similar (hence the possibility of combining results). The difference in the removal of orthophosphates depended on the form of used DWTS (dried or hydrated), especially in the range of doses of 4–13 mgFe/mgP. Fig. 7 presents the combined results of primarily hydrated sludge I and II, depending on the initial concentration of orthophosphates in the reject water. The increase of the dose of the adsorbent, which in this case is iron sludge, and the decrease of the initial concentration of orthophosphates in the reject water caused decrease of the specific orthophosphates removal efficiency referred to the iron added.

The increase of initial concentration caused an incremented contact of PO_4^{-3} ions with active centres of iron sludge. Due to the small possibility of contact at low concentration of orthophosphates, adsorption has not reached saturation state. The amount DWTS added to the solution determined the number of binding sites available for adsorption. However, the removal efficiency of orthophosphate decreased with increasing dose of adsorbent. Such result may be caused by competitive reactions to adsorption of orthophosphates.

3.3 Impact of pH on the effectiveness of orthophosphates removal using DWTS

Efficiency of orthophosphate removal from reject water was highly influenced by changes in pH (Fig. 8). Experiments conducted both for control samles within the pH range of 3–9.4 and, under the same conditions, after dosing iron sludge. The results showed that only pH control of reject water and its mixing for 2 h may influence changes of reject water quality. This effect was observed at pH = 7.6 and 9.4, for which along with the pH increase, the concentration of orthophosphates decreased from the initial value of 19.63 mg P/l to 10.62 and 2.7 mg P/l. Dosage of DWTS to reject water increased the efficiency of the process, especially at low pH = 3. In case of lack of any other interactions in reject water, which were observed for the higher pH values, the initial concentration of orthophosphates at pH = 3, without addition of DWTS, was at the level of 20.26 mg P/l.

Under such conditions, DWTS showed maximum sorption potential against PO_4^{-3} which allows obtaining final concentration of orthophosphate lower than 1 mg P/l. This process could be the result of previously described:

– impact of generated at low pH positive hydroxo complexes of iron ions. Such complexes constitute the active sorption surface for PO_4^{-3} ions;

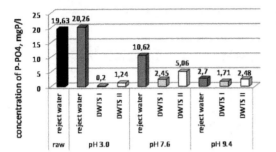

Figure 8. Impact of initial pH value in reject water on concentration of orthophosphates. The dose of primarily hydrated DWTS—3 g/l, reaction time—2 hours.

– increased migration of calcium ions from DWTS into supenatant at low pH. Calcium ions react with orthophosphates.

4 CONCLUSIONS

The study results indicate that the iron sludge obtained from underground water treatment (DWTS) may be used in practise for pre-treatment of reject water, especially for removal of orthophosphates. Since the largest specific removal efficiency of orthophosphates from reject water was observed at the highest tested initial concentration of PO_4^{-3} ions (74 mgP/l) and for primarily hydrated sludge, it is possible to use iron sludge, without additional processes, for removal of orthophosphates from effluent from sewage sludge digestion, where the highest concentrations of orthophosphates are generated. There is no process justification for drying of DWTS and dosing it into reject water in the powdery form. Dosage of iron sludge in the dry form significantly decreases the efficiency of process. Moreover, additional significant costs are required to conduct drying in a technical scale.

The study results also indicate the buffer properties of DWTS and occurrence of reactions affecting the decrease of conductivity of reject water. No differences in the process efficiency of sludge I and II were obtained, despite certain differences in their chemical composition. The only feature which makes the difference is the form of used DWTS: dried—with a lower physicochemical activity and primarily hydrated. The research was also a verification of the results obtained using model solution of orthophosphates presented in the previously published article (Piaskowski 2013). The acquired dependencies were also confirmed, in a wide range, also for reject water. However, for primarily hydrated sludge the specific orthophosphates removal efficiency at an initial concentration of 74 mgP/l was in case of real reject water almost twice as high as in the model solution. That indicates that, despite the apparent lower efficiency of iron sludge compared to commercial reagents, they may be successfully used in the pre-treatment of reject water. Such activity allows obtaining two effects: reduction of DWTS disposal costs and a decrease of costs of phosphorus removal from reject water.

ACKNOWLEDGEMENTS

The scientific research was financed from the resources of the National Science Centre of Poland (the research project No. N523739240).

REFERENCES

Battistoni, P., Pavan, P., Prisciandaro, M. & Cecchi, F. 2000. Struvite crystallization: a feasible and reliable way to fix phosphorus in anaerobic supernatants. *Water Res.* 34: 3033–3041.

Guan, XH., Chen, GH. & Shang, C. 2005. Re-use of water treatment works sludge to enhance particulate pollutant removal from sewage. *Water Res.* 39: 3433–3440.

Hill, C.B. & Khan, E. 2008. A comparative study of immobilized nitrifying and co-immobilized nitrifying and denitrifying bacteria for ammonia removal from sludge digester supernatant. *Water Air Soil Poll.* 195: 23–33.

Klinksieg, K., Dockhorn, T., Dichtl, N. & Osterloh, W. 2007. Verwertung von Schlaemmen aus der Trinkwasseraufbereitung in kommunalen Klaeranlagen—Ein Praxisbeispiel. *Mull und Abfall* 9: 433–439.

Masłoń, A. &Tomaszek, J.A. 2009. Anaerobic ammonium nitrogen oxidation in Deamox process, *Environ. Prot. Eng.* 35: 123–130.

Peng, Y., Zhang, L., Zhang, S., Gan, Y. & Wu, C. 2012. Enhanced nitrogen removal from sludge dewatering liquor by simultaneous primary sludge fermentation and nitrate reduction in batch and continuous reactors. *Bioresource Technol.* 104: 144–149.

Piaskowski, K. 2009. Generation and management of wastewater from ground water treatment. *Arch. Environ. Prot.* 35: 35–46.

Piaskowski, K. 2013. Orthophosphate removal from aqueous solutions using drinking-water treatment sludge. *Water Sci. Technol.* 68: 1757–1762.

Razali, M., Zaho, Y.Q. & Bruen, M. 2007. Effectiveness of a drinking-water treatment sludge in removing different phosphorus species from aqueous solution. Sep. *Purif. Technol.* 55: 300–306.

Song, X., Pan Y., Wu Q., Cheng Z. & Ma W. 2011. Phosphate removal from aqueous solutions by adsorption using ferric sludge, *Desalination* 280: 384–390.

Yang L., Wei J., Zhang Y., Wang J. & Wang D. 2014. Reuse of acid coagulant-recovered drinking waterworks sludge residual to remove phosphorus from wastewater. *Appl. Surf. Sci.* 305: 337–346.

Zeng L., Li X. & Liu J. 2004. Adsorptive removal of phosphate from aqueous solutions using iron oxide tailings, *Water Res.* 38: 1318–1326.

Environmental Engineering V – Pawłowska & Pawłowski (Eds)
© 2017 Taylor & Francis Group, London, ISBN 978-1-138-03163-0

Magnetic separation of submicron particles from aerosol phase

A. Jaworek, A. Marchewicz, T. Czech, A. Krupa & A.T. Sobczyk
Institute of Fluid Flow Machinery, Polish Academy of Sciences, Gdansk, Poland

K. Adamiak
Department of Electrical and Computer Engineering, University of Western Ontario, London, Ontario, Canada

ABSTRACT: Power plants or solid waste incinerators are the main anthropogenic sources of air contaminants. Bag filters, absorbers, scrubbers or electrostatic precipitators are used for the removal of PM10 and larger particles. PM2.5 particles are difficult to remove by these methods. Recently, attention was drawn towards magnetic separation as an effective method for these particles removal because magnetic properties of submicron particles can be superparamagnetic that facilitates their motion towards the magnetic poles. In this paper, a quadrupole type magnetic separator for the collection of fly ash particles has been investigated. The trajectories of ferromagnetic particles were analyzed in order to determine the collection efficiency of particles by the magnetic separator. The results show that ferromagnetic particles larger than 1 micron can be removed with efficiency higher than 99% in 7-stack quadrupole system, while smaller particles require longer stacks. The collection efficiency was higher for smaller space between the stacks.

Keywords: magnetic separation, submicron particles, PM2.5, magnetic quadrupole

1 INTRODUCTION

Power industry is the largest anthropogenic emitter of PM10 and gaseous contaminants. Electrostatic precipitators and bag filters are the most frequently used devices for the removal of fly ash particles produced in coal fired boilers. Although mass collection efficiency of those devices is higher than 99.9%, in the flue gases there still remain particles smaller than 2.5 μm, which are classified as PM2.5. Main components of fly ash leaving coal fired boiler are SiO_2 (roughly 1/2), Al_2O_3 (1/4) and Fe_2O_3 (1/8). Other compounds found in fly ash in amounts larger than 1 mg/g each, are CaO, Na_2O, MgO, K_2O, TiO_2 (Coles et al. 1979, Smith et al. 1979, Ylätalo & Hautanen 1998, Hower & Robertson 2004). The percentage of those compounds can vary depending on the coal mine and the combustion conditions. Because of the presence of iron containing particles, some of them can be separated from the non-magnetic fractions. The percentage of toxic elements (As, Cd, Hg, Ni, Zn, Pb, Cr, Sr, Be, V, U) is significantly higher in submicron particles (PM1) than in larger ones (Shu et al. 2001). It was found that trace elements abundant in nanometer particles adhere to the larger particles or are fused with them. Up to 20% of PM2.5 particles are of magnetic properties. The magnetic fractions are mainly composed of magnetite and hematite fused with quartz or mullite (Lu S. et al. 2009, Dudas & Warren 1987). Other elements, which are of 4–6 times higher concentration than in non-magnetic fractions, are Cd, Pb, Cr, Cu, Mn, Mo, Se, Sn, V, and Zn (Vassilev et al. 2004a,b, Lu S. et al. 2009). These elements greatly contribute to contamination of air, soil, and surface and ground water, but their affinity to magnetic fractions is favorable to their removal by magnetic forces. Metals adsorbed on the particle surfaces are leached by rains or in contact with surface water and are mobilized into aquatic environment (Jankowski et al. 2006, Bednar et al. 2010, Cowan et al. 2013). Magnetic particles, which are carriers of heavy metals, sediment to soil, and can be found in tree leaves (Lu S. et al. 2009, Strelets and Vatin 2015).

Recovery of trace elements from waste is drawing increasing attention of engineers in order to abate environmental impact of different emission sources, such as energy boilers, solid waste incinerators, sea vessel engines etc. Recently, in order to achieve this goal, an interest was turned to magnetic separation, a method suggested in 1970s (Watson 1973, Sakata et al. 1976, Ariman et al. 1979, Gooding & Drehmel 1979, Mendrela et al. 1985). Two types of devices utilizing magnetic forces to separate fine magnetic particles are used: wet and dry separators. Wet separators remove the particles from a liquid conveying them. This method is used for enrichment of materials from ores or for water purification. Dry magnetic filters or separators are used for the collection of fine para- or ferromagnetic particles from fluidized beds or from aerosol phase (Sakata et al. 1976, Rossier et al. 2012, Zyryanov et al. 2011).

Magnetic filters are made from ferromagnetic (usually steel) wires, rods, wool or beads packed in a canister and embedded in a strong external magnetic field (usually generated by a large magnetic coil), which magnetizes the wires and creates a highly non-uniform magnetic field near their surface that attracts the particles towards them due to induced magnetic moment (Watson 1973, Gooding & Drehmel 1979, Arajs et al. 1985, Zarutskaya et al. 1998, Li Y.W. et al. 2007a,b, Kumar & Biswas 2005, Alvaro et al. 2007, Nakai et al. 2011). Magnetic separators are made in form of stacks of permanent magnets arranged co-linearly or transversally to the flowing gas. In this case, the particles are drawn towards magnetic gaps between permanent magnets forming the stack due to induced magnetic dipole moment. Permanent magnets or conventional coils can produce magnetic field with the magnetic flux density much lower than 1 T. In order to increase the magnetic force acting on a particle, the gradient of magnetic field is maximized. Such devices are called High Gradient Magnetic Separators (HGMS). The magnetic field can be further increased by using superconducting coils. Those devices are called Superconducting Coil HGMS (SCHGMS).

Magnetic separation has been widely used for separation and concentration of mining ores and wastes, for the removal of magnetic particles from ores or coal, or recently, as materials presorting technology in Refuse Derived Fuels (RDF) or Municipal Solid Waste (MSW) incinerators (Watson 1973, Chang et al. 1998, Lopez-Delgado et al. 2003, Svoboda & Fujita 2003, Watson & Beharrell 2006, Padmanabhan & Sreenivas 2011, Xia et al. 2015). Those processes proceed for large particles (tens of µm) or for grains of mm in size.

Application of magnetic aggregation and magnetic separation for gas cleaning from micron or smaller particles, for example, fly ash from coal fired power plants or ashes from waste incinerators is a subject of recent investigations (Lu S. et al. 2009, de Boom et al. 2011, Cheng et al. 2014).

In this paper, a quadrupole type magnetic separator has been considered. The paper analyzes the trajectories of ferromagnetic particles leading to particles deposition onto magnetic poles. Experimental results on the speciation of various elements along the separator, and on magnetic poles are also presented. Morphology of separated particles and their elemental analysis was carried out by SEM and EDS methods, respectively.

2 MAGNETIC SEPARATION

It is known from the literature that magnetic properties of submicron particles (0.1–1 µm) are different from the bulk material and their properties can be changed to superparamagnetic, which facilitates their aggregation in the aerosol phase and affects their motion towards the magnetic poles (Zachariah et al. 1995, Relle et al. 1999). Most of the rare earth oxides and actinide compounds are paramagnetic. A strong dependence of magnetic force on the superparamagnetic particle results in increased collection efficiency for particles <100 nm, as compared to the larger ones (Zarutskaya & Shapiro 2000, Nübold et al. 2003). As a recovery technique, magnetic separation is particularly suitable for solid waste incinerators, rich in heavy metals or other trace elements, among them para- and ferromagnetic (Chang et al. 1998, Cheng et al. 2014).

Quadrupole magnetic separator presented in this paper has been designed for the removal of ferro- and paramagnetic particles from flue gases. The separator consists of four magnetic stacks arranged on a square grid and placed co-linearly with the flowing gas (Fig. 1). Neodymium-iron-boron magnets in each magnetic stack are arranged in such a way that like poles of two neighbour permanent magnets in stack are facing each other, and are separated by a magnetic spacer. In the experiments, commercial stacks, model 25 × 225/2 × M6w/N manufactured by ENES Magnesy Paweł Zientek, Warszawa (Poland), were used. The diameter of these stacks was 25 mm and their length 225 mm. The magnetic flux density at the spacer surface was about 0.65 T.

Particle trajectory in a magnetic field **H** can be determined from the following vector differential equation of motion (Zarutskaya et al. 1998, Zarutskaya & Shapiro 2000):

$$m_p \frac{d\mathbf{w}}{dt} = m_p \mathbf{g} + \frac{3\pi \eta_g d_p}{C_c}(\mathbf{u}-\mathbf{w}) + \mu_0 \nabla[\mathbf{m} \cdot \mathbf{H}] + \mathbf{F_B}$$

(1)

where **m** = magnetic moment of the particle governed by the angular motion equation;

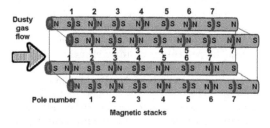

Figure 1. Permanent magnets configuration in quadrupole magnetic separator.

m_p = particle mass; η_g = dynamic viscosity of the gas; r_p = particle radius; C_c = Cunningham slip correction factor; **u** = gas velocity; **w** = particle velocity; μ_0 = magnetic permeability of the free space ($\mu_0 = 4\pi \times 10^{-7}$ [Vs/Am]); **g** = gravitational constant; $\mathbf{F}_B$ = stochastic Brownian force on the particle.

The rotation of the particle is:

$$J_p \frac{d^2\phi}{dt^2} = -\frac{1}{8\pi\eta_g r_p^3}\frac{d\phi}{dt} + \mu_0 \nabla(\mathbf{m} \times \mathbf{H}) + \mathbf{T}_B \quad (2)$$

where J_p is moment of inertia of particle, ϕ is the angle of particle rotation and $\mathbf{T}_B$ is the stochastic Brownian moment of the particle.

The magnetic moment **m** of a spherical particle of diameter d is:

$$\mathbf{m} = \mathbf{M}\frac{\pi d_p^3}{6} \quad (3)$$

where **M** = magnetization.

The particle trajectories along one of the stacks of magnets determined numerically with the commercial software COMSOL multiphysics (version 5.0) are presented in Figure 2, for two particle sizes 0.5 μm (Fig. 2a) and 5 μm (Fig. 2b), for the gas face velocity of 0.5 m/s. The distance between the neighbor magnet's axis equaled to 55 mm, therefore the distance between the nearest magnets surfaces was 30 mm. The assumed mass density of the particles was 2200 kg/m³, and their permeability 100 μ_0. Magnetic moment of particles smaller than 1 μm decreases with particle size decreasing that results in lower magnetic force on the particles.

Thus, smaller particles can pass through the magnet stack without being deposited (Fig. 2a), and lower collection efficiency is obtained. Larger particles are predominantly deposited onto the magnet's pole-spacers as shown in Fig. 2b. The collection efficiency of ferromagnetic particles is also higher for short distances between magnetic stacks because of higher gradient of magnetic field.

3 EXPERIMENTAL SET-UP

The measurements were carried out in an experimental stand shown schematically in Figure 3. The magnetic separator consisted of four magnets arranged in quadrupole configuration (cf. Fig. 1). The space between the neighbor magnet surfaces was 30 mm or 6.3 mm. The quadrupole separator with magnets spaced to 30 mm was placed in a PMMA channel of square cross section of 160 × 160 mm, while that with magnets spaced at 6.3 mm in 63 × 63 mm channel. Para- and ferromagnetic particles flowing through the separator

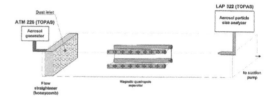

Figure 3. Experimental setup from investigation of PM2.5 particles deposition in quadrupole magnetic separator.

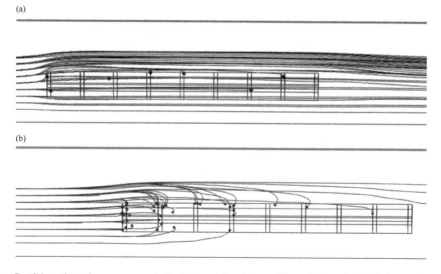

Figure 2. Particle trajectories near magnets for 0.5 μm (a) and 5 μm (b) particles and 0.5 m/s face velocity (magnet separation 30 mm, particle permeability 100 μ_0).

were attracted and deposited onto the poles of the magnetic stacks A flow straightener was mounted at the inlet of the channel to eliminate air vortices. The airflow through the channel was forced by a suction fan placed at its outlet.

The experiments were carried out for fly ash particles collected from third section of an electrostatic precipitator in a coal fired power plant. The particles were dispersed in methanol forming a colloidal suspension of mass concentration lower than 0.5%, and stirred for 1 hour. The suspension was sprayed by Aerosol Generator ATM 226 (TOPAS) and injected to the channel at its inlet. In ATM 226, the nozzle atomizes particle suspension towards a baffle on which larger particles are separated and the smaller ones flow outside the device to the channel. After spraying, the solvent evaporates and only PM2.5 particles are obtained. The concentration of particles was measured at the outlet of the channel using Aerosol Particle Size Spectrometer LAP 322 (TOPAS). This instrument is using 90° light scattering for particles detection. LAP 322 is able to measure aerosols in concentrations <10^4 particles/cm^3, the measuring range is: 0.2–40 μm. Accuracy of LAP 322 meets ISO 21501-1 standard. The air flow velocity in the channel was measured with a hot wire anemometer TSI model 8455, which operates within a range of 0.125–50 m/s with ±2% accuracy. The air velocity was set to 1 m/s or 0.5 m/s.

4 RESULTS

A photograph of magnetic separator taken from the experimental channel is shown in Figure 4. The dust deposited onto the magnet poles can be easily noticed as dull rings formed on the spacers. Inset shows a zoomed photograph, in which particles deposited between the magnetic poles are also visible. SEM micrographs of fly ash particles collected from the poles number 1 (first one from the inlet), 4 (in the middle of the stack), and 7 (the last pole i this stack) of one of the stacks (cf. Fig. 1) are shown in Figure 5. Most of the particles are spherical, that is characteristic of fly ash from coal fired power plants, of the size < 5 μm. The particles are mostly composed of aluminosilicate glass, and are covered with much smaller particles of a size of 10–40 nm.

A comparison of cumulative size distributions of fly ash particles collected from 1st, 4th, and 7th poles with those entering the magnetic separator is presented in Figure 6. The mean size of particles collected from the poles of magnetic stacks is typically about or smaller than 1 μm. Each size distribution was determined from the analysis of about 1000 particles from SEM micrographs. SEM micrograph also indicated that the particles are preferentially deposited along the circumference of magnetic spacer forming parallel conglomerates with free space between them (Fig. 7). This effect is not understood and requires further investigations if it is caused by material properties of the spacer, local deformation of magnetic field by the first few deposited particles, or is only caused by agglomeration of particles in the aerosol phase.

The EDS spectra of fly ash particles entering the separator are shown in Figure 8. The comparison of atomic percentage of the most abundant elements detected by means of EDS analysis is shown in Figure 9. From these measurements it can be inferred that regardless of small differences in size distribution of particles collected at various poles along the magnetic stack, the elemental composition of those particles does not differ significantly, and differences are within statistical error of the method of EDS analysis. The content of iron, which is responsible for magnetic properties of deposited particles, varies between 4.5 wt.% at first pole and 8.8 wt.% at the 4th pole (Fig. 9). Comparing those numbers with iron content in the inlet particles, the magnetic particles enrichment by the quadrupole separator can be estimated at a level of 2.

5 COLLECTION EFFICIENCY

The collection efficiency of magnetic separator for ferromagnetic particles has been determined from numerical simulation of the particles at different poles of magnetic stack. Results of numerical simulations are illustrated in Figures 10 and 11. For large space between magnetic stacks, equal to

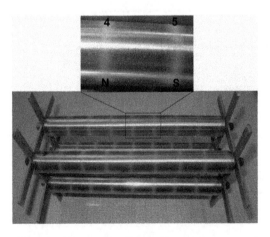

Figure 4. Photograph of magnetic separator removed from experimental channel. Above: Magnification of photograph of magnet between 4th and 5th poles, showing particles deposited between these poles.

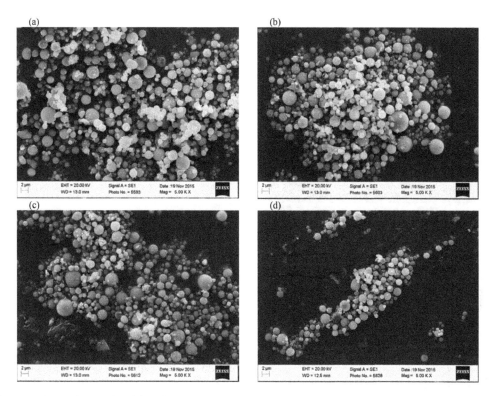

Figure 5. SEM micrographs of fly ash particles collected from 1st (a), 4th (b) and 7th (c) pole-spacer of magnetic separator and between 4th and 5th pole-spacers (d).

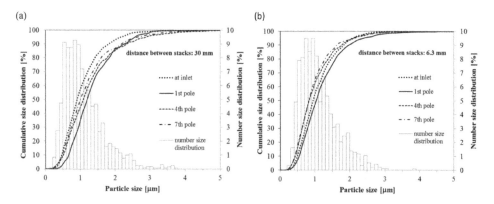

Figure 6. Comparison of cumulative size distribution of fly ash particles entering the magnetic separator with those collected from 1st, 4th, and 7th poles for distance between magnetic stacks: 30 mm (a) and 6.3 mm (b). The histogram presents number size distribution of inlet particles.

30 mm, the cumulative collection efficiency does not exceed 50% along the distance of entire stack. The results of numerical simulations for a slower particle velocity (0.5 m/s) and the distance between the surfaces of permanent magnets stack reduced to 6.3 mm are illustrated in Figure 12. For those distances and flow rates, the cumulative collection efficiency of ferromagnetic particles along the distance of entire stack exceeds 50% and can approach 100%, as illustrated in Figure 12.

141

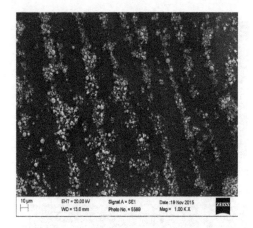

Figure 7. SEM micrograph of fly ash collected from 1st pole spacer. Particles are preferentially deposited along the circumference of magnetic spacer forming parallel conglomerates.

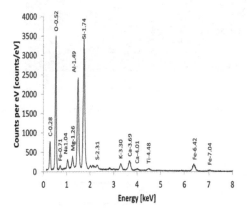

Figure 8. EDS spectrum of particles entering the magnetic separator.

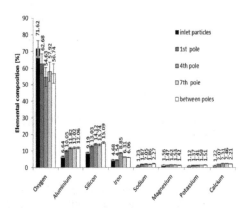

Figure 9. Comparison of elemental composition for particles collected from various places of magnetic separator.

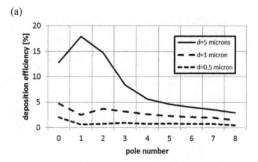

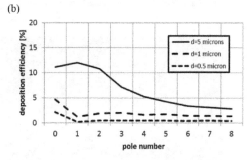

Figure 10. Deposition efficiency of magnetic particles of various sizes at different magnet poles. Inlet flow velocity 0.5 m/s (a) and 1 m/s (b) (section 0 is the surface of stack facing the flow). Space between stacks equals 30 mm.

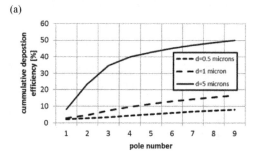

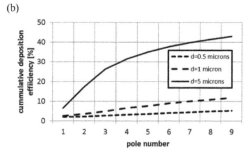

Figure 11. Cumulative deposition efficiency of magnetic particles of various sizes at different magnet poles. Inlet flow velocity 0.5 m/s (a) and 1 m/s (b) (section 0 is the surface of stack facing the flow). Space between stacks equals 30 mm.

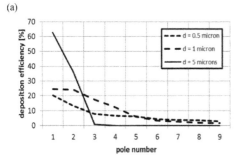

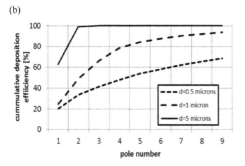

Figure 12. Deposition efficiency of magnetic particles of various sizes at different magnet poles (a) and cumulative deposition efficiency. Inlet flow velocity 0.5 m/s. Space between stacks equals 6.3 mm.

6 CONCLUSIONS

Magnetic separation of ferromagnetic particles from fly ash has been considered in this paper. Magnetic separator, in the form of quadrupole configuration with magnetic stacks co-linear with gas flow, is able to remove fine particles of magnetic properties. This type of device can be used for the removal of magnetic PM2.5 particles from contaminated gases after bag filter or the last stage of electrostatic precipitator in power plants or solid waste incinerators. Potentially, it can also be used for the separation of magnetic particles which contain higher amounts of rare earth or transition elements. The collection efficiency of ferromagnetic particles by the separator, determined from numerical simulation of their trajectories, can approach 100%. The elemental composition of particles deposited on subsequent pole-spacers do not differ significantly, and differences are within statistical error of the method of EDS analysis.

ACKNOWLEDGEMENTS

The work was supported in part by the Institute of Fluid Flow Machinery, Polish Academy of Sciences within the project O1/T3/Z4 and in part by the Natural Sciences and Engineering Research Council (NSERC) of Canada.

REFERENCES

Alvaro, A., Rodriguez, J.M., Augusto, P.A. & Estevez, A.M., 2007. Magnetic filtration of an iron oxide aerosol by means of magnetizable grates. *China Particuology* 5: 140–144.

Arajs, S., Moyer, C.A., Aidun, R. & Matijevic, E., 1985. Magnetic filtration of submicroscopic particles through a packed bed of spheres. *J. Appl. Phys.* 57(1): 4286–4288.

Ariman, T., Ojalvo, M.S. & Drehmel, D.C., 1979. Novel concepts, methods and advanced technology in particulate/gas separation. *Report on a Workshop. J. Air Poll. Contr. Assoc.* 29(8): 818–822.

Bednar, A.J., Chappell, M.A., Seiter, J.M., Stanley, J.K., Averett D.E., Jones W.T., Pettway B.A., Kennedy A.J., Hendrix S.H. & Steevens J.A., 2010. Geochemical investigations of metals release from submerged coal fly ash using extended elutriate tests. *Chemosphere* 81: 1393–1400.

Chang, Y.-H., Chen, W.C. & Chang, N.-B., 1998. Comparative evaluation of RDF and MSW incineration. *J. Hazardous Materials* 58: 33–45.

Cheng, M.-D., Allman, S.L., Ludtka, G.M. & Avens, L.R., 2014. Collection of airborne particles by a high-gradient permanent magnetic method. *J. Aerosol Sci.* 77: 1–9.

Coles, D.G., Ragaini, R.C., Ondov, J.M., Fisher, G.L., Silberman D. & Prentice B.A., 1979. Chemical studies of stack fly ash from a coal-fired power plant. *Envir. Sci. Technol.* 13(4): 455–459.

Cowan, E.A., Seramur, K.C. & Hageman, S.J., 2013. Magnetic susceptibility measurements to detect coal fly ash from the Kingston Tennessee spill in Watts Bar Reservoir. *Environmental Pollution* 174: 179–188.

De Boom, A., Degrez, M., Hubaux, P. & Lucion, C., 2011. MSWI boiler fly ashes: Magnetic separation for material recovery. *Waste Management* 31: 1505–1513.

Dudas, M.J. & Warren, C.J., 1987. Submicroscopic model of fly ash particles. *Geoderma* 40: 101–114.

Gooding, C.H. & Drehmel, D.C., 1979. Application of high gradient magnetic separation to fine particle control. *J. Air Pollution Control Assoc.* 29(5): 534–538.

Hower, J.C. & Robertson, J.D., 2004. Chemistry and petrology of fly ash derived from the co-combustion of western United States coal and tire-derived fuel. *Fuel Proc. Technol.* 85: 359–377.

Jankowski, J., Ward, C.R., French, D. & Groves, S., 2006. Mobility of trace elements from selected Australian fly ashes and its potential impact on aquatic ecosystems. *Fuel* 85: 243–256.

Kumar, P. & Biswas, P., 2005. Analytical expressions of the collision frequency function for aggregation of magnetic particles. *J. Aerosol Sci.* 36(4): 455–469.

Li, Y.W., Zhao, C.S. & Wu, X., 2007a. Aggregation experiments on fine fly ash particles in uniform magnetic field. *Powder Technol.* 174: 93–103.

Li, Y.W., Zhao, C.S. & Wu, X., 2007b. Aggregation mechanism of fine fly ash particles in uniform magnetic field. *Korean J Chem Eng* 24(2): 319–327.

Lopez-Delgado, A., Peña, C., Lopez, V. & Lopez, F.A., 2003. Quality of ferrous scrap from MSW incinerators: a case study of Spain. Resources, *Conservation and Recycling* 40: 39–51.

Lu, S.G., Chen, Y.Y., Shan, H.D. & Bai, S.Q., 2009. Mineralogy and heavy metal leachability of magnetic fractions separated from some Chinese coal fly ashes. *J. Hazardous Materials* 169: 246–255.

Mendrela, E.A., Adamiak, K. & Mendrela, E.M., 1985. Magnetic separator for volatile dust. *Magnetic Separation News* 1: 185–194.

Nakai, Y., Senkawa, K., Mishima, F., Akiyama, Y. & Nishijima, S., 2011. Study on interparticle interaction for dry HGMS system using pneumatic conveyance. *Physica* C 471: 1533–1537.

Nübold, H., Poppe, T., Rost, M., Dominik, C. & Glassmeier, K.-H., 2003. Magnetic aggregation II. Laboratory and microgravity experiments. *Icarus* 165: 195–214.

Padmanadhan, N.P.H. & Sreenivas, T., 2011. Process parametric study for the recovery of very-fine size uranium values on super-conducting high gradient magnetic separator. *Advanced Powder Technology* 22: 131–137.

Relle, S., Grant, S.B., Tsouris, C. & 1999. Diffusional coagulation of superparamagnetic particles in the presence of an external magnetic field. *Physica A* 270, 427–443.

Rossier, M., Schreier, M., Krebs, U., Aeschlimann, B., Fuhrer, R., Zeltner, M., Grass, R.N., Günther, D. & Stark, W.J., 2012. Scaling up magnetic filtration and extraction to the ton per hour scale using carbon coated metal nanoparticles. *Separation and Purification Technology* 96: 68–74.

Sakata, S., Yoshikawa, A. & Tasaki, A., 1976. Magnetic separation of aerosol particles from air flow. *Japan. J. Appl. Phys.* 15, No.10, 2017–2018.

Shu, J., Dearing, J.A., Morse, A.P., Yu, L. & Yuan, N., 2001. Determining the sources of atmospheric particles in Shanghai, China, from magnetic and geochemical properties. *Atmosph. Environ.* 35: 2615–2625.

Smith, R.D., Campbell, J.A. & Nielson, K.K., 1979. Characterization and formation of submicron particles in coal-fired plants. *Atmosph. Environ.* 13: 607–617.

Strelets, K. & Vatin, N. 2015. Dust Features Used to Calculate Dust Removal Performance in Cyclones.

Rocznik Ochrona Środowiska/Annual Set the Environment Protection, 17(1): 104–112.

Svoboda, J. & Fujita, T., 2003. Recent developments in magnetic methods of material separation. *Minerals Eng.* 16: 785–792.

Vassilev, S.V., Menendez, R., Borrego, A.G., Diaz-Somoano, M. & Martinez-Tarazona, M.R., 2004a. Phase-mineral and chemical composition of coal fly ashes as a basis for their multicomponent utilization. 2. Characterization of ceramic cenosphere and salt concentrates. *Fuel* 83: 585–603.

Vassilev, S.V., Menendez, R., Borrego, A.G., Diaz-Somoano, M. & Martinez-Tarazona, M.R., 2004b. Phase-mineral and chemical composition of coal fly ashes as a basis for their multicomponent utilization. 3. Characterization of magnetic and char concentrates. *Fuel* 83: 1563–1583.

Watson, J.H.P., 1973. Magnetic filtration. *J. Appl. Phys.* 44: 4209–4213.

Watson, J.H.P. & Beharrell, P.A., 2006. Extracting values from mine dumps and tailings. *Minerals Eng.* 19: 1580–1587.

Xia, W., Xie, G. & Peng, Y., 2015. Recent advances in beneficiation for low rank coals. *Powder Technology* 277: 206–221.

Ylätalo, S.I. & Hautanen, J., 1998. Electrostatic precipitator penetration function for pulverized coal combustion. *Aerosol Sci. Techn.* 29(1): 17–30.

Zachariah, M.R., Aquino, M.I., Shull, R.D. & Steel, E.B., 1995. Formation of superparamagnetic nanocomposites from vapor phase condensation in a flame. *Nanostruct. Materials* 5, 383–392.

Zarutskaya, T., Lekhtmakher, S. & Shapiro M., 1998. Monte-Carlo simulation of particle stochastic trajectories: A model for magnetic filtration of nanometer aerosols. *J. Aerosol Sci.* 29, Suppl. 1: 1113–1114.

Zarutskaya, T. & Shapiro, M., 2000. Capture of nanoparticles by magnetic filters. *J. Aerosol Sci.* 31(8): 907–921.

Zyryanov, V.V., Petrov, S.A. & Matvienko, A.A., 2011. Characterization of spinel and magnetospheres of coal fly ashes collected in power plants in the former USSR. *Fuel* 90: 486–492.

Environmental Engineering V – Pawłowska & Pawłowski (Eds)
© 2017 Taylor & Francis Group, London, ISBN 978-1-138-03163-0

Treatment of wastewater from textile industry in biological aerated filters

J. Wrębiak, K. Paździor, A. Klepacz-Smółka & S. Ledakowicz
Faculty of Process and Environmental Engineering, Lodz University of Technology, Lodz, Poland

ABSTRACT: The biodegradation of industrial textile wastewater was investigated in the up-flow Biological Aerated Filters (BAFs). Four support materials (ceramsite, beechwood shavings, Intalox saddles, RK BioElements) and various Hydraulic Retention Times (HRTs) were used. The results obtained from the experiments revealed the ability of the microorganisms to adapt to changes of wastewater composition. The continuous process in BAFs with ceramsite and beechwood at HRT equal to 72 h led to a good removal performance of organic carbon compounds—approx. 82% for COD and TOC, while 98% was obtained for BOD_5. Decolorization efficiencies varied between 52 and 75%. As a result, the effluents met the local discharge standards for organic carbon and total nitrogen concentrations. Furthermore, SEM microphotographs of all support materials are presented. The performed investigations confirmed that biofilm carriers type as well as HRT are crucial factors to achieve effluent quality requirements.

Keywords: biodegradation, textile wastewater, industrial effluent, aerobic sewage purification, up-flow biological aerated filters

1 INTRODUCTION

The textile industry, as one of the most water consuming branches, uses plenty of surfactants, vast quantities of inorganic compounds and a wide selection of dyes and pigments. The number of chemicals that may regularly or temporarily occur in the effluent is difficult to assess. Currently the Society of Dyers and Colourists jointly with the American Association of Textile Chemists and Colorists collected in Colour Index 27,000 individual products under 13,000 Colour Index™ Generic Names, where ca. 45% are still available on the market (Zollinger 2003). Around 3600 colorants are applied in textile industry. Generally, commercial colorants are not just 100% pure substance responsible for the color. They usually contain additional substances that improve the product application properties; e.g. its dispersibility, flow or flocculation resistance. Moreover, some colorants, especially reactive dyes, require massive quantity of electrolytes to bond the fibre. In consequence, apart from colorant, more than 8000 chemicals may appear in dying bath in the including various processes of textile finishing (Handa 1991). However, most researchers still concentrate on decolorization as a main problem of textile wastewater treatment (Ali 2010, Kazemi et al. 2016, Sharma et al. 2011, Gupta & Suhas 2009, Vignesh et al. 2014) and little is said about

how to treat effluent in order to gain purification level ready to discharge into watercourses or reuse in dye mill (He et al. 2013, López-Grimau et al. 2011, Vandevivere 1998, Zou 2015).

Nowadays, the wastewater reuse is a precondition for sustainable development. Thus, various biological, chemical, physical and physicochemical methods, such as activated sludge process, fixed film bioreactors, adsorption, advanced oxidation processes and separation are applied for textile wastewater treatment (Fu et al. 2011). However, closing the water cycle cannot be done, regardless of the effort and cost needed. Most physicochemical methods are not only expensive but also generate wastes that are more difficult to dispose than raw wastewater, which in ordinary way, is channelled off into sewage treatment plant. Additionally, some chemical methods present the complication associated with the possible toxicity of degradation products (Li et al. 2015). Activated carbon has limited application. Filtration, as well as electrochemical processes and coagulation, are of doubtful value owing to the cost of treating the sludges and number of restrictions concerning their disposal.

The biodegradation is considered to be the cheapest and environmentally-friendly method (Imran et al. 2015). However, by biological treatment alone it is almost unattainable to treat the effluents of textile industry in order to be able to release them to the environment, all the more to reuse such water in the industry. Thereupon, additional advanced steps are needed to achieve these objectives. Biodegradation is universally followed by other methods, like coagulation–flocculation or adsorption on activated carbon. In France "activated sludge" is the most commonly utilized method, applied as a part of integrated textile effluents treatment systems (Allegre et al. 2006).

Immobilization of microorganisms on different carriers is a promising technique with reference to biological treatment of wastewater (Khan et al. 2010). Biological Aerated Filter (BAF) is used for physical filtration of suspended solids and simultaneously aerobic biodegradation. Several natural and artificial filtering media have been reported. Among them were: volcano rock (Li et al. 2015), schist, slag, haydite (Yu et al. 2008), expanded clay (Wu et al. 2016), lava (Wang et al. 2006), expanded polystyrene (Ryu et al. 2008), tuff, activated charcoal, coco slag, gravel (Melián et al. 2008), zeolite, sand (Chang et al. 2002), plastics and ceramic particles (Wenyu et al. 2007). After running a period of some weeks, a biocoenosis is anchored on these filter media. Biological aerated filters have the following advantages:

– relatively low floor space requirements and shorter biodegradation times of wastewater due to higher biomass concentrations;
– elimination of the final sedimentation tank;
– carbon removal, nitrification and solids filtration can be accomplished in a single unit;
– beneficial conditions for slowly multiplying bacteria and for higher microorganisms;
– higher resistance to toxicants;
– synergy between biodegradation and adsorption may occur

(Rouette 2001, Wenyu et al. 2007).

A wastewater treatment method should be adjusted as much as possible to the pollution in order to achieve the highest purification efficiency. Textile wet finishing includes several unit processes. In the Table 1 a typical dyeing process is outlined. Fabric pre-treatment is done to remove impurities before dyeing, while wash off is to reduce salt and unfixed dye concentration. Excluding fabric finishing (drying and softening), the whole wet process takes 8–12 hours and contributes 8–11 baths with different concentration of pollutants.

A sequence of emitted baths from exhaust dyeing slightly differs for light and dark shades of cotton knitted fabric. Light shades need bleaching, whereas dark ones require more dye and salt and extra rinsing. In the Table 2 there are typical amounts of input chemicals distinguished between light, medium and dark shades.

Table 1. Wet finishing process stages for cotton knitted fabric with reactive dyestuffs (Rouette 2001).

	Pretreatment	Dyeing	Wash off
Time [hour]	2–3	2–3	4–6
Number of baths	2–3	1	5–7
Bath temperature [ºC]	98	60 or 80	Hot, warm and cold
Effluent load	COD, alkali, salt	Color, alkali, salt	Color, salt

Table 2. Typical input chemicals for exhaust dyeing of cotton knitted fabric with reactive dyestuffs (EU 2003).

Input chemicals [g · kg⁻¹ textile]	Shade		
	Light	Medium	Dark
Dyestuff	0.5–4	5–30	30–80
Org. auxiliary	0–30	0–30	0–35
Salt	90–400	600–700	800–2000
Inorganic auxiliary	50–250	30–150	30–150

As far as the characteristics of the discharged water are concerned, some examinations were carried out beforehand for individual baths of typical dyeing processes (Bemska et al. 2012, Wrębiak et al. 2014). Explicit differences between baths occurred among the following parameters: COD, pH, conductivity and absorbance.

The results clearly indicate that high-, medium- and low-loaded baths are discharged during exhaust dyeing process. This shows the importance of separating the different streams in order to allow recycling of the low-loaded baths and more effective treatment of the rest. Such a conduct is required by European Union law and expounded in Reference Document on Best Available Techniques for the Textiles Industry (EU 2003). This leads us to the conclusion that the problems associated with textile effluents are not restricted to the dyes themselves. There are issues associated with the high levels of detergents, inorganic salts and other auxiliaries required by certain textile dyeing processes. Eventually, each specific process presents its own set of environmental issues which need to be addressed by the simultaneous development of new products, process improvements and effluent treatment methods.

This paper presents the results of investigations on the industrial textile wastewater biodegradation by means of up-flow Biological Aerated Filters (BAFs). The experiments were performed with four different biofilm carriers and Hydraulic Retention Times (HRTs).

2 MATERIALS AND METHODS

2.1 Industrial textile wastewater

Effluent undergoing biodegradation was taken from an industrial dye house (Z.W. Biliński Sp. J.). It is one of the biggest dye mills in Poland, which processes mostly cotton knitted fabric with reactive dyestuffs at 40–60°C. The discharged water was divided into two streams and averaged in an expansion tank for 24h. The experiments were conducted on a stream consisting mainly of washing baths and rinses. This stream was chosen for biological treatment as less toxic (Bemska et al. 2012, Wrębiak et al. 2014).

2.2 Experimental set-up

The three parallel up-flow BAF reactors were made of glass pipe, each with a diameter of 15 cm and a height of 86 cm. Apparatus was equipped with inlet and outlet pipes, a reservoir for wastewater storage and a pump for attaining a continuous feed. Volumetric flow rates were set between 5 and 15 $dm^3 \cdot d^{-1}$, and consequently, Hydraulic Retention Times (HRTs) equalled: 72, 48 and 24 h. Compressed air was also introduced from the bottom of bioreactor with volumetric flow around 1 $dm^3 \cdot min^{-1}$. All experiments were performed at an ambient temperature of 22 ± 4°C. Columns, which provided a volume of 15 dm^3 were equipped with perforated plates, placed 8 cm above the bottom and 5 cm below the top. As biofilm support the following materials were used:

a. ceramic *Intalox* saddles; specific surface area 254 $m^2 \cdot m^{-3}$
b. RK BioElements Medium (injection molded in Polypropylene); specific surface area 750 $m^2 \cdot m^{-3}$
c. ceramsite (expanded clay); density: 282 $g \cdot dm^{-3}$, maximum diameter: 15.5 ± 3.1 mm, minimum diameter: 10.8 ± 2.0 mm
d. beech shavings; average size: 15 × 50 mm, 1 mm thick

In the Table 3 there are SEM microphotographs of all used support materials.

No external inoculum was introduced into the reactors. At start-up, the wastewater was recirculated with addition of glucose (1 $g \cdot dm^{-3}$). As a result, the autochthonous microorganisms created the biofilm. The influence of Hydraulic Retention Time (HRT) on the biodegradation efficiency was examined. Experiments took 4 months for 72 h HRT, 2 months for 24 h HRT and one month for 48 h HRT which is still being in operation. The samples of raw, as well as treated wastewater, were taken every 3 ÷ 5 days. Backwash was not conducted.

2.3 Analytical methods

The following analytical methods were applied:
– pH (WTW meter Multi 720);
– conductivity (WTW meter Multi 720);
– spectrophotometric analysis (according to PN-EN ISO 7887:2002);
– five-day biochemical oxygen demand, BOD_5 (dilution method, standard method—APHA, 1992);
– chemical oxygen demand, COD (standard dichromate method, spectrophotometer DR 5000, Hach-Lange);
– total organic carbon, TOC (analyser IL550TOC-TN, Hach-Lange);
– total nitrogen, TN (analyser IL550TOC-TN, Hach-Lange);
– total phosphorus, TP (spectrophotometer DR 5000, Hach-Lange).

The spectral absorption coefficient was calculated at wavelengths of 436 nm, 525 nm and 620 nm, according to Eq. (1).

$$SAC\,(m^{-1}) = \frac{A}{d} \times f. \qquad (1)$$

where A is the measured absorbance at a defined wavelength; d is the cuvette's diameter (mm); f is the factor for the conversion (f is 1000).

Table 3. SEM microphotographs of support materials before immobilisation of microbial cells in BAFs (magnification × 500).

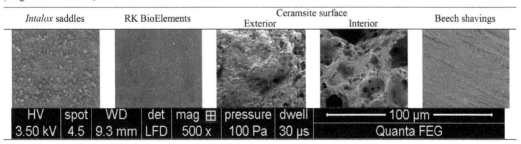

The organic loading rate (OLR, kg · m^{-3} · d^{-1}) was determined by:

$$OLR = \frac{HLR \times COD}{1000\ mg \cdot dm^{-3} \times 1\ m^3 \cdot kg^{-1}} \qquad (2)$$

where COD is the chemical oxygen demand (mgO$_2$ · dm^{-3}) and HLR is the hydraulic loading rate (m^3 · m^{-3} · d^{-1}).

3 RESULTS AND DISCUSSION

3.1 Quality evaluation of the wastewater

The wastewater was taken from the dyeing factory 17 times during 6 months of conducted experiment. Mean values and Standard Deviations (SD) of measured parameters for the used stream are presented in Table 4. COD influent values fluctuated from 523 to 1370 mg O$_2$ · dm^{-3}. The BOD$_5$: COD ratio, which is used as an indicator of the biode-

Table 4. Mean values and Standard Deviations (SD) of pollution parameters for raw wastewater.

		Mean value	SD
pH [−]		8.58	1.26
Conductivity [mS · cm^{-1}]		3.11	0.85
BOD$_5$ [mgO$_2$ · dm^{-3}]		296	90
COD [mgO$_2$ · dm^{-3}]		804	246
BOD$_5$/COD [−]		0.39	0.09
TOC [mgC · dm^{-3}]		296	107
TN [mgN · dm^{-3}]		16.48	14.07
TP [mgP · dm^{-3}]		4.97	1.56
SAC [m^{-1}]	436 nm	17.96	6.76
	525 nm	14.18	6.55
	620 nm	10.39	7.19

gradability, was in the range of the biodegradable wastewater, and contrary to other values, it underwent slight changes. High deviations among BOD$_5$, COD, TOC, TN, TP and spectrophotometric measurements are inevitable and arise from the nature of the dying process, which lasts quite a long time (Table 1). Only much longer wastewater averaging time may equalize all values, but this can cause preliminary biodegradation, especially when wastewater is warm. As the concentration of pollutants changed in raw wastewater, the Organic Loading Rate (OLR) varied from 0.54 to 0.87 kg COD m^{-3} · d^{-1} (24 h HRT); from 0.29 to 0.64 kg COD m^{-3} · d^{-1} (48 h HRT) and from 0.11 to 0.36 kg COD m^{-3} · d^{-1} (72 h HRT).

3.2 Effect of HRT on the treatment efficiency

The significant variations of OLR were easily tolerated by biomass. After a temporary deterioration, a stable degradation of the wastewater was soon observed in the system. Independently of the filling material used, the carrier efficiency of the biodegradable carbon organic compounds removal (measured as BOD$_5$) was very high—over 97% at HRT equal to 72 h (Table 5). The highest reduction of COD level was obtained for ceramsite – 85%. Process conducted basing on beech shavings obtained a slightly worse result – 83%. The worst result had experiment with RK BioElements—only 79%. Similar efficiencies in COD removal observed Chang et al. (2002) for BAF filled with zeolite (88%) and sand (75%). Finally, BOD$_5$, COD, TOC, TN concentrations of ceramsite and beechwood effluents were maintained within legalized range. On the other hand, the effluent from BAF filled with RK BioElements exceeded the permissible limits of COD, TOC and TP content.

Table 5. Characteristics of the BAF effluents for different support materials at 72 h HRT.

		RK BioElements		Ceramsite		Beech shavings		Legislation (Decree 2014)
		Effluent	Removal [%]	Effluent	Removal [%]	Effluent	Removal [%]	
pH [−]		8.27 ± 0.14	−	8.39 ± 0.20	−	8.35 ± 0.17	−	6.5–9.0
Conductivity [mS · cm^{-1}]		2.72 ± 0.34	−	3.01 ± 0.60	−	2.98 ± 0.57	−	n.a.
BOD$_5$ [mgO$_2$ · dm^{-3}]		8.95 ± 3.54	97.3 ± 0.7	4.67 ± 1.60	98.3 ± 0.7	5.22 ± 1.89	98.0 ± 0.7	25
COD [mgO$_2$ · dm^{-3}]		140 ± 39	79.4 ± 5.4	92 ± 25	85.2 ± 5.6	101 ± 21	82.6 ± 5.8	125
BOD$_5$/COD [−]		0.06 ± 0.02	85.9 ± 4.2	0.06 ± 0.03	86.6 ± 7.0	0.06 ± 0.02	88.1 ± 3.8	n.a.
TOC [mgC · dm^{-3}]		39.33 ± 14.52	83.8 ± 3.7	27.71 ± 13.73	87.6 ± 4.3	29.17 ± 5.63	86.7 ± 4.3	30
TN [mgN · dm^{-3}]		4.55 ± 1.66	55.2 ± 17.7	4.30 ± 2.34	57.7 ± 16.3	2.42 ± 0.64	76.0 ± 6.2	30
TP [mgP · dm^{-3}]		3.96 ± 1.72	18.2 ± 9.6	4.33 ± 1.54	8.9 ± 7.4	3.24 ± 1.81	27.4 ± 35.2	3
SAC [m^{-1}]	436 nm	6.71 ± 2.85	59.4 ± 11.2	5.32 ± 0.95	64.7 ± 11.2	5.97 ± 1.63	56.8 ± 18.0	n.a.
	525 nm	4.98 ± 3.18	58.3 ± 19.4	3.37 ± 1.27	70.0 ± 14.1	3.51 ± 1.87	66.4 ± 15.6	n.a.
	620 nm	2.86 ± 1.36	62.7 ± 15.0	2.00 ± 1.00	73.4 ± 12.9	2.17 ± 1.91	70.7 ± 10.0	n.a.

However, none of the experiments resulted in the required total phosphorus concentration level, all effluents exceeded 3 mgP · dm⁻³. In fact, TP removal was negligible. Only at the very beginning of the experiment for beech shavings TP concentration declined to 0.5 mgP · dm⁻³. This can be explained by the phenomenon of biofilm formation i.e. rapid proliferation of biomass phase and perhaps partly by sorption on the chips. In fact, no excess biomass has been taken away from the reactors, thus there was no phosphorus removal after initial period.

Decolorization was noticed for each BAF effluent—from 57% to 73%—and was stable for RK BioElements and ceramsite. The BAF with beechwood improved color removal degree month by month; however, the average values were similar to the afore-mentioned ones (Fig. 1).

Some of the results obtained for BAF filled with beechwood revealed phenomena not occurring in the other cases of used beds. The highest reduction of total nitrogen was probably a result of both adsorption of textile dyes (rich in nitrogen) on shavings, and nitrification plus denitrification processes which might have occurred in anoxic areas clogged by biofilm. However, the mean SAC reduction for the beech during the first two months of experiment was lower than for the ceramsite, 54.6% vs 64.0% respectively. Decolorization at the beech BAF gradually increased week by week, and finally reached 74.7% (mean value) during the third month (Table 6). It is likely that this phenomenon was not directly connected with the improvement of the decolorization process but with inevitable, gradually lessening release of colorants originated from beech shaves. Figure 2 presents the first week of the experiment. It is stated to illustrate the simultaneous extraction of beech originated colorants and decrease of textile dye concentration in effluent. There is a significant increase of absorb-

Table 6. Spectral absorption coefficient of effluents after biodegradation in BAF at 72 h HRT measured in third month from the beginning of the process.

		Beech shavings	
		Effluent	Removal [%]
SAC [m⁻¹]	436 nm	5.14 ± 1.29	69.6 ± 8.0
	525 nm	2.53 ± 0.98	77.3 ± 9.5
	620 nm	1.24 ± 0.45	77.1 ± 7.5

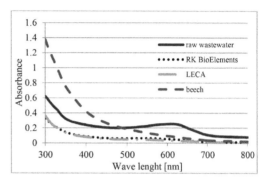

Figure 2. UV/Vis spectra of initial and treated wastewaters at HRT 72 h (first week of the process).

ance between 300–450 nm and wastewater decolorization effect is observed between 500–800 nm.

The color of beechwood is determined by many chemical compounds, i.e. cellulose, hemicelluloses, lignin and extractives that provide brownish hue. The last ones which take only a negligible percentage of wood weight are crucial. Beech extractive compounds consist mainly of polyphenols, especially flavonoids. Catechin is the dominant phenolic compound in beechwood (Vek et al. 2013). Both catechin, and other phenols owe the color to benzene ring, not to double nitrogen bound. Hence the conclusion that TN reduction better reflects textile dye lessening than spectrophotometric measurements. BAF filled with beech shavings remove 20% more TN than those with other carriers.

It is widely assumed that if BOD$_5$ reduction is close to 100%, the biodegradation process is complete and further treatment by this method is inefficient. Although the usage of the ceramsite and the beech shavings resulted in the satisfactory reduction levels, the process took a long time and influent OLR amounted only 0.22÷0.23 kg COD m⁻³ · d⁻¹. Thus, the experiment was repeated with two modifications: HRT was switched to 24 h and RK BioElements were changed to ceramic *Intalox* saddles.

Despite the shortening of HRT three times biodegradable organic compounds were still very efficiently removed: 94–96% (Table 7). The average COD reductions after biodegradation at 24 h HRT

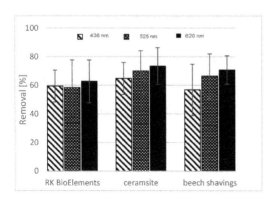

Figure 1. Color removal for different support materials at 72 h HRT.

Table 7. Characteristics of the BAF effluents for different support materials at 24 h HRT.

	Ceramic Effluent	Removal [%]	Ceramsite Effluent	Removal [%]	Beech shavings Effluent	Removal [%]	Legislation (Decree 2014)
pH [−]	8.43 ± 0.18	–	8.51 ± 0.19	–	7.97 ± 0.27	–	6.5–9.0
Conductivity [mS · cm^{-1}]	4.16 ± 1.2	–	4.04 ± 0.98	–	3.34 ± 0.18	–	n.a.
BOD$_5$ [mgO$_2$ · dm^{-3}]	12.51 ± 6.48	93.8 ± 2.7	10.54 ± 3.70	94.5 ± 2.3	7.85 ± 2.45	96.2 ± 0.9	25
COD [mgO$_2$ · dm^{-3}]	229 ± 44	60.4 ± 9.6	170 ± 33	70.0 ± 7.3	177 ± 23	68.3 ± 2.6	125
BOD$_5$/COD [−]	0.05 ± 0.02	85.5 ± 6.3	0.06 ± 0.02	81.0 ± 7.9	0.04 ± 0.01	88.8 ± 2.0	n.a.
TOC [mgC · dm^{-3}]	57.91 ± 8.86	66.5 ± 7.1	45.26 ± 7.74	73.4 ± 5.8	55.68 ± 2.82	68.0 ± 1.8	30
TN [mgN · dm^{-3}]	5.69 ± 1.78	43.4 ± 15.7	6.07 ± 1.86	42.8 ± 15.7	3.37 ± 0.43	75.6 ± 9.8	30
TP [mgP · dm^{-3}]	7.09 ± 1.92	0	6.13 ± 1.71	0	2.54 ± 1.45	26.8 ± 13.9	3
SAC [m^{-1}] 436 nm	10.00 ± 2.92	58.2 ± 6.9	9.47 ± 3.3	60.7 ± 8.7	6.75 ± 1.24	68.0 ± 8.6	n.a.
525 nm	5.22 ± 2.00	59.4 ± 10.3	4.89 ± 2.18	62.2 ± 12.1	3.00 ± 0.58	73.1 ± 7.7	n.a.
620 nm	3.62 ± 1.50	54.5 ± 10.0	3.68 ± 1.85	54.6 ± 14.9	1.58 ± 0.62	75.1 ± 11.5	n.a.

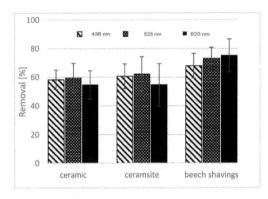

Figure 3. Color removal for different support materials at 24 h HRT.

were as follows: about 60% for *Intalox* saddles, 70% for ceramsite and 68% for beech shavings. Similar reductions occurred for TOC. It turned out that the kind of support, as well as HRT used, had the influence on decolorization—the best color removal was obtained for beechwood (77% at 525 and 620 nm by HRT equal to 72 h), the worst for ceramic and ceramsite (54.5%) at 620 nm by 24 h HRT (Figs. 1, 3).

The BAFs with ceramsite and beechwood were selected as the best operating bioreactors but 24 h for hydraulic retention time was not enough to meet the requirements for COD and TOC content. Thus experiment was repeated for these two supports and HRT was set at 48 h.

The received results, as expected, were between those previously described. The degree of BOD$_5$ removal was again very high: 96 and 97% (Table 8). Beech shavings achieved slightly better results for all parameters. For this application, TOC concentration mean value in effluent was 28.38 mgC · dm^{-3}, though temporarily exceed 30 mgC · dm^{-3}. Its color was also intangible (all SAC values below 5 m^{-1}). Unfortunately, raw wastewater introduced to BAFs operating at 48 h had very low color, thus decolorization capabilities are not fully demonstrated (Fig. 4). Irrespectively of these facts, both BAFs has insufficient COD concentration in effluents.

To sum up, at each HRT the required BOD and TN levels were obtained—below 25 mgO$_2$ · dm^{-3} and 30 mgN · dm^{-3}, respectively. However, COD and TOC concentrations did not always stay below the permissible level for effluent intended to be discharged into surface water. Only experiments conducted with ceramsite and beech shavings at HRT equal to 72 h fulfilled these requirements. Due to the fact that the presented experiments were conducted on the raw industrial textile wastewater, significant values of standard deviations were observed—for all parameters measured. Despite considerable fluctuations of analyzed pollution indicators, high removal efficiencies were obtained for each filling. Especially for BOD, but also COD, TOC, TN and color. The best purification outcome was observed after treatment in BAF with ceramsite at 72h HRT—effluent average values of BOD$_5$, COD and TOC were 4.67 mgO$_2$ dm^{-3}, 92 mgO$_2$ dm^{-3} and 27.71 mgC dm^{-3}, respectively.

Treating textile wastewater He *et al.* (2013) gained similar purification levels as for BOD$_5$ (≤ 7.6 mgO$_2$ dm^{-3}) and lower COD values (≤ 45 mgO$_2$ dm^{-3}) after integrated ozone biological aerated filters and membrane filtration. However, their biochemical and chemical oxygen demands of influent were lower: in the range of 12.6–23.1 mgO$_2$ dm^{-3} and 82–120 mgO$_2$ dm^{-3}, respectively. Liu et al. (2008) achieved average COD equal to 31.2 mgO$_2$ dm^{-3} after tertiary treatment of textile wastewater with combined media biological aer-

Table 8. Characteristics of the BAF effluents for different support materials at 48 h HRT.

		Ceramsite		Beech shavings		Legislation (Decree 2014)
		Effluent	Removal [%]	Effluent	Removal [%]	
pH [−]		8.32 ± 0.28	–	8.28 ± 0.14	–	6.5–9.0
Conductivity [mS · cm^{-1}]		2.41 ± 0.47	–	2.45 ± 0.51	–	n.a.
BOD$_5$ [mgO$_2$ · dm^{-3}]		9.43 ± 0.44	95.8 ± 1.3	6.89 ± 2.77	97.0 ± 1.4	25
COD [mgO$_2$ · dm^{-3}]		164 ± 23	76.6 ± 4.1	144 ± 31	77.8 ± 2.5	125
BOD$_5$/COD [−]		0.07 ± 0.01	81.3 ± 3.9	0.05 ± 0.02	86.4 ± 6.0	n.a.
TOC [mgC · dm^{-3}]		30.21 ± 6.75	85.3 ± 1.9	28.38 ± 6.61	84.8 ± 2.1	30
TN [mgN · dm^{-3}]		4.63 ± 3.03	65.4 ± 8.6	4.07 ± 2.70	71.0 ± 8.7	30
SAC [m^{-1}]	436 nm	6.11 ± 0.73	54.7 ± 5.4	4.77 ± 0.70	56.9 ± 13.0	n.a.
	525 nm	4.05 ± 0.62	51.3 ± 5.5	2.71 ± 0.41	66.1 ± 8.2	n.a.
	620 nm	3.61 ± 0.92	49.8 ± 6.1	1.83 ± 0.78	70.4 ± 7.8	n.a.

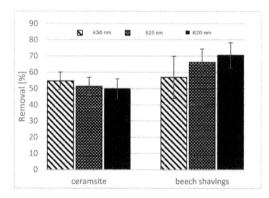

Figure 4. Color removal for different support materials at 48 h HRT.

ated filter. These results and our research confirm that BAFs are useful as for pretreatment, as well as secondary and tertiary wastewater purification.

3.3 SEM analysis of the support materials

In order to better understand the role of supports structure in biofilm formation, the SEM microphotographs were taken. In BAFs the carriers acted as supporting material for microbial growth. The larger active surface a carrier has, the more microbes can aggregate staying attached to the filter and more effective biodegradation is expected. After approximately a month of the biodegradation process, a noticeable biomass coating the support was observed in bioreactors. The deposit on ceramic *Intalox* saddles and RK BioElements was attached only to the carrier surface. On the other hand, the bioreactors with ceramsite and beechwood contained biomass in the entire volume. Furthermore, it was observed that on the expanded clay and beech shavings the biofilm was formed not only by bacteria but also by fungi (fungus mycelium was clearly visible, even small sporocarps). The wood chips were not sterilized before the application and thus fungus autochthonous microflora probably developed and then transferred by itself to ceramsite but did not grow in case of artificial carriers.

As it was presented in the Table 3, the expanded clay has a very porous internal structure, though the external surface is rough with only a few deep holes. It was expected that certain bacteria, especially anoxic, will grow inside the granules. In order to prove this statement, a sample of ceramsite was taken from BAF, after six months of continuous biodegradation. A thick layer of the formed biomass easily set apart from expanded clay. Only the firmly fixed layer of biofilm was documented in the Figure 5. The SEM observation confirmed that internal surface of ceramsite, as well as external, can be occupied by bacteria and higher organisms. Under × 5000 magnification some protozoa and bacteria are visible (Fig. 5), as well as certain examples of diatoms.

Expanded clay is beneficial for microbial growth providing extra internal space to aggregate and form the biofilm matrix. Cells growing within a biofilm have higher chances of adaptation to unfavorable changing concentration of the influent wastewater. Biofilms maintain optimal pH conditions and secure against harmful factors, thus allow cells to improve mineralization processes. Generally, the biofilms are characterized by the three layers that can be distinguished: a top layer the most fragile and easily detached; an intermediate, and a third residual layer remained on the surface, as mentioned by Klepacz-Smółka et al. (2015). The latter layer was documented on SEM photos.

Beech shavings providing irregular surface are also an excellent support for a biofilm formation. The immense amount of biomass was noticed after four weeks of biodegradation process. Additionally, at some underfed areas, it may become a substrate for certain cultures allowing their survival until con-

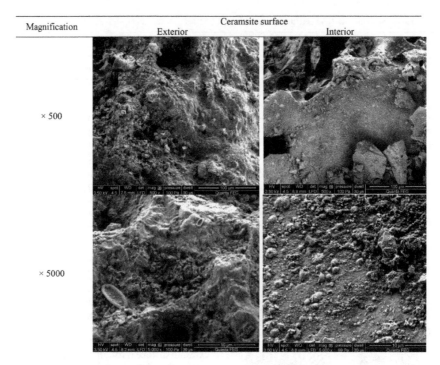

Figure 5. SEM microphotographs of ceramsite external and internal surface with the biofilm.

centration of mineralized pollutant grove. On the other hand, flavonoids protect wood against instant biodegradation, thanks to its antiseptic properties.

4 CONCLUSIONS

The biodegradation of the separated textile stream in BAF reactors operating at HRT equal to 72 h resulted in the fulfilling of the requirements of the Polish legislation for the influents to surface water as for BOD_5 and TN values. However, among the four tested biofilm supports i.e.: polypropylene, ceramic, ceramsite and beechwood only the latter two fulfilled the legal discharge requirements with COD and TOC.

Nevertheless, TP concentration periodically exceeded the acceptable range—a little reduction of phosphorus was noticed. That can be explained by the fact that surplus biomass was almost not produced at all. The low organic content of the influent stream could be efficiently degraded by the attached-growth biomass operating without backwash over several months of treatment. The fixed beds made of natural materials allowed for the development of more sophisticated biofilms: combination of bacteria and fungus consortium. Generally, the results demonstrated that the biodegradation by means of BAFs is a suitable and highly efficient method of treating the separated stream from dye mill.

Apart from the beech shavings, there were no significant differences between the decolorization efficiencies obtained for the used fillings. An average color removal varied between 52 and 69% for *Intalox* saddles, RK BioElements, and expanded clay. The best decolorization efficiencies were obtained after treatment in BAF filled with the beech shavings—up to 75%. Beechwood decreased color slightly better than the other fillings probably due to adsorption and biological dye degradation, what was proved by higher TN reduction—up to 76%, while the other filters 43 ÷ 65%.

ACKNOWLEDGEMENTS

This research was supported by the National Centre of Research and Development (NCBiR) in Poland within research project PBS2/A9/22/2013. The authors wish to thank PhD Jan Sielski for his valuable help and SEM microphotographs.

REFERENCES

Ali, H. 2010. Biodegradation of Synthetic Dyes—A Review. *Water Air Soil Pollut.* 213: 251–273.
Allegre, C., Moulin, P.; Maisseu, M. & Charbit, F. 2006. Treatment and reuse of reactive dyeing effluents. *J. Mem. Sci.* 269:15–34.

APHA, 1992. Standard Methods for the Examination of Water and Wastewater, 18th edn. American Public Health Association (APHA), Washington, DC.

Bemska, J., Bilińska, L. & Ledakowicz, S. 2012. Analiza ścieków włókienniczych pod kątem wyboru najlepszej metody ich oczyszczania. In Zbigniew Hubicki (ed.), *Nauka i przemysł—metody spektroskopowe w praktyce, nowe wyzwania i możliwości*. 706–709 Lublin: Uniwersytet Marii Curie-Skłodowskiej.

Bilinska, L., Gmurek, M. & Ledakowicz, S. 2015. Application of advanced oxidation technologies for decolorization and mineralization of textile wastewaters. *J. Adv. Oxid. Technol.* 18(2): 185–194.

Chang W.S., Hong, S.W. & Park, J. 2002. Effect of zeolite media for the treatment of textile wastewater in a biological aerated filter. *Process Biochemistry* 37:693–698.

Christie, R.M. 2001. *Colour Chemistry*. Cambridge: Royal Society of Chemistry.

Decree 2014, Rozporządzenie Ministra Środowiska z dnia 18 listopada 2014r. w sprawie warunków, jakie należy spełnić przy wprowadzaniu ścieków do wód lub do ziemi, oraz w sprawie substancji szczególnie szkodliwych dla środowiska wodnego (Dz. U. 2014 poz. 1800).

European Commission, Integrated Pollution Prevention and Control (IPPC), Reference Document on Best Available Techniques for the Textiles Industry 2003.

Fu, Z., Zhang, Y. & Wang, X. 2011. Textiles wastewater treatment using anoxic filter bed and biological wriggle bed-ozone biological aerated filter. *Bioresour. Technol.* 102: 3748–3753.

Gupta, V.K. & Suhas 2009. Application of low-cost adsorbents for dye removal—A review. *J. Environ. Management* 90: 2313–2342.

Handa, B.K. 1991. *Treatment and recycle of wastewater in industry*. Nagpur: National Environmental Engineering Research Institute.

He, Y., Wanga, X., Xu J., Yan J., Ge Q., Gu X. & Jian L. 2013. Application of integrated ozone biological aerated filters and membrane filtration in water reuse of textile effluents. *Bioresour. Technol.* 133:150–157.

Imran, M., Crowley, D.E., Khalid, A., Hussain, S., Mumtaz, M.W. & Arshad, M. 2015. Microbial biotechnology for decolorization of textile wastewaters. *Rev. Environ. Sci. Biotechnol.* 14:73–92.

Kazemi, S.Y., Biparva, p. & Ashtiani, e. 2016. *Cerastoderma lamarcki* shell as a natural, low cost and new adsorbent to removal of dye pollutant from aqueous solutions: Equilibrium and kinetic studies. *Ecological Engineering* 88: 82–89.

Khan, R. & Banerjee, U.C. 2010. Decolorization of azo dyes by immobilized bacteria. In Erkurt, H. A. (ed.), *Biodegradation of azo dyes*. Springer Berlin Heidelberg.

Klepacz-Smółka, A., Sójka-Ledakowicz, J. & Ledakowicz, S. 2015. Biological Treatment of Post-Nanofiltration Concentrate of Real Textile Wastewater. *Fibres & Textiles in Eastern Europe*, 23, 4(112): 138–143.

Li, X., Shi, H., Li, K. & Zhang, L. 2015. Combined process of biofiltration and ozone oxidation as an advanced treatment process for wastewater reuse. *Front. Environ. Sci. Eng.* 9(6): 1076–1083.

Liu, F., Zhao, C., Zhao, D. & Liu, G. 2008. Tertiary treatment of textile wastewater with combined media biological aerated filter (CMBAF) at different hydraulic loadings and dissolved oxygen concentrations. *J. Hazard. Mater.* 160: 161–167.

López-Grimau, V., Gutierrez-Bouzan, M., Valldeperas, J. & Crespi, M. 2011. Reuse of the water and salt of reactive dyeing effluent after electrochemical decolorisation. *Color. Technol.* 128: 36–43.

Melián, J.A.H., Ortega Méndez, A., Araña, J., Díaz, O.G. & Rendón, E.T. 2008. Degradation and detoxification of formalin wastewater with aerated biological filters and wetland reactors. *Process Biochem.* 43: 1432–1435.

Rouette, H.K. 2001. *Encyclopedia of Textile Finishing*. Springer-Verlag Berlin, Heidelberg: Woodhead Publishing.

Ryu, H.D., Kim, D., Lim, H.E. & Lee, S.I. 2008. Nitrogen removal from low carbon-to-nitrogen wastewater in four-stage biological aerated filter system. *Process Biochem.* 43: 729–735.

Sharma, P., Kaur, H., Sharma, M. & Sahore, V. 2011. A review on applicability of naturally available adsorbents for the removal of hazardous dyes from aqueous waste. *Environ. Monit. Assess.* 183:151–195.

Vandevivere, P.C., Bianchi, R. & Verstraete, W. 1998. Treatment and Reuse of Wastewater from the Textile Wet-Processing Industry: Review of Emerging Technologies. *J. Chem. Technol. Biotechnol.* 72: 289–302.

Vek, V., Oven, P. & Poljanšek, I. 2013. Quantitative HPLC Analysis of Catechin in Wound Associated Wood and Knots of Beech. *Wood Industry* 64(3): 231–238.

Vignesh, A., Siddarth, A.S., Gokul, O.S. & Babu, B.R. 2014. A novel approach for textile dye degradation by zinc, iron–doped tin oxide/titanium moving anode. *Int. J. Environ. Sci. Technol.* 11: 1669–1678.

Wang, C., Li, J., Wang, B. & Zhang, G. 2006. Development of an empirical model for domestic wastewater treatment by biological aerated filter. *Process Biochem.* 41: 778–782.

Wenyu, X., Li, Z. & Jianjun, C. 2007. Treatment of slightly polluted wastewater in an oil refinery using a biological aerated filter process. *Wuhan University J. of Natural Sci.* 12(6): 1094–1098.

Wrębiak, J., Bilińska, L., Paździor, K. & Ledakowicz, S. 2014. Ocena biodegradowalności wyodrębnionych strumieni ścieków z farbiarni. *Przegląd Włókienniczy— Włókno Odzież Skóra* 5: 46–49.

Wu, Q., Li, W.T., Yu, W.H., Li, Y. & Li, A.M. 2016. Removal of fluorescent dissolved organic matter in biologically treated textile wastewater by ozonation-biological aerated filter. *J. Taiwan Inst. Chem. E.* 59: 359–364.

Yu, Y., Feng, Y., Qiu, L., Han, W. & Guan, L. 2008. Effect of grain-slag media for the treatment of wastewater in a biological aerated filter. *Bioresource Technol.* 99: 4120–4123.

Zollinger, H. 2003. *Color Chemistry: Syntheses, Properties, and Applications of Organic Dyes and Pigments*. Zurich: Verlag Helvetica Chimica Acta.

Zou, X. 2015. Combination of ozonation, activated carbon, and biological aerated filter for advanced treatment of dyeing wastewater for reuse. *Environ. Sci. Pollut. Res.* 22: 8174–8181.

Environmental Engineering V – Pawłowska & Pawłowski (Eds)
© 2017 Taylor & Francis Group, London, ISBN 978-1-138-03163-0

Hydroecological investigations of water objects located on urban areas

W. Wójcik
Faculty of Electrical Engineering and Computer Science, Lublin University of Technology, Lublin, Poland

V.P. Osypenko
Institute of Hydrobiology of National Academy of Sciences of Ukraine, Kyiv, Ukraine

V.V. Osypenko
National University of Life and Environmental Sciences of Ukraine, Kyiv, Ukraine

V.I. Lytvynenko
Kherson National Technical University, Kherson, Ukraine

N. Askarova & M. Zhassandykyzy
Kazakh National Research Technical University After K.I. Satpaev, Almaty, Kazakhstan

ABSTRACT: The results of the annual dynamics investigations of the major hydrochemical properties of the Lybid River (right tributary of Dnipro River) and Gorikhovatsky Pond are presented. The pH, dissolved oxygen concentration and degree of oxygen saturation of water, bichromate and permanganate oxidizabilities of water are defined. Their correlation with the seasonal dynamics of concentrations of such dissolved organic matters as humic acid, fulvic acid, carbohydrate and protein are considered. It has be noted that the ecological features of the chemical composition of water are largely caused both by natural processes and anthropogenic pressure.

Keywords: hydrochemical investigations, humic acid, fulvic acid, carbohydrate, protein, seasonal dynamics

1 INTRODUCTION

Currently, the city water bodies and watercourses are often used not only as a source of drinking and industrial water but also as a means for utilization of both drain and waste water of domestic and industrial origin. The network of city collector sewer drain realizes their runoff that leads to permanent pollution of water objects. The typical examples of this environmental situation are the Lybid River, which is the right tributary of the Dnipro River and flows in the south-western part of Kyiv and Gorikhovatsky pond, located in the city park of recreation.

Experts estimate that the overall quality of water in Lybid is three to five times worse than in the river, which is typical for rivers flowing through the large industrial centers (Zaverukha et al., 2008). The contribution of the organic components of urban waste in their pollution is very high, but the Dissolved Organic Matter (DOM) of Lybid River is practically unexplored. From the ecological aspect, they are not only the health indicators of

water quality, but also the criteria for the operation of hydrobiocenoses, biological processes in which occurring simultaneously with the absorption and the release of organic compounds. The fact that Lybid River flows into the Dnipro River and the water resources of the Dnipro basin constitute about ¾ of the Ukraine water resources makes the this ecological situation very serious.

The Gorikhovatsky Pond is a part of a network of artificial water bodies in the Gorikhovatka riverbed which is one of the river Lybid tributaries. It is located in the downstream part of the river and, therefore, it is potentially experiencing the greatest anthropogenic pressure. A significant contribution to pollution of this object makes its closely spaced highway, residential area and catering and leisure. In addition, the water level gradually falls because the metabolites of aquatic organisms settle to the bottom, creating swampy sediments.

Consequently, a large and increasingly negative role in shaping the quality of the water environment is played not only by natural but also anthropogenic factors, especially in aquatic ecosystems located near and within settlements. Depending on the influence of any factor to forming the component composition of dissolved organic matter it is possible to evaluate both the chemical and biological state of the reservoir or watercourse. The objective of the present work was to study the annual and seasonal dynamics of the total dissolved organic matter content, concentration their components—humic acid, fulvic acid, carbohydrate and protein—in connection with some hydrochemical characters also pH and the concentration of oxygen dissolved in water.

2 MATERIALS AND METHODS

The samples of water for investigating total hydrochemical indicators were taken in 2014 in monthly intervals. In order to study seasonal dynamics of the content Humic Acids (HA), Fulvic Acids (FA), carbohydrates (C) and proteins (P) the samples were taken in February, April, July and October. The samples of water were filtered through the «Synpor» membranous filter (Czech Republic) with the sieve openings 0.4 μm in diameter. Then pH, dissolved oxygen concentration and degree of oxygen saturation, bichromate and permanganate oxidizabilities (BO and PO) of water were conducted following standard procedures (Semenow, 1977).

Fractionation of dissolved organic matter was carried out with the ion-exchange chromatography method using the DEAE (diethylaminoethyl)- and CM (carboxymethyl)-cellulose (Sirotkina et al., 1974). Humic acids and fulvic acids were determined with the spectrophotometry method in

terms of their own coloration at $\lambda = 400$ nm and by their chemical reaction (Perminova, 2000), carbohydrates—using antrone (Semenow, 1977), proteins—by the Folin–Lawry method (Debeyko et al., 1973).

3 RESULTS AND DISCUSSION

As has been noted, the content and dynamics of dissolved organic matter concentration in the water are important hydrochemical indicators of the aquatic environment. Active reaction of water environment (pH) and the concentration of oxygen dissolved in water are abiotic factors that greatly affect the overall content and component composition of dissolved organic matter. In order to describe the overall ecological condition of Lybid River, monthly investigations of the aforementioned hydrochemical indicators have been conducted (Table 1).

As shown in the Table 1, the samples of water from the Lybid River were selected in two sections, i.e. the upper and lower stream of river. The observations of the pH change showed that it ranged from 7.4 to 8.3. Moreover, the lowest pH was noted in the spring and summer, whereas the highest one was noted in the autumn and winter. This dynamic is different from the annual dynamics of pH in the natural water and the streams studied by us earlier (Osypenko et al., 2012). Probably, the high anthropogenic load on Lybid River violates the natural processes of the phytoplankton on regulate the carbonate equilibrium and, consequently, the pH of the water (Denisova et al., 1989). It should be

Table 1. The annual dynamics of indicators pH, dissolved oxygen, degree of oxygen saturation, bichromate and permanganate oxidizabilities in the water of the Lybid River.

Month	pH		O_2, mg/dm^3		Degree of oxygen saturation, %		Permanganate oxidizabilities, mgO/dm^3		Bichromate oxidizabilities, mgO/dm^3	
	1	2	1	2	1	2	1	2	1	2
January	7.7	7.7	11.3	12.1	82.5	90.6	18.2	16.2	30.4	27.4
February	7.6	7.6	10.5	11.6	82.0	100.7	7.7	6.4	43.2	40.3
March	7.6	7.6	10.8	11.5	100.2	104.5	8.0	6.4	43.5	38.4
April	7.4	7.5	8.1	8.8	76.9	81.8	9.6	9.1	53.8	51.2
May	7.6	7.6	7.8	7.9	88.4	89.5	14.4	14.7	36.0	38.4
June	7.6	7.5	8.3	8.4	87.0	89.7	13.1	11.8	29.1	27.9
July	7.6	7.5	7.0	7.2	79.8	82.1	13.1	12.8	33.4	27.4
August	7.7	7.7	7.8	8.1	86.5	89.9	12.5	10.9	35.4	35.0
September	8.3	8.3	8.5	9.0	88.2	93.2	10.5	9.6	65.3	54.4
October	8.2	8.2	9.1	9.6	82.1	82.2	7.4	6.4	25.6	23.0
November	–	–	–	–	–	–	–	–	–	–
December	8.3	8.3	9.5	9.7	82.2	82.4	7.0	6.5	33.3	29.1

Note: 1, 2 – numbers of the sites where the water samples were taken.

156

Table 2. The seasonal dynamics of the components content dissolved organic matters in the water from the Lybid River.

Season	Humic acids, mg/dm^3	Fulvic acids, mg/dm^3	Carbo-hydrates, mg/dm^3	Proteins, mg/dm^3
Winter	0.20	5.6	1.28	0.38
Spring	0.30	8.6	1.71	0.59
Summer	0.46	7.3	2.43	0.72
Autumn	0.07	6.3	0.57	0.23

noted that even in natural water bodies these factors can exert an adverse effect on the development of aquatic plants and animals.

The oxygen regime of Lybid River during the year was characterized by sufficiently high content of dissolved oxygen in water ranging from 7.0 mg/dm^3 (79.8% saturation) in July to 12.1 mg/dm^3 (90.6% saturation) in January. However, this oxygen regime shows more likely about the hydrology watercourse features than about the satisfactory conditions.

Throughout the year, the total content of organic matters in the water were monitored in terms of BO and PO of water. Their values varied in fairly wide ranges without clear intra-annual regularities. It should be noted that often been observed a discrepancy in the changes of magnitude of bichromate and permanganate oxidizabilities. Thus, in September, the indicator of bichromate oxidizabilities that normally displays the presence in water the hard oxidizable organic substances was the highest (65.3 mg O/dm^3), whereas the quantity was substantially reduced. The maximum indicators for permanganate oxidizabilities (18.2 mg O/dm^3), which mainly reflect the presence of easily oxidizable organic compounds has come from January, in the absence of ice. The bichromate oxidizabilities of this month was lower than the average annual value. The minimum values of bichromate and permanganate oxidizabilities observed in October amounted to 23.0 and 6.4 mg O/dm^3, respectively.

As for the ratio of the bichromate and permanganate oxidizabilities quantities, in some months the value of bichromate oxidizabilities was 5.5 to 6.0 times higher than the value of permanganate oxidizabilities. The average annual bichromate to permanganate oxidizabilities averaged 3.5. In natural waters, this ratio of bichromate and permanganate oxidizabilities may indicate a low degree of transformation of organic matter and predominance in waterways aliphatic autochthonous dissolved organic matter (Vasylchuk, 2011). In this case, we can also speak poorly-oxidizable contaminants naturally occurring both during spring floods and torrential rains, and through the sewers, coming into contact with the water runoff.

The content of individual groups of organic substances during the year varied in the intervals: humic acids—from 0.07 to 0.46 mg/dm^3, fulvic acids—from 5.60 to 8.60 mg/dm^3, carbohydrates—from 0.57 to 2.43 mg/dm^3, proteins—from 0.23 to 0.72 mg/dm^3 (Table 2). The seasonal dynamics of the main components concentration of dissolved organic matter are mostly consistent with traditional changes in their content in water (Bashenkhaeva, 2006). The continuous process of metabolism in the cells hydrobionts includes both a release of the products of their vital activity into environment and the uptake of organic nutrients for their growth and reproduction. Thus, it is possible to assess the processes of production and decomposition registered in natural waters using seasonal dynamics of these basic groups of organic matter—humic acids, fulvic acids, carbohydrates and proteins.

It is known, that the formation of the water chemical composition is greatly affected by how the hydrological and hydrobiological factors are associated with the life of aquatic organisms (Klochenko et al., 2014; Sakevich, 1985). Thus, the concentrations of humic acids, carbohydrates and proteins have been increased from winter to spring, reaching the maximum values of 0.46, 2.43 and 0.72 mg/dm^3, respectively in the summer due to the intensive development of phytoplankton and zooplankton in the water and isolating the products of their metabolism. Only the content of fulvic acids had a maximum value (8.6 mg/dm^3) in spring. It could be explained by of large amounts in water humus substances which appeared in the spring flood season. In autumn, due to the attenuation of vital processes of aquatic organisms and settling of organic products on the bottom, decreases have been observed in the content of humic acids, carbohydrates and proteins – 6.6, 4,3 and 3.1 times, respectively. fulvic acids concentration decreased slightly, possibly because they come with an additional surface runoff and groundwater during long autumn rains.

Among the features pertaining to the seasonal distribution of the various components of dissolved organic matter in the water from the Lybid River, it should be noted that the changes in fulvic acids content were less pronounced as compared to other organic studied substances. From the literature it is known that the products of wastewater entering the river during a year, can serve as an additional source of water humus, which also includes fulvic acids, i.e. the components with low molecular weight and more mobile than humic acids ones in the ecosystems organic compounds (Manka et al., 1974).

As it has been said, the list of important abiotic factors that determine the general content and component composition of dissolved organic

Table 3. The annual dynamics of indicators temperature, pH, dissolved oxygen, degree of oxygen saturation, bichromate and permanganate oxidizabilities in the water of the Gorikhovatsky Pond.

Month	Water temperature °C	pH	O_2, mg/L	Oxygen saturation %	Permanganate oxidizabilities mgO/L	Bichromate oxidizabilities mgO/L
January	1	7.5	5.0	35.5	8.3	36.5
February	2	7.2	4.1	29.0	8.4	46.1
March	14	7.7	11.0	106.4	8.8	58.9
April	10	7.6	10.2	90.8	9.7	61.4
May	23	7.8	8.7	100.3	12.2	43.6
June	20	7.9	8.9	96.3	13.8	45.8
July	25	7.3	5.0	58.9	16.0	45.6
August	22	7.9	5.5	61.8	16.6	43.5
September	17	8.0	4.3	43.7	14.1	52.6
October	6	8.2	4.8	38.3	10.9	56.1
November	–	–	–	–	–	–
December	2	7.6	5.6	40.9	6.4	49.3

matter in water include the temperature mode and pH, as well as the concentration of oxygen dissolved in water. The summarized results of the annual study of above-mentioned indicators concerning the dynamics in water Gorikhovatsky Pond are presented in the Table 3.

Important abiotic factors that determine the total content and component composition of dissolved organic matter in water include the temperature mode, pH and the dissolved oxygen concentration. The summarized results of the annual study pertaining to the dynamics of these indicators in the water from the Gorikhovatsky pond are presented in the Table 3.

Monitoring of the pH in water showed that it ranged from 7.2 (February) to 8.2 (October). The study of the content of dissolved oxygen in water showed that it varied widely: 4.1 mg/dm³ (29.0% saturation) - 11.0 mg/dm³ (106.4% saturation). In general, the oxygen regime of the water pond can be assessed as unsatisfactory. Six months within a year, i.e. in January, February, July, September, October and December, the degree of water saturation with oxygen does not exceed 50.0%.

The annual average concentration of oxygen in water was only 6.6 mg/dm³ (62.9% saturation). Observation of the dynamics of its content in the water showed that the studied parameters decreased during the freezing of the pond (February), during intense "bloom" of water (July) and further slow decomposition of the accumulated biomass of phytoplankton (September, October).

Table 4 shows the seasonal dynamics of the separate components of dissolved organic matter in water—humic acids, fulvic acids, carbohydrates and proteins. They are connected in a certain way with indicators PO and BO, which reflect both

Table 4. The seasonal dynamics of the components content dissolved organic matters in the water of the Gorikhovatsky Pond.

Season	Humic acids, mg/dm³	Fulvic acids, mg/dm³	Carbo-hydrates, mg/dm³	Proteins, mg/dm³
Winter	0.30	6.0	1.91	0.47
Spring	0.45	9.4	2.71	0.70
Summer	0.61	9.7	3.43	0.83
Autumn	0.40	20.3	2.57	0.47

the hydrological and biological processes that take place in the pond.

Thus, the permanganate oxidizabilities indicators ranged from 6.4 mg O/dm³ (December) to 16.6 mg O/dm³ (August). The minimum and maximum concentrations of carbohydrates and proteins, which belong to the group of easily oxidized dissolved organic matter have also been observed in winter and summer respectively. For carbohydrates, these concentrations were 1.91 (winter) and 3.43 (summer) mg/dm³; for proteins – 0.47 (winter) and 0.83 (summer) mg/dm³.

Bichromate oxidizabilities values varied from 36.5 (January) to 61.4 (April) mg O/dm³. During the year, a high bichromate oxidizabilities was observed in March and April (58.9 and 61.4 mg O/dm³), as well as in September and October (52.6 and 56.1 mg O/dm³). Analyzing the seasonal concentration of compounds with a low capacity for oxidation of humic acids and fulvic acids, it should be noted that they increased in spring to 0.45 and 9.4 mg/dm³ respectively. Among the features concerning the seasonal distribution of humus substances in the water of the Gorikhovatsky Pond, the maximum concentration of fulvic acids occurred in autumn (20.3 mg/L).

In contrast to the appearance of humic substances in water from surface and underground outflows as a result of spring floods, such uncharacteristically high concentration of fulvic acids in the autumn can be explained by the slow decomposition of excessive phytoplankton biomass and the formation of the water humus (Novikov, 2000). A relatively high concentration of carbohydrates (2.57 mg/dm^3), which are the basic building blocks of phytoplankton cells, supports this conclusion. From the literature it is known that these substances may accumulate in water when the mass of algae dies off (Sakevich, 1985). Low oxygen content in the water at this time, as noted above, dull greenish colour and odour of water indicate stagnation in the reservoir. We cannot include the accumulation of organic residues in the Gorikhovatsky Pond water as a result of their supply from the other ponds that are located upstream (Osypenko, 2014).

4 CONCLUSIONS

The ecological features of the chemical composition of water in Kyiv water objects are largely determined both by natural processes and anthropogenic pressure, which is manifested not only in the direct entering into the water products of human activity, but also indirectly, as a result of changes in the physical, chemical and biological factors in the aqueous environment. Monitoring of annual and seasonal changes of the total and components content dissolved organic matter in the water, including humic acids, fulvic acids, carbohydrates and proteins, along with other hydrochemical and hydrobiological indices makes it possible to assess the ecological state of water objects.

REFERENCES

Bashenkhaeva, N.V., Sinyukovich, V.N., Sorokovikova, L.M. & Khodzher, T.V. 2006. Organic matter in the water of the river Selenga. *Geographiya i prirodniye resursy* 1: 47.
Debeyko, Y.V., Ryabov, A.K. & Nabivanets B.I. 1973. Direct photometric determination of dissolved proteins in natural waters. *Hydrobiological J.* 9(6): 109.
Denisova, A.N., Timchenko, V.M., Nakhshina, E.P. et al. 1989. *Gidrologiya i gidrokhimiya Dnepra i ego*

vodokhranilisch (Hydrology and hydrochemistry of the Dnipro and its reservoirs). Kyiv: Naukova Dumka Press.
Dziadak, B. 2013. Short and long-term data prediction for water quality estimation. *Przeglad Elektrotechniczny* 89(6): 278–280.
Klochenko, P. D., Shevchenko, T. F., Vasilchuk, T. A., Osypenko, V. P., Yevtukh, T. V., Medved, V. A. & Gorbunova Z. N. 2014. On the ecology of phytoepiphyton of water bodies of the Dnieper River basin. *Hydrobiological J.* 50(3): 41.
Manka, J., Robhum, M., Mandelbaum, A. & Bortinger A. 1974. Characterization of organics in secondary effluents. *Environ. Sci. Technol.* 8(12): 1017.
Novikov, M.A. & Kharlamov, M.N. 2000. Transabiotic factors in the aquatic environment (review). *Zhurnal obschey biologii.* 61(1): 22.
Osypenko, V. P. 2014. Seasonal and spatial changes in the content and molecular mass distribution of carbohydrates in the surface water. *Hydrobiological J.* 50(5): 89.
Osypenko, V.P., Vasylchuk, T.A. & Yevtukh, T.V. 2012. Seasonal dynamics of basic groups of organic matter in different water bodies. *Gidrologiya, gidrokhimiya i gidroekologiya (Hydrology, hydrochemistry and hydroecology)* 1(26): 134.
Perminova, I.V. 2000. Analiz, klassifikatsiya I prognoz svoystv gumusovykh kislot (Analisis, classification and prediction of the properties of humic acid). *Author's abstract of Doctoral thesis.* Moscow.
Rakhmetullina, S., Turganbayev, Y. & Gromaszek, K. 2012. Application of variational data assimilation algorithms in the ecological monitoring system. *IAPGOS* 2012(4): 33–35.
Sakevich, A.I. 1985. *Ekzometabolity presnovodnykh vodorosley (Exometabolites of limnetic algae)*. Kyiv: Naukova Dumka Press.
Semenow, A.D. 1977. *Rukovodstvo po khimicheskomu analizu poverkhnostnykh vod sushi (Manual on the chemical analysis of surface waters of land)*. Leningrad: Gidrometeoizdat.
Sirotkina, I.S., Varshal, G.M., Lurye, Y.Y. & Stepanova N.P. 1974. Use of cellulose sorbents and sefadexes in the systematic analysis of organic matter in natural waters. *Zhurnal analiticheskoy khimii (Analytical chemistry J.)* 29(8): 1626.
Vasylchuk, T.A., Osypenko, V.P. & Yevtukh T.V. 2011. The peculiarities of the migration and distribution of main groups of organic matter in the water of the Kyiv reservoir depending on the oxygen regime. *Hydrobiological J.* 47(2): 97.
Wojcik, W., Bieganski, T., Kotyra, A., Smolarz, A., Wojcik, J. & Janoszczyk, B. 1997. Application of algorithms of forecasting in the optical fibre coal dust burner monitoring system. *Proc. of SPIE* 3189: 100–109.
Zaverukha, N.M., Serebryakov, V.V. & Skiba, Y.A. 2008. *Osnovy ekologii (Foundations of ecology)*. Kyiv: Caravela Press.

Environmental Engineering V – Pawłowska & Pawłowski (Eds)
© 2017 Taylor & Francis Group, London, ISBN 978-1-138-03163-0

An influence of sludge compost application on heavy metals concentration in willow biomass

D. Fijałkowska, L. Styszko & B. Janowska
Faculty of Civil Engineering, Environmental and Geodetic Sciences, Koszalin University of Technology, Koszalin, Poland

ABSTRACT: The aim of the study was to determine the contents of heavy metals in biomass of willow fertilized with compost from municipal sewage sludge. Following objects were drawn on the big plots: (a) without fertilization, (b) fertilized with compost (10 t dm·ha^{-1}), (c) fertilized with compost and mineral nitrogen (90 kg N·ha^{-1}) and (d) fertilized with compost and mineral nitrogen (180 kg N·ha^{-1}). Within big plots, nine clones of willow were drawn. The highest contents of heavy metals in willow shoots were determined on object (a) for cadmium and manganese and (b) for copper, cobalt and zinc. The lowest were noted on objects (a) for cobalt, (b) for manganese, (c) for cadmium and (d) for copper and zinc. The highest accumulation of metals was noted for the following clones: 1054 and 1052 (cadmium), 1033 (copper), 1018 (lead), 1054 (zinc), 1052 (manganese), and the lowest respectively: 1013, 1054, 1052, 1013 and 1018.

Keywords: heavy metals, *Salix viminalis*, clones, fertilization, compost, sewage sludge, mineral nitrogen

1 INTRODUCTION

Heavy metals are naturally present in the Earth's crust. They include, among others: Cadmium (Cd), Copper (Cu), Chromium (Cr), Cobalt (Co), Nickel (Ni), lead (Pb), Zinc (Zn) and Manganese (Mn). Natural phenomena and human activity in a variety of industries and municipal economy are sources of environmental pollution with heavy metals. Sewage sludge produced during wastewater treatment may be used as an organic fertilizer in agriculture and for restoration of soilless and degraded areas. Several wastewater treatment plants produce compost from municipal sewage sludge for commercial purposes. Koszalin University of Technology had purchased compost from sewage sludge and applied it for fertilization of energetic willow cultivation on sandy soil in Kościernica experimental field. That particular compost (produced by composting plant in Sianów) is certified for use in agriculture. It is a suitable fertilizer for use in cultivation of energetic willow due to high content of biogenic elements and alkali metals. It is also characterized by a low heavy metal content and very low odour nuisance (Czechowska-Kosacka et al. 2015a, b, Sobczyk et al. 2015, Starzyk et al. 2015). According to our previous studies, a dose of 10 Mg of compost from municipal sludge per hectare had an influence on the content of alkali (1.5 fold) and no influence on heavy metals content in layer of soil up to the depth of 90 cm in the first year after its application (Fijałkowska et al. 2010).

The application of compost during cultivation of energetic plants was easier and more socially acceptable than the use of wastewater or sewage sludge due to lower odour nuisance.

Plants which may effectively absorb heavy metals from the soil should be characterized by: rapid growth, high biomass yield, ease of harvest, deep root system and accumulation of large amounts of those metals in the overground parts (Karczewska 2008). The effectiveness of phytoextraction of heavy metals from soil depends on the amount of water passing through the plant per unit of time. Therefore, it is considered that willow is well suited for this purpose, and it has a high water demand (Żurek & Majtkowski 2009). Among the species of willow, *Salix viminalis* can absorb 217 g·ha^{-1} of cadmium from soil and *Salix caprea* has the ability to absorb $2,340$ g·ha^{-1} of zinc, 76 g·ha^{-1} of copper, 242 g·ha^{-1} of lead and 41 g·ha^{-1} of cadmium (Żurek & Majtkowski 2009). Heavy metals content in plants varies depending on their ability to move from soil to shoots. The arrangement of soil to shoots transportation of heavy metal ions is in accordance with their decreasing mobility: Cd > Zn > Ni > Cu > Pb (Starck 2002).

Selected clones of common osier (*Salix viminalis*) with a relatively high tolerance to pollution, large phytoextraction capacities and satisfactory yield of biomass are recommended for cultivation on soils of medium and heavy pollution with heavy metals (Kabała et al 2010). The wood of common osier, cultivated for energy purposes, contains small amounts of heavy metals in dry mass (70–140 mg·kg^{-1} of Zn, 7–10 mg·kg^{-1} of Cu, 5–11 mg·kg^{-1} of Ni, 1–3 mg·kg^{-1} of Pb, 0.1–0.3 mg·kg^{-1} of Cr and 0.3–0.6 mg·kg^{-1} of Cd) (Kabała et al 2010, Kaniuczak et al. 2000).

Willow has a strong ability to absorb and accumulate zinc and cadmium from soil. Wood of willow cultivated on soil heavily contaminated by emissions from smelting plants may contain up to 4000 mg·kg^{-1} of Zn, 64 mg·kg^{-1} of Cd, 20 mg·kg^{-1} of Cu and 10 mg·kg^{-1} of Pb (Mathe-Gaspar & Anton 2005). Contents of zinc and cadmium in roots and overground parts of willow are usually similar, and content of copper and lead may be 10–30 times higher in roots than in wood of shoots (Mathe-Gaspar & Anton 2005). 40-fold differences in the ability of absorption and accumulation of cadmium between species and willow clones may occur (Greger & Landberg 1999). The use of willow for phytoremediation may be possible only up to a certain level of soil contamination, due to dying of rootstocks and shortening period of their effective yielding. During cultivation of willow on contaminated soil about 180 g·ha^{-1} of Cu, 10.8 g·ha^{-1} of Cd and 2,700 g·ha^{-1} of Zn is absorbed yearly (Kabała et al. 2010).

Studies of Sieciechowicz (2015) show that the lowest cadmium content in biomass of *Salix vinimalis* was found on plantation with mineral fertilization and the highest—on objects not fertilized, while the average copper content in biomass from objects with mineral fertilization was approx. 23.4% higher than its content in plants cultivated on not fertilized soil. The highest contents of copper and cadmium were found in leaves of *Salix viminalis* (respectively 1.76 mg·kg^{-1} dm and 12.73 mg·kg^{-1} dm), lower in roots and the lowest in culms.

The aim of the study was to determine the contents of heavy metals in biomass of willow fertilized with compost from municipal sewage sludge.

2 MATERIALS AND METHODS

The field experiment with bushy willow was established using randomized sub-blocks in a dependant system in 2006 on an experimental field in Kościernica (N: 54°9'42.86", E: 16°27'8.36"). It was established on light soil of IVb-Va bonitation class of good rye complex, with the granulometric composition of light loamy sand up to a depth 100 cm and light loam, below 100 cm. Experiment was conducted through the years 2007–2009. Before experiment, the content of macro elements was: N – 0.03%, C – 0.71%, while the content of assimilable forms in 100 g of soil amounted to: P – 0,007%, K – 0,003% and Mg – 0,003% and pH$_{KCl}$ equalled to 5.1 in soil profile 0–90 cm (Styszko et

al. 2010). The level of groundwater was 940–980 cm below soil surface (measurements performed every 2 months in March-November periods, in the years 2008–2010) (Styszko & Fijałkowska 2015).

On big plots (sub-blocks of the first order) the following objects were drawn: (a) object without fertilization, (b) fertilized with compost from sewage sludge (10 t dm·ha^{-1}), (c) fertilized with compost from sewage sludge (10 t dm·ha^{-1}) and mineral nitrogen (90 kg N·ha^{-1}) and (d) fertilized with compost from sewage sludge (10 t dm·ha^{-1}) and mineral nitrogen (180 kg N·ha^{-1}). Within big plots, the following nine clones of willow (sub-blocks of the second order) were drawn: 1047, 1054, 1023, 1013, 1052, 1047D, 1056, 1033 and 1018, which were planted in three repetitions. The area of a small plot equalled to 34.5 m^2.

The compost produced only from sewage sludge was obtained from Sianów Waste Recovery Center. Compost was characterized by the following parameters: pH$_{KCl}$ – 6.63, dry mass content – 68.42%, and in dry mass following contents were noted: organic matter – 39.06%, N – 1.75% P – 1.60% K – 0.112%, Ca – 3.426% and Mg – 0.325%. Compost was applied in the spring of 2006, and mineral fertilizer was applied each year before the start of willow vegetation in the years 2006–2009.

In the spring of 2006, before the application of compost, soil in a layer 0–90 cm contained the following quantities of heavy metals per 1 kg of dry mass: chromium – 10.2 mg, nickel – 12.3 mg, cadmium – 0.5 mg, copper – 12.6 mg, cobalt – 8.0 mg, lead – 17.8 mg, iron – 19.3 mg, zinc – 48.9 mg and manganese – 646.6 mg (Fijałkowska et al. 2010). Dose of 10 t dm·ha^{-1} of compost introduced into the soil the following amounts of heavy metals: chromium – 160.6 g, nickel – 70.6 g, cadmium – 11.2 g, copper – 627.3 g, cobalt – 313.2 g, lead – 120.9 g, iron – 96,875.0 g, zinc – 3652.4 g and manganese – 2618.0 g (Fijałkowska et al. 2010).

Each of the was mowed separately in the following years. Growing shoots were mowed after the second, third and fourth growing season, following the application of compost. Samples of shoots were collected on the day of mowing (time I) and after several months of seasoning (time II). Shoots were dried, crushed and ground to a particle size of 0.2 mm.

Samples of biomass were mineralized using the mixture of acids: 65% HNO$_3$ + 70% HClO$_4$ + 30% H$_2$O$_2$ and microwave energy (Mega Milestone 1200 apparatus). The content of heavy metals (Cd, Cu, Cr, Co, Ni, Pb, Zn and Mn) was determined using Flame Atomic Absorption Spectrometry (FAAS) on iCE 3500 Series (Thermo Scientific) spectrometer. For low concentrations of metals (cadmium, lead) atom trap was used in order to lower the detection level (Janowska & Szymański 2009, Janowska 2013).

Vegetation of willow clones started in the 2nd to 3rd ten-day period of April each year. During all the experiment years (2006–2009) precipitation in the period January-December was higher than 753 mm, and in the period April-October, precipitation ranged from 459 mm (in 2008) to 654 mm (in 2007). The highest precipitation was 1,062 mm in 2007, which was considered as a very wet year. The year 2008, with precipitation of 855 mm, was considered as considerably wet, and the years 2006 and 2009, with precipitation 753 mm and 787 mm respectively, were considered as wet. Characteristics of hydrothermal conditions basing only on annual precipitation proved to be insufficient for willow growing, because there were periods extremely and very dry, even in the years considered as wet. Using Sielianinow hydrothermal coefficient it was proven that extremely dry and very dry periods occurred in July 2006, May 2008 and in April 2009 (Styszko et al. 2010).

ANOVA analysis was conducted in order to estimate the difference between the results obtained for the biomass harvested from particular plots. Percentage structure of variance components was estimated, and the significance of the results was assessed using F test. The years of sampling harvest (R) were random factors and constant factors were: two periods of biomass samples collection (I, II) (A), four fertilization combinations (a, b, c, d) (B) and nine clones of willow (9) (C). Statistical analysis was conducted using STATISTICA 9.

3 RESULTS AND DISCUSSION

The method of variance components was used to determine the influence of variability of main factors and their interactions on the total variability for the period of shoots biomass sampling, fertilizer combinations, clones of willow on the content of heavy metals (Cd, Cu, Cr, Co, Ni, Pb, Zn and Mn) in the dry mass of shoots (Table 1). The

analysis showed that the total variability of content of heavy metals in willow shoots was very low for cobalt, chromium, nickel, cadmium and copper, average for lead, high for manganese and very high for zinc. After summing proportional participation of factors in effects of interactions, it was found that most of that variation was attributed to random factors and their interactions (chromium, lead, nickel and cobalt), fertilizer combinations (copper and zinc) and willow clones (manganese and cadmium).

The average content of heavy metals for the given levels of examined factors, limits of detection of heavy metals for used methodology of its determination and number of analyses in which heavy metals were not detected with available methods, are given in Table 2. During the statistical analyses, the value of 0.00 mg·kg^{-1} of dry mass was used in cases when heavy metals were not detected. The highest number of analyses below the limit of detection was noted for cobalt (47.7%), chromium (30.1%) and cadmium (19.9%) and the lowest for lead (8.8%) and copper (2.8%). For zinc and manganese, no such cases were observed (Table 2). That is why during statistical analysis of chromium and cobalt in willow biomass, the obtained average content of those metals was below the detection limit.

Change of sampling collection period from I (directly after harvest) to II (after 4–6 months of biomass curing on the field) does not significantly differentiate content of cadmium, copper, chromium, nickel, zinc and manganese in the shoots of willow, but it decreased the content of lead and increased the content of cobalt. Such phenomenon may be explained by a higher leaching of lead compounds than cobalt compounds during biomass curing.

Fertilizer combinations had no significant effect on the content of chromium, nickel and lead in the shoots of willow. In case of other heavy metals, a

Table 1. Influence of examined factors on total variability of heavy metals content in the biomass of willow shoots.

Metal	Total variability	Sum of variability of main factors and their interactions [%]				
		A	B	C	R	sum
Cadmium (Cd)	0.6187	3.43	29.28	34.49	32.8	100.0
Copper (Cu)	1.0498	3.53	65.63	9.94	20.9	100.0
Chromium (Cr)	0.3394	2.80	1.05	1.05	95.1	100.0
Cobalt (Co)	0.1557	16.55	17.10	12.35	54.0	100.0
Nickel (Ni)	0.5667	9.26	13.32	4.22	73.2	100.0
Lead (Pb)	5.2843	9.10	5.05	8.35	77.5	100.0
Zinc (Zn)	444.5341	2.76	47.37	33.77	16.1	100.0
Manganese (Mn)	219.2992	4.96	20.67	38.97	35.4	100.0

Designations of sums of main factors and their interactions:
A—period of biomass samples collection; B—fertilizer combinations; C—willow clones; R—years of harvest.

Table 2. Content of heavy metals in willow shoots obtained during vegetation in the years 2007–2009.

Factor	Factor levels	Content in the willow biomass [mg·kg^{-1} dm]							
		Cd	Cu	Cr	Co	Ni	Pb	Zn	Mn
Period of sample collection (A)	I	0.99	2.41	0.00	0.05	0.58	3.80	62.08	46.14
	II	1.21	2.55	0.17	0.20	0.34	2.83	61.92	47.81
	LSD$_{0.05}$	0.27 n.i.	0.16 n.i.	0.26 n.i.	0.11*	0.28 n.i.	0.88*	2.03 n.i.	4.35 n.i.
Fertilizer combinations (B)	a[1]	1.66	2.82	0.00	0.06	0.51	3.11	72.22	52.22
	b[2]	1.16	3.22	0.24	0.30	0.72	3.50	72.54	42.60
	c[3]	0.77	2.52	0.00	0.07	0.25	3.71	59.15	44.58
	d[4]	0.81	1.38	0.11	0.08	0.36	2.94	44.08	48.50
	LSD$_{0.05}$	0.38***	0.23***	0.37 n.i.	0.16*	0.39 n.i.	1.24 n.i.	2.87***	6.16*
Willow clones (C)	1047	1.07	2.57	0.00	0.20	0.38	3.27	61.73	45.22
	1054	1.61	2.01	0.07	0.09	0.55	2.87	72.98	55.91
	1023	1.16	2.72	0.09	0.15	0.49	3.27	68.06	44.52
	1013	0.47	2.31	0.00	0.10	0.39	3.40	46.19	43.49
	1052	1.61	2.31	0.16	0.19	0.53	2.80	72.12	58.40
	1047D	1.51	2.47	0.27	0.17	0.58	3.45	65.54	48.28
	1056	0.59	2.47	0.07	0.05	0.48	3.65	48.90	45.22
	1033	1.12	2.78	0.00	0.08	0.36	2.96	71.92	48.85
	1018	0.75	2.70	0.13	0.11	0.39	4.20	50.56	32.89
	LSD$_{0.05}$	0.20***	0.26***	0.28 n.i.	0.14 n.i.	0.29 n.i.	0.57***	4.49***	4.07***
Average from the experiment		1.10	2.48	0.09	0.13	0.46	3.32	62.00	46.98
Limit of detection		0.09	0.12	0.18	0.24	0.18	0.30	0.03	0.09
Number of analyses below limit of detection [%]		19.9	2.8	30.1	47.7	28.2	8.8	0.0	0.0

LSD$_{0.05}$ – the least significant difference for α = 0.05.
Significance at: n.i. – not significant difference; *α = 0.05; **α = 0.01; ***α = 0.001.
a[1] – without fertilization; b[2] – compost from sewage sludge (10 t dm·ha^{-1}); c[3] – compost from sewage sludge (10 t dm·ha^{-1}) and mineral nitrogen (90 kg N·ha^{-1}); d[4] – compost from sewage sludge (10 t dm·ha^{-1}) and mineral nitrogen (180 kg N·ha^{-1}).

specific diversification of their content was noted (Table 2). Fertilization with compost from municipal sewage sludge (object (b)) did not cause a significant increase of zinc content in the shoots of willow. On the other hand, the content of copper and cobalt increased, and cadmium and manganese decreased. Application of mineral fertilization with nitrogen at objects where compost was used, did not cause a change of manganese content, but the content of cadmium, copper, cobalt and zinc decreased significantly. The content of cadmium and cobalt was not significantly lower at objects with nitrogen dose 180 kg N·ha^{-1} (object (d)) compared to objects with dose 90 kg N·ha^{-1} (object (c)), but the content of copper and zinc was significantly lower at object (d) than (c).

Chromium, cobalt and nickel content in shoots was not significantly different among willow clones. In the case of other heavy metals, the differences were significant (Table 2). The highest contents of metals were found in following clones: cadmium in clones 1054 and 1052 (1.61 mg·kg^{-1} dm),

copper in clone 1033 (2.78 mg·kg^{-1} dm), lead in 1018 (4.20 mg·kg^{-1} dm), zinc in 1054 (72.98 mg·kg^{-1} dm) and manganese in 1052 (58.40 mg·kg^{-1} dm). The lowest content of these elements was noted, respectively: for cadmium—in clone 1013 (0.47 mg·kg^{-1} dm), copper—in 1054 (2.01 mg·kg^{-1} dm), lead—in 1052 (2.80 mg·kg^{-1} dm), zinc – 1013 (46.19 mg·kg^{-1} dm) and manganese—in 1018 (32.89 mg·kg^{-1} dm). The differences between the highest and lowest vales of particular heavy metals content were varied and amounted to 27.7% for copper, 33.3% for lead, 36.7% for zinc, 43.7% for manganese and 70.8% for cadmium.

Charts (Figs. 1–6) show the impact of fertilizer combinations on the content of heavy metals in shoots of nine willow clones.

Charts (Figs. 1–6) show the impact of fertilizer combinations on the content of heavy metals in shoots of nine willow clones.

The highest cadmium content in shoots of all willow clones was noted on objects without fertilization (a[1]).

Fertilization with compost (b²) caused a slight decrease of its content. For clones 1013 and 1018, the decrease was significant (61.3–80.0%) (Fig. 1). Additional fertilization with nitrogen (c³ and d⁴) caused a very strong decrease of cadmium content in shoots (48.5–85.5%). In case of clones 1013 and 1018, the differences among combinations b², c³ and d⁴ were not significant.

The highest copper content was noted in clones 1047, 1054, 1023, 1013, 1052 and 1047D at objects fertilized with compost (b²) and in clones 1056, 1033 and 1018 at objects without fertilization (a¹). The lowest concentration of copper in all clones was noted on plots with additional nitrogen fertilization with the dose of 180 kg N·ha⁻¹ (d⁴) (Fig. 2).

The highest nickel content in the shoots of willow was found on objects without fertilization (a¹) in clones: 1047, 1056, 1033 and 1018 and on objects fertilized only with compost (b²) in clones: 1054, 1023, 1013, 1052 and 1047D. The lowest concentration of Ni was observed on objects c³ in clones 1052, 1047D, 1056, 1033 and 1018 (decreased by 58.8–93.2%) and on objects d⁴ in clones: 1047, 1054, 1023 and 1013 (Fig. 3).

Changes of lead content in willow shoots were not similar to the pattern observed in the case of cadmium, copper, nickel and zinc (Fig. 4). The highest lead content was detected on the objects fertilized with compost (b²) in clones: 1054, 1013, 1047D and 1018, fertilized with compost and nitrogen with dose of 90 kg N·ha⁻¹ in clones: 1047, 1052, 1047D, 1056 and 1033 and fertilized with compost and nitrogen with dose of 180 kg N·ha⁻¹ in clone 1023. Differences between the highest and the lowest lead content in the shoots, depending on willow clone, were from 19.8% to 64.2%.

The highest zinc contents in willow shoots were found on objects without fertilization (a¹) in clones: 1047, 1054, 1023, 1013, 1047D and 1033 and on objects fertilized with compost (b²) in clones 1052 and 1018. The lowest zinc concentration was noted on objects fertilized with compost and nitrogen with dose of 90 kg N·ha⁻¹ (c³) in clone 1013 (16.7% lower than in object a¹) and fertilized with compost and nitrogen with dose of 180 kg N·ha⁻¹ (d⁴) in clones: 1047, 1054, 1023, 1052, 1047D, 1056, 1033 and 1018 (32.8–52.5% lower than in object a¹) (Fig. 5).

Manganese content in willow shoots, like in the case of lead, was changing differently than in the

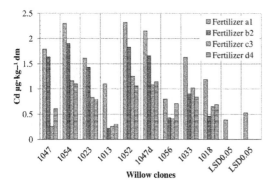

Figure 1. Impact of fertilization on the content of cadmium in biomass of willow shoots.

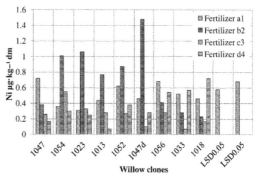

Figure 3. Impact of fertilization on the content of nickel in biomass of willow shoots.

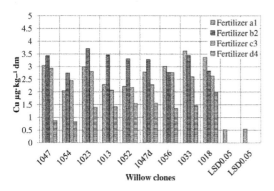

Figure 2. Impact of fertilization on the content of copper in biomass of willow shoots.

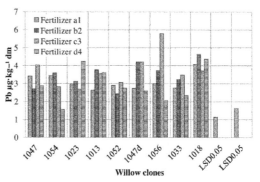

Figure 4. Impact of fertilization on the content of lead in biomass of willow shoots.

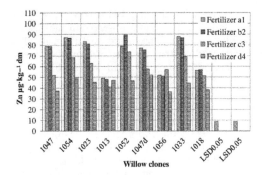

Figure 5. Impact of fertilization on the content of zinc in biomass of willow shoots.

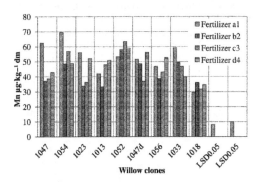

Figure 6. Impact of fertilization on the content of manganese in biomass of willow shoots.

case of cadmium, copper, nickel and zinc (Fig. 6). The highest manganese content was found on objects without fertilization (a[1]) in clones: 1047, 1054, 1023 and 1033, fertilized with compost (b[2]) in clone 1018, fertilized with compost and nitrogen with the dose of 90 kg N·ha^{-1} (c[3]) in clone 1052 and fertilized with compost and nitrogen with the dose of 180 kg N·ha^{-1} in clones: 1013, 1047D and 1056. The differences between the highest and lowest values of manganese content in willow shoots, depending on willow clone, ranged from 15.9% to 40.7%.

Our experiments were conducted on sandy soil, in which contents of heavy metals were within the limits set by relevant Ministry of Environment Regulation (2002). as soil not contaminated with those metals (Fijałkowska et al. 2010). Dose of 10 t·ha^{-1} of dry mass of compost introduced into soil following amounts of heavy metals: chromium – 160.6 g, nickel – 70.6 g, cadmium – 11.2 g, copper – 627.3 g, cobalt – 313.2 g, lead – 120.9 g, zinc – 3,652.4 g, manganese – 2,618.0 g (Fijałkowska et al. 2010). According to the certificate of compost as organic fertilizer, dose of 10 t·ha^{-1} can be applied once every five years.

In our studies analysis of shoots biomass during first four year rotation has shown that the average content of following metals per 1 kg of dry mass was: chromium – 0.09 mg, nickel – 0.46 mg, cadmium – 1.1 mg, copper – 2.48 mg, cobalt – 0.13 mg, lead – 3.32 mg, zinc – 62.0 mg and manganese – 46.98 mg. Moderately lower contents of nickel, copper and zinc, and slightly higher contents of chromium, cadmium and lead than reported in the literature for wood of common osier cultivated for energy purposes were found (Kabała et al 2010, Kaniuczak et al. 2000).

On average, significantly highest content of heavy metals per 1 kg of dry mass was found on objects: "a" (without fertilization) for cadmium (1.66 mg·kg^{-1}) and manganese (52.22 mg·kg^{-1}), "b" (fertilized with compost) for copper (3.22 mg·kg^{-1}), cobalt (0.30 mg·kg^{-1}) and zinc (72.54 mg·kg^{-1}), and significantly lowest on objects: "a" for cobalt (0.60 mg·kg^{-1}), "b" for manganese (42.60 mg·kg^{-1}), "c" (fertilized with compost and 90 kg·ha^{-1} of N) for cadmium (0.77 mg·kg^{-1}) and "d" (fertilized with compost and 180 kg·ha^{-1} of N) for copper (1.38 mg·kg^{-1}) and zinc (44.08 mg·kg^{-1}). Also significant variability of average content of heavy metals in the shoots of willow clones (cadmium – 70.8%, copper – 27.1%, lead – 33.3%, zinc – 36.7% and manganese – 25.5%) and for interaction between clones and combinations of fertilizers were found.

On average, the highest amounts of many heavy metals were accumulated per 1 kg of dry mass of following willow clones: cadmium – 1054 and 1052 (1.61 mg·kg^{-1}), copper –1033 (2.78 mg·kg^{-1}), lead – 1018 (4.20 mg·kg^{-1}), zinc – 1054 (72.98 mg·kg^{-1}) and manganese – 1052 (58.40 mg·kg^{-1}) and the lowest amounts: cadmium – 1013 (0.47 mg·kg^{-1}), copper – 1054 (2.01 mg·kg^{-1}), lead – 1052 (2.80 mg·kg^{-1}), zinc – 1013 (46.19 mg·kg^{-1}) and manganese – 1018 (32.89 mg·kg^{-1}). Differences between extreme contents of metals among fertilizer combinations were for the following genotypes: cadmium—from 48.5% in clone 1033 to 85.5% in clone 1047, copper—from 40.9% in clone 1018 to 80.5% in clone 1047, nickel—from 58.8% in clone 1056 to 90.9% in clone 1013, lead—from 19.8% in clone 1018 to 64.2% in clone 1056, zinc—from 16.7% in clone 1013 to 52.5% in clone 1047 and manganese—from 15.9% in clone 1052 to 40.7% in clone 1047.

4 CONCLUSIONS

The most important conclusions which may be drawn on the basis of the conducted experiments on the impact of compost from municipal sewage sludge on the content of heavy metals in willow shots are as follows:

Establishment of plantation of willow on land set aside, with low groundwater level, requires assumption that its practical importance expressed with diverse heavy metal content in the biomass is determined by: date of sampling (aging time of shoots on the field after mowing) and correctly balanced organic and mineral fertilization adapted to the requirements of a clone.

Biomass obtained from shoots of 9 clones of willow Salix viminalis cultivated on a light soil fertilized with compost from sewage sludge in dose of 10 t dm·ha^{-1} contained lower content of nickel, copper and zinc, and a slightly higher content of chromium, cadmium and lead than the values given in the literature for average conditions of field cultivation.

The highest average contents of heavy metals in dry mass of shoots was found on plots with no fertilization—for cadmium and manganese, and fertilized with compost—for copper, cobalt and zinc, and the lowest—on plots not fertilized—for cobalt, fertilized with compost—for manganese, fertilized with compost and 90 kg N·ha^{-1}—for cadmium and fertilized with compost and 180 kg N·ha^{-1}—for copper and zinc.

The use of mineral fertilizer containing nitrogen significantly decreases the content of heavy metals in the biomass of willow shoots.

Significant differences in the average content of heavy metals in shoots of Salix viminalis clones were observed. The highest amounts of metals were accumulated by following clones: cadmium – 1054 and 1052, copper – 1033, lead – 1018, zinc – 1054 and manganese – 1052. The lowest amounts were accumulated by following clones, respectively: 1013, 1054, 1052, 1013 and 1018.

A peculiar reaction of clones to fertilization with compost from municipal sludge and additional fertilization with nitrogen in the aspect of accumulation of heavy metals in shoots was proven.

Seasoning of shoots in the field for 4–6 months had a slight impact on the content of heavy metals.

ACKNOWLEDGEMENTS

The scientific work financed from funds for science in the years 2008–2011 as a research project N N523 4162 35.

REFERENCES

Czechowska-Kosacka, A., Cao, Y. & Pawłowski, A. 2015a. Criteria for Sustainable Disposal of Sewage Sludge. *Annual Set the Environment Protection (Rocznik Ochrona Środowiska)* 17: 337–350.
Czechowska-Kosacka, A., Cel. W., Kujawska, J., Wróbel, K. 2015b. Alternative Fuel Production based on Sewage Sludge Generated in the Municipal Wastewater Treatment. *Annual Set the Environment Protection (Rocznik Ochrona Środowiska)*, 17(1): 246–255
Fijałkowska, D., Janowska, B. & Styszko L. 2010. Influence of amendment of short-rotation willow plantation in the vicinity of Koszalin with sewage sludge compost

on modifications of content of some metals in the soil. *Fresenius Environmental Bulletin* 19(2a): 1–3.
Greger, M. & Landberg, T. 1999. Use of willow in phytoextraction. *Int. J. Phytoremed.* 1: 115–123.
Janowska, B. 2013. Humic Acids Complexes with Heavy Metals in Municipal Solid Wastes and Compost. In Pawłowski, Dudzińska & Pawłowski (eds) *Environmental Engineering IV:* 249–255. London: Taylor & Francis.
Janowska, B. & Szymański, K. 2009. Transformation of selected trace elements during the composting process of sewage sludge and municipal solid waste. *Fresenius Environmental Bulletin* 18(7): 1110–1117.
Kabała, C., Karczewska, A. & Kozak, M. 2010. Suitability of energy plants for reclamation and development of degraded soils (in Polish). *Zesz. Nauk. UP Wroc. Rol.,* XCVI Nr 576: 97–118.
Kaniuczak, J., Błażej, J. & Gąsior, J. 2000. Content of trace elements in various clones of wicker. Vol. I. Content of iron, manganese, copper and zinc (in Polish). *Zesz. Probl. Post. Nauk. Rol.* 472: 379–385
Karczewska, A. 2008. *Protection of soil and reclamation of degraded areas (in Polish).* Wrocław: Wydawnictwo Uniwersytetu Przyrodniczego.
Mathe-Gaspar, G. & Anton, A. 2005. Study of phytoremediation by use of willow and rape. *Acta Biolog. Szeged* 49(1–2): 73–74.
Regulation of the Polish Minister of Environment from September 9, 2002 on soil quality standards and ground quality standards (in Polish). Journal of Laws 2002 No. 165 item 1359.
Sieciechowicz, A. 2015. The impact of varied fertilisation on the concentration of cadmium and copper in organs of willow trees (Salix viminalis). *Civil and Environmental engineering Reports* 16(2): 181–194.
Sobczyk, W., Sternik, K., Sobczyk E.J. & Noga, H. 2015. Rating of Yielding of Willow Fertilized with Sewage Sludge. *Annual Set the Environment Protection (Rocznik Ochrona Środowiska)* 17: 1113–1124.
Starck, Z. 2002. Physiological basis of plant productivity (in Polish). In Kopcewicz, J. & Lewak S. (eds) *Fizjologia roślin:* 679–706. Warszawa: Wydawnictwo PWN.
Starzyk, J., Czekała, J., Wolna-Maruwka, A., Swędrzyńska, D. 2015. Zmiany stanu mikrobiologicznego kompostów na bazie kory sosnowej z dodatkiem preparatu Efektywne Mikroorganizmy (EM), zielonej masy roślin i mocznika. *Annual Set the Environment Protection (Rocznik Ochrona Środowiska)*, 17(2): 1512–1526.
Styszko, L., Fijałkowska, D. & Sztyma-Horwat M. 2010. Shrubby willow crop in the four-year cycle of cultivation in light soil in Pomerania. In M. Jasiulewicz (ed) *Regional and Local biomass Potential*: 159–190. Koszalin: Polish Economics Association, Koszalin University of Technology.
Styszko, L. & Fijałkowska, D. 2015. Successive effect of fertilization with compost from municipal sewage sludge and frequency mowing on architecture of energy willow field in subsequent years of cultivation (in Polish). *Zeszyty Problemowe Postępów Nauk Rolniczych* 582: 73–80.
Żurek, G. & Majtkowski W. 2009. Alternative plants in phytoextraction of heavy metals at contaminated areas (in Polish). *Problemy Inżynierii Rolniczej* 2009(3): 83–89.

Environmental Engineering V – Pawłowska & Pawłowski (Eds)
© 2017 Taylor & Francis Group, London, ISBN 978-1-138-03163-0

An influence of municipal sewage sludge and mineral wool application on sorption properties of coarse-grained soil

S. Wesołowska, S. Baran, G. Żukowska, M. Myszura & M. Bik-Małodzińska
Institute of Soil Science, Environment Engineering and Management, University of Life Sciences in Lublin, Lublin, Poland

A. Pawłowski & M. Pawłowska
Faculty of Environmental Engineering, Lublin University of Technology, Lublin, Poland

ABSTRACT: A large part of Polish soils are formed at coarse-grained parent material which determines low resistance to degradation and productivity. Supplementation with exogenous organic matter and nutrients allow to improve the soil quality. The aim of the study was to evaluate the effect of municipal sewage sludge and spent mineral wool addition to sandy loam soil on its physicochemical properties. A pot experiment showed that sorption properties depended mainly on the addition of the sludge, and on its doses. Application of the sludge at a dose 10 Mg · ha^{-1} significantly increased the content of base cations and a base saturation of the adsorption complex, but it did not have a significant effect on the hydrolytic acidity and cation exchange capacity. The addition of sludge at a dose 100 Mg · ha^{-1} significantly influenced all the examined properties. Addition of mineral wool at doses from 200 to 800 m^3 ha^{-1}, significant differed base saturation of the soil adsorption complex.

Keywords: soil reclamation, sewage sludge, mineral wool, waste land use

1 INTRODUCTION

Soil is an essential and multifunctional element of terrestrial ecosystems, and its main role is to create conditions for plant growth. The solid phase of the soil has the ability to capture the ions or particles from an aqueous solution and store them. This phenomenon, called "sorption" determines the bioavailability of plant nutrients and protects soil organisms against toxic effects of harmful substances, such as heavy metals (Adriano, 2001). Sorption properties of soils regulate leaching of nutrients from the soil and determine the effectiveness of fertilization. Sorption capacity of the soil depends mainly on grain size composition and origin of the soil. A large content of sand, low content of colloids, including the clay minerals and humic substances, high acidity, which are characteristic for most Polish soils, classifies most of them into "susceptible to degradation" category (Baran et al. 2014, Dąbkowska-Naskręt et al. 2001). Low sorption capacity determines their low production potential. It is estimated that the production potential of 1 hectare of average Polish soil corresponds to the potential of 0.6 ha of arable land in the European Union (Skłodowski & Bielska, 2009). This is due to the fact that more than 40% of Polish soils are formed on sands or loamy sands,

have alluvial genesis or belong to mountain soil, which are characterized by poor agricultural suitability (Krasowicz et al. 2011).

High quality of the soil is determined by an appropriate system of physical, physicochemical and biological properties (Carter, 2002). Increase in the content of organic matter in poor-quality soils is one of the ways leading to a radical improvement of their properties. Stabilized municipal sewage sludge can be an effective source of exogenous organic matter for the soils. It improves porosity, water retention, sorption capacity of the soil and supplies the nutrients for plant growth (Żukowska et al. 2002). It was used as a supplement improving the parameters of biofilter beds (Hort et al. 2009, Pawłowska et al. 2011) or soil-like materials formed on the base of different waste (Kujawska et al. 2014).

Currently, the land use of municipal sewage sludge is a preferred option in sludge management, apart from the use in energy-production sector (Pawłowska & Siepak 2006, Heidrich et al. 2008, Cao & Pawłowski 2012, Czechowska-Kosacka et al. 2015), because it allows the use of their fertilizing properties, resulting from the significant content of humic substances, nitrogen and other plant nutrients (Siuta 1998). The land use of sewage sludge relies on its applications to arable soils in order to replenish the nutrients or to degraded soils and soilless land for its restoration, on supporting soils susceptible to wind and water erosions or production of compost for the above-mentioned purposes (Singha & Agrawal 2008).

Beside the sewage sludge, other types of waste may also be used as a source of nutrients or a means of improving the physicochemical properties of the low-quality soil. Increasing in the effectiveness of soil fertilizing can be achieved through the combined application of sewage sludge and other wastes (Baran et al. 2002, Gondek & Filipek-Mazur 2006), for example, spent mineral wool derived from the crops under controlled environment.

The mineral wool is an inert material that does not pose threat to the environment. However, spent mineral wool may contain nutrients that were not taken up by the plant under cultivation, especially magnesium, calcium, nitrogen, phosphorus and potassium. The used wool is also enriched with the organic substances derived from plant residues, which were grown on it (Baran 2008). High porosity, with a predominance of pores capable of retaining water, is an additional advantage of mineral wool. Oświęcimski (1996) showed that plant roots found the optimal conditions for the growth in the artificial substrate formed from mineral wool, and yield of vegetables and ornamental plants were very high.

Application of spent mineral wool in combination with sewage sludge to the poor-quality soil, degraded or soilless lands is a potential method of its management. The aim of the study was to determine the effect of municipal sewage sludge and mineral wool addition to sandy loam soil on the formation of its physicochemical properties, such a pH_{H_2O}, pH_{KCl}, cation exchange capacity (T or CEC), total exchangeable bases (S), percent base saturation (V), and hydrolytic acidity (Hh).

2 MATERIALS AND METHODS

The study was conducted as a pot experiment in an experimental greenhouse of the University of Life Sciences in Lublin. The pots with a capacity of 12 dm^3 were filled with loamy sand. The soil was acidic. It has low content of nitrogen and available forms of phosphorus, potassium, and microelements. The sorption capacity and organic matter content was also low. Two kinds of waste materials were added to the soil in order to improve its properties. These included: municipal sewage sludge (code 19 08 05) from the municipal wastewater treatment plant in Stalowa Wola and spent mineral wool (code 02 01 83) (Reg. of Polish Minister of Environment, O.J. 2014 item 1923) from horticultural company in Niemce near Lublin. Selected properties of the soil and waste components used in the experiment are presented in Table 1.

The components were mixed with different configurations and amounts. Variants of soil substrates studied in the experiment are shown in Table 2. Soil without additives (control I) and the soil fertilized with mineral form of nitrogen, phosphorus and potassium were used as a control samples. The sewage sludge was applied to the soil in doses of 10 and 100 Mg · ha^{-1} (corresponding to 3.4 and 34 g of dry mass of sludge on 1 kg of soil) and the mineral wool in doses of 200, 400 and 800 m^3 · ha^{-1} (corresponding to 80, 160 and 320 cm^3 of wool on 1 kg of soil).

Soil with mineral wool at doses of 200, 400 and 800 m^3 · ha^{-1} enriched with N, P, K fertilizer were also examined. In further variants of the experiment, different combinations of the mixture of soil, mineral wool (in doses of 200, 400 and 800 m^3 · ha^{-1}) and sewage sludge (at a fertilizing dose equal to 10 Mg · ha^{-1}) and a reclamation dose (equal to 100 Mg · ha^{-1}) were analyzed.

The mixture of plant species usually used in soil reclamation (Table 3) was sown into all the pots. The experiment was performed for 3 growing seasons. Soil samples were collected with soil probe from the entire depth of the pots four times during the experiment: at the beginning of the study and every year, at the end of the growing season (in October).

Table 1. Selected properties of soil, municipal sewage sludge and spent mineral wool used in the experiment.

Parameter (unit)	Soil	Mineral wool	Sewage sludge
Grain size:			
sand 2.0–0.05 mm (%)	90	–	–
silt 0.05–0.002 mm (%)	4	–	–
clay 0.002 (%)	6	–	–
pH H$_2$O (–)	5.6	6.9	6.8
pH KCl (–)	5.0	6.6	6.4
Hydrolytic acidity, Hh (cmol(+).kg^{-1})	3.20	3.82	4.50
Total exchangeable bases, S (cmol(+).kg^{-1})	2.22	57.04	50.04
Cation exchange capacity T (cmol(+).kg^{-1})	5.52	60.86	54.54
Percent base saturation, V (%)	40.2	93.7	91.7
N$_{total}$ (g.kg^{-1})	0.50	4.0	28.0
TOC (g.kg^{-1})	5.80	28.5	193.8
C:N (–)	11.6	7.1	6.9
P$_{available}$ (mg.kg^{-1})	1.15	22.99	60.40
K$_{available}$ (mg.kg^{-1})	2.98	33.00	17.02
Cu (mg.kg^{-1})	2.33	42.75	139.0
Zn (mg.kg^{-1})	9.14	133.50	935.0
Pb (mg.kg^{-1})	6.28	35.50	29.2
Cd (mg.kg^{-1})	0.17	0.02	3.45
Cr (mg.kg^{-1})	10.89	18.50	26.7
Ni (mg.kg^{-1})	6.30	9.30	55.1
Hg (mg.kg^{-1})	0.02	0.02	0.45

Table 2. Composition of soil substrate used in the experiment.

No.	Experimental variants
1.	Soil without additives (control I)
2.	Soil + mineral fertilization with NPK: 80, 40 and 60 kg · ha^{-1}, i.e. 0.27, 0.14 and 0.21 g/pot (control II)
3.	Soil + sewage sludge (10 Mg · ha^{-1})
4.	Soil + sewage sludge (100 Mg · ha^{-1})
5.	Soil + mineral wool (200 m^3 · ha^{-1})
6.	Soil + mineral wool (400 m^3 · ha^{-1})
7.	Soil + mineral wool (800 m^3 · ha^{-1})
8.	Soil + mineral wool (200 m^3 · ha^{-1}) + NPK
9.	Soil + mineral wool (400 m^3 · ha^{-1}) + NPK
10.	Soil + mineral wool (800 m^3 · ha^{-1}) + NPK
11.	Soil + mineral wool (200 m^3 · ha^{-1}) + sewage sludge (10 Mg · ha^{-1})
12.	Soil + mineral wool (400 m^3 · ha^{-1}) + sewage sludge (10 Mg · ha^{-1})
13.	Soil + mineral wool (800 m^3 · ha^{-1}) + sewage sludge (10 Mg · ha^{-1})
14.	Soil + mineral wool (200 m^3 · ha^{-1}) + sewage sludge (100 Mg · ha^{-1})
15.	Soil + mineral wool (400 m^3 · ha^{-1}) + sewage sludge (100 Mg · ha^{-1}
16.	Soil + mineral wool (800 m^3 · ha^{-1}) + sewage sludge (100 Mg · ha^{-1})

Table 3. Composition of plant species used in the experiment.

No.	Species	Share [%]
1.	Meadow fescue (*Festuca pratensis*)	41.2
2.	Red fescue (*Festuca rubra*)	19.2
3.	Perennial rye-grass (*Lolium perenne*)	14.7
4.	Italian rye-grass (*Lolium multiflorum*)	12.4
5.	Orchard grass (*Dactylis glomerata*)	6.5
6.	Red clover (*Trifolium pratense*)	6.0

Following physicochemical parameters were determined or calculated in the averaged soil samples: pH value (measured potentiometrically in H_2O and in 1 mol · dm^{-3} KCl), hydrolytic acidity, Hh (by Kappen method, in 1 mol · dm^{-3} CH_3COONa), total exchangeable bases, S (using Pallmann method, in 0.5 mol · dm^{-3} NH_4Cl at pH 8.2), cation exchange capacity, T (as a sum of Hh and S) and percent base saturation of the sorption complex (V), the total organic carbon, TOC (by Tiurin method modified by Simakow), total nitrogen content (by Kjeldhal method, using 1002 Kjeltech distillation unit) and the content of available phosphorus and potassium (by Egner-Riehm method), available magnesium (by Sachtschabel) method). The analysis of variance (ANOVA) for the factors listed in Table 4 was carried out using the STATISTICA 9.0 software.

Table 4. Factors evaluated in ANOVA test.

	Combinations
A.	Control samples
1.	Soil without additives
2.	Soil + NPK: 80, 40 and 60 kg · ha^{-1}, i.e. 0.27, 0.14, 0.21 g/pot
B.	Reclamation variants
1.	Soil + sewage sludge (10 Mg · ha^{-1})
2.	Soil + sewage sludge (100 Mg · ha^{-1})
3.	Soil + mineral wool
4.	Soil + mineral wool + NPK
5.	Soil ++ mineral wool + sewage sludge (10 Mg · ha^{-1})
6.	Soil + + mineral wool + sewage sludge (100 Mg · ha^{-1})
II	Period
1.	Start of the experiment
2.	After the first harvest
3.	After the second harvest
4.	After the third harvest

3 RESULTS

3.1 *The influence of the waste additives on soil pH*

Measurements of the actual and exchangeable acidity in the various seasons of the experiment showed that the introduction of additives into the soil influenced the acid ions concentration in soil water (Table 5).

Mineral fertilization (N, P, K) resulted in a slight decrease in pH compared to the soil without additives (control I). The addition of sewage sludge contributes to the greater reduction of acidity than the addition of mineral wool. The increase in sludge dose from 10 to 100 Mg · ha^{-1} induced the increase in exchangeable acidity of the soil (in the range of pH 0.8 to 1.1). The increase was maintained throughout the whole experiment, although the changes in actual acidity did not follow the trends in changes of pH$_{KCl}$. Sewage sludge used in combination with mineral wool caused further reduction of acidification, visible especially after application the greater dose (100 Mg · ha^{-1}) of sludge. The addition of mineral wool, compared to the control sample (I) contributed to the increase of exchangeable acidity of the soil, but the effect of NPK fertilization in combination with this waste was low.

The changes in pH values were time-dependent. The most evident increase in the exchangeable acidity was noted in the last year of the study (Table 5).

3.2 *The influence of the waste additives on hydrolytic acidity (Hh) of the soil*

In the case of soils to which the sewage sludge or mineral wool were introduced, a significant decrease in hydrolytic acidity was observed (Table 6, Fig. 1).

Table 5. Changes in pH value in examined soil.

Experimental variants	pH							
	in H$_2$O				in KCl			
	Period				Period			
	I	II	III	IV	I	II	III	IV
Soil without additives (control I)	5.2	5.3	5.2	5.6	4.5	4.5	4.3	4.3
Soil + mineral fertilization with NPK: 80, 40 and 60 kg · ha^{-1}, i.e. 0.27, 0.14, 0.21 g/pot (control II)	5.2	5.1	5.0	5.8	4.5	4.3	4.1	4.7
Soil + sewage sludge (10 Mg · ha^{-1})	6.1	5.0	5.0	6.3	5.6	4.5	4.5	5.6
Soil+ sewage sludge (100 Mg · ha^{-1})	6.3	5.4	4.6	5.5	6.5	5.5	4.7	5.0
Soil + mineral wool (200 m^3 · ha^{-1})	5.6	5.7	5.3	5.9	4.8	4.8	4.6	5.2
Soil + mineral wool (400 m^3 · ha^{-1})	5.4	5.8	5.2	6.9	4.7	4.9	4.8	5.8
Soil + mineral wool (800 m^3 · ha^{-1})	5.7	5.8	5.5	6.6	4.9	5.0	4.9	5.7
Soil + mineral wool (200 m^3 · ha^{-1}) + NPK	5.3	5.3	5.3	6.8	4.5	4.5	4.5	5.5
Soil + mineral wool (400 m^3 · ha^{-1}) + NPK	5.5	5.7	5.2	6.7	4.6	4.8	4.6	5.6
Soil + mineral wool (800 m^3 · ha^{-1}) + NPK	6.5	5.8	5.5	7.0	5.8	4.9	4.8	5.8
Soil + mineral wool (200 m^3 · ha^{-1}) + sewage sludge (10 Mg · ha^{-1})	5.7	5.4	5.3	5.6	4.8	4.7	4.6	5.5
Soil + mineral wool (400 m^3 · ha^{-1}) + sewage sludge (10 Mg · ha^{-1})	5.7	5.4	5.3	7.1	5.0	4.9	4.7	6.0
Soil + mineral wool (800 m^3 · ha^{-1}) + sewage sludge (10 Mg · ha^{-1})	5.7	5.9	5.6	6.6	5.0	5.3	5.0	5.8
Soil + mineral wool (200 m^3 · ha^{-1}) + sewage sludge (100 Mg · ha^{-1})	6.3	5.6	5.0	6.5	6.2	5.4	4.7	5.7
Soil + mineral wool (400 m^3 · ha^{-1}) + sewage sludge (100 Mg · ha^{-1})	6.4	5.9	5.5	6.7	6.3	5.7	5.2	5.8
Soil + mineral wool (800 m^3 · ha^{-1}) + sewage sludge (100 Mg · ha^{-1})	6.1	5.5	4.9	6.8	5.9	5.3	4.6	6.2

Comments: I—start of the experiment; II—after the first harvest; III—after the second harvest; IV—after the third harvest.

Mineral fertilization with N, P and K resulted in a non-significant increase in hydrogen ions concentration in the soils. The changes in Hh values were time-dependent. Gradual reduction in the H$^+$ concentration in the soil water and in the sorption complex, statistically significant in the final stage of the experiment, was observed. The changes in Hh values in the control samples (I and II) during the course of the experiment were irregular and lower than in amendment soil.

Taking into account the hydrolytic acidity in tested soil materials and assuming its value measured in control sample I (soil without additives) as 100%, the following order was established:

soil + NPK (101.6%) > soil (100%) > soil + sewage sludge 10 Mg · ha^{-1} (90.9%) > soil + mineral wool (86.0%) > soil + mineral wool + NPK (82.4%)
> soil + sewage sludge 10 Mg · ha^{-1} + mineral wool (75.9%) > soil + sewage sludge 100 Mg · ha^{-1} (71.3%) > soil + sewage sludge 100 Mg · ha^{-1} + mineral wool (66.4%)

Sewage sludge used in higher, "reclamation" dose (100 Mg · ha^{-1}), decreased the concentration of hydrogen ions in the soil more effectively (by 19.6%) than the "fertilizing" dose (10 Mg · ha^{-1}). This pattern was also confirmed, but less intensive (decrease by 9.5%) in the samples in which the sewage sludge were applied simultaneously with mineral wool.

3.3 The influence of the waste additives on total exchangeable bases (S) in the soil

Content of base cations in the soil (control I) was low, but it was increased significantly under the influence of high doses of sewage sludge, as well as a mixture of sewage sludge with mineral wool (Table 7, Fig. 2). The sludge used in the "reclamation" dose (100 Mg · ha^{-1}) caused the increase in average alkaline cation content by 2.82 cmol (+) · kg^{-1}, i.e. by 66.5% as compared to a soil amended with lower dose of the sludge (10 Mg · ha^{-1}) and by 5.44 cmol (+) · kg^{-1}, i.e. 335% as compared to

Table 6. Hydrolytic acidity (H_h, cmol(+) $\cdot$ kg^{-1}) of soils under examinations.

Experimental variant	Period				Average
	I	II	III	IV	
Soil without additions (control I)	3.30	2.85	2.85	3.30	3.07
Soil + mineral fertilization with NPK: 80, 40 and 60 kg $\cdot$ ha^{-1}, i.e. 0.27, 0.14, 0.21 g/pot (control II)	3.20	3.15	2.85	3.30	3.12
Soil + sewage sludge (10 Mg $\cdot$ ha^{-1})	3.00	3.15	2.55	2.45	2.79
Soil + sewage sludge (100 Mg $\cdot$ ha^{-1})	1.80	2.55	2.25	2.15	2.19
Soil + mineral wool (200 m$^3 \cdot$ ha^{-1})	3.00	2.85	2.25	2.85	2.74
Soil + mineral wool (400 m$^3 \cdot$ ha^{-1})	2.95	2.70	2.25	2.55	2.61
Soil + mineral wool (800 m$^3 \cdot$ ha^{-1})	2.85	2.55	2.40	2.45	2.56
Soil + mineral wool (200 m$^3 \cdot$ ha^{-1}) + NPK	3.00	3.00	2.40	2.45	2.71
Soil + mineral wool (400 m$^3 \cdot$ ha^{-1}) + NPK	2.85	2.70	2.25	2.15	2.49
Soil + mineral wool (800 m$^3 \cdot$ ha^{-1}) + NPK	2.55	2.45	2.40	2.15	2.39
Soil + mineral wool (200 m$^3 \cdot$ ha^{-1}) + sewage sludge (10 Mg $\cdot$ ha^{-1})	2.70	2.70	2.25	2.15	2.45
Soil + mineral wool (400 m$^3 \cdot$ ha^{-1}) + sewage sludge (10 Mg $\cdot$ ha^{-1})	2.65	2.50	2.10	2.00	2.31
Soil + mineral wool (800 m$^3 \cdot$ ha^{-1}) + sewage sludge (10 Mg $\cdot$ ha^{-1})	2.40	2.30	2.25	2.00	2.24
Soil + mineral wool (200 m$^3 \cdot$ ha^{-1}) + sewage sludge (100 Mg $\cdot$ ha^{-1})	2.00	2.15	2.10	2.00	2.06
Soil + mineral wool (400 m$^3 \cdot$ ha^{-1}) + sewage sludge (100 Mg $\cdot$ ha^{-1})	1.95	2.10	2.10	2.00	2.04
Soil+mineral wool (800 m$^3 \cdot$ ha^{-1}) + sewage sludge (100 Mg $\cdot$ ha^{-1})	2.00	2.05	2.05	2.00	2.02
Average values					
Soil + sewage sludge (10 Mg $\cdot$ ha^{-1})	3.00	3.15	2.55	2.45	2.79
Soil + sewage sludge (100 Mg $\cdot$ ha^{-1})	1.80	2.55	2.25	2.15	2.19
Soil + mineral wool	2.93	2.70	2.30	2.62	2.64
Soil + mineral wool + NPK	2.80	2.72	2.35	2.25	2.53
Soil + sewage sludge (10 Mg $\cdot$ ha^{-1}) + mineral wool	2.58	2.50	2.20	2.05	2.33
Soil + sewage sludge (100 Mg $\cdot$ ha^{-1}) + mineral wool	1.98	2.10	2.08	2.00	2.04
Average value for period	2.64	2.61	2.33	2.37	2.49
NIR$_{0.05}$ for the period			0.19**		
NIR$_{0.05}$ for the particular variant			0.53**		

Comments: I—start of the experiment; II—after the first harvest; III—after the second harvest; IV—after the third harvest, **differences significant at $P \leq 0.01$.

control sample I. The highest increase in total exchangeable bases was observed in the mixture of soil with sewage sludge (100 Mg $\cdot$ ha^{-1}) and mineral wool (800 m$^3 \cdot$ ha^{-1}). The S value in last research term was by 6.38 cmol (+) $\cdot$ kg^{-1} higher than the value measured in control sample I.

Considering the duration of the experiment, significant changes in the value of total exchangeable bases were observed only in variants involving sewage sludge. Sludge used at a "reclamation" dose (100 Mg $\cdot$ ha^{-1}), as the only additive increased the total exchangeable bases significantly in the II, III and IV research periods, while in the case of combined application of sewage sludge and min-

eral wool—significant increase was noted only in II and IV periods.

Taking into account the content of total exchangeable bases in tested soil materials and assuming its value measured in control sample I (soil without additives) as 100% the following order were established:

soil (100.0%) < soil + NPK (112.3%) < soil + mineral wool (154.9%) < soil + mineral wool + NPK (158.6%)

<soil + sewage sludge 10 Mg $\cdot$ ha^{-1} + mineral wool (208.0%) < soil + sewage sludge 10 Mg $\cdot$ ha^{-1} (261.7%)

<soil + sewage sludge 100 Mg $\cdot$ ha^{-1} (435.8%)

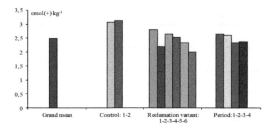

Figure 1. Influence of sewage sludge and mineral wool addition on hydrolytic acidity of sandy soil (average values for factors evaluated in ANOVA test). Symbols: Control samples: 1-soil without additives, 2-soil + NPK (80, 40, 60 kg ha^{-1}); Reclamation variants: 1-soil + sewage sludge (10 Mg · ha^{-1}), 2-soil + sewage sludge (100 Mg · ha^{-1}), 3-soil + mineral wool, 4-soil + mineral wool + NPK, 5-soil + mineral wool + sewage sludge (10 Mg · ha^{-1}), 6-soil + mineral wool + sewage sludge (100 Mg · ha^{-1}); Research terms: 1-start of the experiment; 2-after the first harvest; 3-after the second harvest; 4-after the third harvest.

<soil + sewage sludge 100 Mg · ha^{-1} + mineral wool (490.1%)

3.4 The influence of the waste additives on cation exchange capacity (T or CEC) in the soil

The control sample I (soil without additives) was characterized by a low sorption capacity, equal to 4.69 cmol(+) · kg^{-1}. The sewage sludge applying increased the value of CEC, but only in the variant with the addition of 100 Mg · ha^{-1} the increase was significant (Table 8, Fig. 3).

Cation exchange capacity of the soil sorption was influenced both by the additive in the sewage sludge and mineral wool. Sludge added in a dose of 100 Mg · ha^{-1} caused an increase in CEC value of tested soil by 2.73 cmol (+) · kg^{-1}, ie. bt 41.8%, compared to a dose of 10 Mg · ha^{-1}. When the mineral wool was applied together with sludge CEC value increased by 4.27 cmol (+) · kg^{-1}, i.e. by 74.8%.

Considering the CEC variation in particular research period, significant growth of analyzed parameter in soil amended with sewage sludge in fertilizing "dose" (10 Mg · ha^{-1}) was observed only in III period. On the other hand, the addition of mineral wool to this mixture has not influenced CEC value significantly.

Sewage sludge used in a "reclamation" dose 100 Mg · ha^{-1} gradually increased cation exchange capacity of the soil, but significant differences in the values of CEC were found only in III and IV research period—when the sludge was used as a sole additive; and in II, III and IV periods—when it was used with mineral wool.

Changes in the cation exchange capacity of the soil caused by the NPK fertilization, as well as by the addition of mineral wool, proved to be insignificant in comparison to the soil without additives (control I). Additionally, the changes observed in these substrates in successive research seasons were also insignificant.

Taking into account the changes in cation exchange capacity in tested soil materials and assuming the value measured in control sample I (soil without additives) as 100% the following order was established:

soil (100.0%) < soil + NPK (105.3%) < soil + mineral wool + NPK (108.7%) < soil + mineral wool (109.8%)

<soil + sewage sludge 10 Mg · ha^{-1} + mineral wool (121.7%) < soil + sewage sludge 10 Mg · ha^{-1} (139.2%)

<soil + sewage sludge 100 Mg · ha^{-1} (197.2%)
<soil + sewage sludge 100 Mg · ha^{-1} + mineral wool (212.8%)

3.5 The influence of the waste additives on percent base saturation (V) in soil

The low level of saturation of the soil sorption complex with base cations slightly increased under the influence of mineral NPK fertilization (Table 9, Fig. 5). However, a significant growth of base saturation was observed under the influence of the both waste additives. The most visible impact was noted upon the addition of the sludge, similarly as in the case of the previously described sorption properties of the soil. When the sludge was used as sole additive at higher dose (100 Mg · ha^{-1}), the value of V increased about 2.2 times, while at a lower dose (10 Mg · ha^{-1}), by 1.8 times. In the case when sludge was applied together with mineral wool, these indicators amounted to were 2.3 and 1.7, respectively. It should be emphasized that the sewage sludge applied at a "reclamation" dose allowed to achieve a percent base saturation value corresponding with the standards for the soil of very good quality.

Base saturation of the soil sorption complex in samples with addition of a lower dose of the sludge was reduced significantly in III and IV research periods, while in the samples containing the higher sludge dose, it exhibited a significant decrease in the period II, but after this time it clearly increased.

Taking into account the percent base saturation in the tested soil materials and assuming the value measured in control sample I (soil without additives) as 100%, the following order was established:

soil (100%) < soil + NPK (106.4%) <soil + mineral wool (141.4%) <soil + mineral wool + NPK (146.2%)

<soil + sewage sludge 10 Mg · ha^{-1} + mineral wool (171.1%) <soil + sewage sludge 10 Mg · ha^{-1} (188.0%).

Table 7. Total exchangeable bases (S, cmol(+) · kg^{-1}) of soils under examinations.

Experimental variant	Period I	II	III	IV	Average
Soil without additions (control I)	1.55	1.65	1.62	1.66	1.62
Soil + mineral fertilization with NPK: 80, 40 and 60 kg · ha^{-1}, i.e. 0.27, 0.14, 0.21 g/pot (control II)	1.60	1.70	1.81	2.19	1.82
Soil + sewage sludge (10 Mg · ha^{-1})	4.10	4.58	4.36	3.93	4.24
Soil + sewage sludge (100 Mg · ha^{-1})	5.40	6.56	7.81	8.49	7.06
Soil + mineral wool (200 m^3· ha^{-1})	2.27	2.34	2.43	2.52	2.39
Soil + mineral wool (400 m^3· ha^{-1})	2.35	2.42	2.58	2.85	2.55
Soil + mineral wool (800 m^3· ha^{-1})	2.40	2.50	2.67	2.90	2.61
Soil + mineral wool (200 m^3· ha^{-1}) + NPK	2.28	2.38	2.53	2.55	2.43
Soil + mineral wool (400 m^3· ha^{-1}) + NPK	2.38	2.48	2.60	2.95	2.60
Soil + mineral wool (800 m^3· ha^{-1}) + NPK	2.44	2.58	2.77	2.98	2.69
Soil + mineral wool (200 m^3· ha^{-1}) + sewage sludge (10 Mg · ha^{-1})	3.27	3.00	3.23	3.42	3.23
Soil + mineral wool (400 m^3· ha^{-1}) + sewage sludge (10 Mg · ha^{-1})	3.35	3.42	3.28	3.55	3.40
Soil + mineral wool (800 m^3· ha^{-1}) + sewage sludge (10 Mg · ha^{-1})	3.40	3.50	3.47	3.60	3.49
Soil + mineral wool (200 m^3· ha^{-1}) + sewage sludge (100 Mg · ha^{-1})	5.66	7.52	9.00	9.45	7.90
Soil + mineral wool (400 m^3· ha^{-1}) + sewage sludge (100 Mg · ha^{-1})	5.74	8.00	8.33	9.62	7.92
Soil + mineral wool (800 m^3· ha^{-1}) + sewage sludge (100 Mg · ha^{-1})	5.73	8.83	8.21	9.25	8.00
Average values					
Soil + sewage sludge (10 Mg · ha^{-1})	4.10	4.58	4.36	3.93	4.24
Soil + sewage sludge (100 Mg · ha$^-$)	5.40	6.56	7.81	8.49	7.06
Soil + mineral wool	2.34	2.42	2.56	2.75	2.51
Soil + mineral wool + NPK	2.36	2.48	2.63	2.82	2.57
Soil + sewage sludge (10 Mg · ha^{-1}) + mineral wool	3.34	3.30	3.32	3.52	3.37
Soil+sewage sludge (100 Mg · ha^{-1}) + mineral wool	5.71	8.11	8.51	9.44	7.94
Average value for period	3.37	3.90	4.11	4.43	3.95
NIR$_{0.05}$ for the period			0.66**		
NIR$_{0.05}$ for the particular variant			1.80**		

Comments: as shown under Table 6.

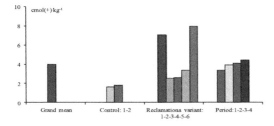

Figure 2. Influence of sewage sludge and spent mineral wool on total exchangeable bases in sandy soil (average values for factors evaluated in ANOVA test). Symbols as shown in Fig. 1.

< soil + sewage sludge 100 Mg · ha^{-1} (220.1%) < soil + sewage sludge 100 Mg · ha^{-1} + mineral wool (228.8%).

4 DISCUSSION

Studies have shown that the application of sewage sludge as the only supplement – as well as in combination with mineral wool – to acidic (pH 5.2) sandy soil had a significant influence on its physicochemical properties, as evidenced by the results of cluster analysis of sorption properties (Fig. 5).

Table 8. Cation exchange capacity (T, cmol(+) · kg^{-1}) of soils under examinations.

Experimental variant	Period I	II	III	IV	Average
Soil without additions (control I)	4.85	4.50	4.47	4.96	4.69
Soil + mineral fertilization with NPK: 80, 40 and 60 kg · ha^{-1}, i.e. 0.27, 0.14, 0.21 g/pot (control II)	4.80	4.85	4.66	5.49	4.94
Soil + sewage sludge (10 Mg · ha^{-1})	6.10	6.73	6.91	6.38	6.53
Soil + sewage sludge (100 Mg · ha^{-1})	7.20	9.11	10.1	10.6	9.25
Soil + mineral wool (200 m^3 · ha^{-1})	5.27	5.19	4.68	5.37	5.13
Soil + mineral wool (400 m^3 · ha^{-1})	5.30	5.12	4.83	5.40	5.16
Soil + mineral wool (800 m^3 · ha^{-1})	5.25	5.05	5.07	5.35	5.17
Soil + mineral wool (200 m^3 · ha^{-1}) + NPK	5.28	5.38	4.93	5.00	5.14
Soil + mineral wool (400 m^3 · ha^{-1}) + NPK	5.23	5.18	4.85	5.10	5.09
Soil + mineral wool (800 m^3 · ha^{-1}) + NPK	4.99	5.03	5.17	5.13	5.08
Soil + mineral wool (200 m^3 · ha^{-1}) + sewage sludge (10 Mg · ha^{-1})	5.97	5.70	5.48	5.57	5.68
Soil + mineral wool (400 m^3 · ha^{-1}) + sewage sludge (10 Mg · ha^{-1})	6.00	5.92	5.38	5.55	5.71
Soil + mineral wool (800 m^3 · ha^{-1}) + sewage sludge (10 Mg · ha^{-1})	5.80	5.80	5.72	5.60	5.73
Soil + mineral wool (200 m^3 · ha^{-1}) + sewage sludge (100 Mg · ha^{-1})	7.66	9.67	11.1	11.45	9.96
Soil + mineral wool (400 m^3 · ha^{-1}) + sewage sludge (100 Mg · ha^{-1})	7.69	10.1	10.43	11.62	9.96
Soil + mineral wool (800 m^3 · ha^{-1}) + sewage sludge (100 Mg · ha^{-1})	7.73	10.88	10.26	11.25	10.02
Average values					
Soil + sewage sludge (10 Mg · ha^{-1})	6.10	6.73	6.91	6.38	6.53
Soil + sewage sludge (100 Mg · ha$^-$)	7.20	9.11	10.06	10.64	9.25
Soil + mineral wool	5.27	5.12	4.86	5.37	5.15
Soil + mineral wool + NPK	5.17	5.20	4.98	5.08	5.10
Soil + sewage sludge (10 Mg · ha^{-1}) + mineral wool	5.92	5.81	5.53	5.57	5.71
Soil + sewage sludge (100 Mg · ha^{-1}) + mineral wool	7.69	10.22	10.60	11.44	9.98
Average value for period	6.01	6.51	6.44	6.80	6.44
NIR$_{0.05}$ for the period	0.77				
NIR$_{0.05}$ for the particular variant	2.10**				

Comments: as shown under Table 6.

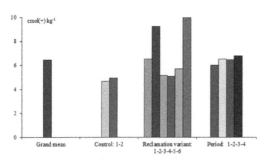

Figure 3. Influence of sewage sludge and spent mineral wool on cation exchange capacity in sandy soil (average values for factors evaluated in ANOVA test). Symbols as shown in Fig. 1.

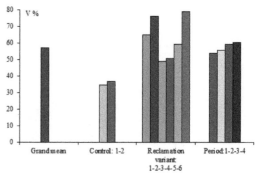

Figure 4. Influence of sewage sludge and spent mineral wool on percent base saturation in sandy soil (average values for factors evaluated in ANOVA test).

Table 9. Percent base saturation V (% of CEC) of soils under examinations.

Experimental variant	Period I	II	III	IV	Average
Soil without additions (control I)	31.96	36.67	36.24	33.47	34.58
Soil + mineral fertilization with NPK: 80, 40 and 60 kg · ha^{-1}, i.e. 0.27, 0.14, 0.21 g/pot (control II)	33.33	35.05	38.84	39.89	36.78
Soil + sewage sludge (10 Mg · ha^{-1})	67.21	68.10	63.10	61.60	65.00
Soil + sewage sludge (100 Mg · ha^{-1})	75.00	72.01	77.63	79.79	76.11
Soil + mineral wool (200 m^3 · ha^{-1})	43.07	45.09	51.92	46.93	46.75
Soil + mineral wool (400 m^3 · ha^{-1})	44.34	47.27	53.42	52.78	49.45
Soil + mineral wool (800 m^3 · ha^{-1})	45.71	49.50	52.66	54.21	50.52
Soil + mineral wool (200 m^3 · ha^{-1}) + NPK	43.18	44.24	51.32	51.00	47.43
Soil + mineral wool (400 m^3 · ha^{-1}) + NPK	45.51	47.88	53.61	57.84	51.21
Soil + mineral wool (800 m^3 · ha^{-1}) + NPK	48.90	51.29	53.58	58.09	52.96
Soil + mineral wool (200 m^3 · ha^{-1}) + sewage sludge (10 Mg · ha^{-1})	54.77	52.63	58.94	61.40	56.94
Soil + mineral wool (400 m^3 · ha^{-1}) + sewage sludge (10 Mg · ha^{-1})	55.83	57.77	60.97	63.96	59.63
Soil + mineral wool (800 m^3 · ha^{-1}) + sewage sludge (10 Mg · ha^{-1})	58.62	60.34	60.66	64.29	60.98
Soil + mineral wool (200 m^3 · ha^{-1}) + sewage sludge (100 Mg · ha^{-1})	73.89	77.77	81.08	82.53	78.82
Soil + mineral wool (400 m^3 · ha^{-1}) + sewage sludge (100 Mg · ha^{-1})	74.64	79.21	79.87	82.79	79.13
Soil+mineral wool (800 m^3 · ha^{-1}) + sewage sludge (100 Mg · ha^{-1})	74.13	81.16	80.02	82.22	79.38
Average values					
Soil + sewage sludge (10 Mg · ha^{-1})	67.21	68.10	63.10	61.60	65.00
Soil + sewage sludge (100 Mg · ha$^-$)	75.00	72.01	77.63	79.79	76.11
Soil + mineral wool	44.38	47.29	52.67	51.30	48.91
Soil + mineral wool + NPK	45.86	47.80	52.84	55.64	50.54
Soil + sewage sludge (10 Mg · ha^{-1}) + mineral wool	56.41	56.92	60.19	63.22	59.18
Soil + sewage sludge (100 Mg · ha^{-1}) + mineral wool	74.22	79.38	80.32	82.51	79.11
Average value for period	53.79	55.69	59.22	60.35	5.73
NIR$_{0.05}$ for the period			2.01**		
NIR$_{0.05}$ for the particular variant			5.48**		

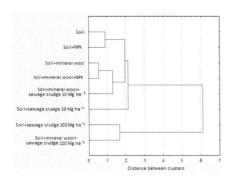

Figure 5. Dendrogram presenting the results of cluster analysis for sorption properties of soil materials examined in the experiment.

The greatest similarity has been shown between both the controls samples (soil without additives and the soils fertilized with mineral form of N, P, and K), and between the both soil mixtures with mineral wool (unamended and amended with N, P, and K). It means that NPK fertilization had no influence on sorption properties of the soil.

The mixtures of soils with sewage sludge in dose 100 Mg · ha^{-1} differed from other tested materials to the greatest degree. Examined soils mixtures differed in regard to acidity. Waste-derived additives increased the pH values, both of these measured in water and KCl. Sewage sludge addition exerted the most beneficial effect on the reduction of acidification. It increased the pH$_{KCl}$ value of the soil of about 0.8–1.1, and the higher increase was observed at the "reclamation" dose ($100 \, \text{Mg} \cdot \text{ha}^{-1}$) of sewage sludge. In the case of simultaneous application of sewage sludge in the mentioned amount and mineral wool, a further increase in pH value (of about 0.5) was found. The obtained results comply with the observations of Roig et al. (2012) and Ociepa et al. (2008), who showed a beneficial effect of sewage sludge addition on the quality of acidic soils, consisting of increasing the pH value.

Our experiment showed that both the examined additives decreased the hydrolytic acidity of soil, which initially was equal to $3.07 \, \text{cmol}(+) \cdot \text{kg}^{-1}$. However, the decrease in acidity caused by the addition of sewage sludge at a dose of $10 \, \text{Mg} \cdot \text{ha}^{-1}$ was negligible. The addition of the sludge at a dose of $100 \, \text{Mg} \cdot \text{ha}^{-1}$ just resulted in a significant decrease of H^+ ions concentration in the soil to the value $2.19 \, \text{cmol}(+) \cdot \text{kg}^{-1}$. Mineral wool application caused further decline in the value of Hh to $2.04 \, \text{cmol}(+) \cdot \text{kg}^{-1}$. This was the lowest value hydrolytic acidity observed throughout the experiment.

The phenomenon of decline in the Hh value in the soil after the introduction of sewage sludge is not consistent with the results obtained by Stańczyk-Mazanek et al. (2013), which showed that the addition of sewage sludge at a dose of $100 \, \text{Mg} \cdot \text{ha}^{-1}$ to the soil with initial Hh equal to $3.70 \, \text{cmol}(+) \cdot \text{kg}^{-1}$ resulted in an almost 2-fold increase in the value of the parameter, already after the first year of the experiment. Differences in chemical composition of sewage sludge caused by dewatering method may be the reason of the divergence between these results and obtained in our experiment. The use of aluminum salts as a coagulant could explain such a significant increase in hydrolytic acidity caused by the addition of sludge. However, the authors do not provide information on this subject. The increase in Hh value in sandy soils amended with sewage sludge was observed also by Iżewska (2007) and Gondek (2009).

From a practical point of view, a very important feature of soils is a cation exchange capacity underlying the retention capacity of the positive ions. This capacity determined the fertility of soils and their ability to buffer against the action of chemical agents. The sandy soil used in the experiment was characterized by a low CEC averaging $4.69 \, \text{cmol}(+) \cdot \text{kg}^{-1}$. Studies have shown that an addition of sewage sludge played a crucial role in the growth of CEC value of tested soil mixtures. The addition of mineral wool raised CEC value only to a slight degree. The highest, over 2-fold increase in sorption capacity of soil compared to the control sample I, was observed in the mixture with higher dose of sludge and the highest dose of mineral wool. A significant increase in CEC values in soils amended with high doses of sewage sludge was also found by other researchers (Soon 1981, Kladivko & Nelson 1979, Mitchel 1978).

The adsorption complex of the soils mixed with sewage sludge was saturated by bases to a great extent.

Percent base saturation of the soil amended with higher dose of sewage sludge amounted to 76.1%, and it was approximately 2.2 times higher than the V value found in the control soils. Addition of mineral wool with the highest dose ($800 \, \text{m}^3 \cdot \text{ha}^{-1}$) to this mixture increased base saturation of adsorption complex only by about 3.3%. Thus, the sewage sludge was the main reason for the soil enrichment with the base cations. The total exchangeable bases in the control sample I was $1.62 \, \text{cmol}(+) \cdot \text{kg}^{-1}$, and the addition of sewage sludge at a dose $100 \, \text{Mg ha}^{-1}$ increased the value of this parameter to $7.06 \, \text{cmol}(+) \cdot \text{kg}^{-1}$, i.e. more than four times. A similar effect of the sandy soil amendment with sewage sludge was observed by Stanczyk-Mazanek et al. (2013). They applied sludge at a dose of 100 Mg of ha^{-1} and obtained an increase in total exchangeable bases from 2.49 to $10.62 \, \text{cmol}(+) \cdot \text{kg}^{-1}$.

5 CONCLUSIONS

Pot experiment carried out in a greenhouse over a three-year period showed that the addition of municipal sewage sludge and/or mineral wool influences the physicochemical properties of acid, coarse-grained soil that was poor in organic matter, as well as silt and clay fractions. An analysis of pH, hydrolytic acidity, cation exchange capacity, total exchangeable bases, and percent base saturation determined for the soil and it mixture with sewage sludge and spent mineral wool indicates that changes in these properties depended mainly on the addition of the sludge and on its doses. Application of the sludge at a "fertilizing" dose ($10 \, \text{Mg ha}^{-1}$) significantly increased the amount of base cations in the soil and a base saturation of the adsorption complex, but it did not have any significant effect on hydrolytic acidity and cation exchange capacity. The addition of sludge at a "reclamation" dose ($100 \, \text{Mg} \cdot \text{ha}^{-1}$) significantly changed all the examined sorption properties. In the case of addition of mineral wool at all tested doses (200 to $800 \, \text{m}^3 \cdot \text{ha}^{-1}$), significant differences were observed only in regard to base saturation of the soil adsorption complex.

ACKNOWLEDGEMENTS

The study was supported by Polish Ministry of Science and Higher Education, Research Project No BLUE GAS-BG1/SOIL/2013, carried out as a part of joint programme between the National Centre for Research and Development and the Industrial Development Agency JSC.

REFERENCES

Adriano, D.C. 2001. Trace Elements in Terrestrial Environments: Biogeochemistry. Bioavailability and Risk of Metals. Springer—Verlag: New York.

Baran, S. 2008. Możliwości wykorzystania wełny mineralnej Grodan do kształtowania właściwości wodnych gleb i gruntów. *Zeszyty Problemowe Postępów Nauk Rolniczych* 533: 15–19.

Baran, S., Oleszczuk, P. & Żukowska, G. 2002. Zasoby i gospodarka odpadami organicznymi w glebie. *Acta Agrophysica* 73: 17–34.

Baran, S., Bielińska, E.J., Smal, H., Wójcikowska-Kapusta, A., Paluszek, J., Pranagal, J., Żukowska, G., Chmielewski, S.Z., & Futa, B. 2014. *Innowacyjne Metody Ochrony i Rekultywacji Gleb.* Monografie Komitetu Inżynierii Środowiska PAN, tom 120: 15–45.

Cao, Y. & Pawłowski, A. 2012. Sewage sludge-to-energy approaches based on anaerobic digestion and pyrolysis: Brief overview and energy efficiency assessment. *Renewable & Sustainable Energy Reviews* 3(16); 1657–1665

Carter, M.R. 2002. Soil quality for sustainable land management: organic matter and aggregation interactions that maintain soil functions. *Agronomy Journal* 94: 38–47.

Czechowska-Kosacka, A., Cao, Y. & Pawłowski, A. 2015. Criteria for sustainable disposal of sewage sludge. *Rocznik Ochrona Środowiska—Annual Set The Environment Protection* 1(17); 337–350.

Dąbkowska-Naskręt, H., Jaworska, H. & Kobierski, M. 2001. Cation exchange capacity of clay rich soils in relations to mineralogical composition and organic matter content. *Acta Agrophysica* 50: 113–118.

Epstein, E., Taylor, J.M. & Chaney, R.L. 1976. Effects of sewage sludge and sludge compost applied to soil on same soil physical and chemical properties. *Journal of Environmental Quality* 5: 422–426.

Gondek, K. 2009. Aspekty nawozowe i środowiskowe przemian i dostępności dla roślin wybranych pierwiastków w warunkach nawożenia różnymi materiałami organicznymi. *Zeszyty Naukowe Uniwersytetu Rolniczego im. Hugo Kołłątaja w Krakowie* 452: 329.

Gondek, K & Filipek-Mazur, B. 2006. Ocena efektywności nawożenia osadami ściekowymi na podstawie plonowania roślin i wykorzystania składników pokarmowych. *Acta Scientiarum Polonorum Formatio Circumiectus* 5: 39–50.

Heidrich, Z., Kalenik, M., Podedworna, J. & Stańko, G. *Sanitacja wsi.* Warszawa 2008.

Hort, C., Gracy, S., Platel, V. & Moynault, L. 2009. Evaluation of sewage sludge and yard waste compost as a biofilter media for the removal of ammonia and volatile organic sulfur compounds (VOSCs). *Chemical Engineering Journal* 152 (1): 44–53.

Iżewska, A. 2007. Wpływ nawożenia obornikiem, osadem ściekowym i kompostem z osadów ściekowych na właściwości gleb. *Zeszyty Problemowe Postępów Nauk Rolniczych* 518: 85–92.

Kladivko, E.J. & Nelson, D.W. 1979. Changes in soil properties from application of anaerobic sludge. *Journal of the Water Pollution Control Federation* 51: 325–332.

Krasowicz, S., Oleszek, W., Horabik, J., Dębicki R., Jankowiak J., Stuczyński T. & Jadczyszyn J. 2011. Racjonalne gospodarowanie środowiskiem glebowym Polski. *Polish Journal of Agronomy* 7: 43–58.

Kujawska J., Pawłowska M., Czechowska-Kosacka A. & Cel W. 2014. Wykorzystanie przefermentowanych osadów ściekowych z komunalnych oczyszczalni ścieków w procesie unieszkodliwiania odpadów wiertniczych. *Inżynieria i Ochrona Środowiska* 4 (17): 583–595.

Mitchel, M.J., Hartenstein, R., Swift, B.L., Neuhauser, E.F., Abrams, B.I., Mulligan, R.M., Brown, B.A., Craig, D. & Kaplan, D. 1978. Effects of different sewage sludges on same chemical and biological characteristics of soil. *Journal of Environmental Quality* 7: 551–559.

Ociepa, A., Pruszek, K., Lach, J. & Ociepa, E. 2008. Influence of long-term cultivation of soils by means of manure and sludge on the increase of heavy metals content in soils, Ecological Chemistry and Engineering S 15(1): 103–109.

Oświęcimski, A. 1996. Aktualne tendencje w wykorzystaniu podłoży nieorganicznych w uprawach pod osłonami. *Zeszyty Problemowe Postępów Nauk Rolniczych* 429: 9–13).

Pawłowska, M., Rożej, A. & Stępniewski, W. 2011. Effect of bed properties on methane removal potential in aerated biofilter—model studies. *Waste Management* 31(5): 903–913.

Pawłowska, M. & Siepak, J. 2006. Enhancement of methanogenesis at a municipal landfill site by addition of sewage sludge. *Environmental Engineering Sciences* 23(4): 673–679.

Regulation of the Polish Environment Minister of 9 December 2014 about the catalogue of waste; Official Journal of 2014, item 1923.

Roig, N., Sierra, J., Nadal, M., Marti, E., Navalon-Maadrigal, P., Schuhmacher, M. & Domingo, J.L. 2012. Relationship between pollutant content and ecotoxicity of sewage sludges from Spanish wastewater treatment plants. *Science of the Total Environment* 425: 99–109.

Singha, R.P. & Agrawal, M. 2008. Potential benefits and risk of land application of sewage sludge. *Waste Management* 28: 247–358.

Siuta, J. 1998. Warunki i sposoby przyrodniczego użytkowania osadów ściekowych. W: Materiały Międzynarodowego Seminarium Szkoleniowego nt. Podstawy oraz praktyka przeróbki i zagospodarowania osadów. Kraków. Wyd. LEM s.c.

Skłodowski, P. & Bielska, A. 2009. Właściwości i urodzajność gleb Polski – podstawą kształtowania relacji rolno – środowiskowych. *Woda-Środowisko-Obszary Wiejskie* 9(28): 203–214.

Soon, Y.K. 1981. Solubility and Sorption of Cadmium in Soils Amended with Sewage Sludge. *Journal of Soil Science* 32: 85–95.

Stańczyk-Mazanek, E., Piątek, M. & Kępa, U. 2013. Wpływ następczy osadów ściekowych stosowanych na glebach piaszczystych na właściwości kompleksu sorpcyjnego. *Rocznik Ochrona Środowiska - Annual Set of the Environment Protection* 15: 2437–2451.

Żukowska, G., Flis-Bujak M. & Baran S. 2002. Wpływ nawożenia osadem ściekowym na substancję organiczną gleby lekkiej pod uprawą wikliny. *Acta Agrophysica* 73: 357–367.

Environmental Engineering V – Pawłowska & Pawłowski (Eds)
© 2017 Taylor & Francis Group, London, ISBN 978-1-138-03163-0

Nitrate monoionic form of anion exchanger as a means for enhanced nitrogen fertilization of degraded soils

M. Chomczyńska
Faculty of Environmental Engineering, Lublin University of Technology, Lublin, Poland

ABSTRACT: The study was performed to test the possibility of applying nitrate monoionic form of anion exchanger for enhanced nitrogen fertilization of degraded soils. To achieve the study aim, a pot experiment with *Dactylis glomerata* L. as a test species was conducted. Thirteen soil series were prepared with increasing doses of nitrogen, applied as $Ca(NO_3)_2 \cdot 4H_2O$ and nitrate form for the test purpose. It was shown that all nitrogen doses introduced into soil in nitrate form influenced plant growth advantageously, significantly increasing the values of vegetative parameters. Nitrogen doses applied in nitrate form showed the same effectiveness, because the values of vegetative parameters obtained in media series with nitrate form additions did not differ significantly. Regarding total dry biomass of plants, it could be stated that the efficiencies of application of nitrate monoionic form (in the presence of an additional Ca source) and calcium nitrate are similar for enhanced nitrogen fertilization of degraded soils.

Keywords: ion exchangers, degraded soils, nitrogen fertilization

1 INTRODUCTION

Under the impact of human activity the soil degradation is progressing rapidly. Nowadays, 0.5–0.7% of soils disappear every year and degraded lands occupy about 2 billions ha in the world (Molo 2010). This threaten sustainable development (Sztumski 2016, Żukowska et al. 2016). To restore the original and/or utilizable properties of degraded lands, technical and biological reclamation is required. Biological restoration involves humus layer forming, as well as scarps and slopes planting. Humus layer forming occurs due to the cultivation of suitable plant mixtures, mainly containing grass and legume species. Scarps or slops planting requires application of appropriate trees, shrubs and perennial herbaceous plants. Plant growth on degraded soils is supported by organic and mineral fertilization. Organic fertilization can be realized using, among others, manure, compost, peat or sludge from biological sewage treatment. Sometimes the afore-mentioned materials —especially sewage sludge should be applied with caution because it can contain large amount of pathogens or increased heavy metals concentrations (Czechowska-Kosacka & Pawłowski 2006). Mineral fertilizing mostly enriches soils with macronutrients, especially with nitrogen, phosphorus and potassium (Maciak 1999, Karczewska 2008). Nitrogen fertilization increases plant yield with a relatively low efficiency, because a considerable fraction (up to two-thirds) of N input accumulates as runoffs (Frink et al. 1999). The resulting nitrate leaching from soil and nitrous oxide emissions in the atmosphere have harmful consequences on the environment (Donner & Kucharik 2008, De Pessemier 2013, Krajewski 2016). It should be also added that while performing soil reclamation, it is recommended to repeat the nitrogen fertilization, which generates additional costs e.g. connected with continued operation of agricultural machineries (Karczewska 2008, Patrzałek 2010). Hence, in order to reduce the afore-mentioned costs on the one hand, and to prevent negative consequences of environment eutrophication on the other hand, it seems reasonable to apply monoionic form of ion exchangers saturated with NO_3^- anions for soil fertilization with higher nitrogen doses. Ion exchangers (ion exchange resins) are synthetic organic polymers with an electrostatic charge that is neutralized by selected counterion of opposite charge (Skogley & Dobermann 1996). Ion exchange resins can be successfully used in soil reclamation due to high ion exchange capacity, relatively high chemical and mechanical stability and sanitary safety (Soldatov 1998, Qian & Schoenau 2002). These materials are also characterized by the ability of gradual nutrient ions release, thus preventing their leaching by rainfall (Soldatov 1988, Chomczyńska & Pawłowski 2003). Therefore, the aim of the present study was:

i. to determine the effect of increasing nitrogen doses applied as nitrate monoionic form on growth of orchard grass (*Dactylis glomerata* L.) —species recommended as a constituent of plant restoration mixtures

ii. to compare the effectiveness of application of monoionic nitrate form and conventional fertilizer (calcium nitrate) for enhanced nitrogen fertilization of degraded soils.

2 MATERIALS AND METHODS

Sandy soil, monoionic nitrate form and calcium nitrate ($Ca(NO_3)_2 4H_2O$) were used as basic materials in the study. Soil was taken from the excavation boundary of the sand mine in Rokitno, situated near Lublin (Eastern Poland). It consisted of the following fractions: sand (2.0–0.05 mm) – 76%; silt (0.05–0.002 mm) – 23%; clay (<0.002 mm) – 1%. The pH value of the soil in KCl solution was 5.53. The contents of macronutrients available for plants in the soil were determined according to Polish standards (Lityński & Jurkowska 1982, Ostrowska et al. 1991). It was characterized by low (P, K, S), very

low (Mg) or insufficient (N, Ca) contents of nutrients in terms of plants requirements—Table 1.

Monoionic nitrate form was prepared on the basis of polyfunctional anion exchanger EDE-10P. The total exchange capacity of anion exchanger in the OH$^-$ form is 10.5 molg^{-1}. It contains four types of functional groups of different basic strength. Nitrate form of EDE-10P was prepared treating its OH-form with HNO_3 solution. The monoionic form contained 2.95 mol of NO_3^- per gram. The studies were performed using orchard grass (*Dactylis glomerata* L. cv. Amera) as the test species. For the purpose of the pot experiment, thirteen series of media were prepared including: the control series (soil alone), four soil series with increasing N doses applied as $Ca(NO_3)_2 4H_2O$, four soil series with increasing N doses applied as monoionic nitrate form and four soil series with increasing N doses applied as monoionic nitrate form and supplemented with increasing calcium doses as $CaCl_2$ (Table 2). The nitrogen doses were: 100, 200, 250 and 300 kg of N per ha, where 100 kg N per ha was the basic dose recommended for fertilizing meadows on mineral soil (Niewiadomski 1983). The experiment started on 21 April 2015. In each

Table 1. Contents of nutrients in the soil.

N-NH$_4$ mg per kg	N-NO$_3$ mg per kg	P$_2$O$_5$ mg per 100 g	K$_2$O mg per 100 g	Mg mg per 100 g	Ca mg per dm^3	S-SO$_4$ mg per 100 g
1.4	<1.37	5.6	<2.5	2.2	303	0.92

Table 2. Characteristics of media series in the pot experiment.

Media series[*]	Soil g per pot	Fertilizer g per pot	Nitrogen g per pot	Calcium g per pot	Pot number
S	470	–	–	–	7
S+1Ca(NO$_3$)$_2$	470	0.6617	0.0785	0.1123	6
S+2Ca(NO$_3$)$_2$	470	1.3234	0.1570	0.2246	6
S+2.5Ca(NO$_3$)$_2$	470	1.6543	0.1963	0.2808	6
S+3Ca(NO$_3$)$_2$	470	1.9851	0.2355	0.3369	6
S+1 m.f.	470	1.9007	0.0785	–	7
S+2 m.f.	470	3.8015	0.1570	–	7
S+2.5 m.f.	470	4.7530	0.1963	–	7
S+3 m.f.	470	5.7022	0.2355	–	7
S+1 m.f.Ca	470	1.9007	0.0785	0.1123	6
S+2 m.f.Ca	470	3.8015	0.1570	0.2246	6
S+2.5 m.f.Ca	470	4.7530	0.1963	0.2808	6
S+3 m.f.Ca	470	5.7022	0.2355	0.3369	6

[*]s–s oil; S+1Ca(NO$_3$)$_2$, S+1m.f., S+1m.f.Ca—soil with single N dose (100 kg of N per ha) applied as $Ca(NO_3)_2 4H_2O$ or nitrate monoionic form (m.f.Ca—supplemented with CaCl$_2$), respectively; S+2Ca(NO$_3$)$_2$, S+2m.f., S+2m.f.Ca—soil with double N dose (200 kg of N per ha) applied as $Ca(NO_3)_2 4H_2O$ or nitrate monoionic form (m.f.Ca—supplemented with CaCl$_2$), respectively; S+2.5Ca(NO$_3$)$_2$, S+2.5m.f., S+2.5m.f.Ca—soil with 2.5-fold N dose (250 kg of N per ha) applied as $Ca(NO_3)_2 4H_2O$ or nitrate monoionic form (m.f.Ca—supplemented with CaCl$_2$), respectively; S+3Ca(NO$_3$)$_2$, S+3m.f., S+3 m.f.Ca—soil with triple N dose (300 kg of N per ha) applied as $Ca(NO_3)_2 4H_2O$ or nitrate monoionic form (m.f.Ca—supplemented with CaCl$_2$), respectively.

pot (of 360 cm³ volume) 51 orchard grass seeds were sown. After 7 days, the number of plants in each pot was standardized to 25. The experiment was carried out in a phytotron with a 13/11 light/dark regime. The daytime (7 am-8 pm) air temperature was 25°C while the night-time (8 pm-7 am) air temperature was 16°C. During the experiment the plants were watered with distilled water. The amount of water depended on the current needs of the plants. The experiment was terminated after 42 days from the time of seed sowing. The plant stems were cut down and roots were separated. The wet and dry (105°C) biomass of stems and dry (105°C) root biomass were weighed. Data obtained were subjected to analysis of variance (Statistica 12.5 software). The means were separated using Tukey's test at the 0.05 probability level.

3 RESULTS AND DISCUSSION

The study results are presented in Figures 1–8. It can be seen that all nitrogen additions introduced into soil in nitrate monoionic form affected the plant growth advantageously, significantly increasing the values of vegetative parameters. The wet stem biomass of orchard grass growing in media supplemented with particular doses of nitrate form exceeded the one obtained for the test species growing in the control series (Fig. 1) by 200–238%. The dry stem biomass of plants on soil enriched with N additions as nitrate form was over 2 times higher than that obtained for soil alone (Fig. 2). The dry root biomass of the test species in series with nitrate form additions exceeded the one of plants growing on soil of the control series by 69–94% (Fig. 3). Total dry biomass of plants obtained in series with increasing doses of nitrate form was 2 times greater than that observed for test species growing on soil alone (Fig. 4). The results described above are consistent with the findings of other studies carried out by Kloc & Szwed (1995), Wasąg (2000), Chomczyńska & Pawłowski (2003). The above-mentioned authors also observed a positive effect of monoionic forms additions (containing different macronutrients) on

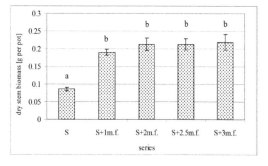

Figure 2. Dry stem biomass of plants in control series (soil) and series with increasing N doses applied as monoionic form; explanations are the same as under Figure 1.

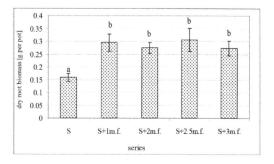

Figure 3. Dry root biomass of plants in control series (soil) and series with increasing N doses applied as monoionic form; explanations are the same as under Figure 1.

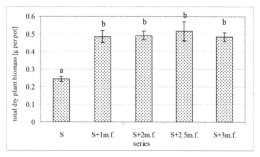

Figure 4. Total dry biomass of plants in control series (soil) and series with increasing N doses applied as monoionic form; explanations are the same as under Figure 1.

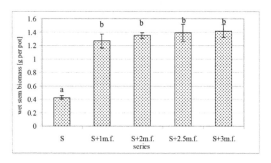

Figure 1. Wet stem biomass of plants in control series (soil) and series with increasing N doses applied as monoionic nitrate form; S—soil, S+1 m.f. – soil+N (100 kg per ha), S+2 m.f. – soil+N (200 kg per ha), S+2.5 m.f. – soil+N (250 kg per ha), S+3 m.f. – soil+N (300 kg per ha); different letters above bars indicate significant differences between mean values; I—standard deviation.

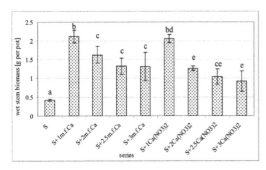

Figure 5. Wet stem biomass of plants in series with increasing N doses applied as Ca(NO₃)₂4H₂O and series with increasing N doses applied as monoionic form supplemented with Ca; S—soil; S+1Ca(NO₃)₂, S+1 m.f.Ca—soil with single N dose (100 kg of N per ha); S+2Ca(NO₃)₂, S+2 m.f.Ca—soil with double N dose (200 kg of N per ha); S+2.5Ca(NO₃)₂, S+2.5 m.f.+Ca—soil with 2.5-fold N dose (250 kg of N per ha); S+3Ca(NO₃)₂, S+3 m.f.+Ca—soil with triple N dose (300 kg of N per ha); different letters above bars indicate significant differences between mean values; I—standard deviation.

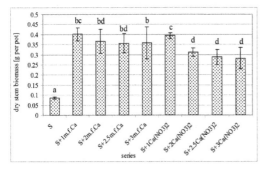

Figure 6. Dry stem biomass of plants in series with increasing N doses applied as Ca(NO₃)₂4H₂O and series with increasing N doses applied as monoionic form supplemented with Ca; explanations are the same as under Figure 5.

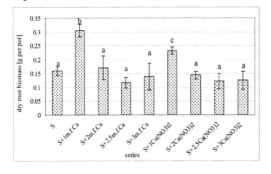

Figure 7. Dry root biomass of plants in series with increasing N doses applied as Ca(NO₃)₂4H₂O and series with increasing N doses applied as monoionic form supplemented with Ca; explanations are the same as under Figure 5.

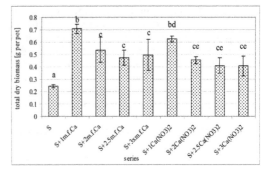

Figure 8. Total dry biomass of plants in series with increasing N doses applied as Ca(NO₃)₂4H₂O and series with increasing N doses applied as monoionic form supplemented with Ca; explanations are the same as under Figure 5.

plant growth, although they did not use increased doses of nutrients in their tests. It is also worthy to point out that some literature date report about a decrease in root biomass of plants or inhibition of root growth under the conditions of enhanced nitrogen fertilization (Lityński & Jurkowska 1998, Fageria & Moreira, 2011, Yu et al. 2014). Obviously, in the present studies—as it was mentioned above—the root biomass in series with additions of nitrate form was much greater than the one in the control series, but root biomass percentage in the total plant biomass in fertilized media series was slightly lower (56–61%) than that determined for soil alone (62%). The lack of a radical decrease in root biomass percentage under the conditions of higher nitrate monoionic form doses can be seen as a positive phenomenon from the soil restoration point of view. A relatively high root biomass percentage (as compared to stem biomass percentage) can indicate an intensified development of grass root systems which favors the stabilization and anti-erosive function of plants also during soil reclamation (Zuazo & Pleguezuelo 2009).

It should be stressed that all nitrogen doses applied in nitrate form showed the same effectiveness. The differences in values of vegetative parameters between media series with nitrate form additions did not exceed 16% and were not statistically significant (Figs. 1–4). In other words, an increase in N additions was not accompanied by a substantial increase in plant biomass. Thus, under the experimental conditions, according to "minimum low" nitrogen at higher doses ceased to be a limiting factor. Nevertheless, plants growing on soil supplemented with increasing N doses in nitrate form were healthy—they did not show any deformations or drying symptoms. It confirms the possibility of application of nitrate monoionic form for enhanced N fertilization to ensure nitrogen reserve in reclaimed soils without negative influence on plant growth.

The tendencies of increasing vegetative parameter values observed for media series supplemented with additions of nitrate form—when compared to the control series—were partially the same as those for the media series where N doses were applied as $Ca(NO_3)_2 4H_2O$ (Figs. 5–8). Namely, all nitrogen doses introduced into soil as conventional fertilizer significantly increased wet and dry stem biomass, as well as total dry biomass of plants (Figs. 5, 6, 8). Dry root biomass in series $S+1Ca(NO_3)_2$ significantly exceeded the one obtained on soil alone, but in series with higher nitrogen doses (200, 250 and 300 kg of N per ha) this parameter did not differ significantly from that in the control series (Fig. 7). All vegetative parameters of the test species growing on soil supplemented with single N dose $(S+1Ca(NO_3)_2)$ were significantly greater than those obtained in the series with higher nitrogen doses (Figs. 5–8). At the same time, there were no significant differences in the values of vegetative parameters between media series enriched with higher N doses (Figs. 5–8). It should be added that plants growing on soil with greater additions of calcium nitrate exhibited drying out of leaf tips or of single shoots which would indicate on disturbances in the water uptake. Disorders in water uptake could result from high osmotic potential of soil solution causing physiological drought phenomenon (Mahajan & Tuteja 2005, Carpici et al. 2010, Carillo et al. 2011, Batool et al. 2014).

It was relevant to compare the application efficiency of nitrate monoionic form and the conventional fertilizer as calcium nitrate. Research in this regard required the use of media series with increasing N doses applied as $Ca(NO_3)_2 4H_2O$ and in nitrate monoionic form, as well as supplemented with calcium because Ca is a macronutrient affecting plant growth (Marschner 2012). The results obtained showed that wet stem biomass of plants growing on soil enriched with double and triple N dose in nitrate form was significantly higher (by 29 and 42%, respectively) than that obtained in series where nitrogen was introduced as $Ca(NO_3)_2 4H_2O$ (Fig. 5). Dry stem biomass of orchard grass obtained in series S+3 mfCa significantly exceeded (by 29%) the one obtained in series $S+3Ca(NO_3)_2$ and dry root biomass of plants growing on soil with single N dose in nitrate form was significantly greater (by 35%) than the one found for soil with single N dose applied as $Ca(NO_3)_2 4H_2O$ (Figs. 6 and 7). However, it should be said that differences between values of total dry plant biomass were not statistically significant for the compared media series with different nitrogen forms dosed (Fig. 8). From a soil remediation point of view, production of dry plant matter is important because it serves as a source material for humus formation. Therefore, the lack of statistically significant differences in total dry biomass allows the efficiencies of application of nitrate monoionic form and calcium nitrate to be determined as similar for enhanced nitrate fertilization of degraded soils. Nevertheless, it would be worth to support this conclusion with further studies, where e.g. the test species could be the other ones recommended as constituents of plant recultivation mixtures.

4 CONCLUSIONS

On the basis of the results presented herein, the following conclusions are offered:

Nitrate monoionic form of anion exchanger seems to be an efficient means for enhanced fertilization of degraded soils—all nitrogen doses introduced as nitrate monoionic form into soil influenced plant growth advantageously, increasing the values of vegetative parameters for orchard grass.

In experimental conditions, all nitrogen doses applied as nitrate monoionic form showed the same effectiveness, i.e. the differences in plant yield between media series with nitrate form additions were not statistically significant.

Regarding total dry biomass of plants, it could be stated that the efficiencies of application of nitrate monoionic form (in the presence of an additional Ca source) and calcium nitrate are similar for enhanced nitrogen fertilization of degraded soils. Nevertheless, the advantage of nitrate form application is gradual release of NO_3^- in exchange with ions of plant metabolites, which may be helpful in reducing eutrophication of the environment.

ACKNOWLEDGEMENTS

This work was supported by grant of Lublin University of Technology, No. S-13/WIŚ/2015

REFERENCES

Batool, N., Shahzad, A., Ilyas, N. & Noor, T. 2014. Plants and salt stress. *International Journal of Agriculture and Crop Sciences* 7 (9): 582–589.

Carillo, P., Annunziata, M.G., Pontecorvo, G., Fuggi, A. & Woodrow, P. 2011. Salinity stress and salt tolerance In A. Shanker (ed.), *Abiotic stress in plants—mechanisms and adaptations*: 22–38. Rijeka: Intech.

Carpici, E.B., Celik, N. & Bayram, G. 2010. The effects of salt stress on the growth, biochemical parameter and mineral element content of some maize (*Zea mays* L.) cultivars. *African Journal of Biotechnology* 9 (41): 6937–6942.

Chomczyńska, M. & Pawłowski, L. 2003. Utilization of spent ion-exchange resins for soil reclamation. *Environmental Engineering Science* 20: 301–306.

Czechowska-Kosacka, A. & Pawłowski, L. 2006. Influence of the addition of solidifying materials on the heavy metals immobilization in sewage sludge. *Rocznik Ochrony Środowiska* 8: 11–26.

De Pessemier, J., Chardon, F., Juraniec, M., Delaplace, P. & Hermans, C. 2013. Natural variation of the root morphological response to nitrate supply in *Arabidopsis thaliana. Mechanisms of Development* 130: 45–53.

Donner, S.D. & Kucharik, C.J. 2008. Corn-based ethanol production compromises goal of reducing nitrogen export by the Mississippi River. *Proceedings of National Academy of Sciences of the United States of America* 105: 4513–4518.

Fageria, N.K. & Moreira, A. 2011. The role of mineral nutrition on root growth of crop plants. In D.L. Sparks (ed.), *Advances in agronomy, vol. 110*: 251–331. San Diego: Academic Press.

Frink, C.R., Waggoner, P.E. & Ausubel, J.H. 1999. Nitrogen fertilizer: retrospect and prospect. *Proceedings of National Academy of Sciences of the United States of America* 96: 1175–1180.

Karczewska, A. 2008. *Soil protection and restoration of degraded lands* (in Polish). Wrocław: Wydawnictwo Uniwersytetu Przyrodniczego we Wrocławiu.

Kloc, E. & Szwed, R. 1995. *Ion exchangers as nutrient carriers. Studies on posibility of ion exchangers application for soil properties improvement* (in Polish). PhD thesis, Lublin: Politechnika Lubelska.

Krajewski, P. 2016. Agricultural Biodiversity for Sustainable Development, *Problemy Ekorozwoju/ Problems of Sustainable Development.* 12(1): 135–141.

Lityński, T. & Jurkowska, H. 1982. *Soil fertility and plant nutrition* (in Polish). Warszawa: PWN.

Maciak, F. 1999. *Restoration and protection of the environment* (in Polish). Warszawa: Wydawnictwo SGGW.

Mahajan, S. & Tuteja, N. 2005. Cold, salinity and drought stresses: an overview. *Archives of Biochemistry and Biophysics* 444: 139–158.

Marschner, P. 2012. *Marschner's Mineral Nutrition of Higher Plants.* San Diego: Academic Press.

Molo, B. 2010. Global problem solving for example of environment protection (in Polish). In E. Cziomer (ed.), *Bezpieczeństwo międzynarodowe w XXI wieku.* Kraków: Krakowskie Towarzystwo Edukacyjne sp. z o.o.

Niewiadomski, W. 1983. *Fundamentals of agrotechny* (in Polish). Warszawa: PWRiL.

Ostrowska, A., Gawliński, S. & Szczubiałka, Z. 1991. *Methods for analysis and properties evaluation of soils and plants* (in Polish). Warszawa: Instytut Ochrony Środowiska.

Patrzałek, A. 2010. Development of plant association and initial soil initiated by grass sowing on mine waste dumping during 30 years (in Polish). *Górnictwo i geologia* 5(4): 191–200.

Qian, P. & Schoenau, J.J. 2002. Practical applications of ion exchange resins in agricultural and environmental soil research. *Canadian Journal of Soil Science* 82: 9–21.

Skogley, E.O. & Dobermann, A. 1996. Synthetic ion-exchange resin: soil and environmental studies. *Journal of Environmental Quality* 25: 13–24.

Soldatov, V.S. 1988. Ion exchanger mixtures used as artificial nutrient media for plants. In M. Streat (ed.), *Ion exchange for industry*: 652–658. London: Ellis Horwood.

Soldatov, V.S., Pawłowski, L., Szymańska, M., Matusevich, V.V., Chomczyńska, M. & Kloc, E. 1998. Ion exchange substrate Biona-111 as an efficient mean of barren grounds fertilization and soils improvement. *Zeszyty Problemowe Postępów Nauk Rolniczych* 461: 425–436.

Sztumski, W. 2016. The impact of sustainable development on the Homeostasis of the social environment and the matter of survival. *Problemy Ekorozwoju/ Problems of Sustainable Development.* 11(1): 41–47.

Wasąg, H., Pawłowski, L., Soldatov, V.S., Szymańska, M., Chomczyńska, M., Kołodyńska, M., Ostrowski, J., Rut, B., Skwarek, A. & Młodawska, G. 2000. *Restoration of degraded soils using ion exchange resins* (in Polish). Raport. Lublin: Politechnika Lubelska.

Zuazo, V.H.D. & Pleguezuelo, C.R.R. 2009. Soil-erosion and runoff prevention by plant covers: A review. In E. Lichtfouse, M. Navarrete, P. Debaeke, S. Véronique, C. Alberola (eds), *Sustainable Agriculture*: 785–811. Springer Netherlands.

Żukowska, G., Myszura, M., Baran, S., Wesołowska, S., Pawłowska, M., Dobrowolski, Ł. 2016. Agriculture vs. Alleviating the Climate Change. *Problemy Ekorozwoju/ Problems of Sustainable Development.* 11(2): 67–74.

Environmental Engineering V – Pawłowska & Pawłowski (Eds)
© 2017 Taylor & Francis Group, London, ISBN 978-1-138-03163-0

The process generation of WWTP models for optimization of activated sludge systems

J. Drewnowski & K. Wiśniewski
Faculty of Civil and Environmental Engineering, Gdansk University of Technology, Gdansk, Poland

A. Szaja & G. Łagód
Faculty of Environmental Engineering, Lublin University of Technology, Lublin, Poland

C. Hernandez De Vega
School of Civil Engineering, Polytechnic University of Valencia, Valencia, Spain

ABSTRACT: The objective of water and wastewater utilities, which are responsible for optimization of wastewater treatment plant operation, is threefold: environmental protection, saving money, and mitigation of pollutant emissions, both to water and air. The performed study aimed to assess the process generation of WWTP models that improve activated sludge systems operation, by means of new developed Mantis2 model. This model accounts for the most prevalent biological and chemical processes occurring in wastewater treatment plants. It integrates the nitrogen, phosphorus, and carbon removal, as well as anaerobic digestion. Mantis2 is based, among others, on ASM2d and UCTADM1 models. In order to conduct evaluation, a WWTP layout was prepared by means of GPS-x software and used in dynamic simulation. The conducted simulations show that optimization of wastewater treatment plant operation can be achieved by changes in operating parameters. It was also demonstrated that energy savings can be obtained in various facilities with AS aeration systems.

Keywords: modelling of WWTP, Mantis2, optimization, activated sludge systems

1 INTRODUCTION

Carrying out computer simulations enables to evaluate wastewater treatment systems efficiency and performance. This includes creating a numerical model of a wastewater treatment plant, simulating its operation under various conditions, and interpreting the obtained results. Proper organization is vital for the creation of a successful model. In order to integrate the computer model with the conditions characterizing the parameters of individual wastewater treatment plant parts, it is necessary to conduct laboratory tests, as well as measurements and calculations. Additionally, a comprehensive analysis of the entire technological system, which includes the specific variables of a modeled object, is required. At present, numerous mathematical models which successfully describe a wastewater treatment plant are available. One of the simplest ones includes Activated Sludge Model (ASM), a biochemical mathematical model which describes the conversion of nitrogen, phosphorus, as well

as organic compounds (Henze et al., 1987). There are also more complex models. For instance, the recently developed Mantis2 model integrates Activated Sludge Model with anaerobic digestion processes, as well as precipitation and/or anammox.

The objective of water and wastewater utilities, which are responsible for optimization of wastewater treatment plant operation, is threefold: environmental protection, saving money, and mitigation of pollutant emissions, both to water and air. The performed study aimed to assess the process generation of WWTP models that improve activated sludge systems operation, by means of new Mantis2 model. In order to conduct evaluation, a wastewater treatment plant layout was prepared with GPS-x ver. 6.4 software and used in a steady state and dynamic simulation. The paper concentrates on the WWTP layout, process definitions, influent wastewater elaboration, control handles, sensors, the benchmarking procedure and criteria of extended mathematical modelling and computer simulation evaluation.

2 MATERIAL AND METHODS

2.1 Study site

The batch reactors, as well as WWTP layouts MUCT and modified Bardenpho system was modelled by means of GPS-x. Figure 1 a–b presents process generation of WWTP models for optimization of activated sludge systems (from popular ASM2d to new, more complex Mantis2) and the example of WWTP layout. Table 1 includes additional information pertaining to the treatment process line, including activated sludge MUCT and modified Bardenpho system with a single bioreactor compartment (volumes of individual compartments, average DO concentrations in the aerobic zone). On the other hand, Table 2 contains information about the average operating conditions of the sample WWTP used in the simulation.

The WWTP layout used in the study was designed in the following way: the wastewater treatment process line begins with a primary settler. Then, a bioreactor was created according to modified Bardenpho system with an additional internal recirculation line from the anoxic/aerobic zone and a final clarifier follows. The sludge from the primary settler is carried to the thickener, whereas the sludge from the final clarifier undergoes dewatering. The thickened sludge is subsequently mixed and fed to the digester and subjected to the digestion process which produces biogas. The digested sludge is eventually dewatered to obtain mud for other purposes. The water separated during thickening and dewatering undergoes treatment in the header plant. For instance, it has allowed greater weight of sediment—both through the reduction of organic matter in the process of stabilization and ultrasonic disintegration, as well as a greater reduction of water contained in the sediment processes of dehydration. The result is a complete disposal and the use of sludge composting process, taking into account both production growth of deposits (additional waste from neighboring municipalities), as well as the expected amount of incoming tailings (fats). The reject water produced during the dewatering processes of raw and digested sludge is recirculated to the head of wastewater train.

Table 1. Characteristics of the wastewater treatment process line including activated sludge MUCT/modified Bardenpho system with a single bioreactor compartment (volumes, aver-age DO concentrations etc.) used for preliminary simulation study in GPS-x.

Parameter	Surface area	Max. volume	Depth	DO concentration
MUCT	m^2	m^3	m	gO_2/m^3
Primary settler	1000	–	3.5	–
Anaerobic	–	3500	4.0	–
Anoxic 1	–	2050	4.0	–
Anoxic 2	–	3930	4.0	–
Anoxic/Aerobic	–	3930	4.0	–
Aerobic 1/2	–	1950	4.0	2.93/2.25
Aerobic 3/4	–	1950	4.0	1.66/1.21
Aerobic 5/6	–	1950	4.0	0.66/0.60
Deoxic	–	1250	4.0	–
Final settler	3020	–	3.0	–
Mod. Bardenpho	m^2	m^3	m	gO_2/m^3
Primary settler	3500	–	3.5	–
Mixing tank	–	720	4.0	
Anaerobic 1	–	720	4.0	–
Anaerobic 2	–	1470	4.0	–
Anoxic	–	1470	4.0	–
Anoxic/Aerobic	–	520	4.0	–
Aerobic 1	–	1080	4.0	2.55/2.13
Aerobic 2	–	1080	4.0	2.20/1.90
Aerobic 3	–	1480	4.0	1.54/1.33
Aerobic 4	–	1480	4.0	0.82/0.43
Final settler	1500	–	3.5	–
Digester	–	4000	–	–

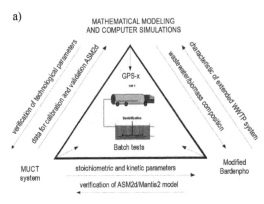

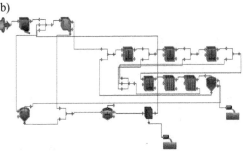

Figure 1. A schema of the process generation of WWTP models for optimization of activated sludge systems (a) including batch reactors and (b) the example of WWTP layout of modified Bardenpho system created in the GPS-x simulator.

Table 2. Characteristics of the average operating conditions as a base point of MUCT/modified Bardenpho plant layout for preliminary simulation study in GPS-x.

Parameter	Unit	MUCT/modified Bardenpho summer 2008/2013
Concentrations in settled wastewater:		
COD	$gCOD/m^3$	695/760
BOD_5	$gBOD/m^3$	357/432
$N_{tot.}$	gN/m^3	65.9/105
$N\text{-}NH_4^+$	gN/m^3	56.1/62.9
$P_{tot.}$	gP/m^3	50.9/13.9
$P\text{-}PO_4^-$	gP/m^3	45.6/13
Concentrations in secondary effluent:		
COD	$gCOD/m^3$	42.1/31.2
$COD_{Soluble}$	$gCOD/m^3$	34.5/30.1
$N_{tot.}$	gN/m^3	10.3/6.7
$N\text{-}NH_4^+$	gN/m^3	0.135/0.07
$N\text{-}NO_3^-$	gN/m^3	7.55/4.69
$P_{tot.}$	gP/m^3	2.53/0.91
$P\text{-}PO_4^-$	gP/m^3	0.812/0.15
Operating parameters:		
Q_{INF}	m^3/d	21568/6594
Q_{MLR1} (anox 1–anaer)/ (aer 4—anaer 2)	m^3/d	40000/18850
Q_{MLR2} (aer—anox 2)	m^3/d	120000/–
Q_{RAS}	m^3/d	3352
Temperature	°C	19/20
SRT	d	21.3/26
Biomass characteristics:		
MLSS	g/m^3	5450/4519
MLVSS/MLSS (i_{VT})	–	0.437/0.742

2.2 Extracting experimental data

The results of batch experiment (concerning, for example the oxygen/ammonia utilization rate—OUR/AUR, nitrogen utilization rate—NUR, as well as P release rate and anoxic/aerobic P utilization rate—PRR and anoxic/aerobic PUR), conducted in a large BNR WWTP (600,000 PE), were obtained by Drewnowski and Mąkinia (2014) and used to create the experimental database. The database was employed for the comparison of model predictions. Moreover, the 96-hour measurement series carried out by Swiniarski et al. (2012) in the full-scale MUCT bioreactor at WWTP plant was also implemented in the calibration/validation process. The full-scale MUCT and modified Bardenpho bioreactor was preliminarily simulated by the ASM2d base model. The performed research allows to assess various biochemical processes and extend optimize modes by using the Mantis2 model. This, in turn, enables to develop the WWTP layout as a base point in this study and simulate the operation of a complex model concerning a full-scale WWTP system.

2.3 Process generation of WWTP models for optimization of activated sludge systems

Apart from the high efficiency of the wastewater treatment process, increasing emphasis is put on the energy costs and additional operational expenses. Numerous methods of reducing the cost of power which can be employed in all types of wastewater treatment plants are already known. Energy, which is used, among others, for the aeration of bioreactors, as well as wastewater pumping system, constitutes a significant share of operational costs.

On the basis of dynamic simulation drawing on the actual conditions in a wastewater treatment plant, energy optimization of the system was performed. Following parameters were analyzed:

- the concentration of oxygen dissolved in nitrification chamber,
- changes in the intensity of internal recirculation from the end of a nitrification chamber to the denitrification chamber.

During the optimization of system energy consumption, the main factor taken into consideration was the efficiency of wastewater treatment, which was to match the wastewater quality parameters laid out in the Regulation of the Minister of Environment of 18 November 2014 (Rozporządzenie MŚ, 2014). The capacity of the presented treatment plant exceeds 100000 ENI, the concentration of total nitrogen in the treated wastewater should not surpass 10 mg/l, while the concentration of total phosphorus should not be greater than 1 mg/l.

The chemical engineering community (Skogestad, 2000 a,b) has drawn the attention to the importance of integrated and plant-wide control strategies. The advantages of these solutions start being appreciated in wastewater industry. Treatment of wastewater in a plant is carried out as an integrated process. The primary and secondary clarification units, anaerobic digesters, thickeners, activated sludge reactors, dewatering systems, storage tanks, etc. have to be operated and controlled jointly, rather than individually. It is necessary to consider all the interactions occurring between the processes (Kelessidis & Stasinakis, 2012; Stefaniak et al., 2014; Werle & Dudziak 2014). Therefore, the IWA Task Group on Benchmarking of Control Strategies for WWTP and the scientific community reached consensus regarding the use of Mantis2, a plant-wide evaluation tool which greatly raised the number of control handles. This, in turn enabled new control possibilities, including examining the impact of sludge digestion regimes and either enhanced control (Vanrolleghem et al., 2010) or biological treatment (Volcke et al., 2006 a,b) of the nitrogen rich digested reject water, as well as the influence of activated sludge control strategies on the sludge line (Jeppsson et al., 2007).

2.4 *Organization of the modeling study procedure*

Mantis2 model accounts for the most prevalent biological, and chemical processes occurring in wastewater treatment plants. It integrates the nitrogen, phosphorus, and carbon removal with the anaerobic digestion. Mantis2 is based, among others, on ASM2d and UCTADM1 models. The computer simulations employing the Mantis2 model were performed with GPS-x ver. 6.4 simulation platform software (Hydromantis, Canada).

The calibration of activated sludge process model was carried out on the basis of batch tests results (summer study period) and 96-hour measurements performed in the full-scale MUCT bioreactor (summer 2008). The obtained results of batch tests with process biomass and settled wastewater from winter sessions were utilized in the comparison of model predictions pertaining to the NO_3-N, NH_4-N, and PO_4-P behavior and also were used further for validated model of modified Bardenpho plant layout. The kinetic and stoichiometric process parameters in ASM2d constituted the basis for further process calibration in the new model, i.e. Mantis2. Additionally, the Nelder-Mead simplex method was employed for numerical optimization and final validation/recalibration of Mantis2 parameters for carrying out a simulation as a base point for modified Bardenpho plant layout which was prepared by using GPS-x software (see details in Table 3).

In order to create a complex wastewater treatment plant model, the primary settler and anaer-

Table 3. List of the kinetic and stoichiometric parameters in ASM2d/Mantis2 adjusted during model calibration in the GPS-x simulator.

Symbol	Unit	Default value	Calibrated ASM2d/Mantis2
Heterotrophic organisms:			
Aerobic heterotrophic Y on soluble substrate	gCOD/gCOD	0.666	0.63/0.625
Anoxic heterotrophic Y on soluble substrate	gCOD/gCOD	0.533	0.63/0.625
Maximum specific growth rate on substrate	1/d	3.2	3/6
Saturation/inivition coefficient for Sac	g/COD/m^3	5	4
Reduction factor for denitrification on nitrate-N	–	0.32	0.8
Aerobic heterotrophic decay rate	1/d	0.62	0.3/0.4
Phosphorus accumulating organisms:			
Aerobic yield on PAO growth	gCOD/gCOD	0.639	0.625
Anoxic yield on PAO growth	gCOD/gCOD	0.511	0.625
PHA storage yield	gP/gCOD	0.4	0.5
Saturation coefficient of PAO for Sac	gCOD/m^3	4	2
Rate constant for storage of poly-phosphate	gP/gPAO/d	1.5	2.5
Aerobic decay coefficient for PAO	1/d	0.2	0.18
Anoxic reduction factor for decay rate	–	0.9	0.6
Poly-P lysis coefficient	1/d	0.2	0.1
Ammonia-oxidizing Organisms:			
Oxygen saturation for ammonia oxidizer	gO$_2$/m^3	0.25	0.22/0.2
Ammonia oxidizer aerobic decay rate	1/d	0.17	0.14/0.12
Anoxic reduction factor for decay rate	–	0.5	1.0/1.5
Nitrite-oxidizing organisms:			
Nitrite saturation coefficient for nitrite oxidizer	gN/m^3	0.5	0.34/0.4
Oxygen saturation for nitrite oxidizer	gO$_2$/m^3	0.68	0.1
Anoxic reduction factor for decay rate		0.5	0.4/0.3
Hydrolysis:			
Hydrolysis rate constant for Xs	1/d	3	2.5
Saturation coefficient for particulate COD	–	0.1	0.2
Anoxic hydrolysis reduction factor	–	0.28	0.6
Anaerobic hydrolysis reduction factor	–	0.4	0.1

obic digestion chamber were added to the AS bioreactor line. The simulations were conducted at steady-state conditions. However, in order to calibrate the entire plant layout for simulation study in GPS-x software, dynamic conditions were employed as well.

3 RESULTS AND DISCUSSION

The paper demonstrates the possibility of utilizing the wastewater treatment plant model as a tool for optimizing plant operation. Figure 2 a–b presents the obtained sample results, as well as model predictions of batch experiments (Drewnowski & Mąkinia, 2013, 2014). Initially, the model was calibrated with yearly average primary effluent data (including, for instance COD, NH_4, NO_3, PO_4) and run at steady state conditions. In order to check whether the model was calibrated properly, the steady state results were compared with the yearly average MLSS, %VSS and effluent parameters (including TSS, NO_3, NH_4). The obtained comparison results of the measured (actual) and predicted (model) concentrations of batch tests parameters were presented in Figure 2 (a, b). The kinetic and stoichiometric process parameters of ASM2d constituted the basis for further process calibration in Mantis2 model, which was used to simulate modified Bardenpho plant layout prepared by means of GPS-x software (See Figure 3, where calibrated data were made bold).

The mentioned data and information were employed in the development and calibration of a complex Mantis2 WWTP model. The process was modelled in order to improve the knowledge about its operation and to derive key model parameters which could be applied in predictive process models. Changes in the concentration of phosphorus and nitrogen were observed during the performed simulation both in biological tanks, as well as the treatment plant influent. The obtained model predictions were fitted to the measured NURs. This required adjusting two parameters which were based on selected kinetic ASM2d coefficients, i.e. the hydrolysis rate constant (k_h) and the maximum growth rate of heterotrophs (Y_H) (See Figure 2a). During anoxic P uptake, no additional modifications were required for calibration of NUR in the anoxic phase of the PRR/anoxic PUR test. The calibration of PUR and PRR tests was carried out with the following parameters: saturation coefficient for PAOs with respect to S_A ($K_{SA,PAO}$),

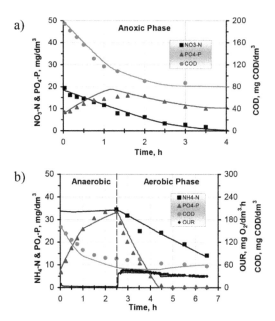

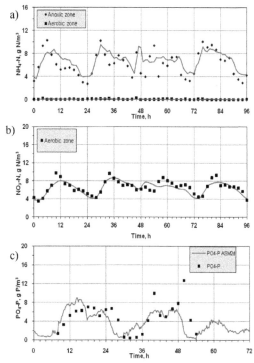

Figure 2. Sample results of different process rate measurements and model ASM2d predictions of batch tests: (a) conventional NUR experiments, (b) PRR/aerobic PUR experiments (Drewnowski and Mąkinia, 2013 and 2014).

Figure 3. Sample results of different process rate measurements and model ASM2d/Mantis2 predictions of the modified Bardenpho bioreactor simulation study in full-scale plant layout: (a) anoxic/aerobic zone of $N-NH_4$, (b) aerobic zone of $N-NO_3$, (c) aerobic zone of $P-PO_4$.

saturation coefficient for PAOs with respect to polyphosphate (K_{PP}), the rate constant for storage of PHA (q_{PHA}), anaerobic hydrolysis reduction factor (η_{fe}) and saturation coefficient for particulate COD (K_X).

The calibration of nitrification process, which was based on the data obtained from PRR and aerobic PUR batch tests, was carried out by means of kinetic parameters which encompassed the NH_4-N saturation coefficient ($K_{NH4,A}$) and the maximum growth rate of autotrophs (μ_A) (See Figure 2b). In order to precisely match the kinetic and stoichiometric parameters, the process calibration in Mantis2 model, used for the simulation of WWTP layout with full-scale MUCT bioreactors prepared with GPS-x software, was initiated with the calibrated results of a series of batch tests with the process biomass (ASM2d). Afterwards, the previous MUCT bioreactor model (Swinarski et al., 2012) provided the basis for further process calibration, which was conducted with the aid of Nelder-Mead simplex optimization techniques, in order to estimate the Mantis2 kinetic and stoichiometric parameters (see details in Table 3). Having calibrated the ASM2d/Mantis2 model, the actual behaviours of NO_3-N, NH_4-N, and PO_4-P in the anoxic/aerobic tank of full-scale modified Bardenpho bioreactor (See Figure 3), were matched by the GPS-x simulation with high accuracy.

The calibrated and/or validated plant-wide Mantis2 model enabled to perform simulations of scenarios which may impact the wastewater treatment plant process optimization and operation. In order to present the prospective application of simulation software for predicting the WWTP energy output, the following scenarios were taken into account:

- varying concentration of the oxygen dissolved in the activated sludge reactor nitrification zones,
- impact of changes in dissolved oxygen concentration on ammonium nitrogen and nitrate nitrogen in relation to electricity consumption,
- impact of changes in the flow rate of internal recirculation on the concentration of nitrate nitrogen in relation to electricity consumption.

The changes of dissolved oxygen concentration in the nitrification chamber were analyzed in relation to the power consumption of the aerating device. While carrying out computer simulations, different dissolved oxygen concentrations in activated sludge chambers were assumed, ranging from 0.5 to 2 m/l. The changes in ammonium nitrogen (NH_4-N) and nitrate nitrogen (NO_3-N) were presented in Figure 4 (a, b).

On the basis of the dynamic modelling results it was ascertained that the dissolved oxygen content in the range of 0.5 to 2 mg/l is correct in relation to the quality of effluent wastewater. Nevertheless, during the inflow of the highest pollutant load, the ammonia nitrogen concentration increases significantly. In order to determine the optimal concentration of dissolved oxygen in relation to system energy efficiency, an analysis was conducted, as shown in Figure 5. It depicts the relation between

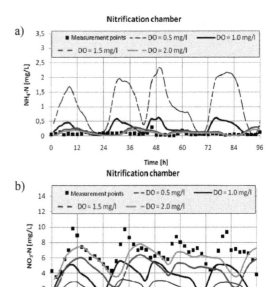

Figure 4. Changes in the concentration of ammonia nitrogen (a) and nitrate nitrogen (b) in the nitrification chamber at various concentrations of dissolved oxygen based on the modified Bardenpho bioreactor simulation study in full-scale plant layout (extended studies under realistic process conditions adapted from Wiśniewski, 2015).

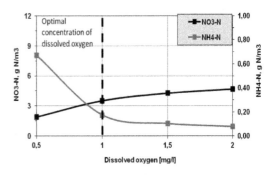

Figure 5. Analysis of changes in dissolved oxygen concentration on ammonia nitrogen and nitrate nitrogen based on the modified Bardenpho bioreactor simulation study in full-scale plant layout (extended studies under realistic process conditions adapted from Wiśniewski, 2015).

the increase of energy consumption of the aeration device, and the change of ammonia nitrogen and nitrate nitrogen in the effluent. On the basis of the analysis it can be stated that the most optimal concentration of dissolved oxygen approximates 1 mg/l, with high efficiency of the process maintained. The total energy consumption can be reduced as low as 14% by assuming the concentration of oxygen equal to 1 mg/l in relation to 2 mg/l in the nitrification chamber.

The next step towards optimization of wastewater treatment is analysing the impact of internal recirculation intensity changes from the end of a nitrification chamber to the denitrification chamber. During the simulation, a constant concentration of dissolved oxygen was assumed (1 mg/l).

The simulation was carried out with varying intensities of internal recirculation, ranging from the 50% to 300% of the influent. The obtained simulation results are presented in Figure 6.

On the basis of the simulation results it can be observed that the changes in recirculation significantly impact the concentration of nitrate nitrogen in the effluent. Lowering the flow intensity deteriorates the efficiency of the process. With the recirculation at 50%, the concentration of nitrate nitrogen is at the limit of acceptable total nitrogen (<10 mg/l). In order to determine the optimal flow intensity in relation to energy efficiency, an analysis was conducted. See Figure 7 for details.

The commonly assumed internal recirculation in a wastewater treatment plant amounts to about 300% of influent wastewater. Therefore, by assuming the internal recirculation at 200% of influent waste-water, up to 30% of energy can be saved, simultaneously deteriorating the total nitrogen concentration in the effluent by about 15%. Nevertheless, the concentration of total nitrogen remains below the limit (<10 mg/l).

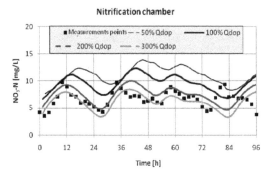

Figure 6. Changes in the flow rate in the internal recirculation at steady oxygen concentration of 1.0 mg/L based on the modified Bardenpho bioreactor simulation study in full-scale plant layout (extended studies under realistic process conditions adapted from Wiśniewski, 2015).

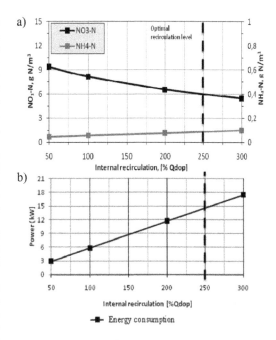

Figure 7. Analysis of changes in the flow rate of internal recirculation (a) on the concentration of N-NH$_4$, N-NO$_3$ (b) in relation to electricity consumption based on the modified Bardenpho bioreactor simulation study in full-scale plant layout (extended studies under realistic process conditions adapted from Wiśniewski, 2015).

By comparing the performed computer simulations with the actual operating conditions of a wastewater treatment plant, it is possible to reduce the operating expenses by as much as 20% (data calculated with GPS-X 6.4 software). The simulation results acquired with a computer model were similar to the values of treated wastewater parameters, obtained through the treatment process in a WWTP.

Wastewater treatment practice has now progressed to the point where the removal of organic matter and nutrient removal by biological nitrification and denitrification and biological phosphorus removal, can be accomplished in a single system. The non-linear dynamics and properties of these biological processes are still not very well understood (Meirlaen et al., 2001; Vanhooren et al., 2003). As a consequence, a unique model cannot always be identified. This contrasts with traditional mechanical and electrical systems where the model can be uniquely derived from physical laws. Also, the calibration of wastewater treatment models is particularly hard: many expensive experiments may be required to accurately determine model parameters (Szaja et al., 2015). Although all the limitations and difficulties stated above, modelling and simulation of wastewater treatment is considered useful (Henze et al., 2000; Korniluk et al., 2008). Models are valuable tools to summarise and increase the understanding of complex interactions in biological systems. More quantitatively, they can be used to predict the dynamic response of the system to various disturbances (Cooney & McDonalds, 1995 Vanhooren et al., 2003).

Despite their promising properties described above, the practical use of dynamic modelling of wastewater treatment could still be rather limited (Morgenroth et al., 2000). Especially the labour and cost intensive calibration of WWTP models is considered hard to accomplish in practical situations. New methodologies are being developed to overcome this bottleneck (Petersen et al., 2001). In some cases, e.g. when modelling biofilm wastewater treatment, the development of models that are able to describe the wastewater treatment system well enough to correctly predict system responses, but that at the same time have a feasible complexity for simulation, is also a challenge (Vanhooren et al., 2003). The application of good practice in WWTP models is still under study process and this field of science need to be extended step by step, e.g. in cases:

- predict dynamic responses of the system to influent variations so as to develop strategies to optimise treatment plant operation—either off-line or with on-line real-time simulations that are used for control and optimisation (Dochain & Vanrolleghem 2001)
- operators might be interested in using models for finding answers to practical questions (Swinarski et al., 2012),
- integrate multiple processes, e.g. nowadays, the removal of organic matter, nitrogen and phosphorus is accomplished in a single system (Henze et al., 2000; Vanhooren et al., 2003),
- design WWTP by evaluating data from pilot-scale reactors and predict performance of full-scale plants (Takors et al., 1997, Vanhooren et al., 2003),
- mathematical modelling and computer simulations are used as a tool to help for greater understanding of the interactions between the above-mentioned processes in lab, pilot and full-scale activated sludge systems (Makinia 2010).

4 SUMMARY AND CONCLUSION

The performed process of optimization, which was conducted on the basis of biological system physiology, is required in order to fully utilize the existing plant facilities. The lack of adequate tools which would enable simulation of biological processes has hampered the universal adoption of this approach. However, Mantis2 model and GPS-x simulation platform proved to be suitable for this purpose. Utilizing the wastewater treatment plant model as a tool for its optimization was described in a simulation of a three-phase system in Bardenpho process equipped with pre-denitrification chamber (similarly to the JHB system). During the simulation, changes in nitrogen and phosphorus concentration were observed both in the biological chambers and in the plant effluent. Additionally, the results of optimization of biological chambers aeration system and the impact of internal recirculation on energy consumption were presented as well. The simulations were carried out with varying concentrations of oxygen dissolved in nitrification phase, and with different flow intensities of internal recirculation from the nitrification chamber to the denitrification chamber. The concentrations of dissolved oxygen amounted to 0.5–2 mg/l. On the other hand, the range of internal recirculation intensities corresponded to the internal flow intensities equal to 50–300% of influent wastewater. The total phosphorus concentration was not presented, as it did not exceed 1 mg/l in any simulation.

While conducting calibration, the kinetic parameters of ASM2d and Mantis2 simulation models were matched to the actual data from MUCT and Bardenpho process at both WWTP layouts. Having calibrated the two afore-mentioned models, it was possible to accurately predict the principal process rates (including PUR, NUR, as well as anoxic/aerobic PUR, and OUR/AUR) in one- or two-phase batch tests, as well as for the simulation of complex WWTP layout as a full-scale MUCT/ Bardenpho process bioreactors, as modelled by GPS-x software. Additionally, in order to confirm the modeling results obtained initially, continuous feedback between the real-time measurements (including the behaviour of NH_4-N, NO_3-N, and PO_4-P) conducted in the treatment plant and the employed simulation model was taken into consideration. It was demonstrated that a properly calibrated model can be used for the assessment of operational changes and upgrades, including the process changes and capacities. The model simulations become increasingly accurate the better calibrated they are. The knowledge of historical periods corresponding to various operating conditions, as well as sampling and remaining methods of optimization, aids in matching and adjusting the model performance to an actual, full-scale treatment plant.

Simulations performed by means of computer software can be used to predict the energy consumption for different treatment strategies, as well as chemical and biological processes. The conducted simulations show that optimization of wastewater treatment plant operation can be achieved by changes in sample operating parameters. It was also demonstrated that energy savings can be obtained in various facilities with AS aeration systems.

REFERENCES

Cooney, M.J. & McDonalds, K.A. 1995. Optimal dynamic experiments for bioreactor model discrimination. Appl. Microbiol. Biotechnol. 43, 826–837.

Cullen, A.C. & Frey, H.C. 1999. Probabilistic techniques in exposure assessment. A handbook for dealing with variability and uncertainty in models and inputs. Plenum, New York.

Dochain, D. & Vanrolleghem, P.A. 2001. Modelling and Estimation in Wastewater Treatment Processes. IWA Publishing, London, UK. ISBN 1-900222-50-7.

Drewnowski, J. & Makinia, J. 2013. Modeling hydrolysis of slowly biodegradable organic compounds in biological nutrient removal activated sludge systems. Wat. Sci. Tech. 67(9): 2067–2074.

Drewnowski, J. & Mąkinia, J. 2014. The role of biodegradable particulate and colloidal organic compounds in biological nutrient removal activated sludge system. Internat. J Environ. Sci. Technol. 11(7): 1973–1988.

GPS-X 6.4. 2014 User's guide and technical reference. Hydromantis Environmental Software Solutions, Inc. Hamilton, Ontario in Canada.

Gujer, W. 2006. Activated sludge modelling: past, present and future. Wat. Sci. Tech. 53 (3): 111–119.

Henze, M., Grady Jr, C.P.L., Gujer, W., Marais, G. & Matsuo, T. 1987. Activated sludge model No. 1. Scientific and Technical Report No. 1. IAWQ, London.

Henze, M., Gujer, W., Mino, T. & van Loosdrecht, M.C.M. 2000. Activated Sludge Models ASM1, ASM2, ASM2d and ASM3. Scientific and Technical Report No. 9., IWA Publishing, London, UK.

Jeppsson, U., Pons, M.N., Nopens, I., Alex, J., Copp, J.B., Gernaey, K.V., Rosen, C., Steyer, J.P. & Vanrolleghem. P.A. 2007. Benchmark Simulation Model No 2 – general protocol and exploratory case studies. Wat. Sci. Tech. 56 (8): 287–295.

Kelessidis, A. & Stasinakis, A.S. 2012. Comparative study of the methods used for treatment and final disposal of sewage sludge in European countries. Waste Manag. 32(6): 1186–1195.

Korniluk, M., Montusiewicz, A., Piotrowicz, A., & Łagód G. 2008. Simulation of wastewater treatment systems with membrane separation. Proc ECOpole. 2(1): 41–45.

Mąkinia, J. 2010. Mathematical Modelling and Computer Simuation of Activated Sludge Systems, IWA Publishing, London, 2010.

Meirlaen, J., Huyghebaert, B., Sforzi, F., Benedetti, L. & Vanrolleghem, P.A. 2001. Fast, simultaneous simulation of the integrated urban wastewater system using mechanistic surrogate models. Wat. Sci. Tech. 43(7): 301–310.

Morgenroth, E., van Loosdrecht, M.C.M. & Wanner, O. 2000. Biofilm models for the practitioner. Wat. Sci. Tech. 41(4–5): 509–512.

Petersen, B., Gernaey, K., Henze, M. & Vanrolleghem, P.A. 2001. Evaluation of an ASM1 model calibration procedure on a municipal-industrial wastewater treatment plant. J. Hydroinformatics. 4 (1): 15–38.

Rozporządzenia Ministra Środowiska z dnia 18 listopada 2014 r. w sprawie warunków jakie należy spełnić przy wprowadzaniu ścieków do wód lub do ziemi, oraz w sprawie substancji szczególnie szkodliwych dla środowiska wodnego (Dz.U. 2014 poz. 1800).

Skogestad, S. 2000a. Plantwide control: the search for the self-optimizing control structure. J Proc. control. 10(5): 487–507.

Skogestad, S. 2000b. Self-optimizing control: the missing link between steady-state optimization and control. Comput. Chem Eng. 24(2): 569–575.

Stefaniak, J., Zelazna, A. & Pawlowski, A. 2014. Environmental assessment of different dewatering and drying methods on the basis of life cycle assessment. Wat Sci Tech. 69(4): 783–788.

Swinarski, M., Mąkinia, J., Czerwionka, K., Chrzanowska, M. & Drewnowski, J. 2012. Modeling external carbon addition in combined N-P activated sludge systems with an extension of the IWA activated sludge models. Wat. Environ. Res. 84(8): 646–655.

Szaja, A., Aguilar, J.A. & Łagód, G. 2015. Estimation of chemical oxygen demand fractions of municipal wastewater by respirometric method—case study. Ann. Set Enviro Prot. 17(1): 289–299.

Takors, R., Wiechert, W. & Weuster-Botz, D. 1997. Experimental design for the identification ofm macrokinetic models and model discrimination. Biotechnol. Bioeng. 56: 564–576.

Vangheluwe, H.L. 2000. Multi-Formalism Modelling and Simulation. PhD Thesis, Faculty of Sciences. Ghent University. Ghent.

Vanhooren, H., Meirlaen, J., Amerlinck, Y., Claeys, F., Vangheluwe, H. & Vanrolleghem, P.A. 2003. WEST: modelling biological wastewater treatment. J. Hydroinformatics. 5(1): 27–50.

Vanrolleghem, P.A., Corominas, L. & Flores-Alsina, X. 2010. Real-time control and effluent ammonia violations induced by return liquor overloads. Proc. WEFTEC2010, New Orleans, Louisiana, USA, 2–6 Oct. 2010.

Vanrolleghem, P.A. & Van Daele, M. 1994. Optimal experimental design for structure characterization of biodegradation models: on-line implementation in a respirographic biosensor. Wat. Sci. Tech. 30(4): 243–253.

Volcke, E.I.P., van Loosdrecht, M.C.M. & Vanrolleghem, P.A. 2006a. Controlling the nitrite: ammonium ratio in a SHARON reactor in view of its coupling with an Anammox process. Wat. Sci. Tech. 53(4–5): 45–54.

Volcke, E.I.P., van Loosdrecht, M.C.M. & Vanrolleghem, P.A. 2006b. Continuity-based model interfacing for plant-wide simulation: a general approach. Wat. Res., 40: 2817–828.

Werle, S. & Dudziak, M. 2014. Influence of wastewater treatment and the method of sludge disposal on the gasification process. Ecol. Chem. Eng S. 21(2): 255–268.

Wiśniewski, K. 2015. Model oczyszczalni ścieków jako narzędzie do optymalizacji procesów biologicznych. in: Interdyscyplinarne zagadnienia w inżynierii i ochronie środowiska. Tom 5. ed. Praca zbiorowa pod red. Jacka Wiśniewskiego, Małgorzaty Kutyłowskiej i Agnieszki Trusz-Zdybek Wrocław: Oficyna Wydawnicza Politechniki Wrocławskiej, 2015, 460–467.

Environmental Engineering V – Pawłowska & Pawłowski (Eds)
© 2017 Taylor & Francis Group, London, ISBN 978-1-138-03163-0

Oxidation of organic pollutants in photo-Fenton process in presence of humic substances

R. Świderska-Dąbrowska & K. Piaskowski
Faculty of Civil Engineering, Environmental and Geodetic Sciences, Koszalin University of Technology, Poland

R. Schmidt
Urban Water Supply and Sanitation in Koszalin, Poland

ABSTRACT: Influence of Humic Acids (HA) in concentration of 10–50 mg/l on efficiency of phenol removal in photo-Fenton process was evaluated in this work. Natural zeolite, surface modified with Fe(II) ions, was used as a catalyst of oxidation reaction. The results of the study show that the efficiency of phenol removal decreases along with the increase of concentration of HA. Humic acids form stable complexes with ions of iron (III), which decreases the concentration of Fe (III) available for the photo-Fenton reaction. It was also found that the presence of HA affects the selectivity of the catalyst of Fenton reaction. In that case the preferred oxidation product is hydroquinone, whereas for oxidation of phenol without HA, during first stage of reaction, catechol is the dominant intermediate product of the oxidation. The results also indicate a significant impact of solution pH on the efficiency of phenol degradation and type of intermediate products.

Keywords: phenol, humic acids, oxidation, photo-Fenton

1 INTRODUCTION

Humic Substances (HS) are natural macromolecules of an organic acid character. They are produced in the processes of biochemical decomposition of plants and animals remains. Chemical structure of HS is still not completely known. The separation of humic substances into three basic groups: humins, Humic Acids (HA) and Fulvic Acids (FA) is based on their solubility in acids and bases. Fulvic acids are soluble in the whole pH range, humic acids only in alkaline and slightly acidic solution, at pH<2 they precipitate.

Humic substances are important for soil and water environment. They cause yellow to brown colour of natural waters. HS may also form stable complexes with many metal ions, preferably with Fe(III) and Cu(II). Fuentes et al. (Fuentes et al. 2013) indicate that Fe(III) ions are preferably attached to the carboxyl groups in the aliphatic structures of HS. While Cu(II) are complexed by phenol and O-alkyl groups, located on the aromatic parts.

Adsorption properties in relation to many organic compounds, including toxic phthalates, pesticides or PCBs is an important feature of HS, which causes an increase of their solubility in water.

For these reasons, HS may significantly influence the removal efficiency of other organic contaminants from water and wastewater.

Many literature reports describe the impact of HA on the oxidation of organic compounds (Farré et al. 2007, Georgi et al. 2007, He et al. 2010, Kępczyński et al. 2007, Lipczynska-Kochany & Kochany 2008, Romero et al. 2011, Vione et al. 2004). However, it is impossible to define clearly their action. Depending on the source and fraction, and thus the chemical structure of HA and conditions of the oxidation process both improvement and worsening of efficiency of organic contaminants removal was noted.

Lipczynska-Kochany & Kochany (2008) have proven that the efficiency of oxidation of organic compounds (phenol, 2,4-dimethylphenol, benzene, toluene, o-, m- and p-xylene and dichloroethane) in the Fenton process, conducted in the presence of HS (3000 mg/l) and at pH 7 is similar to the efficiency obtained without HA at pH 3.5. Lower pH = 3.5 and presence of HS may impede the oxidation of organic compounds.

Georgi et al. (2007) had similar observations. They obtained comparable benzene oxidation efficiency at pH 5, in the presence of HS (50 mg/l) and at pH 3 for benzene only.

Kępczyński et al. (2007) have shown that the photolysis of phenol which was irradiated with light at 253.7 nm of wavelength, is much less efficient in the presence of HA. However, at a wavelength within range 310–420 nm, the efficiency of the process depends more on the pH of the solution than on the concentration of HA. Alkaline pH is more favourable than neutral.

Literature reports indicate that the interest paid to Advanced Oxidation Processes (AOPs) results from high efficiency of removal of non-biodegradable substances, especially from industrial wastewater. Fenton processes are readily used due to high performance, low costs, and the ease of use. Fenton process does not require any specialized equipment and reagents. It can be run as homogeneous and heterogeneous process. The disadvantage of homogeneous process is the production of deposits with high Fe content. In order to solve this problem, a heterogeneous catalyst is used due to the ease of separation and possibility of re-use. Zeolites, carrying iron ions, are often used as a heterogeneous catalyst. Such catalyst may be re-used with insignificant decrease of efficiency. Studies on the application of heterogeneous Fe-zeolite catalyst during the degradation of organic contaminants were conducted by Blanco et al., Pereira et al., among others (Aleksić et al. 2010, Blanco et al. 2014, Pereira et al. 2012). Zeolite of Y type, modified with Fe was used. Results indicate that the efficiency of organic substances removal is high and it significantly depends on catalyst and H_2O_2 doses, initial pH and UV radiation (Blanco et al. 2014).

The conditions of Fenton process are optimized by the change of catalyst and H_2O_2 doses in order to minimize their use.

The aim of the research presented in the paper was to explain the mechanism of phenol removal from model solution in the presence of humic acids, including identification of intermediate oxidation products of photo-Fenton process.

2 MATERIAL AND METHODS

Following materials were used in the experiments: Humic Acid (HA) (Aldrich Chemical Co., Switzerland), phenol (Chempur, pure), hydrogen peroxide 30% (POCh, Gliwice, pure), sodium sulfate (POCh, Gliwice, pure), modified natural zeolite.

HA solution was prepared by dissolving 250 mg of HA in 0.1 M NaOH and then deionized water was added up to 1 l (TOC = 87.31 mgC/l).

2.1 Preparation of Fe-zeolite

Natural zeolite from Slovakia, of particle size 0.25–0.50 mm, contains 84% of clinoptilolite, 8% of cristobalite, 4% each of feldspar and illite, and traces of silica and carbonate minerals. During the first stage, the natural zeolite was transformed into a hydrogenous form by shaking it in a threefold volume of 5% HCl (H-zeolite) for 2 hours. It was subsequently modified with Fe (II) ions by co-precipitation (Fe-zeolite). The zeolite was stirred with 0.05 M $FeSO_4$ solution for 4 hours at pH 3 and at a temperature of 50°C. Subsequently, the pH was increased to 9.0 with 25% $NH_{3(aq)}$, and the solution was stirred for another hour. The zeolite was then washed with deionised water and dried at a temperature of 105°C. The procedure was repeated three times. Each time the content of iron in the zeolite was higher. During the last stage of modification, the zeolite was calcined for 2 hours at a temperature of 450°C.

2.2 Characteristics of Fe-zeolite

The content of elements present in structural space of natural zeolite decreased after transforming it into a hydrogenous form. It also caused an increase of the specific surface area of the zeolite and volume of its pores, which could be blocked by the removed ions (Na^+, K^+, Ca^{2+} and Mg^{2+})—see Table 1. Modification of zeolite with Fe(II) ions increased its concentration up to 2.94% of the zeolite mass, which corresponds to 4.21% Fe_2O_3. The surface of zeolite and its pores were covered by active iron(III) oxides in the forms: FeO, α-Fe_2O_3, β-Fe_2O_3 and polymorphous variations.

2.3 Methods

The photo-Fenton process was conducted in a photo-reactor (a 0.75 l cylindrical Pyrex container) placed on a magnetic stirrer. A Heraeus TQ 150 Z1 lamp with a power of 150 W, emitting radiation of higher intensity within the wavelength range of 400–450 nm, was used as the radiation source. Lamp was placed in glass cover which blocked radiation below 300 nm. The experiment was conducted at a constant temperature of 22 ± 1°C. A 1 l of model aqueous solution, containing phenol (PhOH) at a concentration of 200 mg/l and variable concentration of HA (0, 10, 20 and 50 mg/l) was pumped into the reactor from the bottom upwards. Circulation ensured proper aeration and better mixing of the Fe-zeolite catalyst, dosed in the amount of 2 g/l. After 15 minutes of UV lamp heating, a 30% solution of hydrogen peroxide was added to the reactor in the amount of 1 g/l; this started the process of phenol oxidation (t = 0). The experiment was conducted with no pH correction of the phenol solution at its initial value of about pH 5–6.

The oxidation process was conducted for 4 hours, and the samples for determination of pH,

Table 1. Parameters characterizing structure of zeolites.

Zeolite type	BET surface area [m^2/g]	Total volume of PSD-DST pores [cm^3/g]	Total volume of pores—one point method [cm^3/g]	Volume of micropores [cm^3/g]	Volume of mezopores [cm^3/g]
Natural zeolite	28.2	0.09	0.09	0	0.066
H-zeolite	81.5	0.12	0.14	0.017	0.072
Fe-zeolite	211.2	0.16	0.22	0.051	0.082

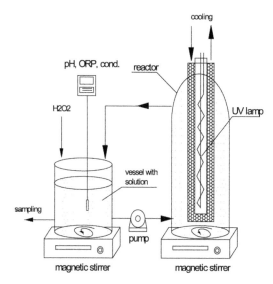

Figure 1. Test stand for experiments on organic contaminants oxidation in the photo-Fenton reaction.

solution temperature, total organic carbon (TOC), phenol and selected products of its decomposition, absorbance at 420 nm and total iron concentration were taken every 30 minutes. Addition of solid Na$_2$SO$_3$ stopped the process of oxidation.

Phenol, hydroquinone, catechol, o-quinone, glyoxylic acid and formic acid were determined on a liquid chromatographer (HPLC) manufactured by Varian. The organic substances were separated on an RP 18 column, 150 mm long, with a grain diameter of 3 μm, using isocratic elution with the acetonitrile water as a mobile phase. TOC was determined on a Shimadzu TOC/VCPH with IR detection. Analytical determination of Fe was conducted on a Varian Spectr AA 20 plus atomic absorption spectrometer.

2 RESULTS AND DISCUSSION

Results of experiments indicate that the increase of the oxidation reaction time causes the decreases of concentration of phenol (from 200 mg/l to 24 mg/l after 210 minutes of reaction) and intermediate products, such as: hydroquinone, o-quinone, catechol and glyoxylic acid, and formic acid are formed—see Figure 2.

Concentration of the intermediate products: hydroquinone, o-quinone and catechol increases up to the 150th minute of the process. Subsequently, their ring systems are destroyed, and in a mixture of products, mainly carboxylic acids are obtained.

The pH of reaction environment has a significant impact on the catalytic oxidation of phenol, the reaction rate, intermediate and final products. The pH of the expcrimental system was changing significantly during the reaction (Fig. 3).

Migration of iron from the matrix of the zeolite is most intensive at a pH approx. 3. It amounts to approximately 2 mg/l. Electropositive complexes of FeOH^{2+} and Fe(OH)$_2^+$ type and dissociated Fe(III) ions, which predominate at pH<4, allow occurring of homogeneous Fenton reaction—reactions (1–3) (Lipczynska-Kochany & Kochany 2008).

$$Fe^{3+} + H_2O_2 = HO_2^\bullet + Fe^{2+} + H^+ \quad (1)$$

$$HO_2^\bullet + Fe^{3+} = O_2 + Fe^{2+} + H^+ \quad (2)$$

$$H_2O_2 + Fe^{2+} = OH^\bullet + Fe^{3+} + OH^- \quad (3)$$

Additional UV radiation may cause an increase of the efficiency of mineralization of numerous organic compounds in aqueous solution. Fe(III) compounds are photoreduced to Fe(II), and additional new HO• radicals from H$_2$O$_2$ are produced, according to the following mechanism (reaction 4). Such mechanism is the cause of positive impact of radiation on the rate of degradation of organic material.

$$Fe^{3+} + H_2O + h\nu \rightarrow Fe^{2+} + OH^\bullet + H^+ \quad (4)$$

Decrease of solution pH during oxidation (Fig. 3) was caused by:

– migration of H$^+$ protons from the zeolite matrix, which has been shown during examinations of ZFe zeolite washout (Świderska-Dąbrowska et al. 2012),

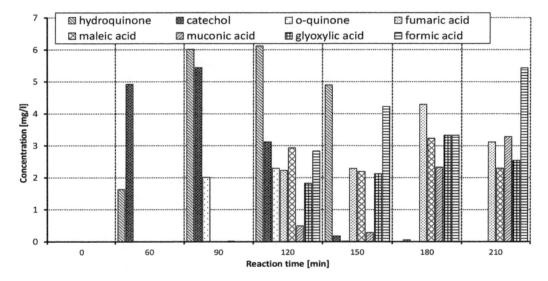

Figutr 2. Impact of reaction time on concentration of products of phenol oxidation in the photo-Fenton process. Concentration of humic acids – 0 mg/l.

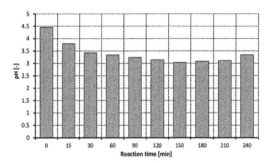

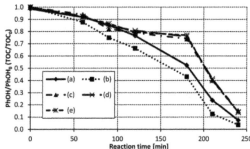

Figure 3. Impact of reaction time on pH of phenol solution (without humic acids) during photo-Fenton process. Initial pH of solution – 5.5.

Figure 4. Comparison of efficiency of phenol oxidation with and without presence of HA; (a) – TOC/TOC$_0$ ratio for phenol solution without HA, (b) – PhOH/PhOH$_0$ ratio for phenol solution without HA, (c) – PhOH/PhOH$_0$ ratio for phenol solution with HA – 10 mg/l, (d) – PhOH/PhOH$_0$ ratio for phenol solution with HA – 20 mg/l, (e) – PhOH/PhOH$_0$ ratio for phenol solution with HA – 50 mg/l.

– acidic properties of hydrogen peroxide and its catalytic decomposition,
– organic acids—intermediate products of oxidation.

Introduction of HA into the reaction solution causes change of the mechanism of phenol oxidation in the photo-Fenton process. The results show that the efficiency of the process decreases along with increasing concentration of HA (Fig. 4).

Increment of HA concentration up to 50 mg/l causes a 40% decrease of phenol removal efficiency, after 180 minutes of reaction time—Figure 4(a) versus (c), (d) and (e). Comparing changes of concentrations of phenol and TOC (sum of phenol and intermediate products of oxidation) in the solution without HA (Fig. 4(a) and (b)), an increase of percentage share of oxidation products in the total content of organic substances may be observed, along with reaction time (up to 120 minutes). This indicates degradation of phenol. Comparison of Figures 2–4 indicates that the removal of phenol is more efficient at a pH approx. 3 than at a pH approx. 5.0. The hydroxyl radical "attacks" the aromatic ring of the phenol, which may be oxidized to catechol, hydroquinone and benzoquinone, and subsequently to carboxylic acids. Different mechanisms of phenol degradation take place during photo-oxidation reaction, depending on the solution pH and the concentration of HS. It is possible that the hydroxyl radicals may "attack" not only the aromatic ring of phenol but also alkyl groups of intermediate products.

Despite a lower efficiency of the oxidation process in the presence of HA, an increase of total Fe concentration in the solution was observed at the same time. After 210 minutes of reaction time and initial concentration of HA 50 mg/l, the concentration of total Fe was higher than 3 mg/l—Figure 5.

Such phenomenon can be explained by the migration of iron compounds from zeolite matrix to the solution due to friction of Fe-zeolite and also formation of stable Fe-HA complexes, which limit the amount of Fe(III) ions involved in the Fenton reaction and thereby limit the amount of generated hydroxyl radicals. On the other hand phenol in the solution (aromatic compound with hydrophilic group—OH) causes an increase of the complexes solubility.

The formation of organic complexes with iron is also confirmed by measurements of the absorbance of post-reaction solution at a wavelength of 420 nm (Fig. 6).

Higher intensity of yellow colour of the solution along with increasing concentration of HA and duration of the oxidation reaction was observed. Absorbance of solution, measured at 420 nm, increased proportionally to the initial concentration of HA, up to 120th minute of reaction time. This is probably caused both by formation of colour intermediate products of phenol oxidation, e.g. o-quinone, and the presence of Fe-HA complexes in the solution. Between 120 and 240 minute, a rapid decrease of absorbance of the solution, to the value below its initial level, was observed. This may prove that the degradation of Fe-HA complexes takes place. As a result, concentration of Fe(III) ions in the solution increases and these are required for generation of next hydroxyl radicals, according to reaction (4). Consequently, intensity of the oxidation of phenol after t = 180 minutes increased, which is represented by a clear inflection of the curve (e) at this point—see Figure 4.

Selectivity of Fe-zeolite catalyst was changing during oxidation of phenol at a constant concentration of 200 mg/l, in the presence of HA at

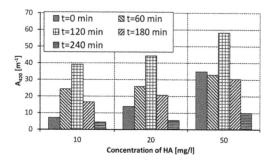

Figure 6. Changes of absorbance (at 420 nm) of oxidized phenol solution with various concentration of HA.

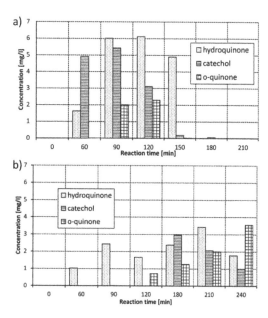

Figure 7. Impact of HA on concentration of selected products of phenol oxidation in photo-Fenton process: (a) phenol concentration 200 mg/l, (b) phenol concentration 200 mg/l + HA concentration 50 mg/l.

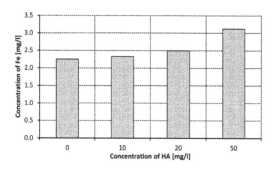

Figure 5. Impact of HA on concentration of Total iron in the solution after 210 minutes of oxidation.

various concentrations (10–50 mg/l) (Fig. 7). After 60 minutes of oxidation of phenol without HA, two main intermediate products of its decomposition were present in the solution: catechol and hydroquinone. Concentration of catechol was more than twice as high as the one of hydroquinone. Its maximum concentration was determined in the sample taken after 90 minutes of oxidation. After this time, the concentration of catechol was decreasing, o-quinone and then simpler organic compounds, such as glyoxylic acid or formic acid were formed, according to the pattern presented in Fig. 8. While hydroquinone decomposed to p-quinone (pathway B, Fig. 8) and subsequently to carboxylic acids containing 1–2 C atoms. In the case

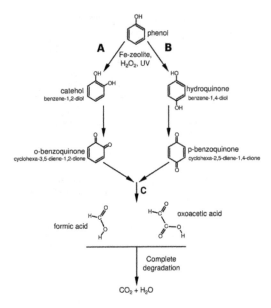

Figure 8. Pattern of phenol oxidation in the photo-Fenton process.

of oxidation of phenol in the solution containing HA at concentration of 50 mg/l, only the formation of hydroquinone was observed during first 90 minutes of reaction. Its concentration was almost two times lower than in the case of the oxidation of phenol only. Catechol was found in sample taken after 180 minutes of the photo-Fenton process. Humic acids inhibit the degradation of phenol, and in particular the formation of catechol.

3 CONLUSIONS

Results of the study indicate that humic acids reduce the efficiency of the oxidation of phenol in the photo-Fenton process. Removal of phenol after 240 minutes of oxidation was over 10% lower in the presence of HA at concentration of 50 mg/l than in case of oxidation without HA. Humic acids form stable complexes with ions of iron (III), which decreases the concentration of Fe(III) available for the photo-Fenton reaction. It was also found that the presence of humic acids affects the selectivity of the catalyst of Fenton reaction. In that case the preferred oxidation product is hydroquinone, whereas for oxidation of phenol without HA, during first stage of reaction, catechol is the dominant intermediate product of the oxidation.

Research results allow to draw conclusions that pH of solution is a significant factor limiting photo-Fenton process. In the studied range of pH changes, the process of degradation of phenol was not so efficient. Results also indicate that the presence of HA inhibits the process of its degradation at pH approx. 3.

The results of presented studies may be used in practice. Photo-Fenton may be used for treatment of wastewater, e.g. leachate from landfills.

Additional advantages of using the Fe-zeolite catalyst for the treatment of wastewater in the process of photo-Fenton is that it can be used multiple times without introducing additional substances to the environment.

REFERENCES

Aleksić, M., Kušić, H., Koprivanac, N., Leszczynska, D. & Božić, A.L. 2010. Heterogeneous Fenton Type Processes for the Degradation of Organic Dye Pollutant in Water—The Application of Zeolite Assisted AOPs. *Desalination* 257: 22–29.

Blanco, M., Martinez, A., Marcaide, A., Aranzabe, E. & Aranzabe, A. 2014. Heterogeneous Fenton Catalyst for the Efficient Removal of Azo Dyes in Water. *American Journal of Analytical Chemistry* 5: 490–499.

Farré, M.J., Doménech, X. & Peral, J. 2007. Combined photo-Fenton and biological treatment for Diuron and Linuron removal from water containing humic acid, *Journal of Hazardous Materials* 147: 167–174.

Fuentes, M., Olaetxea, M., Baigorri, R., Zamarreño, A.M., Etienne, P., Laîné, P., Ourry, A., Yvin, J.-C. & Garcia-Mina, J.M. 2013. Main binding sites involved in Fe(III) and Cu(II) complexation in humic-based structures. *Journal of Geochemical Exploration* 129: 14–17.

Georgi, A., Schierz, A., Trommler, U., Horwitz, C.P., Collins, T.J. & Kopinke, F.-D. 2007. Humic acid modified Fenton reagent for enhancement of the working pH range. *Applied Catalysis B: Environmental* 72: 26–36.

He, D., Guan, X., Ma, J., Yang, X. & Cui, C. 2010. Influence of humic acids of different origins on oxidation of phenol and chlorophenols by permanganate. *Journal of Hazardous Materials* 182: 681–688.

Kępczyński, M., Czosnyka, A. & Nowakowska, M. 2007. Photooxidation of phenol in aqueous nanodispersion of humic acid. *Journal of Photochemistry and Photobiology A: Chemistry* 185: 198–205.

Lipczynska-Kochany, E. & Kochany, J. 2008. Effect of humic substances on the Fenton treatment of wastewater at acidic and neutral pH. *Chemosphere* 73: 745–750.

Pereira, M.C., Oliveira, L.C.A. & Murad, E. 2012. Iron oxide catalysts: Fenton and Fentonlike reactions—a review. *Clay Minerals* 47(3): 285–302.

Romero A., Santos A., Cordero T., Rodríguez-Mirasol J., Rosas J.M. & Vicente F. 2011. Soil remediation by Fenton-like process: Phenol removal and soil organic matter modification. *Chemical Engineering Journal* 170: 36–43.

Świderska-Dąbrowska R., Schmidt R. & Nowak R., 2012. Influence of the Calcination Temperature of Iron Modified Zeolite on Its Chemical Stability, (in Polish), *Monografia Komitetu Inżynierii Środowiska PAN*, Vol. 99: 307–318

Vione, D., Merlo, F., Maurino, V. & Minero, C. 2004. Effect of humic acids on the Fenton degradation of phenol. *Environ Chem. Lett.* 2: 129–133.

Environmental Engineering V – Pawłowska & Pawłowski (Eds)
© 2017 Taylor & Francis Group, London, ISBN 978-1-138-03163-0

Gas sensors array as a device to classify mold threat of the buildings

Z. Suchorab, H. Sobczuk, Ł. Guz & G. Łagód
Faculty of Environmental Engineering, Lublin University of Technology, Lublin, Poland

ABSTRACT: Mold threat is an important problem in building industry. It is one of the most important features that cause Sick Building Syndrome occurrence. In many circumstances this problem is caused by the increase of moisture in the building envelopes but also excessive indoor air humidity. Problem of mold threat is often connected with introduction of new technologies and materials applied for the external walls of the buildings and improper performance of natural ventilation. These factors create suitable conditions for growth of mold. E-nose technique is currently often applied to evaluate air quality in environmental measurements. This paper presents the possibility of classifying mold threat in indoor air of different rooms using low-selective Metal Oxide Semiconductors (MOS) gas sensor array being the element of an e-nose device. Conducted investigation confirmed that e-nose supported by statistical data processing enables to distinguish air from different rooms affected with mold at a satisfactory level.

Keywords: e-nose, MOS sensors, gas sensors array, mold threat, buildings

1   INTRODUCTION

1.1   Mold threat in housing sector

WHO (World Health Organization) reports that indoor air quality influences human health more significantly than outdoor air. This is mainly because nowadays people spend about 90% time inside the closed rooms (Guo et al., 2004). Scientific examinations confirm that indoor air contamination is higher compared to outdoor air (Połednik 2013; Dudzińska et al., 2009). This symptom is popularly called SBS (Sick Building Syndrome). Its visible effects could be the following: mucosa inflammation, asthma, laryngeal and bronchial inflammation. Other symptoms are: migraine, exasperation, concentration disorders leading to a decrease of motivation and effectiveness (Wargocki, 2005).

Microbial contamination is a significant problem of indoor air quality. It ought to be mentioned that fungous sporangia always appear in the air which is a normal situation and the quality and quantity of microorganisms is similar between indoor and outdoor conditions. Problem of moisture in building materials is common in civil engineering (Barnat-Hunek & Smarzewski 2015; Franus et al., 2016). Moist building materials are good nourishment, which combined with high air humidity intensifies the development of microorganisms. Inside the rooms which are threatened with moisture problem, the quantity and quality of microbes is significantly different. Humidity influences the microbial composition of the indoor

air (Table 1) where different species of fungus are sorted into three groups in relation to air humidity (Samson et al., 1994; Żukiewicz-Sobczak et al., 2012). Intervals of humidity that are the most favorable for expansion of the microbes are different for particular species, anyhow it is a general rule that air humidity above 75% is a conducive condition for mold to grow and contaminate indoor air (Pasanen et al., 1994).

According to information in (Eggleston & Bush, 2001), maximum concentration of fungal spores can even exceed 1000 units/m^3. Other sources like (Książek, 2010) state that the maximum allowable quantity of mold concentration equals 50 units/m^3 or 150 units/m^3 in the case of mixture of different species without pathogens. In the case of species as *Cladosporium* and *Alternaria*, maximum allowable spore concentration equals 300 units/m^3. It is important that the presence of the following species: *Aspergillus fumigatus* and *Stachybotrys atra* is considered unacceptable (Jagjit & Jagjit, 1994).

A major factor, decreasing indoor air quality due to mold presence includes fungal metabolites as Microbial Volatile Organic Compounds (MVOC), mycotoxins, allergens, spores and fungi fragments. MVOCs are the main marker showing the mold threat in the indoor environment. They are released by mold along with its expansion. Their composition is characteristic for mold odor. They can lead to various disorders due to olfactory reaction. Mycotoxins are toxic substances which are generated by various microbes (Frąc et al., 2016). They

Table 1. Presence of particular species of fungi depending on air humidity (Żukiewicz-Sobczak et al., 2012).

Relative air humidity	Microorganisms
High, above 90%	*Asperigulus fumigatus*
	Trichoderma
	Exophiala
	Stachybotrys
	Phialophora
	Fusarium
	Ulocladium
	Rhodotorula
	Actinomycetes
Average, 85% 90%	*Aspergillus versicolor*
Low, below 85%	*Aspergillus versicolor*
	Eurotium
	Wallemia
	Penicillia

are mostly produced by *Stachybotrus, Fusarium* and *Aspergillus versicolor* (Samson et al., 1994). Other toxins generated by mold are the following: alpha-toxins, ochratoxin A, trichothecenes and fumonisins (Dutkiewicz & Górny, 2002; Wiszniewska et al., 2004; Żukiewicz-Sobczak et al., 2012).

1.2 Techniques of mold threat estimation

1.2.1 Microbiological methods
Microbiological techniques of mold threat estimation can be divided into traditional (mycological) and molecular methods.

Mycological estimation relies on macroscopic and microscopic observations of the collected samples, moisture determination of the contaminated envelopes, evaluation of ventilation and other sanitary installations performance. It relies on determination of the presence of mold fungi in air samples or swabs, breeding the colonies on agar nourishment, calculation of colonies number and determination of species or genus (Blicharska et al., 2015). Among the traditional techniques, the following can be mentioned:

- sedimentation method by Koch,
- filtration method by filters that capture living and dead fungi,
- cascade impaction method that relies on the flow of contaminated air by small openings.

Traditional microbiological techniques are time consuming, and can last even for 43 days.

Another attempt in detection of fungal contamination is using PCR (Polymerase Chain Reaction) method. This is a molecular microbiological method. It relies on amplification of DNA (deoxyribonucleic acid) fragments. PCR method is expensive and time-consuming, anyhow it offers unequivocal results.

1.2.2 Chemical techniques
Chemical techniques enable to detect selected markers being the products of metabolites as mycotoxins, volatile organic compounds etc. Among them, the most popular are (Wady et al., 2005; Ozonek et al., 2008):

- GC-MS (Gas Chromatography—Mass Spectrometry),
- HPLC-MS (High Performance Liquid Chromatography—Mass Spectrometry).

Both methods require a complicated sample preparation process which involves a lot of manual labor.

1.2.3 Methods of early detection
Early detection techniques enable quick estimation of the possibility of mold threat. They are used to check the status of building contamination and select the proper objects for further investigations with the traditional methods. The most common technique among early detection methods is gas sensor array evaluation, commonly called e-nose.

This technique has been previously used in many fields, mainly in stable indoor applications. Many scientific articles about e-nose application show laboratory experiments or indoor examinations. *In-situ* measurements are limited due to unstable external conditions as temperature, air flow, air humidity etc. Good examples include e-nose implementation in the medical industry (Gardner et al., 2000; Bruins et al., 2013; Pennazza et al., 2013), pharmaceutical industry, cosmetology, food industry and other areas (Baldwin et al., 2011; Bonnefille, 2007; Śliwińska et al., 2014; Marini, 2009; Guz et al., 2010; Guz et al., 2015). A comprehensive list of e-nose applications is presented in the following article (Wilson & Baietto, 2009).

In the articles (Paolesse et al., 2006; Magan & Evans, 2000) there is presented an attempt to evaluate grains quality for fungal strike. Other research, conducted by Jonsson (Jonsson et al., 1997) confirmed the possibility of precise classification of mold into the assumed groups: good, moldy and musty, while in the other experiment of classification by fungal strike (Olsson et al., 2000) the e-nose device was even more precise than gas chromatography—among examined 40 samples of stricken air, only 3 were improperly recognized by e-nose technique, while GC-MS techniques incorrectly classify 6 samples.

According to literature report by Pinzari (Pinzari, 2004), the e-nose was also successfully applied for detection of fungus presence in old libraries and archives. It could be also used for mold threat estimation of the indoor rooms and the equipment. The article (Schiffman et al., 2000) presented a gas sensor array consisting of 16 Metal Oxide Semiconductors (MOS), which was able to detect and classify 5 different species of fungus present in indoor rooms, and the level of detection quality was about 96%. Moreover, it enabled to detect 5 of the selected MVOCs emitted by mold. Other examinations for fungus threat detection were conducted using two commercial devices— Kamina and Moses II (Kuske et al., 2005). The results obtained by Moses II device were more precise, anyhow it was not possible to detect the differences between species. The used technique of data processing—PCA (Principal Component Analysis) enabled only general information about mold presence. According to (Kuske et al., 2006) it was proven that having 12 MOS sensors it was possible to distinguish one of four species of mold, and the estimation accuracy was about 80–85%.

Using standard techniques of mold risk evaluation, contamination is expressed by the number of airborne mold spores present in the air or settled dust on surfaces. Chemical methods mostly show mycotoxins and MVOC concentration as mold presence markers. It is an important feature, because mold can grow behind some layers of finishing materials being the barriers for the spores, not for their metabolism products. It commonly occurs that compounds causing negative effects for the people could be present before any visual symptoms of mold presence. MVOCs cannot be used as bioindicators themselves for particular fungus species which produce wide range of MVOC. Alcohols, ketones, terpenes, esters and sulfur compounds can be listed among them (Kuske et al., 2005). Recognition of particular fungus species should consider the whole composition of the compounds that is why e-noses seem to be a perspective solution for that purposes.

It should also be mentioned that there is a serious limiting factor of e-nose use for mold threat detection, i.e. low MVOC concentration together with the presence of disrupting substances. According to the measurements of Laboratory of Air Quality Sciences (US), performed on 600 buildings, it was noticed that average MVOC concentration amounted to $27\ \mu g/m^3$ with the maximal observed values of about $2000\ \mu g/m^3$ (Kuske et al., 2005). According to (Strom et al., 1994) average concentration level of selected 15 types of MVOCs in stricken rooms amounts to between 50 and 150 $\mu g/m^3$, and in non-stricken objects is about 10 $\mu g/m^3$ which can be compared to the concentration in outdoor air and the detection level of most sensors is over a dozen of $\mu g/m^3$.

2 MATERIAL AND METHODS

2.1 *Applied techniques*

Measurements were conducted using the e-nose consisting of eight MOS type, resistance semiconductor sensors. The applied sensors were produced by TGS Figaro, series 2600: TGS2600-B00, TGS2602-B00, TGS2610-C00, TGS2610-D00, TGS2611-C00, TGS2611-E00, TGS2612-D00, TGS2620-C00. They are small in dimensions, with low electric power consumption (about 300 mW). The low cost of sensors, availability and reliability caused that they are used for many implementations, which enables detailed comparisons of the obtained results. Precise specification of the particular sensors is presented in the producer's technical specification (Figaro, 2016). Additionally, the Maxim-Dallas DS18B20 and Honeywell HIH-4000sensors were applied for the measurement temperature and humidity, respectively.

The flux of sampled air was equal to 200 cm^3/min and sampling was conducted close to the building barriers with visible mold bloom. The datasets of readouts obtained from the gas sensor array are multidimensional with the number of dimensions equal to the number of gas sensors. This hinders the readouts analysis and for the extraction of the desirable information additional statistical methods should be applied as PCA—Principal Component Analysis, PLS—Partial Least Squares, FDA—Functional Discriminant Analysis, LDA—Linear Discriminant Analysis or ANN—Artificial Neural Networks (Fu et al., 2007; Smolarz et al., 2012a; Smolarz et al., 2012b).

The PCA method was applied for the data analysis it. PCA enables reduction of the number of variables (dimensions) to a smaller number of uncorrelated variables (dimensions), as well as classification of variables and cases; therefore, it is possible to detect the hidden dependences between in a large-sized data set. For better interpretation, transformed data are presented in new coordinate system, with the decreased number of dimensions. This enables to reduce the non-important information from the point of view of conducted analysis. Axes of the coordinate system are selected to maximize covariance within the data set. Details about PCA technique are presented in (Krzanowski, 2000).

In this research the following reference techniques were applied:

- visual estimation,
- air quality evaluation using human respiratory system,
- moisture determination of the walls surfaces using TDR (Time Domain Reflectometry) method,
- cascade impaction technique conducted with ACI (Andersen Cascade Impactor).

For moisture determination of the threatened envelopes the TDR method with modified moisture sensors—which enable non-invasive moisture determination of building materials—was applied.

This is an indirect, electric method that allows to measure dielectric permittivity of building materials. Physical basics of the technique are presented in the following paper (Suchorab et al., 2008; Skierucha et al., 2012). Mobile FOM (Field Operated Multimeter) manufactured by ETest producer, Lublin, Poland, was applied in the experiment.

Measurement of fungi bioaerosols was conducted using cascade impactor MAS-6 (Staplex Air Sampler). Samples of air were collected on glass Petri plates with the diameter of 90 mm. Nourishment was placed in the segments of impactor with the following opening diameters:

1. $> 7 \mu m$,
2. $7.0–4.7 \mu m$,
3. $4.7–3.3 \mu m$,
4. $3.3–2.1 \mu m$,
5. $2.1–1.1 \mu m$
6. $1.1–0.65 \mu m$

Air flow was provided by air pump (Staplex Model EC-1HDA Econometric Air Sampler) with constant flow 28.3 dm^3/min. It was applied Sabourand agar with Chloramphenicol produced by BTL manufacturer. Air sampling lasted 10 minutes. After sampling the plates were incubated for the period of 7–14 days in the temperature of 28°C. After incubation the grown colonies were counted and their number was corrected according to Andersen correlation table. Results were given in CFU/m^3 (Colony Forming Unit).

2.2 Description of examined objects

Examinations were conducted in five rooms that visually differed in level of mold threat. Their characteristics are presented below:

a. house from the 1990s, basement pantry, cement-lime plaster rendering, high odor nuisance with visible mold bloom, building stricken with capillary rise—moisture readouts using TDR at lower parts exceed 16%$_{vol}$, air relative humidity 70%—high mold threat level
b. house from the 1980s, bedroom in single-family house, building constructed from aerated concrete without thermal insulation, poor ventilation, visible mold bloom on mortar around the blocks of aerated concrete perceptible odor nuisance, TDR moisture readouts in stricken places shown 13%$_{vol}$, air relative humidity reached 60%,
c. house from 1980s, wainscot, perceptible odor nuisance, no visual bloom, moisture of envelopes – 8%$_{vol}$, relative humidity 60% –medium level of threat
d. wardrobe in new block of flats (built in 2011), attic, finishing in gypsum boards, poor ventila-

tion, no window, no odor nuisance, mold bloom visibility—only fine stains, maximum moisture readouts—4%$_{vol}$, relative humidity—60%—low level of threat,
e. bedroom in new block of flats (built in 2011), proper ventilation, no odor nuisance, no visual mold bloom, maximum moisture readouts—below 1%$_{vol}$, relative humidity 45%, no visible mold threat.

Additionally, e-nose readouts were conducted on reference samples of air:

f. air highly stricken by mold (sampled from timber),
g. clean air,
h. synthetic air.

3 RESULTS AND DISCUSSION

The results of mold threat evaluation are presented in the following forms:

- radial diagrams (gas fingerprints) showing the sensors responses in matrix array obtained in particular environment (Fig. 1.),
- PCA diagram showing all measured environments (Fig. 2),
- microbiological valuation using cascade impaction technique in building (b).

As it is presented in Figure 1, responses of particular diagrams differ, anyhow it is hard to separate the differences between air quality measured in threatened rooms. Major differences can be noticed in diagrams "f" and "g" meaning air stricken by mold and clean air respectively.

PCA processing allows for better interpretation of signals read. This is presented on Figure 2. In the diagram there can be distinguished four visible groups: i) samples of pure air—g,h; ii) samples of non-contaminated air—d,e; iii) samples of contaminated air—a,b,c; iv) sample of air stricken by timber mold—f.

The e-nose was able to distinguish the rooms with obvious presence of mold. In these rooms, smell of mustiness was noticed and in two cases (a,b) was visible bloom.

Timber sample, hardly stricken by mold generated significantly different signals, which were easily recognizable by the e-nose. It must be remembered that such level of mold threat is rare in real conditions. The conducted research using e-nose does not allow to detect what kind or species of mold spreads in indoor air, also it does not allow for clear evaluation of the threat level. This information can be obtained using the traditional microbiological evaluations.

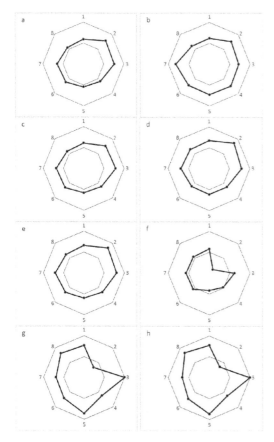

Figure 1. Radial diagrams showing responses of particular sensors in MOS matrix. 1—TGS2600-B00, 2—TGS2602-B00, 3—TGS2610-C00, 4—TGS2610-D00, 5—TGS2611-C00, 6—TGS2611-E00, 7—TGS2612-D00, 8—TGS2620-C00.

For that purpose, room (b) was additionally examined using traditional technique – cascade impaction method. Determination of microscopic fungi in atmospheric air was calculated according to PN-89 Z-04111/03 Standard. Number of fungi units on particular segments of the impactor was the following: 1—10; 2—10; 3—42; 4—113; 5—53; 6—3. Total number of colony forming units was 231 CFU/m^3. The most frequent type of collected mold was *Pencilinium*.

Additionally, particular colonies of *Aspergillus* (also *Aspergillus nigerii*) and *Alternaria* were observed. Comparing to the limits of fungi presence in examined air stated in (Książek, 2010) it should be noticed that the obtained value shows an increased level of mold threat in this room, anyhow it is still below maximum allowable value.

4 CONCLUSION

The conducted investigation confirmed that an e-nose consisting of 8 MOS sensors supported by PCA data processing technique enables satisfactory level of distinction of air from different rooms affected with mold. It would not be able to distinguish air samples which slightly differ in concentration of MVOC, anyhow it enables an objective early detection of contamination. Differences in the achieved readouts were visible when visual or fragrance symptoms were noticed. For more accurate measurement it is still necessary to conduct traditional microbiological determination, for example based on cascade impaction method. Within the microbiological research, it was confirmed that the examined room (b) was significantly stricken by mold. It was also possible to identify the genus of particular fungi.

REFERENCES

Baldwin, E.A., Bai, J., Plotto, A. & Dea, S. 2011. Electronic noses and tongues: applications for the food and pharmaceutical industries. *Sensors* 11(5): 4744–4766.

Barnat-Hunek, D. & Smarzewski, P. 2015. Increased water repellence of ceramic buildings by hydrophobisation using high concentration of organic solvents, *Energ. Buildings.* 103: 249–260.

Blicharska, E., Flieger, J., Oszust, K., Frąc, M., Swieboda, R. & Kocjan, R. 2015. High-Resolution Continuum Source Atomic Absorption Spectrometry with Microwave-Assisted Extraction for the Determination of Metals in Vegetable Sprouts, *Anal. Lett.* 48(14): 2272–2287.

Bonnefille, M. 2007. Electronic noses. Sniffing fast, safe and objective, *Cosmetics* VIII: 9–12.

Bruins, M., Rahim, Z., Bos, A., van de Sande, W.W.J. Endtz, H.P. & van Belkum, A. 2013. Diagnosis of active tuberculosis by e-nose analysis of exhaled air. *Tuberculosis* 93(2): 232–238.

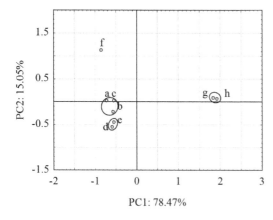

Figure 2. PCA diagram of e-nose readouts.

Dudzińska, M.R., Staszowska, A. & Połednik, B. 2009. Preliminary study of effect of furniture and finishing materials on formaldehyde concentration in office rooms. *Environ. Protect. Eng.* 35(3): 225–231.

Dutkiewicz, J. & Górny, R.L. 2002. Biologiczne czynniki szkodliwe dla zdrowia—klasyfikacja i kryteria oceny narażenia, *Med. Pr.* 53(1): 29–39.

Eggleston, P. & Bush, K. 2001. Environmental allergen avoidance: An overview, *J. Allergy Clin. Immunol.* 107: 403–405.

Figaro, General information for TGS sensors, 2005: 1–12 (available on http://www.figarosensor.com, 2016).

Franus, M., Barnat-Hunek, D. & Wdowin, M. 2016. Utilization of sewage sludge in the manufacture of lightweight aggregate, *Environ. Monit. Assess.* 188(1):10.

Frąc, M., Gryta, A., Oszust, K. & Kotowicz, N. 2016. Fast and Accurate Microplate Method (Biolog MT2) for Detection of Fusarium Fungicides Resistance/ Sensitivity. *Front. Microbiol.* 7(112) 1–16.

Fu, J., Li, G. & Qin, Y. Freeman, W.J. 2007. A Pattern Recognition Method for Electronic Noses Based on an Olfactory Neural Network. *Sens. Actuators B Chem.* 125: 489–497.

Gardner, J.W., Shin, H.W. & Hines E.L. 2000. An electronic nose system to diagnose illness, *Sens. Actuat. B Chem.* 70(1–3): 19–24.

Guo, H., Lee, S.C. & Chan, L.Y. 2004. Indoor air quality investigation AT air-conditioned and non-air-conditioned markets in Hong Kong, *Sci. Total Environ.* 323: 87–98.

Guz, Ł, Łagód, G., Jaromin-Gleń, K., Suchorab, Z., Sobczuk, H. & Bieganowski, A. 2015. Application of gas sensor arrays in assessment of wastewater purification effects. *Sensors*. 15: 1–21.

Guz, Ł., Sobczuk, H. & Suchorab, Z. 2010. Pomiar odorów za pomocą przenośnego miernika z matrycą półprzewodnikowych czujników gazu (In Polish), *Przem. Chem.* 89(4): 378–381.

Jagjit, S. & Jagjit, S. 1994. Building Mycology: Management of Decay and Health, in Buildings, Taylor & Francis.

Jonsson, A., Winquist, F., Schnurer, J., Sundgren, H. & Lundstrom, I. 1997. Electronic nose for microbial quality classification of grains, *Int. J. Food Microb.* 35: 187–93.

Krzanowski, W.J. 2000. Principles of Multivariate Analysis, New York, Oxford, University Press.

Książek, M. 2010. Mykologiczne skutki powodzi cz. 1 (In Polish), *Dachy* 12(132).

Kuske, M., Padilla, M., Romain, A.C., Nicolas, J., Rubio, R. & Marco, S. 2006. Detection of diverse mould species growing on building materials by gas sensor arrays and pattern recognition, *Sens. Actuat. B Chem.* 119(1): 33–40.

Kuske, M., Romain, A-C. & Nicolas, J. 2005. Microbial volatile organic compounds as indicators of fungi. Can an electronic nose detect fungi in indoor environments? *Building and Environment* 40: 824–831.

Magan, N. & Evans, P. 2000. Volatiles as an indicator of fungal activity and differentiation between species, and the potential use of electronic nose technology for early detection of grain spoilage, *J. Stored Prod. Res.* 36(4): 319–340.

Marini, F. 2009. Artificial neural networks in foodstuff analyses: Trends and perspectives A review., *Anal. Chim. Acta.* 635(2): 121–131.

Olsson, J., Borjesson, T., Lundstedt, T. & Schnurer, J. 2000. Volatiles for mycological quality grading of barley grains: determinations using gas chromatographymass spectrometry and electronic nose, *Int. J. Food Microb.* 59: 167–78.

Ozonek, J., Czerwiński, J. & Piotrowicz, A. 2008. Identification of odorous compounds in off-gases from yeast production. *In proc. 7th International Conference on Environmental Engineering, Vilnius, Lithuania,* 261–266.

Paolesse, R., Alimelli, A,. Martinelli, E., Di Natale, C., D'Amico, A., D'Egidio, M.G., Aureli, G., Ricelli, A. & Fanelli, C. 2006. Detection of fungal contamination of cereal grain samples by an electronic nose, *Sens. Actuat. B Chem.* 119(2): 425–430.

Pasanen, A-L., Kalliokoski, P. & Jantunen, M. 1994. *Recent studies of fungal growth on building materials.* In Health Implications of Fungi in Indoor Environments. 485–493. Amsterdam, Netherlands, Elsevier Science.

Pennazza, G., Fanali, C., Santonico, M., Dugo, L,. Cucchiarini, L., Dachà, M., D'Amico, A., Costa, R., Dugo, P. & Mondello, L. 2013. Electronic nose and GC-MS analysis of volatile compounds in Tuber magnatum Pico: evaluation of different storage conditions, *Food Chem.*, 136(2): 668–674.

Pinzari, F. 2004. Electronic nose for the early detection of moulds in libraries and archives, *Indoor Build. Environ.* 13(5): 387–395.

Polish Committee for Standardization. Ochrona czystości powietrza—Badania mikrobiologiczne—Oznaczanie liczby grzybów mikroskopowych w powietrzu atmosferycznym (imisja) przy pobieraniu próbek metodą aspiracyjną i sedymentacyjną (In Polish); PN-Z-04111-03:1989; Polish Committee for Standardization: Warsaw, Poland, 1989.

Połednik B. 2013. Particulate matter and student exposure in school classrooms in Lublin, Poland. *Environ. Res.* 120: 134–139.

Samson, R.A., Flanningan, B., Verhoeff, A.P., Adan, O.C.G. & Hoekstra, E.S. (ed.), 1994. *Health Implications of Fungi in Indoor Environments.* Amsterdam Netherlands, Elsevier Science.

Schiffman, S.S., Wyrick, D.W., Payne, G.A., O'Brian, G. & Nagle, H.T. 2000. Effectiveness of an electronic nose for monitoring bacterial and fungal growth. *In proc. ISOEN 2000,* Brighton, July 20–24: 173–180.

Skierucha, W., Sławiński, C., Szypłowska, A., Lamorski, K. & Wilczek, A. A. 2012. TDR-Based Soil Moisture Monitoring System with Simultaneous Measurement of Soil Temperature and Electrical Conductivity. *Sensors*, 12: 13545–13566.

Śliwińska, M., Wiśniewska, P., Dymerski, T., Namieśnik, J. & Wardencki, W. 2014. Food analysis using artificial senses., *J. Agric. Food Chem.* 62(7): 1423–48.

Smolarz, A., Kotyra, A., Wojcik, W. & Ballester, J. 2012a. Advanced diagnostics of industrial pulverized coal burner using optical methods and artificial intelligence. *Exp. Therm. Fluid Sci.* 43, 82–89.

Smolarz, A., Wojcik, W. & Gromaszek, K. 2012b. Fuzzy modeling for optical sensor for diagnostics of pulverized coal burner. *In Proc. of the Procedia Engineering, 26th European Conference on Solid-State Transducers, Eurosensors 2012*, Krakow, Poland, 9–12 September 2012; pp. 1029–1032.

Strom, G., West, J., Wessen, B. & Palmgren, U. 1994. *Quantitative analysis of microbial volatiles in damp Swedish houses*, In Health implications of fungi in indoor environments. 291–305, Amsterdam, Netherlands, Elsevier.

Suchorab, Z., Jedut, A. & Sobczuk, H. 2008. Water content measurement in building barriers and materials using surface TDR probe. *Proc. ECOpole*, 2(1): 123–126.

Wady, L., Parkinson, D.R. & Pawliszyn, J. 2005. Methyl benzoate as a marker for the detection of mould in indoor building materials. *J. Sep. Sci.* 28(18): 2517–2525.

Wargocki, P. 2005. Measurements of Perceived Indoor Air Quality, *Proc. intern. conf. on energy efficient technologies in indoor environment*. Silesian University of Technology. Gliwice. Poland.

Wilson, A.D. & Baietto, M. 2009. Applications and advances in electronic-nose technologies, *Sensors* 9: 5099–5148.

Wiszniewska, M., Walusiak, J., Guntarowska, B., Żakowska Z. & Pałczyński C. 2004. Grzyby pleśniowe w środowisku komunalnym i w miejscu pracy—istotne zagrożenia zdrowotne (In Polish), *Med. Pr.* 55(3): 257–266.

Żukiewicz-Sobczak, W., Sobczak, P., Imbor, K., Krasowska, E., Zwoliński, J., Horoch, A., Wojtyła, A. & Piątek, J. 2012. Zagrożenia grzybowe w budynkach i w mieszkaniach—wpływ na organizm człowieka (In Polish), *Med. Og. N. Zdrow.* 18 (2): 141–146.

Environmental Engineering V – Pawłowska & Pawłowski (Eds)
© 2017 Taylor & Francis Group, London, ISBN 978-1-138-03163-0

Operation of detention pond in urban area—example of Wyścigi Pond in Warsaw

A. Krajewski, M. Wasilewicz & K. Banasik
Department of Water Engineering, Warsaw University of Life Sciences—SGGW, Warsaw, Poland

A.E. Sikorska
Department of Geography, University of Zurich, Zürich, Switzerland
Department of Water Engineering, Warsaw University of Life Sciences—SGGW, Warsaw, Poland

ABSTRACT: Small detention ponds should perform two major functions, i.e., reduce flood flows and improve water quality by trapping sediment. Such reservoirs, by providing additional benefits, e.g., for sport and recreation, become an increasingly frequent element of city and suburban landscape. However, by practice, while the impact on flood reduction is considered at the design stage of such reservoirs, the aspect of water quality improvement is often undervalued. In this paper results of fields and analytical investigations on the impact of Wyścigi Pond (a detention pond located in the catchment of Służew Creek in Warsaw) on flood flows and on the suspended sediment load reduction are presented. Our investigation shows that Wyścigi Pond impacts flood reduction only to a limited degree, but has a meaningful influence on the sediment load reduction.

Keywords: flood flows, suspended sediment, detention pond, Służew Creek catchment

1 INTRODUCTION

Major consequences of heavy rainfalls are flood flows and increased concentration of suspended solids in rivers. Both of them threaten people's lives and environment. Floods devastate households, infrastructure, crops and monuments, while excessive amounts of suspended solids in rivers cause: silting of valleys, hydraulics structures, reservoirs and also increased flood risk at river mouths. Also other contaminants are transported together with the solids, mainly phosphate compounds (Hejduk 2011, Rodríguez-Blanco et al. 2012), heavy metals (Rivaro et al. 2011, Sikorska et al. 2012), hydrocarbons (Bathi et al. 2012) and some radionuclides (Walling 2006, Porto et al. 2014). Moreover, floods and erosion processes are foreseen to intensify in many regions, due to increase in frequency and intensity of heavy rainfalls resulting from foreseen climate changes (Fiener et al. 2015, Pai et al. 2015).

With a purpose to reduce peak runoff and enhance outflow water quality from urban areas, small reservoirs are more often placed in watercourses (EPA 2014). Due to their relatively small dimensions, they can be well integrated into cityscape or suburban areas. In order to emphasize the main function performed by such reservoirs, they are referred to as sediment detention ponds or dual purpose detention ponds (Haan et al.1994, Luk 1999). Pursuant to EPA (2002) detention ponds are capable of reducing suspended solids by 49% up to 80%, and of reducing total phosphorus level by 20% up to 52%. However, by practice, at the project stage of such reservoirs, while impact on flood reduction is considered, the aspect of water quality improvement is often undervalued (Gradowski & Banasik 2008, Pietrak & Banasik 2009, Waga-Bart 2013).

In this paper, we present results of fields and analytical investigations on impact of the Wyścigi Pond (a detention pond located in the catchment of Służew Creek in Warsaw) on flood flows and on the suspended sediment load reduction.

2 MATERIAL AND METHODS

The catchment of the Służew Creek (Fig. 1), located in the south-west of Warsaw, has been periodically monitored by the WULS Department of Water Engineering since mid-1980s (Banasik 1987, Banasik et al. 1988). The conducted research has been mainly connected with modelling the rainfall-runoff process and, to a lesser extent, the runoff quality (Banasik et al. 2008, 2014, Sikorska & Banasik 2010; Sikorska et al. 2013,

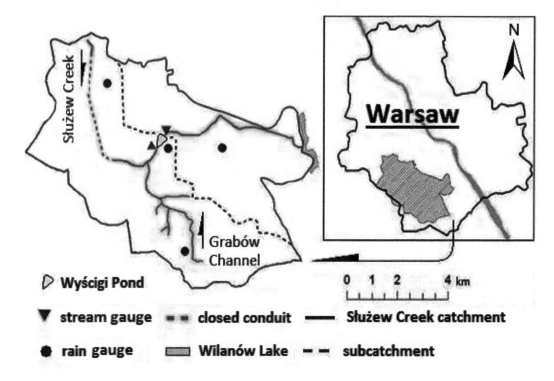

Figure 1. Locality of the Służew Creek catchment.

2015). The area of the entire catchment till its outlet to Wilanów Lake amounts to 54.8 km², whereas the area of the analyzed sub-catchment at the profile Wyścigi Pond is 28.7 km². The catchment is heterogeneous in terms of land development. While the northern part can be characterized by a stronger urbanization, as it is encircled by housing estates and the Chopin Airport, the southern part is less urbanized and dominated by single-family houses, fields, wastelands and woodlands. The Służew stream at the airport is closed in a conduit. As for a typical lowland catchment in Poland, the area of the catchment is very flat without any hills or depressions and with inconsiderable land slopes (less than 0.5%).

The Wyścigi Pond is a low-volume detention pond located directly at the stream. Because of the entire loss of its retention capacity, the pond was renovated in 2007, i.e., its basin was restored and deepened, and new shores were formed. The outflow from the Wyścigi Pond is currently carried by a trapezoid channel. At a normal pool level, the pond surface amounts to 1.3 ha and its volume to 14 500 m³. The Department of Water Engineering has been performing regular acoustic depth measurements of Wyścigi Pond since 2008.

The study catchment is equipped with two stream gauges and four rain gauges, which enable determining the mean areal rainfall within the catchment. The two stream gauging stations are located upstream and downstream of the Wyścigi Pond. Both are equipped with staff gages and water level data loggers. The stations are also adapted for a manual sampling of suspended sediment probes in the main stream. Water level records are verified weekly by staff gage readings. On the basis of the verified water levels, stream flows are estimated according to the established rating curves, which are periodically verified with hydrometric measurements. Manual bathometer allows for taking stream samples during the passage of the flood flows. This device consists of 1-litre container and two pipes. One of them carries water with sediment into a container, while at the same time another pipe discharges air. Shape and dimensions of bathometer are equivalent to that developed by Polish Hydrological Service (Pasławski 1973). The suspended sediment concentration in each sample is next determined by means of the gravimetric analysis.

An example event of 6.05.2015, measured upstream and downstream of Wyścigi Pond is presented in Figure 2. Sedimentgraphs were constructed based on manual samples, taken directly from the stream. The highest value of suspended sediment concentrations in the inflow (204 mg·dm⁻³) were measured prior to the discharge

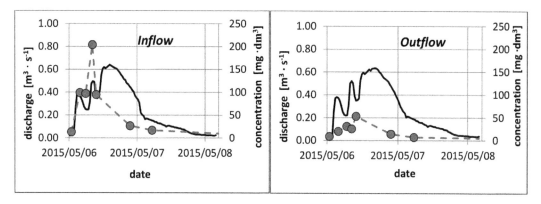

Figure 2. Rainfall—runoff event from 06.05.2015, measured inflow and outflow from the Wyścigi Pond. Solid line is discharge; dashed line with dots is suspended sediment concentration.

peak, which is known as a first flush phenomenon. The suspended sediment concentration measured at low stream flows was close to 5 mg·dm^{-3}. Seven recorded rainfall-runoff-suspended sediment concentration events were used in the further analysis.

3 RESULTS AND DISCUSSION

3.1 Impact of the Wyścigi Pond on flood flows and suspended sediment load reduction during flood events

Since the beginning of its operation, it was foreseen that the pond would cut the peak flows. However, the first investigations conducted by Pietrak & Banasik (2009) have shown that reduction rates of flood hydrographs are insufficient for flood protection. In this paragraph we present the results of field measurements conducted upstream and downstream of the reservoir, as well as the results of analytical investigations. Table 1 shows the characteristics of 7 rainfall—runoff events measured in the catchment of Służew Creek in the period 2014–2015. For the analyzed dataset average, the maximum inflow to Wyścigi Pond was estimated at 0.736 m^3·s^{-1} and was about 3.5 times higher than the mean inflow in the period 2009–2012 (SQ$_{09-12}$ = 0.216 m^3·s^{-1}). Maximum average outflow from reservoir was very close to the inflow (0.710 m^3·s^{-1}). The reduction rate of discharge was ranging from 0.44% to 13.3%. The highest values of reduction rate were observed for lowest inflows. These observations confirm the results of previous studies on flood flows reduction by Wyścigi Pond. The sediment mass outflowing from reservoir was significantly lower than the inflowing one (average reduction rate = 71.1%). This indicates that Wyścigi Pond has a strong impact on removing suspended sediments during flood flows.

3.2 Silting of Wyścigi Pond in the period 2009–2015

Since 2009, the Department of Water Engineering has been performing regular acoustic depth measurements of Wyścigi Pond. Measurements were done 4 times using an echosounder and GPS receiver. On the basis of the digital elevation model of the reservoir, the area and capacity curves were constructed (Fig. 3), while the second volume of sediment deposits was calculated (Table 2). In the period 2009–2015, Wyścigi Pond has lost about 27% of his capacity (from 20000 m^3 to 14 540 m^3). In the first period of operation (2009–11), 1163 m^3 of sediment was trapped by the reservoir. Most of it was settled in the upper part of the pond. The amount of deposits is decreasing over time, which means that the trap efficiency will be expected at a lower ratio in the future.

3.3 Operation of Wyścigi Pond and similar reservoirs

The first results of field and analytical investigations on the impact of the small detention pond, i.e., Wyścigi Pond, on flood flows and on the suspended sediment load reduction were presented in this paper. The reservoir was built in 2007 with the aim of flood protection. Our investigations proved that this main function is not fulfilled; however, the pond has meaningful impact on the sediment load reduction.

Small detention ponds become an increasingly frequent element of city and suburban landscape. The designers and policymakers often emphasize the role of such constructions in flood protection programs. However, in many cases, the size of such reservoir does not correspond to the catchment area (reservoir volume is too low to capture the flood) which inhibits its proper performance (Pietrak & Banasik 2009). Regarding lowland

Table 1. Characteristics of rainfall-runoff events measured in the catchment of Służew Cek.

No	Date	Rainfall [mm]	Inflow characteristics I_{max}–maximum inflow [m^3·s^{-1}]	Y—suspended sediment mass entering reservoir [Mg]	Outflow characteristics O_{max}–maximum outflow [m^3·s^{-1}]	Y_o—suspended sediment mass leaving reservoir [Mg]	Reduction [%] rate for Max. Discharge $\left(\frac{I_{max.}-O_{max.}}{I_{max.}}\right)\cdot 100\%$	Susp. sediment load $\left(\frac{Y-Y_o}{Y}\right)\cdot 100\%$
1	19.04.2014	6.9	0.529	1.95	0.459	0.430	13.3	77.9
2	24.04.2014	23.5	1.440	9.66	1.41	3.22	2.00	66.7
3	06.05.2015	15.8	0.640	2.84	0.633	1.18	1.07	58.3
4	25.07.2015	8.0	0.460	0.744	0.450	0.210	2.73	71.5
5	04.09.2015	5.1	0.358	0.069	0.340	0.005	4.97	92.3
6	06.09.2015	19.0	1.045	2.207	1.040	0.736	0.440	66.6
7	26.09.2015	15.6	0.677	0.604	0.639	0.210	5.53	64.7
range		5.1–23.5	0.460–1.440	0.069–9.66	0.340–1.410	0.005–3.22	0.44–13.3	58.3–92.3
avarage		13.4	0.736	2.58	0.710	0.856	4.29	71.1

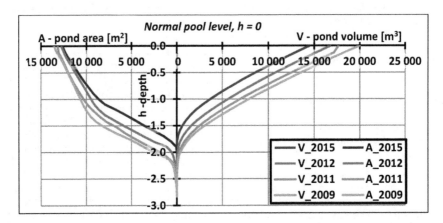

Figure 3. Area and capacity curves of Wyścigi Pond.

Table 2. Volume of sediment deposits in Wyścigi Pond.

Period between measurements	Volume of deposits [m^3] Total	Annual
2009–2011	2 330	1 160
2011–2012	842	842
2012–2015	2 300	766
2009–2015	5 470	911

catchment, the nature of terrain also prevents planning of sufficient reservoir retention.

With this study we would like to point out, that another benefit, i.e., outflow quality improvement, can be provided by such reservoirs. Moreover, smaller reservoirs become silted at a much faster pace than large reservoirs (Madeyski et al. 2008) and consequently, can lose their entire capacity already after few years of their exploitation. Therefore, during the design phase, the maintenance costs to keep the sediment reduction at a required ratio should also be taken into account.

4 CONCLUSIONS

On the basis of the conducted research, following conclusions were made:

i. For seven analyzed rainfall—runoff events, average reduction rate for peak discharge and suspended sediment load equaled 4.3% and 71.7% respectively, the reduction rate of peak discharge was decreasing as the maximum inflow increased, the reduction rate of suspended sediment varied from 58.3% to 92.3%,

ii. In the period 2009–2015, Wyścigi Pond has trapped 5466 m^3 of sediment and has lost about 27% of its capacity, the trap efficiency of reservoir is expected to decrease over time if no maintenance will be undertaken,

iii. Due to low retention volume, reservoirs similar to Wyścigi Pond cannot impact meaningfully the flood flows reduction; therefore, planners and policymakers should consider other benefits provided by detention ponds, i.e., removing excessive amounts of sediments,

iv. Further investigation will be focused on modeling the pond performance, especially by predicting outflow sedimengraphs and the suspended sediment removal.

ACKNOWLEDGMENTS

The study was carried out as part of project no. 2015/19/N/ST10/02665, financed by the National Science Centre, Poland.

REFERENCES

Bathi, J. Pitt, R. & Clark S. 2012. Polycyclic Aromatic Hydrocarbons in Urban Stream Sediments. *Advances in Civil Engineering*, Article ID 372395, 9 pages.

Banasik, K. 1987. Rainfall-runoff conceptual model for an urban watershed. *Proc. XXII Congress of IAHR, and FICUD—Urban Drainage Hydraulics and Hydrology*, 5: 299–300, Lousanne, Swizerland.

Banasik, K., Byczkowski, A. & Ignar, S. 1988. An influence of the method for effective rain determination on the parameters of Nash model for urbanized watershed. *Conference paper of URBAN WATER 88, Hydrological Processes and Water Management in Urban Areas*,17–22, Duisburg-RFN.

Banasik, K., Hejduk, L. & Barszcz, M. 2008. Flood flow consequences of land use Changes in a small urban catchment of Warsaw. *11th International Conference on Urban Drainage* – CD edition. Edynburg, UK.

Banasik, K., Krajewski, A., Sikorska, A. & Hejduk L. 2014. Curve Number estimation for a small urban catchment from recorded rainfall-runoff events. *Archives of Environmental Protection* 40(3): 75–86.

EPA—United States Environmental Protection Agency. 2002. *Urban Stormwater BMP Performance Monitoring*. Office of Water, Washington DC.

EPA—United States Environmental Protection Agency. 2014. *Best Management Practices (BMPs)*. www.epa. gov.

Fiener, P., Neuhaus, P. & Botschek, J. 2013. Long-term trends in rainfall erosivity–analysis of high resolution precipitation time series (1937–2007) from Western Germany. *Agricultural and Forest Meteorology* (171–172): 115–123.

Gradowski, L. & Banasik, K. 2009. Reduction of the flood flow hydrograph by the Berensewicz Pond reservoir on the Służew Creek [in polish]. *Przegląd Naukowy. Inżynieria i Kształtowanie Środowiska* 17, 1(39): 13–25.

Haan C., Barfield B., Hayes J. 1994. D*esign Hydrology and Sedimentology for Small Catchments*. Academic Press, San Diego.

Hejduk, L. 2011. Relation between chosen phosphorus forms and suspended sediment in Zagożdżonka River [in polish]. *Prz. Nauk. Inż. Kszt. Środ.* 54, 20(4): 311–320.

Luk, G. 1999. Evaluation of Dual-Purpose Detention Ponds Design Using the SWMM Model, *Canadian Water Resources Journal*, 24(4): 331–342.

Madeyski, M., Michalec, B. & Tarnawski, M. 2008. *Silting of small water reservoirs and quality of sediments* [in polish]. Infrastruktura i Ekologia Terenów Wiejskich. No 2008/11, Kraków.

Pai, D., Sridhar, L., Badwaik, M. & Rajeevan, M. 2015. Analysis of the daily rainfall events over India using a new long period (1901–2010) high resolution (0.25° × 0.25°) gridded rainfall data set. *Clim Dyn*, 45: 755–776.

Pasławski, Z. 1973. *Methods of River Hydrometry* [in polish]. Wydawnictwo Komunikacji i Łączności, Warsaw.

Pietrak, M. & Banasik, K. 2009. Reduction of the Służew Creek flood flow by small ponds [in polish]. *Przegląd Naukowy Inżynieria i Kształtowani Środowiska*, 3(45): 22–34.

Porto, P., Walling, D. E., Alewell, C., Callegari, G., Mabit, L., Mallimo, N., Meusburger, K. & Zehringer M. 2014. Use of a 137Cs re-sampling technique to investigate temporal changes in soil erosion and sediment mobilisation for a small forested catchment in southern Italy. *Journal of Environmental Radioactivity*, 138: 137–148.

Rivaro, P., Çullaj, A., Frache, R., Lagomarsino, C., Massolo, S., De Mattia, M. & Ungaro, N. 2011. Heavy Metals Distribution in Suspended Particulate Matter and Sediment Collected from Vlora Bay (Albania): A Methodological Approach for Metal Pollution Evaluation. *Journal of Coastal Research*, 58: 54–66.

Rodríguez-Blanco, M. L., Taboada-Castro, M. M., Palleiro-Suárez, L. & Taboada-Castro, M. T. 2012. Phosphorus and Suspended Sediment Loads in Base-Flow and Runoff Events: A Case Study of a Small Stream in NW Spain. *Communications in Soil Science & Plant Analysis*, 43 (1–2): 219–225.

Sikorska, A. E. & Banasik, K. 2010. Parameter identification of a conceptual rainfall-runoff model for a small urban catchment. *Annals of Warsaw University of Life Sciences-SGGW*. Land Reclamation, 42(4): 279–293.

Sikorska, A. E., Scheidegger, A., Chiaia-Hernandez, A.C., Hollender, J. & Rieckermann, J. 2012. Tracing of micropollutants sources in urban receiving waters based on sediment fingerprinting. *9th International Conference on Urban Drainage Modelling*, 9 pp, Belgrade, Serbia.

Sikorska, A. E., Scheidegger, A., Banasik, K., & Rieckermann, J. 2013. Considering rating curve uncertainty in water level predictions. *Hydrol. Earth Syst. Sci.*,17: 4415–4427.

Sikorska, A. E., D. Del Giudice, K. Banasik, Rieckermann, J. 2015. The value of streamflow data in improving TSS predictions—Bayesian multi-objective calibration, *Journal of Hydrology*, 530: 241–254.

WAGA-BART—Zbigniew Bartosik Specjalistycznia Pracownia Projektowa. 2013. Projekt budowy zbiornika retencyjnego na Potoku Służewieckim, Zadanie inwestycyjne: Przywrócenie funkcjin retencyjnej zbiornika retencyjnego Staw Służewiecki [Construction project of Służewiecki Pond]. Warsaw.

Walling, D. E. 2006. Tracing versus monitoring: new challenges and opportunities in erosion and sediment delivery research. In: Owens, P. N., Collins, A. J. (eds.), *Soil Erosion and Sediment Redistribution in River Catchments*: 13–27, CABI, Wallingford.

Environmental Engineering V – Pawłowska & Pawłowski (Eds)
© 2017 Taylor & Francis Group, London, ISBN 978-1-138-03163-0

Hydraulic equations for vortex separators dimensioning

M.A. Gronowska-Szneler & J.M. Sawicki
Faculty of Civil and Environmental Engineering, Gdansk University of Technology, Gdansk, Poland

ABSTRACT: The paper presents a set of hydraulic expressions developed to design vortex separators. These devices are used for gravitational removal of suspensions from wastewater. Methods for vortex separators dimensioning found in the literature could still be improved. Measurements of tangential velocity and liquid residence time conducted on a laboratory vortex tank allowed the authors to formulate a mathematically simple velocity field model of an essential technical advantage. Then, equations describing suspended particle motion in the separator were derived. Finally, a technical procedure for hydraulic design of vortex separators was developed.

Keywords: centrifugal force, storm waste water, vortex separator

1 INTRODUCTION

Vortex separators are designed for the gravitational removal of suspensions from wastewater. In this type of devices, the centrifugal force is the fundamental factor enhancing the process of separation. At present, engineers are becoming increasingly interested in application of these devices along roads and motorways to remove suspensions from rainwater (Andoh & Saul 2003).

The principle of vortex separator operation is analogous to the process of dust removal in a cyclone. However, experiences in cyclone design cannot be applied to vortex separators due to density differences (Gronowska & Sawicki 2011) and a lack of properly described fluid velocity field. Rhodes (2008) proposed to describe tangential velocity inside a separator by an irrational function with velocity value near the outer wall as an important parameter. However, the author did not include the information on how to determine this factor.

Research conducted by Veerapen et al. (2005) leads to a conclusion that influence of the centrifugal force can be neglected. The resulting theoretical description of vortex separator operation was based on the advection-dispersion mass balance of suspension. However, a one-dimensional model of dispersion can be applied provided that the length of the considered liquid stream is greater that the length of the advective subzone LA. For a circular stream with average velocity vz, the length of the stream LA is expressed by G. I. Taylor's (1954) formula:

$$L_A = 0.07 v_z R^2 / K_T \qquad (1)$$

where R—separator radius and KT—transverse dispersion coefficient given by:

$$K_T = 0.041 \sqrt{\lambda} v_z \qquad (2)$$

where λ—Nikuradse coefficient for Darcy-Weisbach formula. For $\lambda = 0.04$ length of the stream is equal to:

$$L_A = 8.6 R \qquad (3)$$

The researchers considered two separate radii: $R_1 = 0.31$ m and $R_2 = 0.75$ m for $L_A = ht$ (tank height) what gave the ratio L_A/R equal to 0.90 for R_1 and $ht_1 = 0.28$ m, and 1.00 for R_2 and $ht_2 = 0.75$ m. Such results indicate that the condition in Equation 1 is not fulfilled and the dispersion model cannot be accepted from the physical point of view. Moreover, Veerapen et.al. (2005) were analyzing fish ponds, where density of suspension ranged from 1060 to 1180 kg/m³, whereas mineral suspension present in storm waste water has the density of circa 2700 kg/m³. With grit density double the value for fish excrements, the centrifugal force may be expected to be greater in magnitude.

Original research of the authors of this paper led to the construction of a formally simple method to design vortex separators—a method consisting

217

of the relations for everyday work of an engineer instead of numerical simulations of separator operation.

2 MATERIAL AND METHODS

2.1 Principle of designing vortex separators

The fundamental design criterion for vortex separators (Rhodes 2008) requires that the flowthrough time of each characteristic suspended particle t_F, i.e. the time of particle motion from $r = R$ to $r = r_w$, is equal to the time necessary for sedimentation of this particle t_s:

$$t_F = H / v_f = t_s \qquad (4)$$

where H—water depth, r—distance from the axis of rotation, r_w—outlet radius, v_f—fluid velocity.

In order to fulfill this condition, a proper model of velocity field inside the separator chamber needs to be determined.

2.2 Liquid flow velocity field

2.2.1 Laboratory examination of the problem

In order to determine the actual velocity distribution in a vortex separator a laboratory test stand was constructed. The stand consisted of a cylindrical chamber with an inlet tangent to the chamber wall (Fig. 1a; $R = 0.40$ m, $r_w = 0.02$ m, inlet diameter $d_{in} = 0.075$ m, inlet height $h_w = 0.20$ m). The inlet was installed at half of the outlet pipe height.

Measurements of liquid tangential velocity u_t were performed by means of a micro-propeller current meter (OTT Hydromet). The flowrate varied from 0.20 L/s to 0.60 L/s. Each series of measurements included a set of 54 measurement points placed evenly within the chamber interior (6 radii every 60°; 3 radial distances: R/4, R/2, 3R/4; 3 measurement verticals: $z = 0.02$ m, $z = 0.10$ m, $z = 0.18$ m).

Liquid residence time within the chamber was determined using a fluorometric method. A known dose of the tracer (Rhodamine WT) in the form of an impulse was injected at the inlet and then its concentration at the outlet pipe was measured (Cyclops-7, Turner Designs, USA).

2.2.2 Results of measurements

Sample results of measurements are presented in Figure 2 (tangential velocity profile according to Equation 6 and ranges of velocity values measured along the depth) and Figure 3 (tracer concentration corresponding to tangential velocity measurements).

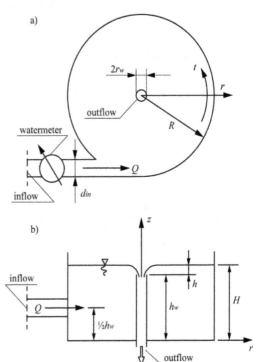

Figure 1. Schematic diagram of vortex separator: a) plan view; b) vertical cross-section.

The following aspects were observed during measurements: tangential velocity profile inside the separator is almost linear; an intensive free surface vortex near the outlet pipe was formed. Mathematical description of these three-dimensional effects can be done by using general equations of fluid mechanics, however, taking into account the requirement of formal simplicity, authors decided to discard them in the final design method.

2.2.3 Tangential velocity model

Results of measurements show that distribution of tangential velocity component u_t is rather symmetrical with respect to z axis and slightly varying along the verticals (Fig. 2). Its radial distribution corresponds to research conducted by Stairmand (1951):

$$u_t(r) r^{1/2} = C \qquad (5)$$

However, the researcher did not provide a way to determine the value of constant C, so the authors of this paper decided to express tangential velocity distribution as a linear function:

$$u_t(r) = A(R - r) \qquad (6)$$

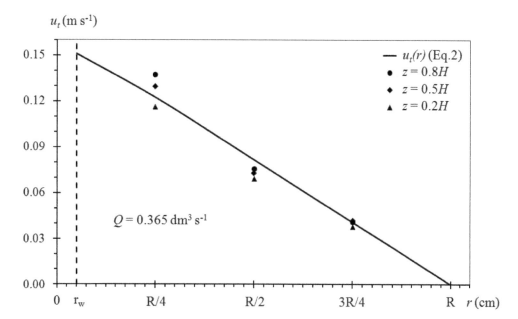

Figure 2. Sample results from tangential velocity measurements.

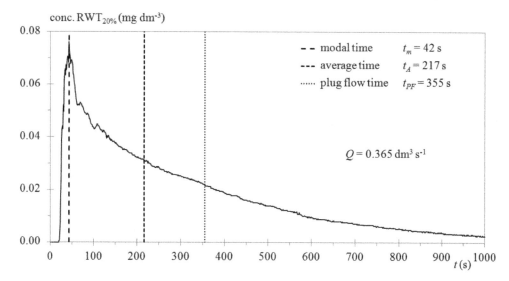

Figure 3. Sample results from tracer measurements.

where coefficient A is calculated from a physically justified relation for steady flow that the energy flux dissipated by the liquid is equal to the energy flux delivered to the device (Sawicki 2004, Slattery 1999). After some transformations, one obtains:

$$A = 6.14Q\left(HR^4 d_m^{\ 4}\right)^{-1/3} \qquad (7)$$

For instance, for the considered model of vortex separator (Fig. 1), this coefficient equals: $A = 0.242$ 1/s for $Q = 0.215$ L/s; $A = 0.411$ 1/s for $Q = 0.365$ L/s; and $A = 0.636$ 1/s for $Q = 0.565$ L/s.

Tangential velocity profiles calculated using Equation 6 show good conformity with the performed measurements (Fig. 2), thus, the linear velocity model u_t was accepted.

2.2.4 *Radial velocity model*
For cylindrical tanks the radial velocity is most commonly described by:

$$u_r(r) = -Q/2\pi H_e r \qquad (8)$$

where H_e is the thickness of the active layer of liquid flow, i.e. a synthetic parameter, which corrects the tank capacity by including the total volume of "dead zones". In an ideal case of plug-flow $H_e = H$ (Rhodes 2008); however, in practice, the thickness of an active layer of flow is somewhere between the values of total water depth H and diameter of the inlet d_{in} (Fig. 1b).

Liquid residence time for a two-dimensional model of vortex separator can be determined from the relation for the plug-flow time:

$$t_k = \int_R^{r_w} dr/u_r(r) = \pi H_e (R^2 - r_w^2)/Q \qquad (9)$$

Calculations of the plug-flow time t_k and empirical determination of corresponding values of the average residence time t_A (defined as the first moment of the residence time distribution curve) indicated that, in general, tA is a little longer than the half of t_k. For instance, calculations for $Q = 0.34$ L/s and $H_e = H = 0.24$ m made from $r = R$ to $r = r_w$ yielded the plugflow time equal to $t_k = 355$ s, whereas the average residence time was $t_A = 217$ s (Fig. 3). In such a situation, to obtain a cohesive and a mathematically simple relation, authors of this paper assumed that the mean depth of the active layer of flow equals (Fig. 1):

$$H_e = 0.5(d_{in} + h_w + h) \qquad (10)$$

Thickness of the water layer overflowing the outlet pipe h, (Fig. 1b) can be calculated from the expression describing a shaft, e.g. "the morning glory overfall" (Gronowska & Sawicki 2014, Nalluri & Featherstone 2001):

$$Q = 2\mu_p 2\pi r_w \sqrt{2g} h^{3/2}/3 \qquad (11)$$

where μ_p—overfall discharge coefficient, g—gravity acceleration.

2.3 Pressure distribution

Research on the laboratory test stand showed that liquid vertical velocity can be ignored. In the prevailing part of the chamber its values are small and have an increasing tendency only near the outflow vortex. This means that vertical pressure has a static character and generates the Archimedes force. On the contrary, radial pressure component needs to be taken into account, as it is responsible for the generation of transverse drift FTD. In the considered case this force can be directly calculated from the relation (Soo 1966):

$$F_{TD} = -V_p \partial p/\partial r \qquad (12)$$

where p—pressure, V_p—particle volume.

Pressure p can be determined from the equation of fluid motion. With the chosen form of velocity vector, this equation takes a simple form without viscous terms (Slattery 1999):

$$\rho_l u_r \partial u_r/\partial r - \rho_l u_t^2/r = -\partial p/\partial r \qquad (13)$$

where ρ_l—liquid density.

2.4 Description of suspended particle motion

2.4.1 Balance of forces acting on a particle

The centrifugal force FC that acts along the radius of curvature of liquid trajectory is, by definition, equal to the product of particle mass ($m_p = r_p V_p$; ρ_p—density of the particle) and liquid angular velocity ($\omega = u_t^2/r$). In the considered case, when tangential velocity is described by Equation 6, one obtains:

$$F_C(r) = \rho_p V_p u_t^2/r = \rho_p V_p A^2 (R - r)^2/r \qquad (14)$$

By substituting the pressure derivative according to Equation 13, the transverse drift FTD given by Equation 12 acquires the form:

$$F_{TD}(r) = \rho_l V_p u_r \partial u_r/\partial r - \rho_l V_p u_t^2/r \qquad (15)$$

Further substitution of radial velocity as in Equation 8 and tangential velocity as in Equation 6 yields:

$$F_{TD}(r) = -\rho_l V_p A^2 (R - r)^2/r - \rho_l V_p Q^2/4\pi^2 H_e^2 r^3 \qquad (16)$$

In comparison to the first term on the right side, the second term on the right is small. For the following technically justified values of parameters: flowrate Q in the range from 0.01 m³/s to 0.02 m³/s, tangential velocity $u_t = 0.30$ m/s, active flow layer $H_e = 1.00$ m, separator radius R in the range from 1.00 m to 3.00 m and outlet radius $r_w = 0.20$ m, the ratio between the two terms exceeds 30 and the second term can be omitted.

While determining trajectory of motion, one should take the drag force into account. Depending on the values of the Reynolds number there are two possibilities: Stokes' formula for $Re < 1$ and Newton's formula for bigger values of Re (e.g. Soo 1966). The considered case displays a significant disproportion between intensity of tangential flow and intensities of flows in other directions. For tangential velocity around 0.30 m/s and particles 0.10 mm in diameter

Reynolds number is equal to $Re_t = 30$, whereas, for vertical and radial motion with flow velocities 0.0067 m/s and 0.0050 m/s, values of the Reynolds number are $Re_z = 0.67$ and $Re_r = 0.50$, respectively. Taking these sets of values into account, the authors of this paper decided that the drag force can be expressed by Stokes' formula (Soo 1966):

$$F_S = 3\pi\mu\left(u_r - v_r\right)d_p \qquad (17)$$

where μ—dynamic viscosity coefficient, v_r—particle radial velocity. This formula is especially convenient as it allows to obtain an analytical solution to describe the time of particle motion. Application of Newton's formula would require the usage of numerical methods resulting in a solution hardly of any use in further analytical considerations.

2.4.2 Equations of suspended particle motion

For the considered suspended particle (coordinate system as in Fig. 1b), Newton's second law has the following forms (Soo 1966) for radial, tangential and vertical directions, respectively:

$$\left(\rho_p + \alpha_s\rho_l\right)V_p dv_r / dt = \\ \left(\rho_p - \rho_l\right)V_p A^2 \left(R - r\right)^2 / r - 3\pi\mu d_p\left(u_r - v_r\right) \qquad (18)$$

$$\left(\rho_p + \alpha_s\rho_l\right)V_p dv_t / dt = 3\pi\mu d_p\left(u_t - v_t\right) \qquad (19)$$

$$\left(\rho_p + \alpha_s\rho_l\right)V_p dv_z / dt = \left(\rho_l - \rho_p\right)V_p g - 3\pi\mu d_p v_z \qquad (20)$$

where α_s—associated mass coefficient, and $\alpha_s = 0.50$, d_p—particle diameter. These relations, together with the equation of particle trajectory (where **rc**—particle radius vector, **v**—particle velocity vector):

$$d\mathbf{r}_c / dt = \mathbf{v} \qquad (21)$$

create a closed system of equations of suspended particle motion inside a vortex separator.

2.4.3 Determination of suspended particle trajectory

In order to solve the system of Equations 18–21, initial conditions are required. This information can be formulated in diverse ways, so that various variants of separator operation can be analyzed. These equations, due to their complexity, should be solved using computer software and CFD methods (e.g. Dyakowski et al. 1999, Gronowska et al. 2013). In the paper these equations were considered in their specific forms.

2.4.4 Simplification of equations of suspended particle motion

Simplification of the system of Equations 18–21 can be based on the fact that particles suspended in wastewater quickly adapt their velocity to the surrounding conditions. This statement is commonly used in engineering (e.g. McGaughey 1956). As a result, inertial terms present in equations of particle motion can be omitted and the vertical velocity component is equal to particle free sedimentation velocity:

$$v_f = 0.035\left(\rho_p - \rho_l\right)d_p^2 g/\mu \qquad (22)$$

Tangential velocity component is equal to tangential velocity of liquid flow:

$$v_t(r) = u_t(r) \qquad (23)$$

Radial velocity component is equal to the sum of radial flow velocity and a factor resulting from the combined influence of the centrifugal force and the transverse drift:

$$v_r(r) = \left(-Q/2\pi H_e r\right) + \left(0.035A^2\left(\rho_p - \rho_l\right)d_p^2/\mu\right) \\ \times\left((R - r)^2/r\right) \qquad (24)$$

2.4.5 Estimation of particle residence time

For particles with resultant motion towards the chamber center and that are to be removed from the separator, residence time t_F cannot be shorter than the time they take to settle down on the device bottom t_s as given in Equation 4.

Taking into account Equation 21, relation 24 can be transformed into the form:

$$dr/dt = \left(N(R - r)^2/r\right) - M/r \qquad (25)$$

where:

$$N = 0.035A^2\left(\rho_p - \rho_l\right)d_p^2/\mu, \quad M = Q/2\pi H_e \qquad (26)$$

Integration of Equation 25 from the initial time $t = 0$ (particle in location $r = R$) to a chosen time (particle located between $r = R$ and $r = r_w$) yields:

$$t(r) = \frac{1}{2N}\ln\left|1 - \frac{(R - r)^2}{M/N}\right| + \frac{R}{2N\sqrt{M/N}}\ln\left|\frac{1 + \frac{R - r}{\sqrt{M/N}}}{1 - \frac{R - r}{\sqrt{M/N}}}\right| \qquad (27)$$

This relation allows to calculate the particle residence time inside the separator t_F for $r = r_w$. Under the action of the centrifugal force, particle residence time tF is longer than the residence time of the carrier liquid tk. As Equation 27 has a rather inconvenient form to be used in technical analysis, authors decided to make use of some simplifications. Both logarithmic terms of Equation 27 can be replaced by their expansions into power series according to:

$$\ln(1-e^2) = -\left(e^2 + \frac{1}{2}e^4 + \frac{1}{3}e^6 + ...\right) \quad (28)$$

$$\ln\left(\frac{1+e}{1-e}\right) = 2\left(e + \frac{1}{3}e^3 + \frac{1}{5}e^5 + ...\right) \quad (29)$$

that are valid for:

$$|e| = \left|\frac{R-r}{\sqrt{M/N}}\right| < 1 \quad (30)$$

When $e \ll 1$, the first terms of the power series are sufficient to replace the logarithmic functions in Equation 27. Some algebraic operations on Equations 25 and 30 give:

$$t(r) = \pi H_e (R^2 - r^2)/Q \quad (31)$$

For $r = r_w$ the time of particle motion given by Equation 31 is equal to the time of liquid flow t_k according to Equation 9. This means that the longer the suspension residence time in relation to the carrier liquid flow time (so the bigger the value of e in Eq. 29), the stronger the influence of the centrifugal force on vortex separator efficiency of operation. As such, parameter e can be used as an approximate, but a convenient algebraic measure of the centrifugal force influence on separator operation. In order to find the lower evaluation of parameter e, courses of two functions were analyzed:

$$\Delta e_1 = \ln(1-e^2) + e^2 \quad (32)$$

$$\Delta e_2 = \ln\left(\frac{1+e}{1-e}\right) - 2e \quad (33)$$

These functions describe the accuracy of transformation of Equation 27 into Equation 31. Figure 4 shows that the choice of $e = 0.30$ as the limit value was an arbitrary decision. However, bigger values of e result in the relative difference $\Delta e/e$ greater than 10%. This means that the carrier liquid residence time and the suspended particle residence time start to differ when the centrifugal force measurably influences the process of separation. Therefore, substituting Equation 26 into 30, authors proposed the following preliminary design criterion for vortex separators:

$$\left[Q(R-r_w)^2 \Delta\rho \, d_p^2 \, H_e/\mu \, H^{2/3} \, R^{8/3} \, d_{in}^{8/3}\right] > 0.3 \quad (34)$$

Relation 34 combines seven variables that characterize operation of a vortex separator (H, R, r_w, d_{in}, Q, d_p, r_p). Knowing the remaining six values, this equation can be used to calculate each one of the seven variables. As such, Equation 34 plays the role of a "closing design criterion" that is applied appropriately to a given problem.

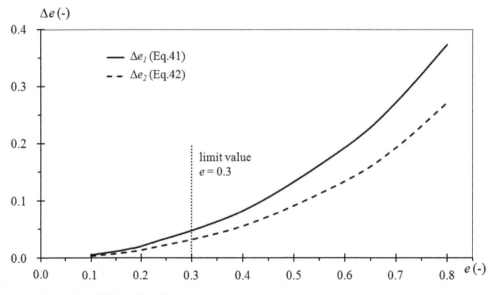

Figure 4. Estimation of limit values of parameter e.

3 RESULTS AND DISCUSSION

The combination of a significant number of variables makes the proposed design criterion difficult for direct analysis. In order to recognize its significance, Equation 34 should be simplified. By assuming the following proportions between variables: $R \gg r_w$; $\Delta\rho = 1700$ kg/m^3; $d_p = 0.20$ mm; $H_e = 0.50$ H; $d_{in} = 0.20$ H; $\mu = 0.001$ kg/ms; criterion becomes a relation that binds H [m], R [m] and Q [m^3/s] together (Fig. 5):

$$H < 2.45 Q^{3/7} R^{-2/7} \tag{35}$$

This relation firstly shows an increase in separator diameter requires a decrease in separator height, and vice versa. Secondly, circulative motion is generated by the energy of the incoming liquid stream tangent to the tank outer wall. If the cubic capacity of the device was too high, the energy of the incoming liquid could be insufficient to generate circulation. This means that vortex separators are rather unsuitable for larger wastewater treatment plants. However, they may successfully serve their purpose as local objects, e.g. being a part of smaller rainwater sewerage systems.

4 CONCLUSION

Producers of devices for local storm waste water treatment are interested in applying principles of centrifuges and cyclones to the technology of suspension removal. In order to answer their need, existing possibilities should be corrected and supplemented with rational methods—methods that are formally simple and sufficiently detailed from physical point of view.

The paper presents a relatively simple two-dimensional algebraic model of liquid flow in a vortex separator, that allowed to formulate equations of trajectories of characteristic suspended particles (Eqs. 18–21). Simplification of these equations yielded practical relations 24 and 27 that can be used to calculate particle velocity and its residence time within the system. General criterion for vortex separators design requires the times of flow and sedimentation to be equal (Eq. 4). Relation 34 can be used to evaluate whether the centrifugal force has a significant influence on the particle residence time.

Concluding, rational design process of a vortex separator can be conducted on three levels of precision:

– simplified level - preliminary calculations of the object characteristics basing on Equation 34;

– technical level - calculations of the object dimensions from conditions given by Equations 4 and 27—particle residence time takes the centrifugal force into account;
– comprehensive level - optional numerical simulations of suspension separation using Equations 18–21.

ACKNOWLEDGMENTS

This paper has been prepared as a part of the Scientific Project N N523 554738, financed by the Polish National Centre of Science.

REFERENCES

Andoh, R.Y.G. & Saul, A.J. 2003. The use of hydrodynamic vortex separators and screening systems to improve water quality. *Water Sci. and Techn.* 47(4): 175–183.

Dyakowski, T., Nowakowski, A.F., Kraipech, W. & Williams., R.A. 1999. A three dimensional simulation of hydrocyclone behaviour. *Proc. 2nd Int. Conf. on CFD in the Minerals and Process Industries, Melbourne, 6–8 December 1999.* CSIRO Australia.

Gronowska, M.A. & Sawicki, J.M. 2011. Study of rotational separators operation and design. In J.M. Sawicki & K. Weinerowska-Bords (eds), *Technical Progress in Sanitary Engineering; 12th Int. Symp. Water Management and Hydraulic Engineering, Gdansk, 5–8 September 2011.* Gdansk: Gdansk University of Technology Publishers.

Gronowska, M.A. & Sawicki, J.M. 2014. Discharge coefficient of shaft overfall for rotational separator (in Polish). *Maritime Eng. Geotech.* 35(1): 11–13.

Gronowska, M.A., Sawicki, J.M. & Zima, P. 2013. Motion of suspended particles in vortex separator. In A. Soltesz, D. Barokova, M. Orfanus & M. Holubec (eds), *Book of Abstracts; 13th Int. Symp. Water Management and Hydraulic Engineering, Bratislava, 9–12 September 2013.*

McGaughey, P.H. 1956. Theory of sedimentation. *J. AWWA* 48(4): 437–448.

Nalluri, C. & Featherstone, R.F. 2001. *Civil engineering hydraulics.* New Delhi: Wiley-Blackwell.

Rhodes, M. 2008. *Introduction to particle technology.* 2nd Edition, Chichester: John Wiley & Sons Ltd.

Sawicki, J.M. 2004. Aerated grit chambers hydraulic design equation. *J. Environ. Eng.* 130(9): 1050–1058.

Slattery, J.C. 1999. *Advanced transport phenomena.* Cambridge: Cambridge University Press.

Soo, L. 1966. *Fluid Dynamics of multiphase systems.* London: Blaisdell Publ. Comp.

Stairmand, C.J. 1951. The design and performance of cyclone separators. *Trans. Inst. Chem. Eng.* 29: 356–373.

Veerapen, J.P., Lowry, B.J. and Couturier, M.F. 2005. Design methodology for the swirl separator. *Aquacultural Eng.* 33: 21–45.

Environmental Engineering V – Pawłowska & Pawłowski (Eds)
© 2017 Taylor & Francis Group, London, ISBN 978-1-138-03163-0

Gaseous fuel production from biomass using gasification process with CO_2 emission reduction

S. Werle
Institute of Thermal Technology, Silesian University of Technology at Gliwice, Gliwice, Poland

ABSTRACT: The paper presents experimental results of the gasification process of the various types of biomass: *Miscanthus* x *giganteous*, sewage sludge and algae. Experiments were conducted on laboratory installation equipped with a fixed bed reactor. Combustible properties of the analyzed fuels were determined. Influence of the biomass type and gasification process parameters on the properties of the produced gas were analyzed. Experiments were performed for the wide range of the air ratio λ. It was equal to 0.12, 0.14, 0.16, 0.18, 0.23 and 0.27. Results show that gasification process parameters strongly influence the produced gas quality. For all analyses biomasses, there is an optimal value of the air ratio in which the produced gas is characterized by the highest values of the lower calorific value. The produced fuel is a lean gas which is a good source of the primary energy use in transformation processes into final form of energy.

Keywords: gasification, *Miscanthus* x *giganteous*, sewage sludge, algae, fixed bed reactor

1 INTRODUCTION

Thermal processes of the biomass utilization, understood as a renewable organic-rich material, become very popular in the world (Mc Namee, 2015). The main reason of this includes climate regulations on carbon dioxide emissions and limited reserves of fossil fuel (Pieńkowski 2012, Dasgupta 2011, Hoedl 2011, Pawłowski 2015, Lindzen 2015). Biomass is carbon-based and composed of a mixture of organic molecules containing hydrogen, usually including atoms of oxygen, often nitrogen and also small quantities of other atoms, including alkali and heavy metals. There are five main groups of biomass (Chinnici et al. 2015, Tripathi et al. 2016):

1. Waste biomass from industry (eg. waste from the food industry, etc.)
2. Human and animal waste (eg. manure, sewage sludge, cooked or uncooked food, paper, etc.)
3. Agricultural biomass (dedicated energy crops—eg. grass, various plants, etc.) and residues from agriculture cultivation (eg. straw, shells, etc.)
4. Woody biomass (eg. stems, branches, leaves, bark, lumps, chips of different trees, etc.)
5. Aquatic biomass (eg. algae, plants and microbes found in water, etc.)

Among presented biomass feedstock samples, sewage sludge, algae and dedicated energy crops seem to be the most perspective. The quantity of the sewage sludge production in the world is still increasing.

The main reason of that is the growth of the world population and implementation of the formal regulation in the field of the wastewater treatment and sewage sludge management. For example, in the European Union (EU) the implementation of the Urban Wastewater treatment directive 91/27/EEC led to 50% increase in sludge production by the year 2005, that is 10 million tons annually. Additionally, The EU Landfill Directive 99/13/EC obliges to reduce the biodegradable waste deposits to landfills. Both Directives cause a strong pressure to present new possible ways of sludge management (Werle and Wilk 2011).

Algae are characterized by a number of benefits in comparison to other source of biomass: (1) they grow fast, (2) consume CO_2 effectively, (3) do not compete with food production, (4) can be grown in both: fresh and sea water. Due to this fact, algae have been considered as one of the most promising alternative biomass sources.

Dedicated energy crops have many advantages, including reduced cost of fertilizer, high yield potential on land not suitable for annual crops (e.g. heavy metal contaminated sites, postindustrial landscapes) and broad adaptation to local environment condition (humidity, soil type etc.) (Pogrzeba et al. 2013). They can be able harvested with typical farm equipment and exhibit positive environmental attributes (e.g. improvement of the biodiversity). Among the species of interest, *Miscanthus* x *giganteus* is the most suitable. It gives the highest yield (up to 30 Mg dry mass/ha) of higher heating value equal to 18.5 MJ/kg dry mass).

Thermal processes of biomass utilization consist of combustion, pyrolysis, gasification and co-combustion (Werle 2011, Song et al. 2014). Gasification is the conversion of solid (or liquid) feedstock into useful and convenient gaseous fuel (or chemical feedstock) that can be burned to release energy or used for production of value-added chemicals. A typical gasification process may include the following steps: (i) drying, (ii) thermal decomposition or pyrolysis, (iii) partial combustion of some gases, vapors and char, (iv) gasification of decomposition products. Gasification requires a gasifying medium like steam, air or oxygen to rearrange the molecular structure of the feedstock in order to convert the solid feedstock into gases or liquids. The main reactions involved during the gasification of organic substances are summarized below (Werle 2013):

Boudouard: $C + CO_2 \rightarrow 2CO$ (a)

Water gas (primary): $CO + H_2O \rightarrow CO + H_2$ (b)

Water gas (secondary): $C + 2H_2O \leftrightarrow CO_2 + 2H_2$ (c)

Methanation: $C + 2H_2 \leftrightarrow CH_4$ (d)

Water gas shift: $CO + H_2O \leftrightarrow CO_2 + H_2$ (e)

Steam reforming: $CH_4 + H_2O \leftrightarrow CO + 3H_2$ (f)

Dry reforming: $CH_4 + CO_2 \leftrightarrow 2CO + 2H_2$ (g)

Gasification has attracted attention as one of the most efficient methods for utilizing biomass as CO_2 emission has become an important global issue. Gasification has several advantages over a traditional combustion process. As a consequence of the reducing atmosphere, gasification prevents emissions of sulfur and nitrogen oxides, heavy metals and the potential production of chlorinated dibenzodioxins and dibenzofurans. A smaller volume of gas is produced compared to the volume of flue gas from combustion, because gasification is characterized by an environment containing low levels of the gasification agent. Due to the reducing conditions used for gasification, most of sulfur, nitrogen, chloride and fluoride in biomass samples may be released as H_2S, NH_3, HCl and HF. The presence of these compounds is undesirable as they may be converted into the respective oxides during gas utilization. Therefore, their formation should be monitored and controlled. Biomass can also contain relevant quantities of heavy metals. Some of them may be volatilized to the gas phase at high temperature while other elements may be retained in the solid residue, trapping some of sulfur, nitrogen and chloride introduced by the feedstock (Ptasiński et al. 2007).

2 MATERIALS AND METHODS

2.1 *Materials*

Three samples of the analysed biomasses are presented in Figures 1–3.

The *Laminaria Hyperborea* algae was used in the study. Algae come from Scotland, from Easdale Sound. Easdale Sound is a narrow channel that separates the island of Easdale Island Sail. Algae were brought into powder form. In such form, they are not suitable for use in a fixed bed gasification reactor, and therefore, pellets have been prepared prior to the gasification (Fig. 3). The experimental plot with *Miscanthus* x *giganteus* was located on contaminated agricultural soil in Bytom (south of

Figure 1. Granulated sewage sludge.

Figure 2. Miscanthus x giganteus sample.

Figure 3. Algae (*Laminaria Hyperborea*) pellets.

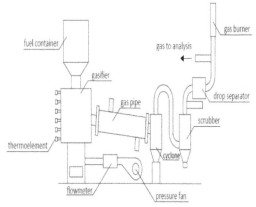

Figure 4. Schematic diagram of the experimental system.

Table 1. Properties of the fuels tested.

Fuel	Sewage sludge	*Miscanthus* × *giganteus*	Algae (*Laminaria Hyperborea*)
% (as received)			
Moisture	5.3	7.6	5.0
Volatile matter	51.5	75.4	47.0
Ash	36.5	1.36	24.7
% (dry)			
C (dry)	31.79	46.6	31.83
H (dry)	4.36	7.16	4.82
O (dry)	20.567	44.73	34.082
N (dry)	4.88	0.16	2.58
S (dry)	1.67	1.35	1.65
LHV, MJ/kg (dry)	12.96	19.45	11.80

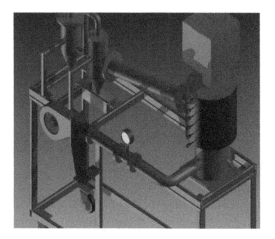

Figure 5. Experimental system visualization.

Poland, Silesia Region). Plots were established in the vicinity of a close-down Pb/Zn/Cd ore mining and processing plant. Before the experiment, the plant has been cut to the required size (Fig. 2).

The analyzed sewage sludge came from a Wastewater Treatment Plant operating in the mechanical and biological system. The sludge produced at the plants was subject to fermentation and then, after being dehydrated, dried in a cylindrical drier on shelves heated up to 260°C. As a result, sludge formed granules (Fig. 3).

The main properties of the analysed fuel are presented in Table 1.

2.2 Apparatus and procedure

The scheme and visualization of the experimental installation (Patent no. P-397225) is presented in Figures 4 and 5. Fuel is fed into reactor from the top (fuel container), while the air was supplied by a pressure fan from the bottom. The temperature of the gasifier interior is measured by six N-type thermocouples located along the vertical axis of the reactor at different six heights: $t_1 = 10$ mm, $t_2 = 60$ mm, $t_3 = 110$ mm, $t_4 = 160$ mm, $t_5 = 210$ mm and $t_6 = 260$ mm above the bottom. The air flow rate supplied into the gasifier is measured by a flow meter. Gasification gas is transported by the pipe. At the outlet of the installation, there is a syngas sampling point where the gas sample is collected and then supplied to CO and H_2 analyzers. The produced fuel is cleaned by a cyclone, scrubber and drop separator. The molar fractions of the main combustible species are measured online at the experimental stand.

The experimental procedure started by turning on the gasifier. The blower was switched on and fuel was placed inside. After approximately 2 hours, the reactor was heated and the experimental measurements started. The air flow rate was adjusted to ensure a specified air ratio. Once syngas production began, the measurements of key variables were taken. During the experiments, the samples were taken twice from the outlet of the gasifier and the outlet of the water scrubber in order to analyze the produced wet gas.

Uncertainties in the measurements of temperature, air mass flow and syngas composition were determined by eq. (1). The total uncertainty in the measurements was determined by using the experimental data and eqs. (1) and (2):

$$\sigma_{x_i}^2 = B_{x_i}^2 + P_{x_i}^2 \quad (1)$$

$$\sigma_r^2 = \sum_{i=1}^{j}\left(\frac{\partial r}{\partial x_i}\right)^2 \cdot \sigma_{x_i}^2 \quad (2)$$

where $r(x_1, x_2, ..., x_j)$ is a function of j measured variables, x_i, σ_r and σ_{xi} are the uncertainties in r and x_i, respectively; and B_{xi} and P_{xi} are the systematic uncertainty and the random uncertainty in the variables x_i, respectively.

The interior temperatures of the reactor were measured by N-type thermocouples to an accuracy of ±0.75% over the measured temperature range. The uncertainty arising from the gasifier operation was assumed to be ±1.00%, and the total uncertainty in the temperature measurements was estimated to be ±1.25% using eq. (1). The air mass flow rate was measured by a flow meter with a full scale value of 4.0 kg/h, a least count of 0.1 kg/h and a specified accuracy of ±2.5%; the total uncertainty in the mass air flow rate into the gasifier was calculated to be ±2.5% from eq. (1). The air ratio λ was calculated using two independent variables (the air mass flow rate and the fuel flow rate) and the minimum theoretical air needed for complete combustion (this variable was calculated on the basis of the fuel composition for which the uncertainty was equal to 0). The uncertainty in λ was equal to 3.50%. The total uncertainty in the gasification gas composition, as measured on-line by the ABB gas analyzer, was equal to ±2.50%. Variable area mass flow meters with an accuracy of +/–3% were used to measure the air/syngas flow. The temperature was measured using N-type thermocouples and an Agilent temperature recorder with an instrument uncertainty of +/–1.5°C. The gas composition was measured by a gas analyzer that was calibrated using known gas mixtures starting from the 72 h to ensure accurate readings.

3 RESULTS

Figure 6 presents the temperature values inside the gasification reactor at six different points located above the gasifier bottom. While analyzing the presented results it can be concluded that

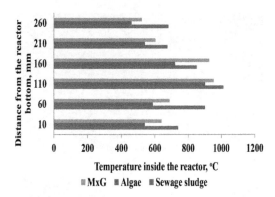

Figure 6. Temperature profiles inside the gasification reactor.

the temperature at T_3 was always the highest temperature in the reactor; thus, T_3 may have been located in the partial oxidation zone, which would have been the hottest area in the fixed bed gasifier. Correspondingly, the other monitoring sites may have been located as follows: T_6 and T_5 in the drying zone, T_4 in the pyrolysis zone, T_2 in the oxidation (combustion) zone and T_1 in the ash zone. It can also be concluded that the sewage sludge gasification process is characterized by the highest value of the measured temperature. This can be explained by the lowest concentration of the oxygen in the sewage sludge in comparison to the rest of analysed fuels.

The thermochemical processes of gasification occur in three major steps. The first step is associated with the initial decomposition of biomass via the thermochemical processes that produce tar, volatiles and char residues. The second step involves reacting the volatiles. The last step comprises the heterogeneous reactions of the remaining carbonaceous residues with the gaseous producer gas and the homogeneous reactions of carbon monoxide, carbon dioxide, hydrogen, vapor and hydrocarbon gases. Figure 7 presents the Lower Heating Value (LHV) of the obtained gas on the air ratio for the biomass samples that were investigated. The lower heating value (LHV) in MJ/m^3_n of the syngas was estimated using the formula given below (Kim et al. 2011).

$$LHV = 0.126 \cdot CO + 0.108 \cdot H_2 + 0.358 \cdot CH_4 \quad (3)$$

Analyzing that figure, it can be concluded that there is an optimal value of air ratio, in which the lower heating value of the obtained gas reaches its maximum. As expected, an increase in the air flow rate caused a decrease in the heating value. A greater amount of oxidizer increases the amounts of noncombustible species and the volumetric frac-

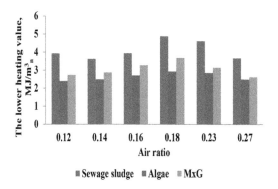

Figure 7. The lower heating value of the gasification gas as a function of the air ratio.

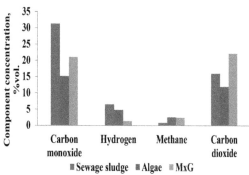

Figure 8. The main component concentration in gasification gas for air ratio 0.18.

tion of nitrogen, thus decreasing the heating value of the obtained gas.

Rapid growth of LHV observed in the value of the air ratio equal to 0.18 is caused by the dominant role of the primary water gas reaction. The reactions that can occur in the gasifier as a result of the gasification agent flow can be categorized as the reaction of gasification agent and carbon in the fuel and the reaction of gasification agent and CO in the gas. The reaction of gasification air and carbon is an endothermic reaction that generates mainly CO, whereas the reaction of gasification air and CO is an exothermic reaction that generates mainly CO_2 (and H_2). When gasification air is fed into the reactor with the fuel, the endothermic reaction of air and carbon occurs first (e.g. primary water gas reaction $C+H_2O \rightarrow CO+H_2$), and the CO in a gaseous state produced from the fuel reacts with the residuals causing other reactions (e.g. water gas shift $CO+H_2O \leftrightarrow CO_2+H_2$). Thus, the composition of H_2, CO and CO_2 (and simultaneously LHV) in the gasification gas changes according to the amount of the air supplied to the reactor.

Analyzing the LHV values, it should be emphasized that regardless of the analyzed feedstock, gasification process gives the opportunity to obtain valuable gaseous fuel. Such fuel can be effectively utilized in boiler, gas turbines or engines (Przybyła et al 2008).

Figure 8 presents the comparison of the main combustible components obtained during the gasification of the analysed feedstock for air ratio equal to 0.18. It can be concluded that sewage sludge is characterized by the highest value of the CO and H_2.

CO_2 shows an inverse relation with CO as the reactions that produce those gases are competing for the same reactants, namely carbon. The concentration of carbon dioxide is generally expected to be a minimum of the optimal air ratio value.

4 CONCLUSIONS

The main conclusions from the study are as follows: (i) Original experimental results on different type of the biomass feedstock gasification are presented in this study. Air gasification of waste, conventional and unconventional biomass were investigated, and a fixed bed reactor was used. (ii) The operating conditions (amount of the gasification agent) of the gasification process greatly influence the syngas composition distribution. (iii) Higher values of the main components (especially C and H) in the sewage sludge plant affect on the increase of the LHV of gasification gas. (iv) Throughout the range of analysed air ratio ($\lambda = 0.12–0.27$), the volumetric fraction of main combustible components of gasification gas (CO and H_2) is higher in the case of the sewage sludge in comparison to other analyzed feedstock. (v) Taking into consideration the lower heating value LHV of the gasification gas, the optimum value of the air ratio in which the LHV takes its maximum value is equal to 0.18. It is irrespective of the biomass type. (vi) The yield of the main producer gas components, CO, H_2 and CH_4, was enhanced by increasing the gasification agent temperature and increasing of the oxygen concentration in the gasification agent.

ACKNOWLEDGMENTS

The paper has been prepared within the frame of the statute research of the Institute of Thermal Technology, Silesian University of Technology and the FP7-People-2013-IAPP (GA No610797) Phyto2Energy Project.

REFERENCES

Chinnici, G., D'Amico, M., Rizzo, A. & Pecorino, B. 2015. Analysis of biomass availability for energy use in Sicily; *Renewable and sustainable Energy Reviews*, 52: 1025–1030.

Dasgupta, P. & Taneja, N. 2011. Low Carbon Growth: An Indian Perspective on Sustainability and Technology Transfer. *Problemy Ekorozwoju—Problems of Sustainable Development*, 6(1): 65–74.

Hoedl, E., 2011. Europe 2020 Strategy and European Recovery. Problemy Ekorozwoju/*Problems of Sustainable Development*, 6(2): 11–18.

Kim, J.W., Mun, T.Y., Kim, J.O. & Kim, J.S. 2011. Air gasification of mixed plastic wastes using a two-stage gasifier for the production of producer gas with low tar and high caloric value. *Fuel*, 90: 2266–2272

Lindzen R. 2010. Global Warming: The origin and nature of the alleged scientific consensus? *Problemy Ekorozwoju/Problems of Sustainable Development*, 5(2), 13–28.

McNamee, P., Darvell, L.I., Jones, J.M. & Williams, A. 2015. The combustion characteristics of high-heating rate chars from untreated and torrefied biomass fuels. *Biomass&Bioenergy*, 82: 63–72

Patent no. P-397225; 2nd of December 2011, Biomass gasification installation particular for sewage sludge.

Pawłowski, L. 2015. Where is the World Heading? Social Crisis Created by Promotion of Biofuels and Nowadays Liberal Capitalism. *Rocznik Ochrona Środowiska/ Annual Set the Environment Protection*, 17(1), 26–39.

Pieńkowski, D. 2012. The Jevons Effect and the Consumption of Energy in the European Union. *Problemy Ekorozwoju/Problems of Sustainable Development*, 7 (1): 105–116.

Pogrzeba M., Krzyżak J. & Sas-Nowosielska A. 2013. Environmental hazards related to *Miscanthus* x *giganteus* cultivation on heavy metal contaminated soil. E3S Web of Conferences 1, 29006.

Przybyla G. & Ziolkowski L. 2008. Analysis of Energy Conversion Process of SI Engine Fuelled with LCV Gas. *Proceedings of the ninth Asia-Pacific international symposium on combustion and energy utilization* Beijing, 113–117

Ptasinski K.J., Prins M.J. & Pierik A. 2007. Exergetic evaluation of biomass gasification. *Energy, 32*,: 568–574.

Song C., Pawłowski A., Ji A., Shan S. & Cao Y. 2014,: Catalytic pyrolysis of rice straw and product analysis. *Environmental Protection Engineering*, 1: 35–43.

Tripathi M., Sahu J.N. & Ganesan P. 2016. Effect of process parameters on production of biochar from biomass waste through pyrolysis: A review. *Renewable and sustainable Energy Reviews*, 55: 467–481.

Werle S. & Wilk R.K. 2011. Analysis of use a sewage sludge derived syngas in the gas industry. *Rynek energii*, 5,: 23–27

Werle S. 2011. Estimation of reburning potential of syngas from sewage sludge gasification process, *Chemical and Process Engineering*, 32: 411–421.

Werle S. 2013. Sewage sludge gasification: theoretical and experimental investigation. *Environmental Protection Engineering*, 2: 25–32.

Environmental Engineering V – Pawłowska & Pawłowski (Eds)
© 2017 Taylor & Francis Group, London, ISBN 978-1-138-03163-0

Methane emissions and the possibility of its mitigation

A. Czechowska-Kosacka & W. Cel
Faculty of Environmental Engineering, Lublin, University of Technology, Lublin, Poland

ABSTRACT: The work characterized the natural and anthropogenic sources of methane, as well as the phases of its generation on a landfill. Two methods of degassing, i.e. passive and active, were described. As it is impossible to directly utilize landfill gas due to its pollution, the steps aiming its treatment—initial and advanced—were discussed as well. Using landfill gas for energy production is the most favourable solution. It contributes to elimination of methane emissions and acquisition of an additional source of energy. Sometimes, the industrial methods of utilizing landfill gas are extremely expensive, both in regard to investment and operation. However, oxidation of methane in soil carried out by methanotrophic bacteria may constitute an alternative solution.

Keywords: methane, landfill gas, energy, treatment, collection

1 INTRODUCTION

The climatic changes caused by the emission of greenhouse gases will lead to serious disturbances in the Earth ecosystems and, consequently, to social unrest which could threaten the realization of intergenerational equity, one of the essential paradigms of sustainable development (Cholewa and Pawlowski, 2009; Pawlowski 2008, 2009 a i b, 2013; Udo et. al. 2010, 2011; Mulia et al., 2016; Sztumski, 2016). It should be noted that Lindzen (2016), an outstanding American climatologist, believes that the climatic changes will not be as disastrous as it is being forecast. Nevertheless, it seems that actions aiming at mitigation of greenhouse gases emission should be taken. Although the role of CO_2 in generating the greenhouse effect is widely discussed, much smaller attention is drawn to the role of methane. Meanwhile, methane is the second greenhouse gas, which corresponds to 20.7% of the greenhouse effect. The share of CO_2 equals 71.6%, whereas nitrous oxide amounts to 6.2% (IPCC-2007; Foster et al. 2007).

The greatest amounts of methane are emitted from natural sources (see Table 1).

The characteristic of anthropogenic sources was presented in Tables 2 and 3. It is forecast (EPA 2012) that the emission of methane to the atmosphere will continue to rise (EPA, 2012, JRC, 2014).

From the point of view of environmental protection, the emission from landfills is the relevant one.

Table 1. Natural methane sources (Kirsche et al., 2013).

Source	Emission Tg/yr
Wetlands	140–280
Geological (including oceans)	30–75
Freshwater	10–70
Wild animals	15
Termites	2–20
Hydrates	1–10
Wildfires	1–5
Permafrost	0–1
Total	199–461

Table 2. Anthropogenic methane sources (Kirsche et al., 2013).

Source	Emission Tg/yr
Fossil fuels industry	85–105
Domestic ruminants	85–95
Waste decomposition	65–95
Landfills	33–44
Biomass burning and biofuels	30–40
Rye cultivation	30–40
Total	295–375

2 METHANE GENERATION IN LANDFILLS

Methane in landfill is produced through the anaerobic decomposition of organic wastes.

Table 3. Global CH_4 emissions by sector as MteqCO$_2$ (Cofala et al. 2006).

Source Category	Sector	1990	2005	2030
Energy	Natural Gas and Oil Systems	1,278.3	1,542.7	1,112.9
	Coal Mining and Activities	529.8	521.6	784.3
	Stationary and Mobile Combustion	221.3	224.3	362.9
	Biomass Combustion	176.3	198.0	230.4
	Other Energy Sources	0.5	0.5	0.5
	Total	2,206.2	2,487.1	2,491.0
Industrial Processes	Other Industrial Processes Sources	7.7	7.5	6.3
	Total	7.7	7.5	6.3
Agriculture	Enteric Fermentation	1,763.9	1,894.3	2,320.5
	Rice Cultivation	480.0	500.9	510.4
	Manure Management	232.7	219.2	252.7
	Other Agricultural Sources	506.6	421.0	421.0
	Total	2,983.2	3 035.4	3,504.6
Waste	Landfilling of Solid Waste	706.1	794.0	959.4
	Wastewater	351.9	476.7	608.8
	Other Waste Sources	13.4	15.2	15.5
	Total	1,071.4	1,285.9	1,583.7
Total		6,268.5	6,815.9	7,585.6

A landfill, due to the composition of deposited waste and occurring processes, can be treated as a bioreactor.

In the surface layer of a landfill, the main processes include biochemical organic matter decomposition in aerobic conditions. On the other hand, in the deeper layers of a landfill, the anaerobic processes—which produce biogas mainly composed of methane and carbon dioxide—become prevalent. The biogas also contains oxygen, nitrogen and trace amounts of organic compounds decomposition products, e.g. hydrogen sulphide, acetaldehyde, ammonia, and ethyl mercaptan.

The process of waste fermentation occurs in five phases which eventually lead to the production of biogas:

I. aerobic
II. acetogenesis
III. unsteady methanogenesis
IV. steady methanogenesis
V. maturation with gradual fading of methanogenesis.

The individual biogas production phases last for different periods of time and are connected with varied waste composition (see Fig. 1)

In the first, i.e. aerobic phase, the aerobic microorganisms decompose part of the organic substance to CO_2 and H_2O in the presence of oxygen. When the oxygen contained in waste is depleted, the anaerobic phase—acetogenesis—follows. With the help of anaerobic organisms, carbohydrates are decomposed, while CO_2 and H_2 produced. In Phase II, the share of N_2 is signifi-

cantly reduced. After approximately 10–50 days, Phase III—unsteady methanogenesis—begins. In Phase III, the methanogenic microorganisms slowly begin processing the fatty acids which were produced earlier into CH_4, H_2O, and CO_2. After about 180–500 days from the beginning of the process, the system stabilizes and Phase IV begins, i.e. steady methanogenesis. During this phase, the production of biogas is observed. In Phase V, the bioprocess ceases (Allen et al. 1997, Galle et al. 2001, Ocieczek & Mniszek 2010, Bialowiec 2011, Bo-Feng et al. 2014, Cai et al. 2014).

The calorific value of landfill gas is related to the methane content, amounting to 20 000 kJ/m^3 on average. The explosivity of landfill gas is connected to the methane content as well. Methane is a combustible gas and forms an explosive mixture with air in 5–15% concentration range. As opposed to other types of biogas, the landfill gas is characterized by a great amount of trace substances, which is illustrated by Table 4 (Parker et al. 2002, Gardziuk 2003, Pawlowska et al. 2008, Staszewska et al. 2011, Klimek 2012).

3 COLLECTION AND TREATMENT OF LANDFILL GAS

The gas produced at each landfill is released into the atmosphere in a natural manner, thus threatening the natural environment. One can distinguish between two methods of degassing: passive and active (see Table 5).

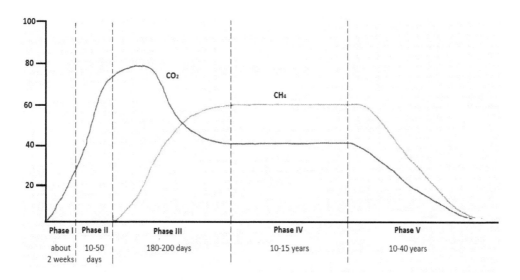

Figure 1. Change in CH_4 and CO_2 content overtime in landfill (Galle et al. 2001, Ocieczek & Mniszek 2010, Kerfoot et al. 2013, Zhu et al. 2013).

Table 4. Composition of biogas produced in municipal waste landfills (Eklund et al. 1998, Ocieczek & Mniszek 2010).

Biogas component	Mean value [%]	Mean values [ppm]
Methane	45	n/a
Carbon dioxide	35	n/a
Nitrogen	15	n/a
Hydrogen	1	n/a
Oxygen	1	n/a
Argon	0.1	n/a
Benzene	0.02	0–15
Toluene	0.02	0–15
Xylene	0.02	0–15
Acetaldehyde	0.015	0–150
Ethyl mercaptan	0.012	0–120
Acetone	0.01	0–100
Hydrogen sulphide	0.01	0–100
Chloro-organic compounds	0.01	0–100
Ammonia	0.01	0–100
Hydrocarbons	0.005	0–50
Ethane	0.003	0–30
Total sulphur	0.003	n/a
Total chlorine	0.002	n/a

The passive degassing system is found in 30.5% of landfills, while the active system was installed in 2.4% of landfills. Mostly non-processed landfill gas is released into the atmosphere.

Passive degassing is the most commonly employed solution. It involves construction of elements, to which the gas flows under its own pressure and is subsequently released into the atmosphere. Such installation does not prevent uncontrolled migration of gas inside and outside the landfill. Part of the gas will flow into the well as a result of pressure differences of landfill biogas, which is higher than the atmospheric pressure, and due to diffusion. This process does not ensure collection of the entire volume of gas generated at a landfill or the control over its emission.

In active degassing systems, gas is pumped from wells. This technology involves linking gas wells by means of gas pipes equipped with sucking, regulating, and pumping modules. Landfill gas is sucked from a landfill owing to a slight vacuum created by this module. In order to prevent the release of landfill gas into the atmosphere, it is necessary to seal the landfill. Gas obtained in this way constitutes a very good fuel used for the production of electric and heat energy (Shin et al. 2005, Sołwiej 2005, Tsai 2007, Solowiej & Neugebauer 2008).

4 TREATMENT OF LANDFILL GAS

Direct use of landfill gas is impossible due to its pollution. Therefore, the gas should undergo treatment, which involves two steps:

I. Initial treatment, consisting in removal of solid and liquid components and drying of gas in the final phase,
II. Advanced treatment involving desulphurization, removal of silicon compounds and other gaseous pollutants, e.g. hydrogen sulphide, ammonia, etc.

Table 5. Landfill degassing.

Source	Passive degassing	Active degassing with energy recovery	Active degassing without energy recovery	No degassing
GIOŚ Report (2010)	42%	n/a	n/a	58
MŚ (2009)	30.5%	1.7	0.7	56.7
GUS, 2008 (Sieja, 2010)	31.5%	5.1	2.1	61.3
GUS, 2004 (Sieja, 2010)	15.8%	3.1	0.9	80.2
GUS, 2000 (Sieja, 2010)	7.6%	1.3	0.7	90.4

Simple physical processes are employed in the initial treatment, resulting in removal or reduction of main pollutants: water (vapour) and solid particles.

The presence of water in the biogas pipeline negatively impacts the efficiency of the system. It causes, among others, reduction of the inside diameter of pipeline, thus increasing the pressure losses in the transit pipeline. Moreover, the nature of the two-phase flow is unstable and may cause vibrations that contribute to the unstable operation of the system. Equally important are the pollutants found in water which may precipitate on pipe walls and cause further pressure losses. Depending on the source of gas and its projected use, the treatment involves the removal of: liquid drops, gas-liquid foam, and non-condensed water vapour.

Dewatering systems, which are integrated with the landfill gas reception system, are used for the removal of liquid drops. In order to prevent the liquid water from migrating in the installation, additional security measures are employed, e.g. drains or wells. Another commonly used solution is a condensate trap, which constitutes the final element of dewatering system and aims to reduce the gas flow rate to the point where precipitation of liquid is possible. This part is found near the gas compressor inlet.

Removal of gas-liquid foam is possible in coalescing filters which are placed in gas pipelines put into a condensate trap and its outlet. These filters cause the foam to sink and prevent its further movement in the system.

Typical processes of initial treatment of uncondensed vapour do not require using special devices for lowering the biogas temperature. Cooling the gas by employing 5–10 meter-long steel pipes with anti-corrosive coating is one of the methods of removing water vapour. It involves achieving the gas temperature equal or lower to the water vapour dew point. This solution causes condensation of water vapour on radiator walls. A drop in the temperature of biogas in the pipelines than supply biogas to power-producing devices is an adverse effect. Therefore, it is necessary to control the temperature in the pipeline.

Advanced treatment constitutes the second step, which enables to obtain gas of higher purity. The processes include physical and chemical treatments which can be jointly subjected to advanced treatment processes.

Removal of hydrogen sulphide from landfill biogas may be conducted through biological, (Zdeb et al. 2009), chemical or physical and the choice depends on the gas composition, purity and volumetric flow rate (see Table 6).

The basic methods for the removal of siloxanes include: absorption on activated carbon, absorption in a liquid mixture of hydrocarbons and cooling of gas with water removal, which eliminates 99% of siloxanes. In a landfill, siloxanes with low molecular mass oxidize and infiltrate into biogas. Afterwards, during energy generation through incineration, they accumulate on the elements connected with the incineration process and/or exhaust.

Remaining biogas pollutants, including ammonia or aromatic hydrocarbons, are present in landfill biogas in low concentrations; thus, it is possible to utilize biogas without the necessity of employing additional treatment systems.

5 MITIGATION OF LANDFILL METHANE EMISSIONS

Natural mechanisms of removing methane from the atmosphere are found in Earth's ecosystems. The greatest amounts (450–620 Tg/yr) of methane are removed in the troposphere through its oxida-

Table 6. Comparison of various hydrogen sulphide removal methods (Shin et al. 2002, Takuwa et al. 2009).

Method	Efficiency	Investment costs
Biological desulphurization	medium	medium
Adding iron (III) chloride	medium	low
Absorption on activated carbon	high	high
Rinsing with water	high	high
Absorption on iron (iii) hydroxide or oxide	high	medium
Absorption on sodium hydroxide	high	medium

tion with OH radicals or chlorine (15–40 Tg/yr). Large amounts of methane are also removed in the stratosphere (15–85 Tg/yr). The uptake of methane by soil (10–45 Tg/yr) is an interesting phenomenon. It involves methanotrofic microorganisms oxidizing methane to carbon dioxide (Kirsche et al. 2013). The most favourable solution is utilizing landfill gas for generation of useful energy, mostly for generation of electricity. It enables achieving two goals simultaneously, i.e. elimination of emissions to air and obtaining an additional source of energy. This requires installation of a landfill gas collection system (see Fig. 2). Before it is used, the collected landfill gas has to be treated and all water, as well as hydrogen sulphide have to be removed (Popov 2005, Pazna et al. 2010, Ajhar et al. 2010, Bergersen et al. 2014, Lantela et al. 2012). Treated landfill gas is most often used for generation of electricity (Dudek et al. 1997, Bova et al. 2006, Musiał & Zajemska 2014, Zalewska-Bartosz 2014, Ahmedet et al. 2015).

Gas for energy generation can be collected during Phase IV, i.e. when its emission is sufficiently high. Due to the decreasing emission, it is impractical to collect methane for energy generation during phase V. It should be noted that attempts were made to intensify methane production by introducing sewage sludge to a landfill (Pawłowska & Siepak 2006, Sanphtoti et al. 2006, Fytili & Zabaniotou 2008, Hao et al. 2008). However, this method can only be applied in certain cases.

Therefore, residual gas emission—which may last for dozens of years—remains (phase V). The industrial methods that could be employed are relatively expensive, both in regard to investment and operation. Hence, the methane oxidation process that is performed by methanotrophic bacteria in soil seems best suited for this purpose (Czepiel et al. 1996, Stepniewski et al. 2006, Abushammala et al. 2014, Sadasivam et al. 2014, Pawłowska et al. 2006, Montusiewicz et al. 2008).

The simplest and cheapest method of eliminating residual methane emission (from phase V) is utilizing a suitable biocover (Streese et al. 2003, Gebert et al. 2003, Wang et al. 2011, 2012, He et al. 2012).

This method enables to remove residual methane for many years without the need for maintenance.

More advanced methods include utilizing biofilters (Hanbricgs et al. 2006, Dever et al. 2011, Caceres et al. 2016, Brandt et al. 2016, Gamez-Cuervo et al. 2016).

Employing one of the above-mentioned methods ensures a virtually complete elimination of methane emissions from landfills.

REFERENCES

Abushammala, M. F. M., Basri, N. E. A., Irwan D. & Younes M. K., 2014. Methane oxidation in landfill cover soils: A review. *Asian Journal of Atmospheric Environment*, 8–1, doi: http://dx.doi.org/10.5572/ajae.2014.8.1.001.

Ahmed, S. I., Johari, A., Hashim, H., Lim, J. S., Jusoh, M, Mat, R. & Alkali, H. 2015. Economic and environmental evaluation of landfill gas utilisation: A multiperiod optimisation approach for low carbon regions. *International Biodeterioration & Biodegradation*, 102: 191–201.

Ajhar, M., Travesset, M., Yüce, S., Melin, T., 2010. Siloxane removal from landfill and digester gas—A technology overview. *Bioresource Technology*, 101(9): 2913–2923.

Allen M. R., Braithwaite, A. & Hills C. C., 1997. Trace organic compounds in landfill gas at seven U.K. waste disposal sites. *Environmental Science & Technology*, 31(4): 1054–1061.

Bergersen, O. & Haarstad, K., 2014. Treating landfill gas hydrogen sulphide with mineral wool waste (MMW) and rod mill waste (RMW). *Waste Management*, 34(1): 141–147.

Bialowiec, A. 2011. Hazardous emissions from municipal solid waste landfills. *Contemporary Problems of Management and Environmental Protection*. 9: 7–28.

Bove, R. & Lunghi, P., 2006. Electric power generation from landfill gas using traditional and innovative technologies. Energy Conversion and Menagement. 47(11): 1391–1401.

Brandt, E. M. F., Duarte F. V., Vieira J. P. R., Melo V. M., Souza C. L., Araújo J.C. & Chernicharo C. A. L., 2016. The use of novel packing material for improving methane oxidation in biofilters. *Journal of Environmental Management*, 182: 412–420.

Cai, B. F., Liu, J. G., Gao, Q. X., Nie, X. Q., Cao D., Liu L. C., Zhou Y. & Zhang Z. S., 2014. Estimation of methane emissions from municipal solid waste landfills in China based on point emission sources. *Advances in Climate Change Research*, 5(2): 81–91.

Cáceres, M., Dorado, A. D., Gentina, J. C. & Aroca G., 2016. Oxidation of methane in biotrickling filters inoculated with methanotrophic bacteria. *Environ Sci-Pollut Res Int.*, DOI: 10.1007/s11356–016–7133-z.

Cholewa, T. & Pawlowski A. 2009. Sustainable use energy in the communal sector. *Rocznik Ochrona Środowiska*, 11: 1165–1177.

Cofala, J., Amann, M., Klimont, Z., Kupiainen, K. & Höglund-Isaksson L., 2006. Scenarios of global anthropogenic emissions of air pollutants and methane until 2030. Final Report for WP3. International Institute for Applied Systems Analysis.

Czepiel, P. M., Mosher, B., Crill, P. M., Harriss R. C., 1996. Quantifying the effect of oxidation on landfill methane emissions. *Journal of Geophysical Research*: *Atmospheres*, 101(D11): 16721–16729.

Dever, S. A., Swarbrick, G. E. & Stuetz, R. M., 2007. Passive drainage and biofiltration of landfill gas: Australian field trial. *Waste Management*, 27(2): 277–286.

Dudek, J. et al., 1997. Wykorzystanie biogazu ze składowisk odpadów komunalnych do celów energetycznych. *Opracowanie Instytutu Górnictwa Naftowego i Gazownictwa*, Kraków.

Eklund, B., Anderson, E. P., Walker, B. L. & Burrows, D. B., 1998. Characterization of landfill gas composition at the fresh kills municipal solid-waste landfill. *Environ. Sci. Technol.*, 32(15): 2233–2237.

EPA, 2012. Summary Report: Global Anthropogenic Non-CO$_2$ GreenhouseGasEmissions: 1990–2030. EPA 430-S-12–002.

Forster, P., Ramaswamy, V., Artaxo, P., Berntsen T., Betts R., Fahey D. W., Haywood J., Lean J., Lowe, D. C., Myhre G., Nganga J., Prinn R., Raga G., Schulz M. & Van Dorland R., 2007. Changes in atmospheric constituents and in radiative forcing. Climate Change 2007. The Physical Science Basis, http://www.ipcc.ch/pdf/assessment-report/ar4/wg1/ar4-wg1-chapter2.pdf.

Fytili, D. & Zabaniotou, A., 2008. Utilization of sewage sludge in EU application of old and new methods—A review. Renewable and Sustainable Energy Reviews, 12(1): 116–140.

Galle, B., Samuelsson, J., Svensson, B. H. & Börjesson, G., 2001. Measurements of methane emissions from landfills using a time correlation tracer method based on FTIR absorption spectroscopy. Environmental Science & Technology, 35(1): 21–25.

Gebert, J., Groengroeft, A. & Miehlich G., 2003. Kinetics of microbial landfill methane oxidation in biofilters. Waste Management, 23(7): 609–619.

Gómez-Cuervo, S., Hernández, J. & Omil, F., 2016. Identifying the limitations of conventional biofiltration of diffuse methane emissions at long-term operation. Environmental Technology, 37(15):1947–1958.

Hao, X., Yang, H. & Zhang, G., 2008. Trigeneration: A new way for landfill gas utilization and its feasibility in Hong Kong. Energy Policy, 36(10): 3662–3673.

Haubrichs, R. & Widmann, R., 2006. Evaluation of aerated biofilter systems for microbial methane oxidation of poor landfill gas. Waste Management, 26(4): 408–416.

He, R., Wang, J., Xia, F. F., Mao, L. J. & Shen D. S., 2012. Evaluation of methane oxidation activity in waste bio-cover soil during landfill stabilization. Chemosphere, 89(6): 672–679.

IPCC, 2007, C|limate Change 2007, Working Group I: The Physical Science Basis

JRC, 2014. Trends in global CO2 emissions, 2014 Report.

Kerfoot, H. B., Hagedornb, B. & Verwiel, M., 2013. Evaluation of the age of landfill gas methane in landfill gas–natural gas mixtures using co-occurring constituents. Environmental Science: Processes & Impacts, 15, 1153–1161.

Kirschke, S. et al., 2013. Global Methane Budget 2013. Three decades of global methane sources and sinks. Nature Geoscience 6: 813–823.

Klimek, P., 2012. Landfill gas to energy projects in Poland update. Global Methane Initiative Agriculture, Municipal Solid Waste, Wastewater Subcommittee Meetings Singapore 2–3 July 2012.

Läntelä, J., Rasi, S., Lehtinen, J. & Rintala, J., 2012. Landfill gas upgrading with pilot-scale water scrubber: Performance assessment with absorption water recycling. Applied Energy, 92, 307–314.

Montusiewicz, A., Lebiocka, M. & Pawlowska, M., 2008. Characterization of the biomethanization process in selected waste mixtures. Archives of Environmental Protection, 34(3): 49–61.

Mulia, P., Behura, A. K. & Kar S., 2016.Categorical imperative in defense of strong sustainability. Problemy Ekorozwoju/Problems of Sustainable Development, 11 (2): 29–36.

Musiał, D. & Zajemska, M., 2014. Ekonomiczno-ekologiczny aspekt wykorzystania biogazu wysypiskowego. Ekonomia i Środowisko, 1: 141–151.

Ocieczek, L. & Mniszek, W. 2010. The analysis of power utilisation of dump gas in the communal waste dump in DąbrowaGórnicza.Zeszyty Naukowe Wyższej Szkoły Zarządzania Ochroną Pracy w Katowicach, 1(6): 80–99.

Panza, D. & Belgiorno, V., 2010, Hydrogensulphider-emoval from landfillgas. Process Safety and Environmental Protection, 88(6): 420–424.

Parker, T., Dottridge, J. & Kelly, S., 2002. Investigation of the composition and emissions of trace components in landfill gas. R&D Technical Report P1–438/TR.

Pawlowska, M., Czerwinski, J. & Stepniewski, W., 2008. Variability of the non-methane volatile organic compounds (NMVOC) composition in biogas from sorted and unsorted landfill material. Archives of Environmental Protection, 34(3): 287–298.

Pawlowska, M. & Siepak J., 2006. Enhancement of methanogenesis at a municipal landfill site by addition of sewage sludge. Environmental Engineering Science, 23(4): 673–679.

Pawlowska, M. & Stepniewski, W., 2006. Biochemical reduction of methane emissions from landfills. Environmental Engineering Science, 23(4): 666–672.

Pawlowski, A., 2008. The role of social science and philosophy in shaping of the sustainable development concept. Problemy Ekorozwoju—Problems of Sustainable Development, 3(1): 7–11.

Pawlowski, A., 2009a. Theorethical aspects of sustainable development concept. Problemy Ekorozwoju/Problems of Sustainable Development, 11 (2): 985–994.

Pawlowski A. 2009b. Sustainable energy as sine qua non condition for the achievement of sustainable development. Problemy Ekorozwoju/Problems of Sustainable Development, 4(2): 9–12.

Pawlowski, A. 2013. Sustainable Development and Globalization. Problemy Ekorozwoju/Problems of Sustainable Development, 8(2): 5–16.

Popov, V., 2005. A new landfill system for cheaper landfill gas purification. Renewable Energy, 30(7): 1021–1029.

Sadasivam, B. Y. & Reddy, K. R. 2014. Landfill methane oxidation in soil and bio-based cover systems: a review. Reviews in Environmental Science and Biol Technology, 13(1): 79–107.

Sanphoti, N, Towprayoon, S, Chaiprasert, P & Nopharatana A., 2006. Enhancing waste decomposition and methane production in simulated landfill using combined anaerobic reactors. Water Sci Technol. 53 (8): 243–51.

Shin, H. C., Park, J. W., Park, K. & Song, H. C., 2002. Removal characteristics of trace compounds of landfill gas by activated carbon adsorption. Environmental Pollution, 119(2): 227–236.

Shin H. C., Park, J. W., Kim, H. S. & Shin E. S., 2005. Environmental and economic assessment of landfill gas electricity generation in Korea using LEAP model. Energy Policy, 33(10): 1261–1270.

Sieja, L. Tworzenie systemu gospodarki odpadami w gminach w oparciu o składowiska odpadów, Szklarska Poręba 2010,

Sołowiej, P., 2005. Zagospodarowanie odpadów komunalnych na terenie wybranej gminy. Inżynieria Rolnicza, 7. Kraków, 297–303.

Sołowiej, P. & Neugebauer M., 2008. Energetyczne wykorzystanie gazu wysypiskowego na podstawie wybranego obiektu. *Inżynieria Rolnicza*, 6(104): 181–185.

Staszewska, E. & Pawlowska M., 2011. Characteristics of emissions from municipal waste landfills. *Environment Protection Engineering* 37(4): 119–130.

Stepniewski W. & Pawlowska, M., 1996. A possibility to reduce methane emission from landfills by its oxidation in the soil cover. Chemistry from the Protection of the Environment 2, *Environmental Science Research, 51*, Plenum Press, New York, 75–92.

Streese, J. & Stegmann, R., 2003. Microbial oxidation of methane from old landfills in biofilters. *Waste Management*, 23(7): 573–580.

Sztumski, W., 2016. The impact of sustainable development on the homeostasis of the social environment and the matter of survival. *Problemy Ekorozwoju/ Problems of Sustainable Development*, 11(2): 41–47.

Takuwa, Y., Matsumoto, T., Oshita, K., Takaoka, M., Morisawa, S. & Takeda N., 2009. Characterization of trace constituents in landfill gas and a comparison of sites in Asia. *Journal of Material Cycles and Waste Management*, DOI: 10.1007/s10163–009-0257-1.

Tsai, W. T., 2007. Bioenergy from landfill gas (LFG) in Taiwan. *Renewable and Sustainable Energy Reviews*, 11(2): 331–344.

Udo, V. & Pawlowski A., 2010. Human Progress Towards Equitable Sustainable Development: A Philosophical Exploration. *Problemy Ekorozwoju—Problems of Sustainable Development*, 5(1): 23–44.

Udo V. & Pawlowski A., 2011. Human Progress Towards Equitable Sustainable Development—Part II: Empirical Exploration. *Problemy Ekorozwoju—Problems of Sustainable Development*, 6(2): 33–62.

Udo, V. & Pawlowski, A. 2010. Human Progress Towards Equitaible Sustainable Development: A Philosophical Exploration. *Problemy Ekorozwoju/ Problems of Sustainable Development*, 5(2): 23–44.

Wang, J., Xia, F. F., Bai, Y., Fang, C. R., Shen, D. S. & He R., 2011. Methane oxidation in landfill waste biocover soil: Kinetics and sensitivity to ambient conditions. *Waste Management*, 31(5): 864–870.

Zaleska-Bartosz, J., 2014. Gaz składowiskowy jako źródło energii. *Nafta-Gaz*, 12, 932–941.

Zhu, H., Letzel, M. O., Reiser, M., Kranert, M., Bächlin, W. & Flassak, T., 2013. A new approach to estimation of methane emission rates from landfills. *Waste Management*, 33(12): 2713–2719.

Zdeb, M. & Pawlowska, M., 2009. An influence of temperature on microbial removal of hydrogen sulphide from biogas. *Rocznik Ochrona Srodowiska*, 11: 1235–1243.

Environmental Engineering V – Pawłowska & Pawłowski (Eds)
© 2017 Taylor & Francis Group, London, ISBN 978-1-138-03163-0

Mitigation of pollutant migration from landfill to underground water and air

K. Szymański & B. Janowska
Faculty of Civil Engineering, Environmental and Geodetic Sciences, Koszalin University of Technology, Koszalin, Poland

A. Czechowska-Kosacka & W. Cel
Faculty of Environmental Engineering, Lublin University of Technology, Lublin, Poland

ABSTRACT: The work characterized selected methods of eliminating methane emissions to the atmosphere. A simplified method of determining the cleaning properties of a filter bed that stimulates the municipal landfill bed was proposed. Studies on the efficiency of leachate filtering were conducted on the example of a several selected pollution indices found in this medium. A function enabling to forecast the mass of pollutants in a filtrate (leachate from a filtering column) depending on the thickness of a filtering layer, mass of pollutants in the model leachate (artificially prepared) and the intensity of supplied leachates was described. The designated regression functions enable qualitative and quantitative estimation of the impact of the analyzed independent variables $(\mathrm{m}_d^{'}, l, \omega)$ on the mass of pollutants flowing out of a layer of medium-grained sand. The obtained results may be used for predicting the degree of soil and groundwater pollution in the potentially affected area of municipal landfill.

Keywords: landfills, pollution, landfill leachates, groundwater, air

1 INTRODUCTION

Landfills generate pollution which constitutes a threat to human health and to our environment. They emit hazardous chemicals that pollute air and water.

Landfills emit toxic gases affecting the quality of air, as well as significant amounts of methane which is a more potent greenhouse gas than carbon dioxide.

Methane is flammable and creates an explosive mixture when mixed with air. Other gases, i.e. ammonia and sulphides—especially hydrogen sulphide—are toxic and responsible for most of the odour. They are formed during microbial decomposition of organic matter deposited in landfill (Montusiewicz et al. 2008, Pawlowska et al. 2008, Scheutz 2009, Bohna et al. 2011, Rachor et al. 2011, Staszewska et al. 2011).

Gases emitted from landfills can be divided into two groups: the first one includes gases that pollute the surrounding air, i.e. ammonia and sulphides, especially hydrogen sulphide. The other group comprises mercaptans, which are toxic and responsible for most of the odours. Hydrogen sulphide constitutes the greatest nuisance; however, it can be oxidized by bacteria (Zdeb et al. 2009).

In literature, methane is given the most attention. It is the second—after CO_2—most abundant gas which influences climatic changes.

The amount of produced methane in a landfill can be increased by adding sewage sludge (Pawłowska & Siepak 2006, Sanphoti et al. 2006, Sandip & Kanchan 2012). After a certain period of time, the emission of methane drops to the level where it is no longer practical to utilize it for energy generation (Themelis & Ulloa 2007, Reichenauer et al. 2011).

However, it is necessary to eliminate methane emissions into the atmosphere. Oxidation process that utilizes microorganisms is employed for this purpose (Pawłowska & Stępniewski 2006, Jennifer et al. 2007). Covering a landfill with a layer of soil, in which methane is oxidized as a result of the activity of methanotrophic bacteria constitutes the simplest method (Stępniewski & Pawłowska 1995, Barlaz et al. 2004, Stern et al. 2006, Bogner et al. 2008, Kujawska & Cel 2016).

Carbon dioxide (70%) is the gas which contributes to the greenhouse effect to a greatest extent, whereas the share of methane amounts to 27%. The characteristics of methane emissions from various anthropogenic sources were presented in Table 1.

Table 1. Estimated global anthropogenic methane emissions by source, 2010.

Enteric fermentation	29%
Oil and gas	20%
Rice cultivation	10%
Landfills	11%
Wastewater	9%
Other Ag sources	7%
Coal mining	6%
Agriculture (Manure)	4%
Biomass	3%
Stationary and mobile sources	1%

Landfills correspond to 17% of emissions. Methane from landfills continues to be emitted for decades. Initially, the concentration of methane in landfill gas approximates 50%. Such gas is suitable for energy generation. The appearing leachates threaten the purity of groundwater (Białowiec 2011).

By infiltrating into the ground and groundwater, pollutants found in landfill leachates may be transmitted over vast distances transmitted. The lack of information pertaining to the migration of these pollutants prevents conducting adequate assessment of their impact on the groundwater located in the potentially influenced area of a landfill (Koda et al. 2009, Garbulewski 2000, Szymanski et al. 2007).

Partial elimination of mineral and organic pollutants occurs already in the landfill bed. (Malina 2015, Wychowanek & Koda 2015). G. Castany distinguishes between five main natural subsoil self-cleaning mechanisms (Castany, 1982):

- physical—miscibility with water, density of leachates, dilutability, thermodynamic processes,
- hydrochemical—chemical stabilization, solubility and precipitation of salt, ion exchange, polar interaction, polarization,
- hydrobiological—biodegradation,
- adsorption and desorption—physical and chemical phenomena,
- hydrodynamic—effective flow rate of leachate stream, contact time with bed, mechanical dispersion.

Biochemical decomposition processes of organic compounds occur in municipal waste. The products of waste mineralization—in the case of aerobic decomposition—mainly include carbon dioxide and water, whereas in the case of anaerobic decomposition—methane and water (Szymański et al. 2005). This decomposition is carried out by microorganisms that activate when the conditions of appropriate temperature, humidity, and the presence of nutrients are met. Depending on the conditions in the landfill, aerobic or anaerobic bacteria prevail (Siebielska & Siodełko 2015).

Leachates that accumulate on the bottom of a landfill migrate through the aeration zone and further on to the saturation zone (aquifer). This creates a contaminated zone in the groundwater stream. Its range depends on the amount and type of pollutants flowing into groundwater, groundwater flow conditions and the landfill exploitation period (Szymanski & Siebielska 2000a, Szymanski & Siebielska 2000b, Talalaj & Dzienis 2007).

The results of previous studies concerning simulation of leachate infiltration through the aeration zone showed that in a porous bed (gravel or/and sand), mechanical cleaning prevails. Biochemical cleaning, the efficiency of which is determined by the aerobic conditions in the aeration zone, takes place to a lesser extent (Koda et al. 2009, Szymanski 1994, Blaszczyk & Gorski 1996). The conducted studies also show that the increase of silt and clay fraction also increases the role of sorption and ion exchange (Bergata et al. 2006). The afore-mentioned studies also made an attempt to evaluate the factors determining the efficiency of leachate cleaning. These factors included the type of bed, its physical state, layer thickness, chemical composition, concentration of pollutants in leachates and their volume, as well as the intensity with which they were supplied to a layer of porous bed. Castrillon et al. 2010 made an attempt of linking the efficiency of pollutants elimination with the thickness layer and the amount of supplied leachates.

2 AIM AND SCOPE OF THE STUDY

Migration of pollutants in the bed, both in the aeration, as well as saturation zone was the subject of numerous papers (Szymanski et al. 2007, Szymanski & Siebielska 2000, Talalaj & Dzienis 2007, Szymanski 1994, Koda 2009). Nevertheless, despite the implementation of ever-improving numerical models, they are still considered unsatisfactory (Dabrowski et al. 2011). The analysis of pollutant migration process is most often limited to the convective migration, with diffusive migration accounts for in calculations (molecular diffusion and hydrodynamic dispersion) for different variants of hydrodynamic area. Due to the biochemical and physicochemical transformations of pollutants found in landfill leachates in the aeration and saturation zone of landfill bed, implementation of study results in practice (Varank et al. 2011, Renolu et al. 2008).

This work made an attempt of estimating the changes in the groundwater quality resulting from the infiltration of landfill leachates. An approximate method of assessing the migration of mass of pollutants from the aeration zone (landfill bed) to the saturation zone (groundwater) was presented. Moreover, an attempt to assess the phenomenon of self-cleaning of water in the aquifer was made on the basis of model studies.

3 MATERIALS AND METHODS

The model studies, simulating infiltration of leachates through the aeration zone were conducted in order to determine the qualitative and quantitative connection between the mass of pollutants flowing into and out of the porous layer of bed, which was non-cohesive. Sieve analysis showed that it was composed of medium grained sand. Its minimum and maximum density, determined in zero humidity conditions, amounted to $(\rho_d)_{max} = 1.79 \cdot 10^3$ kg/m^3, $(\rho_d)_{min} = 1.63 \cdot 10^3$ kg/m^3, respectively. The maximum density with the humidity that characterized the model research (w = 2.3%) was equal to $\rho_{max} = 1.83 \cdot 10^3$ kg/m^3. Individual layers of sand were formed in a uniform manner. Approximately 2.0 kg of sand were added into a column and subsequently compacted with a 1.0 kg tamper. Afterwards, the thickness of a layer was measured and its density calculated. The layers were formed in three model columns by means of artificially prepared leachates

(Table 2), characterized by physicochemical composition that corresponded to the ones found in a municipal landfill in Middle Pomerania as closely as possible (Table 3).

The variable parameters included: intensity of influent leachates ω (volume of leachate per unit of

Table 2. Composition of the leachate prepared for tests in laboratory conditions.

No.	Composition of model solution	Units	Value
1.	NaCl	g/m^3	2040
2.	NH$_4$NO$_3$	g/m^3	360
3.	NaNO$_2$	g/m^3	360
4.	FeCl$_3$.6 H$_2$O	g/m^3	720
5.	FeCl$_2$.4 H$_2$O	g/m^3	600
6.	KH$_2$PO$_4$	g/m^3	120
7.	K$_2$SO$_4$	g/m^3	240
8.	Acetic acid	dm^3/m^3	12
9.	L-Serine	g/m^3	120
10.	DL-Valine	g/m^3	120
11.	Sucrose	g/m^3	120
12.	Na$_2$CO$_3$	g/m^3	120
13.	MgSO$_4$.7H$_2$O	g/m^3	1200
14.	Ninhydrin	g/m^3	120
15.	L-isoleucine	g/m^3	240

Table 3. Chemical composition of the model solution and leachates occurring in the Middle Pomeranian (Poland) landfills.

Index	Unit	Model solution		Stand 1 Without lining (1980s)	Stand 2 With lining (2010s)
		Mean value	Standard deviation		
pH		7.5	0.49	7.0–8.7	7.52
Total hardness	gCaCO$_3$/m^3	18.2	1.40	5.5–25.0	–
Calcium	g/m^3	–	–	90.0–107.0	184.4
Magnesium	g/m^3	20.0	4.24	8.51–32.0	34.8
Manganese	g/m^3	–	–	0.1–4.0	1.24
Total iron	g/m^3	182.43	2.45	2.3–100	10.0
Sulphates	g/m^3	422.8	92.21	66.8–460	24.6
Oxidizability	gO$_2$/m^3	115.2	6.41	100–3100	420.0
COD $_{K2Cr2O7}$	gO$_2$/m^3	5631.71	209.61	469–7761	3191.0
BOD$_5$	gO$_2$/m^3	810.0	189.99	188–4000	680.0
Ammonia nitrogen	g/m^3	47.43	1.31	10–452	785.0
Nitrites	g/m^3	4.95	2.84	0.05–0.2	0.09
Nitrates	g/m^3	18.71	0.90	0.1–10.0	0.80
Total nitrogen	g/m^3	80.0	0.0	32	–
Phosphates	g/m^3	100.0	14.14	0.2–24.0	36.0
Chlorides	g/m^3	1206.99	31.30	58.5–5732	500.0
Dried solid content	g/m^3	3669.23	88.64	628–21350	5216.0
Residue on ignition	g/m^3	2531.33	82.82	4353–16711	2848.0
Loss on ignition	g/m^3	1137.69	33.71	2158–4716	2368.0

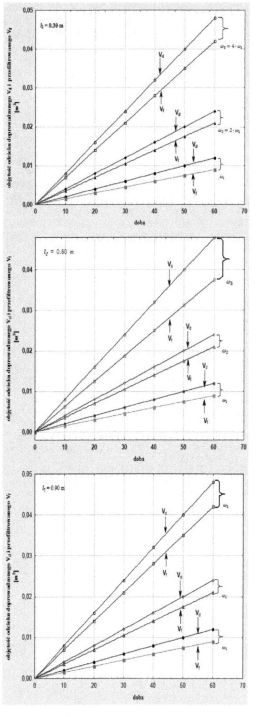

Figure 1. Filtration test results for model landfill leachates on $l_1 = 0.3$ m, $l_2 = 0.6$ m, $l_3 = 0.9$ m thick sand bed (ω – supplied leachate intensity $\frac{m^3}{m^2 \cdot d}$).

area and time), mass of influent pollutants m_d and bed density l. The analysis of changes in the mass of pollutants filtered through the model layer m_f, depending on these parameters, was performed for several essential pollutant indices (Fig. 1). Two indices, i.e. dried solid content (S) and Chemical Oxygen Demand (COD$_{K2Cr2O7}$) were used for the theoretical analysis.

4 RESULTS

Model research was conducted in three series, with varying layer thickness ($l_1 = 0.3$ m, $l_2 = 0.6$ m, $l_3 = 0.9$ m). In each series, leachate was supplied on a daily basis with varying intensity ($\omega_1 = 0.026$ m^3/m^{2}d, $\omega_2 = 0.052$ m^3/m^{2}d, $\omega_3 = 0.104$ m^3/m^{2}d).

Therefore, the leachates introduced to each column were as follows: the first—$v_1 = 195 \cdot 10^{-6}$ m^3, the second—$v_2 = 390 \cdot 10^{-6}$ m^3, and the third—$v_3 = 780 \cdot 10^{-6}$ m^3. Throughout the study, approximately 50 leachate doses were introduced into each column. Determining dried solid content and Chemical Oxygen Demand (COD) in the considered leachates and filtrates was carried out according to the methodology found in valid Polish standards (Kowal & Swiderska-Broz 1996).

The results of studies were illustrated in Figs. 2, 3.

5 DISCUSSION OF RESULTS

5.1 Analytical solution to the pollutant migration process in the bed

The main cleaning processes in the stream of groundwater include mechanical filtering, biochemical transformations, sorption, ion exchange and dilution (Szymanski et al. 2007, Szymanski 1994, Blaszczyk & Gorski 1996). Mechanical filtering, consisting in blocking solid pollutants by the bed, mainly depends on its granulometry and thickness. Sorption, which is a function of grains and soil specific surface area, also depends on its mineral composition. On the other hand, ion exchange—in addition to the afore-mentioned factors—requires humic substances as well. Intensity of biochemical processes, especially in the aeration zone, is dependent on the presence of oxygen, whereas dilution of polluted groundwater is conditioned upon the possibility of replenishing water (infiltration of rainwater) (Koda et al. 2009, Dabrowski et al. 2011, Varank et al. 2011, Zaradny 2008).

The results of conducted model research clearly show that the mass of pollutants filtered through a layer of bed (m_f) depends on the mass

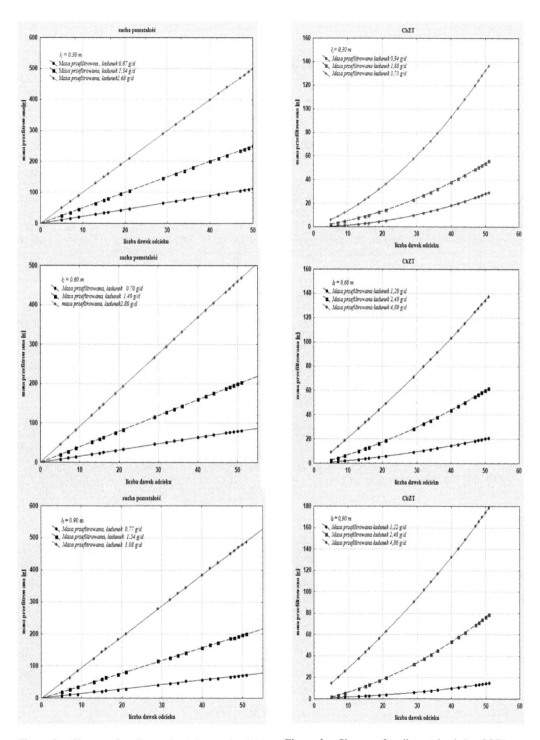

Figure 2. Change of pollutant load (as total solids) contained in filtrates taken from $l_1 = 0.3$ m, $l_2 = 0.60$ m, $l_3 = 0.90$ m thick bed.

Figure 3. Change of pollutant load (as COD) contained in fil-trates taken from $l_1=0.3$ m, $l_2=0.60$ m, $l_3=0.90$ m thick bed.

of influent pollutants (m_d), intensity of influent leachate (ω) and the thickness of a layer (l) (Figs. 1–3).

Statistical methods were used in order to determine the function:

$$m_f = m_f\,(m_d, l, \omega) \qquad (1)$$

Afterwards, analytical solutions were proposed.

5.2 *Multiple regression model*

Multiple regression model was assumed for the description of function (1). It subordinates mean values of a dependent variable to multiple independent variables (Luszniewicz & Slaby 2008). Due to the simplicity of calculations, a linear form of multiple regression was assumed. It was also assumed that the sole variable (m_f) is random, whereas the independent variables (m_d, l, ω) are not, as their values are known (the ones assumed in the study). Such assumption requires introduction of a random component. Granted that the expected value of the random component is equal to zero, a function that subordinates expected values of a dependent variable to the values of independent variables is obtained. The parameters of this function were calculated with least squares method.

Assuming the linear dependency between m_f, m_d, l and ω for a given pollution index, with the accuracy to a random component, it can be formulated as:

$$(m_f)_t = b_1\,(m_d)_t + b_2\,(l)_t + b_3\,(\omega)_t + b_4 + \varepsilon_t \\ (t = 1, 2, 2...., 9), \qquad (2)$$

where ε_t is characterized by normal distribution with unknown variance σ^2.

Calculations of regression parameters and the multiple correlation coefficient were carried out by means of STATSITICA software (Luszniewicz & Slaby 2008). The obtained calculation results for individual pollution indices were presented in Table 4.

The designated regression functions enable estimating the qualitative and quantitative impact of the analyzed independent variables (m_d', l, ω) on the values of pollutant mass flowing out of the layer of sand bed. Fairly uniform values of correlation coefficient indicate high correlation, and thus good coherence with the linear model of m_f' variable empirical values (Table 4). The comparison of linear regression functions of individual pollution indices (Table 4) shows a clear similarity expressed with the dried solid content and COD indices.

Generally, it can be stated that the adopted multiple regression method can be successfully employed for estimation of mass of pollutants flowing out of the incompletely saturated porous bed to the aquifer, for m_d, l and ω values within the range of values assumed in model research. Extrapolation, especially when $m_d \rightarrow 0$, $l \rightarrow 0$ i $\omega \rightarrow 0$, does not ensure correct results.

If the variable Z (corresponding to the mass of pollutant flowing out of the considered layer of sand—m_f) depends on three independent variables x, y, and w, and is designated from the model research, its values $z_{ij} = z_{ij}\,(x)$ are determined for y_i ($i = 1, 2, 3$) and w_j ($j = 1, 2, 3$). It can also be described in the index form as:

$$Z_j = Z_{ij} \cdot D_i, \qquad (3)$$

where:

$$D_1 = \frac{(y - y_2)(y - y_3)}{(y_1 - y_2)(y_1 - y_3)},$$

$$D_2 = \frac{(y - y_1)(y - y_3)}{(y_2 - y_1)(y_2 - y_3)},$$

$$D_3 = \frac{(y - y_1)(y - y_2)}{(y_3 - y_1)(y_3 - y_2)}. \qquad (4)$$

For $j = 1$, the function in the form of equation (3) can be elaborated in the following way:

$$Z_1 = Z_{11}(x)\,D_1 + Z_{21}\,(x)\,D_2 + Z_{31}\,(x)\,D_3 \qquad (5)$$

Table 4. Linear multiple regression functions for leachates supplied to the filter bed.

Pollution index	Regression function $m_f' = m_f'\,(m_d, l, \omega)$	Multiple correlation coefficient (r)
Dried solid content	$m_f' = 2.94\,m_d' - 8.536{,}81\,l + 94.096$, $20\,\omega - 343.76$	0.9996
COD $K_2Cr_2O_7$	$= 0.54\,m_d' + 291.75\,l + 25{,}485.90$ $\omega - 2{,}162.76$	0.9957

*m_f' – mass of filtered pollutant per unit of area [g/m^2].
*m_d' – mass of supplied pollutant per unit of area [g/m^2].

Similarly to the function $Z_j = Z_j(x, y)$, the function Z dependent on three variables $Z = Z(x, y, w)$, can also be described in the index form, where

$$Z = Zj \cdot Kj \qquad (6)$$

$$K_1 = \frac{(w - w_2)(w - w_3)}{(w_1 - w_2)(w_1 - w_3)},$$

$$K_2 = \frac{(w - w_1)(w - w_3)}{(w_2 - w_1)(w_2 - w_3)},$$

$$K_3 = \frac{(w - w_1)(w - w_2)}{(w_3 - w_1)(w_3 - w_2)}. \qquad (7)$$

The extended form of function (6) is as follows:

$$Z(x,y,w) = Z_1(x,y) \cdot K_1 + Z_2(x,y) \cdot K_2 + Z_3(x,y) \cdot K_3 \qquad (8)$$

Calculating the values of this function require prior determination of $Z_{ij} = Z_{ij}(x)$ for the model values y_i and w_j assumed in the model research.

The proposed function (8) formally meets the assumed criteria and can be used for description of the mass of pollutants flowing out of the considered layer of sand (m_f) depending on the mass of supplied pollutants (m_d), thickness of the layer (l) and the intensity of supplied leachates (ω).

Assuming that $Z = m_f$, $x = m_d$, $y = l$, $w = \omega$, the relation between Z and x as well as m_f and m_d can be determined on the basis of model research results. The results of these calculations were presented in Table 5. It contains the regression function equations for the considered pollution indices. The calculations were carried out with least squares method, assuming the power function, linear function, and polynominal function of degree 2. The choice of the most suitable function should be made on the basis of the correlation coefficient value and the function progress in line with the conducted model research. Calculation of regression parameters and the multiple correlation coefficient were carried out with STATISTICA PL software package (Luszniewicz & Slaby 2008). The obtained calculation results for the individual pollution indices were presented in Table 5.

The conducted studies show that all three types of analyzed functions yield very high values of correlation coefficients (Table 5). It was also found that the mass of pollutants in the filtrate can only reach values $m_f \geq 0$. However, there are also cases when soil components are washed from the aeration zone, which results in the parameter $m_f < 0$. It can also be assumed that the mass of pollutants supplied to the saturation zone amounts to $m_d = 0$

Table 5. Regression coefficient between mass of leachate supplied and filtered through medium sand layer ($F = 0,0075\ m^2$).

Pollution index	Thickness l m	Intensity $m^3/m^2/d$	$Z = a\ x^b\ (m_f = a\ m_d{}^b)$			$Z = a\ x + b$ $(m_f = a\ m_{d+}\ b)$			$Z = a\ x^2 + bx + c$ $(m_f = a\ m_d{}^2 + b\ m_{d+}\ c)$			
			a	b	r	a	b	r	a	b	c	r
1	2	3	4	5	6	7	8	9	10	11	12	13
Dried solid content	0.3	0.026	2.860	1.01	0.988	3.17	−3.22	0.987	0.061	1.127	9.415	0.997
		0.052	1.870	1.56	0.947	3.94	−9.93	0.999	0.005	3.653	−6.924	0.999
		0.104	1.440	1.24	0.983	3.84	−2.83	0.000	−0.006	4.539	−18.297	0.999
	0.6	0.026	0.820	1.22	0.983	2.31	−12.96	0.982	0.045	0.230	6.088	0.999
		0.052	1.020	1.24	0.999	3.21	−24.38	0.998	0.010	2.337	−8.461	0.999
		0.104	2.680	1.05	0.999	3.44	−12.83	0.999	0.003	2.974	4.221	0.999
	0.9	0.026	2.700	1.35	0.998	1.48	−6.78	0.992	0.023	0.500	0.500	0.999
		0.052	0.870	1.23	0.996	2.63	−19.56	0.995	0.014	1.455	−3.850	0.998
		0.104	−2.160	1.07	0.994	3.15	−11.22	0.999	−0.001	3.190	−11.763	0.999
COD	0.3	0.026	0.042	1.63	0.925	0.46	−2.77	0.978	0.009	0.049	0.755	0.996
		0.052	0.043	1.60	0.982	0.57	−5.38	0.992	0.003	0.269	−0.629	0.999
		0.104	0.061	1.48	0.993	0.69	−11.02	0.995	0.002	0.409	−2.958	0.999
	0.6	0.026	0.119	1.22	0.942	0.38	−4.07	0.971	0.006	−0.068	2.967	0.999
		0.052	0.060	1.45	0.007	0.59	−11.02	0.997	0.001	0.426	−5.834	0.999
		0.104	0.102	1.32	0.998	0.67	−20.49	0.997	0.001	0.452	−6.601	0.999
	0.9	0.026	0.039	1.30	0.852	0.20	−2.39	0.946	0.005	−0.130	1.748	0.998
		0.052	0.103	1.34	0.979	0.62	−9.48	0.992	0.002	0.353	−3.162	0.995
		0.104	0.344	1.23	0.997	0.69	−6.61	0.998	0.001	0.738	−8.995	0.998

or $m_d > 0$, depending on the intensity of leachate supply and the thickness of filtering layer. In such case, both the linear function, as well as polynomial function of degree 2 (Table 5) lose their validity respectively for:

$$x < \frac{-b}{a} \quad ix < \frac{-b+\sqrt{\Delta}}{2a}$$

Therefore, $m_f = 0$ must be assumed for the range of zero to x. In the case of a linear function, when $b > 0$, and square function, when $c > 0$, it is necessary to correct these functions in such a way, so as to make them pass through the origin of coordinates. This can be done by linking y-axis of the first point, obtained in the study (m_f), with the origin of coordinates. It should be noticed that the correlation coefficients of square functions are in many cases close to uniformity. Nevertheless, the equations where the parameter equals a < 0, may raise concerns. Such functions for high m_d values do not match the expected outcome of the experiment. Taking all of the afore-mentioned aspects into account, the most suitable functions should be chosen from Table 5 for individual pollution indices.

In the research, the values of $y_i = l_i$ amounted to $y_1 = 0.3$ m; $y_2 = 0.6$ m and $y_3 = 0.9$ m, whereas $w_i = \omega_i$ assumed the value 0.026 m³/m²d, $\omega_2 = 0.052$ m³/m²d and $\omega_3 = 0.104$ m³/m²d. Therefore, (4) and (5) can be expressed as:

$$D_2 = \frac{(y-0{,}3)(y-0{,}9)}{-0{,}09},$$

$$D_3 = \frac{(y-0{,}3)(y-0{,}6)}{0{,}18}, \tag{9}$$

$$K_1 = \frac{(w-0{,}052)(w-0{,}104)}{20{,}28 \cdot 10^{-4}},$$

$$K_2 = \frac{(w-0{,}026)(w-0{,}104)}{-13{,}52 \cdot 10^{-4}},$$

$$K_3 = \frac{(w-0{,}026)(w-0{,}052)}{40{,}56 \cdot 10^{-4}}. \tag{10}$$

Afterwards, having selected the most appropriate $z_{ij} = z_{ij}(x)$ functions from Table 5, function (8) can be elaborated for any pollutant index. For instance, in the case of COD, by assuming linear dependencies of these functions, (5) can be presented in the following way:

$$Z_1 = D_1 (0{,}46\ x - 2{,}77) + D_2 (0{,}38\ x - 4{,}07)$$
$$+ D_3 (0{,}20\ x - 2{,}39)$$

$$Z_2 = D_1 (0{,}57\ x - 5{,}38) + D_2 (0{,}59\ x - 11{,}02)$$
$$+ D_3 (0{,}62\ x - 9{,}48) \tag{11}$$

$$Z_3 = D_1 (0{,}69\ x - 11{,}73) + D_2 (0{,}67\ x - 20{,}29)$$
$$+ D_3 (0{,}69\ x - 6{,}61)$$

whereas, function (8) assumes the form

$$Z = Z_1 \cdot K_1 + Z_2 \cdot K_2 + Z_3 \cdot K_3 \tag{12}$$

where the values D_i and K_j are expressed as (4) and (7), respectively. By inserting any of the values $x = m_d$, $y = l$, $w = \omega$ into the above-mentioned equation—provided that they do not exceed the range of adequate values used in model research—it is possible to calculate the value $Z = m_f$, which determines the mass of pollutants filtered through a layer of bed (aeration zone).

6 SUMMARY

Migration of pollutants in a bed is a complex process that is dependent on numerous factors (Janiszewski 2014). Mathematical models describing these processes, drawing on the assumptions of physicochemical hydromechanics that account for the specificity of bed and landfill leachate, do not encompass all of the assumed phenomena, especially the biochemical degradation processes of pollutants contained in leachates. The point of studying the migration process is creating a full theoretical description and indicating the role of individual factors in various hydrogeological conditions. Simultaneously, maximum possible simplification of calculations is achieved by omitting irrelevant components. Theoretical description pertaining to migration of pollutants in groundwater usually assumes that the physicochemical interaction between water and bed occurs in the form of sorption (kinetics of sorption processes are omitted), whereas the interaction between individual groundwater components are not taken into account. Moreover, it is assumed that convective migration is the basic variant. The accuracy of assessment pertaining to the migration of polluted groundwater also depends (to a significant extent) on identifying the water and bed conditions, as well as the chemical composition of leachates. This mainly pertains to the type and properties of bed in the vicinity of pollution source, thickness of aeration zone and aquifer, as well as the direction and flow rate of groundwater. It is known that these values, determined on the basis of geotechnical research, are often assumed with rough approximation. Migration of pollutants has a significant impact on these estimations. The proposed method of estimating the mass of mass of pollutants that infiltrate into groundwater through the aeration zone may prove helpful in solving practical issues (Koda 2011, Koda et al. 2009, Dabrowski et al. 2011, Zaradny 2008).

Bearing in mind that reports pertaining to the impact of municipal landfills on the environment landfills are already in place, it seems reasonable that the impact of municipal landfills on the quality of groundwater should be assessed as well. Such analysis should include both the aeration zone, as well as the saturation zone (aquifer). The proposed method of evaluating changes in the quality of water infiltrating through a modeled aeration zone, enables estimation of the mass of pollutants flowing out from this layer, as well as the intensity of leachate supply. The obtained research results enable assessing both the efficiency of cleaning process in the existing aeration zone, as well as the self-cleaning process occurring in groundwater. This in turn allows predicting the rate of pollution spreading in these waters and the range of polluted zone extending from a landfill in different periods of its exploitation.

REFERENCES

Barlaz M. A., Green R.B., Chanton J. P., Goldsmith C. D., Hater D. R., 2004. Evaluation of a biologically active cover for mitigation of landfill gas emissions. *Environ. Sci. Technol.* 38 (18): 4891–4899.

Bergaya F., Theng B. K. G., Lagaly G., 2006. Handbook of clay science. Elsevier, Amsterdam.

Białowiec A., 2011. Hazardous emissions from municipal solid waste landfills. Contemporary Problems of Management and Environmental Protection. 9: 7–28.

Blaszczyk T., Gorski J., 1996. Odpady a problemy zagrożenia i ochrony wód podziemnych. Biblioteka Monitoringu Środowiska, Warszawa.

Bogner J., Pipatti R., Hashimoto S., Diaz C., Mareckova K., Diaz L., Kjeldsen P., Monni S., Faaij A., Gao Q., Zhang T., Ahmed M. A., Sutamihardja R. T. M., Gregory R., 2008. Mitigation of global greenhouse gas emissions from waste: conclusions and strategies from the Intergovernmental Panel on Climate Change (IPCC) Fourth Assessment Report. Working Group III (Mitigation). Waste Manag Res. 26: 11–32.

Bohna S., Brunkea P., Gebertb J., Jagera J., 2011. Improving the aeration of critical fine-grained landfill top cover material by vegetation to increase the microbial methane oxidation efficiency. Waste Management. 31 (5): 854–863.

Castany G., 1982. Principes et méthodes de l' hydrogeology. Dunod Université, Bordas. Paris.

Castrillón L., Fernández-Nava Y., Ulmanu M., Anger I., Marañón E., 2010. Physico-chemical and biological treatment of MSW landfill leachate. Waste Management. 30: 228–235.

Dabrowski S., Kapuscinski J., Nowicki K., Przybylek J., Szczepanski A., 2011. Metodyka modelowania matematycznego w badaniach i obliczeniach hydrogeologicznych. Poradnik Metodyczny. Poznań.

Garbulewski K., 2000. Dobór i badania gruntowych uszczelnień składowisk odpadów komunalnych. Wydawnictwo SGGW. Warszawa.

Janiszewski A., 2014. Analysis of concentration values and their reductions in relation both to the chosen processes and to the chosen contaminants moving in groundwater. XXI Seminarium Naukowe z cyklu Regionalne Problemy Inżynierii Środowiska, Wydawnictwo ZUT—Szczecin, 37–56.

Jennifer C. Stern J.C., Chanton J., Abichou T., Powelson D., Yuan L., Escoriza S., Bogner J.,2007. Use of a biologically active cover to reduce landfill methane emissions and enhance methane oxidation. Waste Management, 27 (9):1248–58.

Koda E., 2011. Stateczność rekultywowanych składowisk odpadów i migracja zanieczyszczeń przy wykorzystaniu metody obserwacyjnej. Wydawnictwo SGGW. Warszawa.

Koda E., J. Golimowski, E. Wienclaw, 2009. Assessment of the old landfill protection system based on transport modelling and monitoring research. 17th International Conference on Soil mechanics and Geotechnical Engineering. 3: 1977–1980.

Kowal A. L., Świderska—Bróż M., 1996. Oczyszczanie wody. Wydawnictwo Naukowe PWN, Warszawa-Wrocław.

Kujawska J., Cel W., 2016. Mitigation of greenhouse effect by reduction of the methane emissions. Problemy Ekorozwoju/Problems of Sustainable Development 2: 127–129.

Luszniewicz, A., T. Slaby: Statystyka z pakietem komputerowym STATISTICA PL. Teoria i zastosowania. Wydawnictwo C.H. Beck, Warszawa 2008.

Malina G., 2015. Remediacja, rekultywacja i rewitalizacja. Wydawnictwo Polskiego Zrzeszenia Inżynierów Sanitarnych i Techników, Poznań.

Montusiewicz A., Lebiocka M., Pawlowska M., 2008. Characterization of the biomethanization process in selected waste mixtures. Archives of Environmental Protection, 34 (3): 49–61.

Montusiewicz A., Lebiocka M., Pawlowska M., 2008. Characterization of the biomethanization process in selected waste mixtures. Archives of Environmental Protection, 34 (3): 49–61.

Pawlowska M., Czerwinski J., Stepniewski W., 2008. Variability of the non-methane volatile organic compounds (NMVOC) composition in biogas from sorted and unsorted landfill material. Archives of Environmental Protection, 34 (3): 287–298.

Pawlowska M., Siepak J., 2006. Enhancement of methanogenesis at a municipal landfill site by addition of sewage sludge. Environmental Engineering Science, 23 (4): 673–679.

Pawlowska M., Stepniewski W., 2006. Biochemical reduction of methane emissions from landfills. Environmental Engineering Science, 23 (4): 666–672.

Rachor I, Gebert J, Gröngröft A, Pfeiffer EM., 2011. Assessment of the methane oxidation capacity of compacted soils intended for use as landfill cover materials. Waste Management, 31 (5): 833–42.

Reichenauer T. G, Watzinger A, Riesing J, Gerzabek M.H., 2011. Impact of different plants on the gas profile of a landfill cover. Waste Management, 31 (5): 843–53.

Renou S., Givaudan J. G., Poulain S., Dirassouyan F., Moulin P., 2008. Landfill leachate treatment: Review and opportunity. Journal of Hazardous Materials, 150: 468–493.

Sandip T. M., Kanchan C. K., Ashok H B., 2012. Enhancement of methane production and bio-stabilisation of municipal solid waste in anaerobic bioreactor landfill. Bioresource Technology, 110: 10–17.

Sanphoti N, Towprayoon S, Chaiprasert P, Nopharatana A., 2006. Enhancing waste decomposition and methane production in simulated landfill using combined anaerobic reactors. Water Sci Technol. 53 (8): 243–51.

Scheutz Ch., 2009. Microbial methane oxidation processes and technologies for mitigation of landfill gas emissions. Waste Management Research, 27: 409–455.

Siebielska I., Sidelko R., 2015. Polychlorinated biphenyl concentration changes in sewage sludge and organic municipal waste mixtures during composting and anaerobic digestion. Chemosphere, 126: 88–95.

Staszewska E., Pawlowska M., 2011. Characteristics of emissions from municipal waste landfills. Environment Protection Engineering 37 (4): 119–130.

Stepniewski W., Pawlowska M., 1996. A possibility to reduce methane emission from landfills by its oxidation in the soil cover. Chemistry from the Protection of the Environment 2, Environmental Science Research, 51, Plenum Press, New York, 75–92.

Szymanski K., Janowska B., Sidełko R., 2005.: Estimation of bioavailability of copper, lead and zinc in municipal solid waste and compost. Asian Journal of Chemistry, 17 (3): 646–1660.

Szymanski K., Sidelko R., Janowska B., Siebielska I., 2007. Monitoring of waste landfills. Zeszyty Naukowe Wydziału Budownictwa i Inżynierii Środowiska, 23: 75–133.

Szymanski K., Siebielska I., 2000 a. Usage of PTI Method for Evaluation of Landfill Leachte Influence on Underground Waters. Materiały Konferencyjne Sympozjum Forum Chemiczne, Warszawa.

Szymanski K., 1994. Migracja metali ciężkich w strefie aeracji i saturacji podłoża wysypiska. Gospodarka Wodna, 3: 53–56.

Szymański K., Siebielska I., 2000 b. Analityczne aspekty oceny jakości zanieczyszczonych wód podziemnych. Ochrona Środowiska 2000b, 76: 15–18.

Talalaj I. A., Dzienis L., 2007. Changes of groundwater quality in the vicinity of a municipal landfill site. Proceedings of the 2nd National Congress of Environmental Engineering, Taylor & Francis Group, London, 227–231.

Themelis_N. J., Ulloa P. A., 2007. Methane generation in landfills. Renewable Energy, 32: 1243–1257.

Varank G., Demir A., Top S., Sekman E., Akkaya E., Yetilmezsoy K., Bilgili M. S., 2011. Migration behavior of landfill leachate contaminants through alternative composite lines. Science of the Total Environment, 409: 3183–3196.

Wychowanek D, Koda E., 2015. Samooczyszczanie środowiska gruntowo-wodnego w rejonie składowiska odpadów Łubna. Monografia pod redakcją G. Malina, Wydawnictwo Polskiego Zrzeszenia Inżynierów Sanitarnych i Techników, Poznań, 299–308.

Zaradny H., 2008. Modelowanie przepływu wód gruntowych w ujęciu obszarowym. Biuletyn Państwowego Instytutu Geologicznego, 431: 275–285.

Zdeb M., Pawlowska M., 2009. An influence of temperature on microbial removal of hydrogen sulphide from biogas. Rocznik Ochrona Srodowiska, 11: 1235–1243.

Environmental Engineering V – Pawłowska & Pawłowski (Eds)
© 2017 Taylor & Francis Group, London, ISBN 978-1-138-03163-0

Assessment of odour nuisance of wastewater treatment plant

Ł. Guz & A. Piotrowicz
Faculty of Environmental Engineering, Lublin University of Technology, Lublin, Poland

E. Guz
Faculty of Nursing and Health Sciences, Medical University of Lublin, Lublin, Poland

ABSTRACT: Facilities related to wastewater management constitute an emission source of nauseating odours. Despite the activities undertaken in recent years towards limiting the odour nuisance, still the complaints are made about wastewater treatment plants and their negative health effects are reported. In the paper are presented some aspects of the assessment of odour annoyance of WWTPs by means of analytical and sensory methods. Sensory techniques include dynamic olfactometry which is widely described in the literature. It is the method recommended for estimation of the nuisance caused by odours. Analytical techniques mainly include gas chromatography and the use of the specific gas sensors. Gas chromatography is a very precise technique used for qualitative and quantitative analysis, however identification of the chemical composition of odorous gas does not allow to determine its odour nuisance. In contrast, the devices equipped with the array of non-specific gas sensors seem to be very promising.

Keywords: odour nuisance, wastewater treatment plant, emission assessment

1 INTRODUCTION

Sources of odour can be associated both with natural processes such as volcanic eruptions, conflagrations, wetland emissions, degradation of organic matter by microorganisms and with anthropogenic processes related to broadly defined food, chemical and energy industries. In recent years the problem has become urgent because along with the extension of human settlements, residential housing is often built closer and closer to the objects characterised by odour nuisance.

Emission of odours is particularly connected with the operation of facilities related to wastewater management. This includes not only wastewater treatment plants, but also facilities closely linked to WWTPs such as septage dumping station and intermediate pumping stations. Odour emission takes place mostly during every stage of wastewater treatment process. However, it usually pertains to the facilities employed in the primary phase of treatment such as wastewater influent, bar screen, intermediate pumping station of raw wastewater, grit-chamber, grease separator, primary settler, but also suspended growth and attached growth systems (Vincent 2001). Moreover, wastewater sludge treatment processes such as thickening and dewatering can also be characterised by a high odour annoyance.

The main odour sources at 100 German sewage treatment plants were identified by Frechen (1988) by means of questionnaires completed by the plant staff (Fig. 1). Similar results obtained by R. Barczak et al. (2012) are presented in Fig. 2. The odour concentration in different sites of WWTP (Warsaw) was measured using Ecoma TO8 dynamic olfactometer. The results confirm that collection chamber, screen room and sludge processing are main odour sources.

Wastewater flowing into WWTP is very diversified in terms of pollution content. The composition depends mainly on the kind of industrial plants that are connected to the sewerage system. Hence, wastewater treatment plants should continually perform qualitative analyses of incoming sewage in order to determine basic parameters such as chemical oxygen demand (COD), biochemical oxygen demand (BOD), total organic carbon (TOC) and also nitrogen and phosphorus content (Szaja et al. 2015). Often, leachates coming from municipal landfills are introduced into wastewater treatment plant. These leachates contain numerous odour compounds, e.g. dibutyl phthalate, toluene, naphthalene and also trimethylbenzene, hexa tiepane, benzene, chloromethyl-butene, 3-methylbutanoic acid, tetradecane, benzaldehyde, undecane, tetramethylbenzene and many others (Dmochowski et al. 2015).

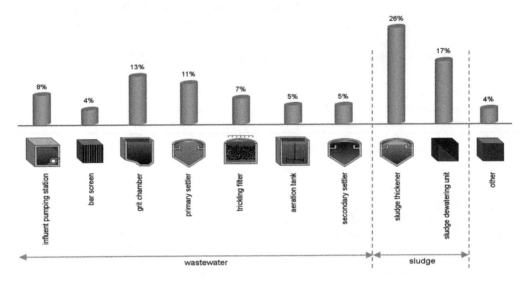

Figure 1. The percentage of respondents chosen among the staff of 100 WWTPs who identified the given facility as nuisance source of odour (Frechen 1998, Gostelow et al. 2001).

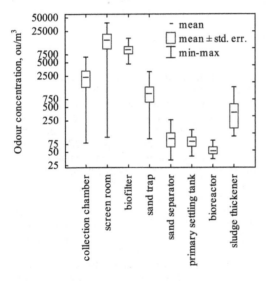

Figure 2. Odour nuisance assessment of WWTP located in Warsaw (Poland) according to data published by R. Barczak et al. (2012).

According to Henry's law, the concentration of volatile compounds in gaseous phase is proportional to the concentration of these compounds in liquid phase. Depending on dimensionless Henry's law constant, the emission can go higher to a greater or lesser extent. For instance, increasing liquid temperature causes the increased emission of volatile compounds. Aside from temperature, other parameters such as liquid turbulence or wind velocity over wastewater surface also significantly affect the phenomenon. For that reason it is difficult to determine the exact composition of polluted air above wastewater. (Sówka et al. 2015).

In gases emitted from wastewater treatment even more than 450 identifiable compounds can be found, with approximately 100 substances being strong odorants (Kośmider et al. 2002). These substances are characterised by a wide range of odours and can be assigned into several main groups (Gostelow et al. 2001):

– Sulfur compounds: hydrogen sulfide (rotten eggs), dimethyl sulfide (rotten, garlic-like), diethyl sulfide (nauseating), diallyl sulfide (garlic-like), carbon sulfides (rotten vegetables), sulfur dioxide (pungent, irritating), methyl mercaptan (rotten cabbage, garlic-like), ethyl mercaptan (rotten cabbage), propyl mercaptan (unpleasant), butyl mercaptan (unpleasant), tert-butyl mercaptan (unpleasant), allyl mercaptan (garlic-like), crotyl mercaptan (skunk, rancid), benzyl mercaptan (unpleasant), thiocresol (skunk, rancid), thiophenol (rotten, nauseating, decomposing);
– Nitrogen compounds: ammonia (pungent, irritating), methylamine (fish-like), dimethylamine (fish-like), trimethylamine (fish-like, ammonia-like), ethylamine (ammonia-like), diethylamine (fish-like), triethylamine (fish-like), cadaverine (rotten meat), pyridine (unpleasant, irritating), indole (fecal, nauseating), skatole (fecal, nauseating);
– Acids: acetic (vinegar-like), butanoic (rancid, sweaty), valeric (sweet);
– Aldehydes and ketones: formaldehyde (irritating, choking), acetaldehyde (fruity, apples),

butyraldehyde (rancid, sweaty), isobutyraldehyde (fruity), isovaleraldehyde (fruity, apples), acetone (fruity, sweet), butanone (green apples).

From among the above listed substances not all of them are toxic, but make a significant contribution towards odour nuisance. The scale of the odour impact is medium with the Odour Emission Rate (OER) of $10^5 \div 10^6$ m³/min and the range of impact accounted for $50 \div 1000$ m (Ozonek et al. 2009a). According to another source the nuisance declines at the distance greater than 900 m from the main emission sources (Szymański et al. 2015). In summer, the range of odour impact extends up to twice as far than in the winter season Therefore, the emission of odours should be determined for various meteorological scenarios. Considering the odour impact, it should be also mentioned that the emission of odorous compounds considerably diminishes the tourist values and economic potential of the adjacent areas (Sówka et al. 2013, Sówka et al. 2015).

In order to legally regulate odour emission issues many standards and regulations were published defining both methodological aspects of the odour nuisance assessment and permissible emission levels together with their duration time. In European countries the first works concerning the standards were started in Germany (VDI 3881), France (AFNOR X-43-101) and the Netherlands (NVN 2820). The European Committee for Standardization (CEN) was also involved in preparation of odour regulations. In 1995 was released the first complete draft of the European olfactometry standard (EN13725) entitled "Air Quality—Determination of Odour Concentration by Dynamic Olfactometry" (Mcginley & Mcginley 2001).

2 HEALTH ASPECTS

Perception of a smell as unpleasant and nuisance-causing depends on individual human characteristics such as olfactory preferences, sex, age and temper (Brudniak et al. 2013). The external factors which can weaken or strengthen the sensation of discomfort resulting from smell perception are also of great importance. In general, the odour impact is the result of several interacting factors referred to as FIDOL: frequency, intensity, duration, offensiveness, and location (Nicell 2009).

Health symptoms caused by ambient odours are explained by means of at least three mechanisms. Primarily, the symptoms can be invoked by contact with odorous compounds at levels enough to trigger irritation. In this case irritation, and not the odour itself, is the cause of health symptoms, while the odour acts as an exposure marker. Secondly, at nonirritant concentrations, health symptoms can appear on account of genetically coded or learned aversions. The third mechanism of arising health symptoms lies in the presence of so called copollutant being one of the components of an odorant mixture (Schiffman & Williams 2005).

The aspect of stimuli generation is very interesting and still not fully known. In the human organism unconditional responses can be produced in the similar manner as saliva production in the presence of a pleasant food aroma (Mcginley & Mcginley, 1999). It was found that the number of health symptoms is not proportional to the distance from the emission source but to the frequency of odour nuisance occurrences (Ames et al. 1991).

Odours are the cause of numerous health afflictions. Individual odorants can exhibit such properties as toxicity, mutagenicity or cancerogenicity (Ozonek et al. 2009a). Exposure to odours may often lead to headache, nausea, reflex nausea, eye irritation, throat irritation, shortness of breath, runny nose, G.I. distress, fatigue, sleep disturbance, inability to concentrate and lack of appetite (Schiffman et al 2004). Moreover, the presence of odours results in mental discomfort or even mental disorders such as depression (Mcginley & Mcginley 1999).

One of the employee groups that is particularly endangered by air pollution is the wastewater treatment staff. They are not only exposed to harmful gases e.g. sulfur dioxide, but may suffer from many disorders including musculoskeletal disorder, various infections such as leptospirosis, hepatitis, helicobacter pylori, and also dermatitis (Rajnarayan 2008).

3 REDUCTION OF ODOUR EMISSION

The reduction of odour nuisance of WWTP primary involves prevention of emission and deodorization of waste gases.

In case of a newly developed treatment plant, the prevention of odour emission should be taken into consideration during the designing stage of treatment process and facilities. In the existing treatment plants, the modernization of facilities and modification of processes can be planned, as well as separation of process from environment by building an airtight hall or hermetic chambers.

On the other hand, deodorization involves removal of malodorous substances from air, transforming odorants into odourless compounds and masking odorants with special compounds changing their smell character. Choosing an efficient deodorization technique is difficult. It mainly consists in reviewing the scientific literature related to the efficiency of techniques applied in plants with a similar profile. Following methods of deodorization can be distinguished (Piecuch et al. 2006):

- adsorption and absorption
- thermal and catalytic burning
- biological treatment of waste gases
- odour neutralization.

Besides the above-mentioned methods, there are several alternative techniques with proven efficiency; for example, hydrodynamic cavitation (Ozonek et al. 2009b).

The volume of waste gas is usually large, and therefore, employing traditional methods of purifying air is technically and economically unjustified. Despite being present in trace concentrations, most volatile compounds are malodorous and have very low odour thresholds, potentially constituting odour nuisance to nearby populations; therefore, it is necessary to eliminate the odour substances to the largest possible degree (Sucker et al. 2008). Among the numerous methods of air purification, an intensive development of biological methods, especially biofiltration, has become noticeable in recent years. In this method, the decomposition of pollutants takes place through the activity of microorganisms colonizing the filter material. Biofiltration is characterised by low exploitation costs and is effective for the decomposition of simple compounds of natural origin such as ammonia and hydrogen sulfide, but often fails in case of mixtures of complex compounds.

In addition to the above-mentioned methods, alternative methods, such as masking-deactivating compounds, are being widely promoted and gain increasingly more supporters. The essence of this solution is spray of special compounds that reduce the intensity of the emitted smell, changing it to a more pleasant one or contain deactivating molecules, blocking smell properties of odorants. Masking the odours is the subject of many studies. Groups of researchers are trying to determine how the essential oils affect the reduction of odour nuisance. It was found that pine essential oil and essential oils obtained from orange skins effectively neutralize unpleasant odours (Kowalczyk et al. 2013, Piecuch et al. 2015). The substrates obtained from rose and mint (Piecuch et al. 2006), extracts from citrus fruits, ginger and carnation (Andriyevska et al. 2008), geranium, caraway, anise, juniper and black cumin (Piecuch et al. 2009), essential oils of lemon (Kowalczyk et al. 2010), as well as grapefruit (Piecuch et al. 2011) all proved to be equally effective.

4 EMISSION ASSESSMENT

Monitoring of odour emissions associated with sewage is not an easy task. It requires the use of many different techniques and numerical simulations conducted by means of special computer software.

The assessment of odour nuisance of an object can be done by measuring the odour concentration, determining the rate of emission and propagation range of odours. The propagation range of odour in the field can be also determined by means of field measurements or by means of a proper mathematical model (Sówka et al. 2015).

The attention was usually drawn to the dominant compounds found in the highest concentrations, such as hydrogen sulfide and ammonia. Apart from these compounds, there is also a group of compounds present at very low concentrations (Szymanski et al. 2015). Therefore the gas samples can be pre-concentrated using a number of techniques such as cryogenic sampling, sorbent tubes and solid phase microexraction (SPME). The advantage of cryogenic sampling is that compounds are not absorbed selectively depending on the fiber material (Muñoz et al., 2010).

4.1 Sensory analysis

Sensory evaluation generally involves describing the character of odours and examining the intensity, as well as the odour threshold by performing serial dilutions of polluted air. Dilution of samples odour may be carried out in a static or dynamic way. Static dilution consists in mixing the sample volume with a specific volume of inert gas. Dynamic dilution involves mixing the flux of sample with the flux of inert gas. The dynamic olfactometer enables automatic dilution of the sample and is one of the most common devices for sensory analysis. The sample is directed to the sniffing ports and is presented to a group of panelists. For example, the Ecoma TO8 dynamic olfactometer allows dilution of the sample in the range from 4 to 65536 ou_E/m^3 with step 2 dilution (Sironi et al. 2010). The measurements of odour threshold, intensity and hedonic tone of the odorant should be performed with appropriately selected evaluation team. Odour threshold can be measured by means of yes/no, as well as the forced choice method. In the case of intensity and hedonic tone assessment, the assessors make the answer on a determined scale, according to the perceived smell stimuli (Sówka et al. 2013).

Field olfactometers, such as NasalRanger, are used in field sensory meanurements. This device allows to obtain dilution to threshold (D/T) ratios of 2, 4, 7, 15, 30 and 60, with an accuracy of 10% (Walgraeve et al. 2015).

Surveys can also be used to evaluate the odour nuisance. The questionnaires with point scale of nuisance are sent to the inhabitants of an area frequently exposed to odour nuisance (Sówka et al. 2011). Point scale should include verbal explanations, such as: 0 (smell imperceptible), 1 (weak),

2 (clear), 3 (strong), 4 (very strong). Monitoring of air quality may also be carried out using scale patterns fragrances which are used to compare the intensity of a stimulus. The scale may be a standard set of aqueous nbutanol solutions with a fixed dilution step. The assessors indicate a pattern with the most similar intensity of odours. The results of evaluations are noted on a proper survey form (Brudniak et al. 2013).

4.2 *Instrumental analysis*

The gas chromatograph is a device capable of separating gaseous sample into individual components and detecting them (Kolb 2006). However determination of chemical composition of odorous gas falls short of its odour nuisance. In order to achieve this, gas chromatography coupled to olfactometry GC-O may be employed. It relies on the division of the carrier gas flowing out of the column and containing eluates into two streams. One of them is directed to the detector of the chromatograph, while the other is headed to the special chromatograph attachment i.e. sniffing port. The advantage of this method is both separation of volatile organic compounds and the classification of smells and the possibility of determining odour intensify by expert panels (Ruth 2001). Measurement participants play a role of additional, very sensitive and selective chromatograph detector. Panelists have to meet the requirement of reproducibility of sensory assessments of odour intensity. In the case of simultaneous assessment of the odour type, the ability to identify the odour and to correctly name the odour is needed. Odour assessments through human detection are recorded during the elution of all sample components. As a result of the measurement so-called olfactogram is obtained and compared with classic chromatogram.

In the literature one can find many successful applications of the GC-O technique, for example, it was used to analyse gas samples emitted from swine barns (Cai 2006) and dairy barns (Zhang et al. 2010).

4.3 *Electronic nose*

An electronic nose (e-nose) can be used to measurement of odorous compounds (Sobczuk & Guz 2011). The e-nose, in great simplification, imitates the human sense of smell. It consist of the array of several nonspecific gas sensor (Craven 1996, Guz et al. 2010). Particular sensors in the array are partially sensitive to miscellaneous groups of chemical compounds; therefore each gas mixture generates different profile of signals. The multidimensional set of signals from all sensors is taken to the analysis. This is comparable to fingerprints in

dactyloscopy, because the probability of the same combination being formed for two different samples is very low; that is why the combination of these signals is popularly called a "gas fingerprint" (Dewettinck et al. 2001).

There are four main groups of sensors: optical, thermal, electrochemical and gravimetric. The most common are resistive and potentiometric gas sensors belonging to the group of electrochemical sensors (Guz et al. 2016).

Using nonspecific gas sensors there is no possibility to precisely distinguish between the different components of the gas mixture. The main purpose of measurement is to identify the general characteristics of the composition of gas. The results of measurement can be also correlated with a number of physicochemical parameters of given gas mixture. Such correlation enables to make concentration assessments of chemical substances, total volatile organic compounds, the concentrations of the fragrance and other parameters of gaseous samples, as well as liquid being in equilibrium with a gas phase. Therefore, the electronic nose can be used e.g. in sewage treatment plants to control the quality of treated waste-water. This versatility of the device combined with a relatively moderate price enables its widespread use in many fields of science and industry (Guz et al. 2015).

The summary of e-nose example application is presented in Table 1.

Table 1. Application of e-nose for evaluation of odours and wastewater parameters.

Description	References
Classification of odour sources, correlation e-nose measurements with odour concentration	Capelli et al. 2008
Classification of wastewater by odours, correlation of e-nose measurements with BOD	Onkal-Engin et al. 2005 Stuetz et al. 1999b
Discrimination between odour samples from different sites in WWTP	Nake et al. 2005
Correlation of odour concentration and e-nose measurements	Stuetz et al. 1999a
Correlation of e-nose measurements with VSS, COD, TSS, turbidity	Dewettinck et al. 2001
On-line detection of pollutants and accidental discharge	Bourgeois et al. 2001 Bourgeois et al. 2002, Bourgeois et al. 2003a, Bourgeois et al. 2003b

5 CONCLUSION

Wastewater treatment plants pose a considerable source of odours, hence proper assessment of the odour nuisance is of great importance. Dynamic olfactometry is the main technique of measuring the odorous pollution from WWTPs in terms of odour annoyance. In turn, gas chromatography is widely used for purposes of qualitative and quantitative analyses. Separate group of devices utilised to measure odorous compounds is constituted by e-nose. Most certainly, many other techniques of measuring volatile air pollutants may be created. Their development may be closely related to genetic engineering and further recognition of the mechanism responsible for the identification of smell with biological olfactory sense. Sensors arrays proved to be an efficient tool and therefore should be developed further.

REFERENCES

Ames, R.G. & Stratton, J.W. 1991. Acute health effects from community exposure to n-propyl mercaptan from an ethoprop (mocap) – treated potato field in Siskiyou County, California. *Archives of Environmental Health* 46(4): 213–217.

Andriyevska, L., Juraszka, B., Kowalczyk, A., Piecuch, T., Pol, K. & Zimoch, A. 2008. Neutralisation of noxious odours by spraying solutions created on the base of extracts from citrus fruits, ginger and carnation. *Rocznik Ochrony Środowiska (Annual Set The Environ ronment Protection)* 10: 707–723 (in Polish).

Barczak, R., Kulig, A. & Szyłak-Szydłowski, M. 2012. Olfactometric methods application for odour nuisance assessment of wastewater treatment facilities in Poland. *Chemical Engineering Transactions* 30: 187–192.

Bourgeois, W., Burgess, J.E. & Stuetz, R.M. 2001. On-line monitoring of wastewater quality: a review. *Journal of Chemical Technology and Biotechnology* 76: 337–348.

Bourgeois, W., Gardey, G., Servieres, M. & Stuetz, R.M. 2003a. A chemical sensor array based system for protecting wastewater treatment plants. *Sensors and Actuators B Chemical* 91: 109–116.

Bourgeois, W., Hogben, P., Pike, A. & Stuetz, R.M 2003b. Development of a sensor array based measurement system for continuous monitoring of water and wastewater. *Sensors and Actuators B Chemical*, 88: 312–319.

Bourgeois, W. & Stuetz, R.M 2002. Use of a chemical sensor array for detecting pollutants in domestic wastewater. *Water Research* 36: 4505–4512.

Brudniak, A., Dębowski, M. & Zieliński, M. 2013. Determination of the odorous influence of selected wastewater treatment plant on area covered by elaboration of spatial development plan. *Rocznik Ochrony Środowiska (Annual Set The Environment Protection)* 15: 1759–1771 (in polish).

Cai, L., Koziel, J., Lo, Y.-C. & Hoff, S.J. 2006. Characterization of volatile organic compounds and odorants associated with swine barn particulate matter using solid-phase microextraction and gas chromatography-mass spectrometry-olfactometry. *Journal of Chromatography A* 1102(1–2): 60–72.

Capelli, S., Sironi, P., Centola, R., del Rosso, R. & Grande, M. 2008. Electronic noses for the continuous monitoring of odours from a wastewater treatment plant at specific receptors: Focus on training methods. *Sensors and Actuators B Chemical* 131: 53–62.

Craven, M.A., Gardner, J.W. & Bartlett, P.N. 1996. Electronic noses—development and future prospects. *Trends in Analytical Chemistry* 15(9): 486–493.

Dewettinck, T., van Hege, K. & Verstraete, W. 2001. The electronic nose as a rapid sensor for volatile compounds in treated domestic wastewater. *Water Research* 35: 2475–2483.

Dmochowski, D., Dmochowska, A. & Biedugnis, S. 2015. Chromatographic analysis of chemical compounds in the leachate from municipal landfill, undergoing electrooxidation. *Rocznik Ochrony Środowiska (Annual Set The Environment Protection)* 17. 1196–1206 (in polish).

Frechen, F.B. 1988. Odour emissions and odour control at wastewater treatment plants in West Germany. *Water Science & Technology* 20: 261–266.

Gostelow, P., Parsons, S.A. & Stuetz, R.M. 2001. Review paper: Odour measurements for sewage treatment works. *Pergtamon* 35(3): 579–597.

Guo, J.S., Abbas, A.A., Chen, Y.P., Liu, Z.P., Fang, F. & Chen, P. 2010. Treatment of landfill leachate using a combined stripping, Fenton, SBR, and coagulation process. *Journal of Hazardous Materials* 178(1–3): 699–705.

Guz. Ł., Łagód, G., Jaromin-Gleń, K., Guz, E. & Sobczuk, H. 2016. Assessment of batch bioreactor odour nuisance using an e-nose. *Desalination and Water Treatment* 57(3): 1327–1335.

Guz, Ł., Łagód, G., Jaromin-Gleń, K., Suchorab, Z., Sobczuk, H. & Bieganowski, A. 2015. Application of gas sensor arrays in assessment of wastewater purification effects. *Sensors* 15(1): 1–21.

Guz, Ł., Sobczuk, H. & Suchorab, Z. 2010. Odor measurement using portable device with semiconductor gas sensors array. *Przemysł Chemiczny* 89, 378–381 (in polish).

Kolb, B. & Ettre, L.S. 2006. *Static Headspace-Gas Chromatography: Theory and Practice*. New Jersey: Wiley-Interscience.

Kośmider, J., Mazur-Chrzanowska, B., Wyszyński, B. 2002. *Odory*. Warszawa: Wydawnictwo Naukowe PWN.

Kowalczyk, A., Kutryn, J. & Piecuch, T. 2010. Neutralization of odours arising during dewatering of municipal sewage sludge in the process of centrifugal sedimentation. *Rocznik Ochrony Środowiska (Annual Set The Environment Protection)* 12: 365–380 (in polish).

Kowalczyk, A., Piecuch, T. & Andriyevska, L. 2013. The use of pine essential oil for masking the odors emitted in the process of mechanical dewatering of municipal sewage sludge. *Rocznik Ochrony Środowiska (Annual Set The Environment Protection)* 15: 807–822 (in Polish).

McGinley, C.M. & McGinley, M.A. 2002. Impact of the new European odor testing standard on wastewater treatment facilities. Water Environment Federation 74th Annual Conference. Atlanta, GA: 13–17.

Mcginley, C.M., & Mcginley, M.A. 1999. The "gray line" between odor nuisance and health effects. Proceedings of Air and Waste Management Association 92nd Annual Meeting and Exhibition St. Louis, Mo: 20–24.

Muñoz, R., Sivret, E.C., Parcsi, G., Lebrero, R., Wang, X., Suffet, I.H., & Stuetz, R.M. 2010. Monitoring techniques for odour abatement assessment. *Water Research* 44(18): 5129–5149.

Nake, A., Dubreuil, B., Raynaud, C. & Talou, T. 2005. Outdoor in situ monitoring of volatile emissions from wastewater treatment plants with two portable technologies of electronic noses. *Sensors and Actuators B Chemical* 106: 36–39.

Nicell, J.A. 2009. Assessment and regulation of odour impacts. *Atmospheric Environment* 43(1): 196–206.

Onkal-Engin, G., Demir, I. & Engin, S.N. 2005. Determination of the relationship between sewage odour and BOD by neural networks. *Environmental Modelling & Software* 20: 843–850.

Ozonek, J., Korniluk, M. & Piotrowicz, A. 2009a. Odour nuisance from animal waste utilization plant. *Rocznik Ochrony Środowiska (Annual Set The Environment Protection)* 11: 1191–1199 (in polish).

Ozonek, J., Szulżyk-Cieplak, J. & Czerwiński, J. 2009b. Reduction of odour emission from wastewater from sugar industry with application of hydrodynamic cavitation. *Rocznik Ochrony Środowiska (Annual Set The Environment Protection)* 11: 1053–1062 (in polish).

Piecuch, T., Andriyevski, B., Andriyevska. L., Juraszka, B. & Kowalczyk, A. 2009. Production and spraying of solutions neutralising noxious odours created on the base of extracts from geranium, caraway, anise, juniper and black cumin. *Rocznik Ochrony Środowiska (Annual Set The Environment Protection)* 11: 607–629 (in polish).

Piecuch, T., Kowalczyk, A. & Dąbrowski, T. 2015. Reduction of odorous noxiousness of sewage treatment plant in Tychowo. *Rocznik Ochrony Środowiska (Annual Set The Environment Protection)* 17: 646–663 (in polish).

Piecuch, T., Kowalczyk, A., Kupś, D. & Gomółka, D. 2011. Method of neutralization of odours arising during mechanical dewatering of municipal sewage sludge. *Rocznik Ochrony Środowiska (Annual Set The Environment Protection)* 13: 747–768 (in polish).

Piecuch, T., Sasinowski, M., Nowak, A., Dąbrowski, J., Kościerzyńska-Siekan, G., Dworaczyk, J. & Zaremba, W. 2006. Production and spraying solutions neutralizing unpleasant smells in the wastewater pretreatment plant in SUPERFISH company. *Rocznik Ochrony Środowiska (Annual Set The Environment Protection)* 8: 239–261 (in polish).

Rajnarayan, R.T. 2008. Occupational health hazards in sewage and sanitary workers. *Indian Journal of Occupational and Environmental Medicine* 12(3): 112–115.

Ruth S.M. 2001. Methods for gas chromatography-olfactometry: a review. *Biomolecular Engineering* 17(4–5): 121–128.

Schiffman, S.S., Walker, J.M., Dalton, P., Lorig, T.S., Raymer, J.H., Shusterman, D. & Williams, C.M. 2004. Potential health effects of odor from animal operations, wastewater treatment, and recycling of byproducts. *Journal of Agromedicine* 9(2): 397–403.

Schiffman, S.S. & Williams, C.M. 2005. Science of odor as a potential health issue. *Journal of Environmental Quality* 34(1): 129–138.

Shicheng, Z., Cai, L., Koziel, J.A., Hoff, S.J., Schmidt, D.R., Clanton, C.J., Jacobson, L.D., Parker, D.B. & Heber, A.J. 2010. Field air sampling and simultaneous chemical and sensory analysis of livestock odorants with sorbent tubes and GC–MS/olfactometry. *Sensors and Actuators B Chemical* 146(2): 427–32.

Sironi, S., Capelli, L., Céntola, P., Del Rosso, R. & Sauro Pierucci, S. 2010. Odour impact assessment by means of dynamic olfactometry, dispersion modelling and social participation. *Atmospheric Environment* 44: 354–360.

Sobczuk, H. & Guz, Ł. 2011. Measurement of odour pollutants with multisensor device with TGS gas sensors. *Rocznik Ochrony Środowiska (Annual Set The Environment Protection)* 13: 1531–1542 (in Polish).

Sówka, I., Nych, A. & Zwoździak, J. 2011. Application of german solutions in odour annoyance evaluation in Poland. *Rocznik Ochrony Środowiska (Annual Set The Environment Protection)* 13: 1275–1288 (in polish).

Sówka, I., Kita, U., Skrętowicz, M. & Nych, A. 2013. The conditions and requirements necessary for the proper functioning of the olfactometric laboratory. *Rocznik Ochrony Środowiska (Annual Set The Environment Protection)* 15: 1207–1215 (in polish).

Sówka, I., Sobczyński, P. & Miller, U. 2015. Impact of seasonal variation of odour emission from passive area sources on odour impact range of selected WWTP. *Rocznik Ochrony Środowiska (Annual Set The Environment Protection)* 17: 1339–1349 (in polish).

Stuetz, R.M., Fenner, R.A. & Engin, G. 1999a. Assessment of odours from sewage treatment works by an electronic nose, H$_2$S analysis and olfactometry. *Water Research* 33: 453–461.

Stuetz, R.M., Fenner, R.A. & Engin, G. 1999b. Characterisation of wastewater using an electronic nose. *Water Research* 33: 442–452.

Sucker, K., Both, R., Bischoff, M., Guski, R. & Winneke, G., 2008. Odor frequency and odor annoyance. Part I: assessment of frequency, intensity and hedonic tone of environmental odors in the field. *International Archives of Occupational and Environmental Health* 81(6): 671–682.

Szaja, A., Aguilar, J.A. & Łagód, G. 2015. Estimation of Chemical Oxygen Demand fractions of municipal wastewater by respirometric method—case study. *Rocznik Ochrony Środowiska (Annual Set The Environment Protection)* 17: 289–299.

Szymański, K., Janowska, B. & Piekarski, J. 2015. Methodological aspects of odorous substances measurement in the vicinity of a sewage treatment plant. *Rocznik Ochrony Środowiska (Annual Set The Environment Protection)* 17: 603–615 (in polish).

Vincent, A.J. 2001. Sources of odours in wastewater treatment. In Stuetz, R.M. & Frehen, F.B. (eds.), *Odours in wastewater treatment: measurement, modeling and control*. London: IWA Publishing.

Walgraeve, Ch., Van Huffel, K., Bruneel, J. & Van Langenhove, H. 2015. Evaluation of the performance of field olfactometers by selected ion flow tube mass spectrometry. *Biosystems Engineering* 137: 84–94.

Xu, Z.Y., Zeng, G.M., Yang, Z.H., Xiao, Y., Cao, M., Sun, H.S., Ji, L.L. & Chen, Y. 2010. Biological treatment of landfill leachate with the integration of partial nitrification, anaerobic ammonium oxidation and heterotrophic denitrification. *Bioresource Technology* 101(1): 79–86.

Environmental Engineering V – Pawłowska & Pawłowski (Eds)
© 2017 Taylor & Francis Group, London, ISBN 978-1-138-03163-0

Air pollution in Poland in relation to European Union

W. Cel
Faculty of Environmental Engineering, Lublin University of Technology, Lublin, Poland

Z. Lenik & A. Duda
Faculty of Fundamental of Technology, Lublin University of Technology, Lublin, Poland

ABSTRACT: The article presents the causes and health effects of excessive atmospheric air pollution in Poland.

Keywords: air pollution, sustainable development, benzo(a)pyrene, PM2.5, PM10

1 INTRODUCTION

In 1983, a report entitled *Chemical Threat to the Environment in Poland* was published (Pawłowski 1999). The report presented an alarming state of environment in Poland. Twenty-seven areas which could be considered as having been struck by ecological disasters were distinguished.

The taken actions have led to a gradual improvement. A significant reduction of pollution emissions occurred in 1990–95. However, it did not result from the implementation of pollution control facilities, but rather from the industrial production collapse due to the introduction of so-called Balcerowicz Plan. As a result, numerous industrial plants went bankrupt. Simultaneously, the Polish market was opened for the import of aging, second-hand automobiles from the entire Europe. Again, this contributed to a continuing deterioration of environment, especially the quality of air.

The number of deaths caused by road traffic accidents is one of the highest in Poland (EU Report 2016). Partly because most of old cars from Europe are now being used in Poland (see Table 1).

This means that indiscriminate introduction of extremely liberal capitalism, without proper connection to the sustainable development, significantly contributed to a degradation of the atmospheric air quality (Pawłowski 2009a, 2009b, 2009c, 2009d, Udo et al. 2010, 2013, Cholewa et al. 2009, Pawłowski 2012, Pawłowski 1999).

According to the data published by European Environmental Agency (EEA 2014), Polish cities are considered among the most polluted ones in Europe and Kraków is one of the most polluted

Table 1. Number of annual deaths caused by road traffic accidents (Greenpeace 2013).

Countries	Number of deaths caused by road traffic accidents
Poland	4572
Germany	4152
Romania	2796
Bulgaria	901
United Kingdom	2337
Czech Republic	901
Greece	1453
France	4273
Slovakia	347
Spain	2605
Italy	4050
Slovenia	171

cities in Europe (Rzeszutek and Bogacki 2016, Szewczyńska et al. 2016).

Table 2 presents the characteristic of average annual concentration of benzo(a)pyrene in individual EU countries in 2012 and Table 3 presents the characteristic of PM_{10} pollution in selected cities in Europe.

1.1 Characteristic of air pollutants emissions in Poland

The impact of individual sectors on the emission of main pollutants into the atmosphere is presented in Table 4. Professional power production contributes mostly to SO_2 emission and together with road transport—to NOx emission. The most

Table 2. Average annual concentration of benzo(a) pyrene in EU countries in 2012 (www.biznesalert.pl/wp-content/uploads/2016/01/PAS-Mapa-BAP-2012.jpg).

	Country	ng/m^3
1	Portugal	>0.01
2	Spain	0.12
3	Netherlands	0.13
4	Malta	0.14
5	Switzerland	0.25
6	Finland	0.25
7	Germany	0.37
8	Great Britain	0.40
9	Cyprus	0.40
10	France	0.47
11	Italy	0.75
12	Latvia	0.80
13	Hungary	1.0
14	Estonia	1.08
15	Slovenia	1.20
16	Czech republic	1.84
17	Poland	5.81

Table 3. Number of days in year (2011) when average daily concentration of PM10 exceeded 35µg/m^3 in selected European cities (ww.nik.gov.pl/aktualnosci).

	City (Country)	
1	Cracow (Poland)	150.5
2	Nowy Sącz (Poland)	126.0
3	Katowice 9Poland)	123.0
4	Sofia (Bulgaria)	122.0
5	Rybnik (Poland)	113.0
6	Poznań (Poland)	88.0
7	Jelenia Góra (Poland)	86.0
8	Bucharest (Romania)	69.0
9	Warsaw (Poland)	57.5
10	Budapest (Hungary)	54.3
11	Rome (Italy)	39.0
12	Prague (Czech republic)	35.3
13	Berlin (Germany)	31.5
14	Lisbon (Portugal)	25.0
15	Amsterdam (Holland)	20.8
16	Vilnius (Lithuania)	19.5
17	Birmingham (UK)	18.0
18	Paris (France)	14.5
19	Madrid (Spain)	6.7
20	Stockholm (Sweden)	1.8

important emitters of NMVOC are road transport and municipal housing sector. Polycyclic aromatic hydrocarbons is emitted by industrial processes (320 t/year), domestic heating (265 t/year), power station (118 t/year), and transport 80 t/year) (Maliszewska-Kordybach 1999). The most harmful PM2.5 particulate matter is emitted in 73% by energy use and supply, 17%—road transport,

Table 4. Characteristic of main pollutants emissions to air (Data from GIOS 2014).

	Thousands tonnes			
	SO$_2$	NO$_x$	PM10	NMVOC
Professional power production	800	300	20	10
Municipal –housing sector	200	40	100	90
Road transport	–	220	20	95

Table 5. Premature deaths attributable to air pollution, ozone (O3), nitrogen dioxide (NO2) in EU countries (EEA-2015).

Country	PM$_{2,5}$	O$_3$	NO$_2$
Austria	6100	320	660
Belgium	9300	170	2300
Bulgaria	14100	500	700
Croatia	4500	270	50
Cyprus	790	40	0
Czech Republic	10400	380	290
Denmark	2900	110	50
Estonia	620	30	0
Finland	1900	60	0
France	43400	1500	7700
Germany	59500	2100	10400
Greece	11100	780	1300
Hungary	12800	610	720
Ireland	1200	30	0
Italy	59500	3300	21600
Latvia	1800	60	90
Lithuania	2300	80	0
Luxembourg	250	10	60
Malta	200	20	0
Netherlands	10100	200	2800
Poland	44600	1100	1600
Portugal	5400	320	470
Romania	25500	720	1500
Slovakia	5700	250	60
Slovenia	1700	100	30
Spain	25500	1800	5900
Sweden	3700	160	10
United Kingdom	37800	530	14100
Albania	2200	140	270
Andorra	60	4	0
Bosnia and Herzegovina	3500	200	70
The Former Yugoslav Republic of Macedonia	3000	130	210
Iceland	100	2	0
Liechtenstein	20	1	3
Monaco	30	2	7
Montenegro	570	40	20
Norway	1700	70	200
San Marino	30	2	0
Serbia	13400	550	1100
Switzerland	4300	240	950
Total	43200	1700	75000

5%—other industrial processes, and 3%—waste (GIOŚ—2014).

The aging second-hand cars constitute another problem. Over 78% of second-hand cars in Poland is 10 or more years old, and 50% out of this figure is 15 or more years old, which results in the highest number of premature deaths caused by road transport in Poland (see Table 1). In recent years, popularization of using biomass for energy purposes creates additional threat of pollutants emission (Gołofit-Szymczak et al. 2016). Waste disposal facilities constitute another significant source of pollutants emitted to air (Cyprowski et al. 2016, Galwa-Widera et al., Pawłowska et al. 2008, Staszewska et al. 2011, Piecuch et al. 2015, Sówka et al. 2016).

2 EFFECT OF AIR POLLUTION ON HEALTH

The studies conducted by WHO indicate a continuous increase of health hazards caused by air pollution. While in 2005, the number of premature

Table 6. Aggregated damage costs, data for European Countries (WHO – 2014).

	Country	Damage costs Million euros
1	Germany	190000
2	Poland	135000
3	United Kingdom	130000
4	France	80000
5	Italy	87000
6	Romania	75000
7	Spain	60000
8	Bulgaria	58000
9	Czech Republic	44000
10	Netherlands	29000
11	Belgium	25000
12	Greece	23000
13	Slovakia	19000
14	Finland	18000
15	Hungary	17000
16	Portugal	17000
17	Sweden	17000
18	Austria	16500
19	Estonia	15000
20	Ireland	15000
21	Norway	15000
22	Denmark	14500
23	Slovenia	14000
24	Lithuania	14000
25	Switzerland	13500
26	Cyprus	1000
27	Luxemburg	1000
28	Malta	1000
29	Latvia	1000

Table 7. Aggregated damage costs normalised against GDP – data for European countries (WHO – 2014).

	Country	EUR/GDP
1	Bulgaria	89000
2	Romania	38000
3	Estonia	31000
4	Poland	25000
5	Czech Republic	20000
6	Slovakia	17000
7	Lithuania	8000
8	Slovenia	8000
9	Hungary	8000
10	Greece	8000
11	Malta	7000
12	Belgium	7000
13	United Kingdom	6000
14	Germany	6000
15	Cyprus	6000
16	Finland	6000
17	Portugal	5000
18	Spain	5000
19	Netherlands	5000
20	Italy	4800
21	France	4000
22	Ireland	4000
23	Austria	3000
24	Sweden	3000
25	Latvia	3000
26	Luxemburg	3000
27	Denmark	2000
28	Norway	2000
29	Switzerland	1000

deaths worldwide caused by air pollution was estimated at 0.8 million, this figure increased to 3.4 million in 2010, and 7 million in 2012 (WHO 2014, Cohen et al. 2004, 2005, Anenberg and Silva et al. 2013). Data for European countries are depicted in Table 5) Approximately 38% of deaths occurred due to ischaemic heart disease, 35% – stroke, 17% – chronic obstructive pulmonary disease, and 10% – acute lower respiratory disease. Negative impact on health translates into greater costs incurred by the economy. Table 6 characterizes the aggregated damage costs by country in 2008–2010, and Table 7 presents the normalized cost by country against GDP. According to the presented data, the highest costs in relation to GDP resulting from air pollution are incurred by Bulgaria, Romania, Estonia, while the fourth place is occupied by Poland.

3 CONCLUSIONS

The main source of PM_{10}, $PM_{2.5}$ and benzo(a) pyrene pollution in Poland are domestic heating stoves and road transport. It results from the common use of solid fuels. Taking into account that out of the 10 cities with the highest PM10 concentration, as many as 6 are found in Poland (Kraków, Nowy Sącz, Gliwice, Zabrze, Sosnowiec and Katowice), there is an urgent need to reduce the use of solid fuels for domestic heating and eliminate old cars.

REFERENCES

Anenberg, S. C. & West, J. J. 2010. The Global Burden of Air Pollution on Mortality: Anenberg et al. respond. *Environ Health Prospect*, 118(10): A424-A425. doi: 10.1289/ehp.1002397R.

Cholewa, T. & Pawłowski, A. 2009. Sustainable Use of Energy in the Communal Sector. *Rocznik Ochrona Środowiska*, 11(2): 1165–1177.

Cohen, A. J., Anderson, H. R., Ostro, B., Pandey, K. D., Krzyzanowski, M., Kuenzli, N., Gutschmidt, K., Pope, C. A., Romieu, I., Samet, J. M. & Smith, K. R. 2004. Mortality impacts on urban air pollution. In *Comparative quantification of health risks: Global and regional burden of disease due to selected major risk factors*, eds. M. Ezzati, A.D. Lopez, A. Rodgers, and C. U. J. L. Murray. Geneva: World Health Organisation.

Cohen, A. J., Anderson, H. R., Ostra, B., Pandey, K. D., Krzyzaowski, M., Kunzli, N., Gutschmidt, K., Pope, A., Romieu, I., Samet J. M. & Smith K. 2005. The Global Burden of Disease due to Outdoor Air Pollution. *Journal of Toxicology and Environmental Health*, Part A, 68: 1–7. doi: 10.1080/15287390590936166.

Cyprowski M., Stobnicka A., Górny R., Gołofit-Szymczak M. & Ławniczek-Wałczyk A. 2016. Aerozole pochodzenia bakteryjnego w pomieszczeniach roboczych zakładu gospodarki odpadami. *Rocznik Ochrona Środowiska* 18(2): 294–308.

EEA 2014 Report: Air pollution fact sheet 2014, Poland.

EEA 2015, Air quality in Europe – 2015 report.

Eurostat 2014, Energy, transport and environment indicators, 2014 edition.

Gałwa-Widera, M. & Kwarciak-Kozłowska, A. 2016. Sposoby eliminacji odorów w procesie kompostowania. *Rocznik Ochrona Środowiska*, 18(2): 850–860.

GIOŚ 2014: Stan Środowiska w Polsce, Raport 2014.

Gołofit-Szymczak, M., Ławniczek-Wałczyk, A., Górny, R., Cyprowski, M. & Stobnicka, A. 2016. Charakterystyka zagrożeń biologicznych występujących przy przetwarzaniu biomasy do celów energetycznych. *Rocznik Ochrona Środowiska* 18, 193–204.

Górny, R. & Gołofit-Szymczak, M. 2016. Zagrożenie środowiskowe powodowane przez włókna szklane. *Rocznik Ochrona Środowiska*, 18, 336–350.

Greenpeace 2013, Annual Report 2013.

KOBiZE 2015. Krajowy Raport Inwentaryzacja 2015, Inwentaryzacja gazów cieplarnianych w Polsce dla lat 1988–2013.

Maliszewska-Kordybach, B. 1999, Sources, concentrations, fata and effects of polycyclic aromatic hydrocarbons (PAHs) in the environment. Part A: PAHs in air., *Polish Journal of Environmental Studies*, 8(3): 131–136.

Pawłowska, M., Czerwiński, J. & Stępniewski, W. 2008. Variability of the non-methane volatile organic compounds (NMVOC) composition in biogas from sorted and unsorted landfill material. *Archives of Environmental Protection* 34(3): 287–298.

Pawłowski, A. 2009a. Teoretyczne uwarunkowania rozwoju zrównoważonego, in: *Rocznik Ochrona Środowiska* 11, 986–994.

Pawłowski, A. 2009b. The Sustainable Development Revolution, in: *Problemy Ekorozwoju/Problems of Sustainable Development* 4(1): 65–76.

Pawłowski, A. 2009c. Sustainable energy as a sine qua non condition for the achievement of sustainable development. *Problemy Ekorozwoju/Problems of Sustainable Development* 4(2): 9–12.

Pawłowski, A. 2009d. Theoretical Aspects of Sustainable Development Concept. *Problemy Ekorozwoju/Problems of Sustainable Development* 11(2): 985–994.

Pawłowski, A. 2013. Sustainable Development and Globalization. *Problemy Ekorozwoju/Problems of Sustainable Development* 8(2): 5–16.

Pawłowski, L. 1990. Chemical threat to the environment in Poland. *Science of The Total Environment* 96(1–2), 1–21.

Pawłowski, L. 2012. Czy liberalny kapitalizm i globalizacja umożliwiają realizację strategii zrównoważonego rozwoju? *Problemy Ekorozwoju/ Problems of Sustainable Development* 7(2): 7–13.

Piecuch, T., Kowalczyk, A., Dąbrowski, T., Dąbrowski, J. & Andrieyevska, L. 2015. Zmniejszenie uciążliwości zapachowych oczyszczalni ścieków w Tychowie. *Rocznik Ochrona Środowiska* 17(1), 646–663.

Rzeszutek M. & Bogacki M. (2016). Ocena modelu dyspersji zanieczyszczeń powietrza OSPM: studium przypadku, Polska, Kraków. *Rocznik Ochrona Środowiska*, 18, 351–362.

Silva, R. A., West, J., Zhang, Y., Anenberg, S. C., Lamarque, J.F., Shindell, D.T., Collins, W.J., Dalsoren, S., Faluvegi, G. & Folberth, G. 2013. Global premature mortality due to anthropogenic outdoor air pollution and the contribution of past climate change, *Environmental Research Letters* 8(3).

Sówka, I. & Grzelka A. 2016b. Zastosowanie krajowych i europejskich rozwiązań w ocenie jakości zapachowej powietrza na obszarach w pobliżu obiektów gospodarowania odpadami. *Rocznik Ochrona Środowiska* 18: 794–802.

Staszewska, E. & Pawłowska, M. 2011. Characteristics of Emissions from Municipal Waste Landfills. *Environmental Protection Engineering* 37(4):119–130.

Szewczyńska, M., Pośniak, M., Dąbrowska J. & Pyrzyńska K. 2016. Badanie rozkładu stężeń wybranych WWA we frakcjach cząstek drobnych emitowanych z silników pojazdów samochodowych. *Rocznik Ochrona Środowiska* 18: 74–85.

Udo, V. & Pawłowski, A. 2010. Human Progress Towards Equitable Sustainable Development: A Philosophical Exploration. *Problemy Ekorozwoju/Problems of Sustainable Development* 5(1): 23–44.

WHO 2014, Global Health Estimates 2013: Deaths by Cause, Age, Sec, by Country, 2000–2012 (provisional estimates). Geneva, World Health Organization, 2014.

WHO 2014, World Health Statistic 2014.

Zdeb, M. Pawłowska, M. 2009. An influence of temperature on microbial removal of hydrogen sulphide from biogas, *Rocznik Ochrona Środowiska* 11: 1235–1243.

Environmental Engineering V – Pawłowska & Pawłowski (Eds)
© 2017 Taylor & Francis Group, London, ISBN 978-1-138-03163-0

The influence of external conditions on the photovoltaic modules performance

A. Zdyb & E. Krawczak
Faculty of Environmental Engineering, Lublin University of Technology, Lublin, Poland

ABSTRACT: Prediction of photovoltaic performance under different outdoor operational conditions in given location is crucial for planning of the installation. This work presents the analysis of the influence of temperature, wind speed and solar irradiation on the PV (Photovoltaic) module efficiency. Since the response to variable environmental conditions differs for each kind of existing PV technology installed outdoors, we performed the analysis for polycrystalline modules as a chosen type of PV material. The results of application of three different mathematical models are shown, and correlated with the data provided by Sunny Portal for the installation facility in Spain.

Keywords: photovoltaics, PV modules, solar irradiation, PV performance

1 INTRODUCTION

Life on the Earth requires continuous supply of energy. Nowadays, our civilization consumes mainly fossil fuel resources that are being rapidly depleted. Simultaneously, according to mainstream opinion CO_2 emission causes unfavorable climatic changes. However, not all scholars share this view. For instance, Lidzen (Lindzen 2010) does not question the essence of greenhouse effect itself, but believes that the forecast changes will be far milder. The idea of sustainable development, formulated in 1987 (Brundland et al. 1987) can be used to integrate all areas of human activity.

The supply of energy is necessary for the further development of our civilization, while the CO_2 emission accompanying the production of energy from fossil fuels may lead to drastic changes in ecosystems on the Earth, possibly threatening the future of civilization. For that reason implementation of sustainable development paradigms requires including this aspects in the development of modern countries (Pawlowski 2009, 2013, Cholewa et al. 2009).

One of the energy sources that would help in implementation of sustainable development is solar energy converted into electricity in solar modules.

PV modules performance parameters, such as the maximum power point, efficiency and temperature coefficient are typically provided by manufacturers. These information is very useful but they are determined in laboratory at Standard Test Conditions (STC) of solar irradiance 1000 W/m^2, Air Mass

(AM) of 1.5 and cell temperature of 25°C. The real external conditions in which modules are expected to work differ significantly from STC parameters values. Modeling of PV modules performance as well as monitoring of working outdoor PV systems is thou necessary since manufacturer information is not reliable to predict PV operation under variable environmental conditions (Olchowik et al. 2004, Olchowik et al. 2006). Most of investigations devoted to the analysis of working PV systems and to prediction of performance of different installations types consider solar irradiation level as the only one or the most important parameter influencing PV operation. Nonetheless, it is worth to notice that only 6–20% of solar radiation illuminating modules is converted into electric energy since most of it causes heating of modules and thereby decreases module efficiency. The issue of wind speed, temperature and other environmental variables is thou addressed in some investigations (Schwingshackl et al. 2013, Dubey et al. 2013, Verhelst et al. 2013, Gökmen et al. 2016). There are number of explicit or implicit mathematical correlations describing temperature of module and its efficiency that include reference state as well as the type of PV technology which is of highly importance in this kind of studies because of the differences in temperature coefficients for various types of materials. Currently, there are several photovoltaic technologies used in modules that are commercially available. The traditional crystalline silicon (c-Si) modules are characterized by the temperature coefficient in the range of 0.4–0.6%°C^{-1}. The values of the temperature coefficient for other technologies are as follows: 0.1–0.2 for the amorphous silicon (a-Si), 0.4–0.45 for polycrystalline silicon (pc-Si), 0.484 for CIS, 0.36 for CIGS and 0.035–0.25 for CdTe (Mattei et al. 2006). In practice, modules that are less sensitive to temperature changes work better in the regions of high ambient temperature but those that are more responsive to temperature changes perform better in low temperature locations.

This work focuses on the analysis of the influence of the ambient temperature, wind speed and solar irradiation on changes in solar modules efficiency. We consider several cases of external conditions: mean values of weather parameters for two different parts of the year in temperate climate and extreme values of this parameters. The type of PV technology used is of fundamental importance to this kind of study so we choose one of them i.e. polycrystalline Si, which is very popular in the PV market. The dependences obtained by different mathematical models are confirmed by the comparison with the performance data of installed PV system.

2 METHODS

Several mathematical models were applied to evaluate the influence of external conditions such as solar irradiation, temperature and wind speed on the PV cells performance. The models and values of various models input parameters are described in details in the next section. All calculations were performed by using MatLab software.

The data of energy production of PV installation, that is analyzed in the second part of this work are provided by Sunny Portal. This open public internet source contains data for small individual PV installations as well as for big plants working in different climate locations. The meteorological data was obtained using Meteonorm 7 software.

3 RESULTS AND DISCUSSION

The performance of solar cell is described by several parameters that can be determined from the I-V characteristic curve. One of them is the short-circuit current (I_{SC}), which is the current through the solar cell when the solar cell is short circuited. Another important parameter is the open-circuit voltage (V_{OC}) observed at open circuit, when no current is flowing. The measure of maximum power from the solar cell and the squareness of the I-V curve is the fill factor:

$$FF = \frac{P_{MP}}{V_{OC}I_{SC}} = \frac{V_{MP}I_{MP}}{V_{OC}I_{SC}}, \tag{1}$$

where P_{MP} – the maximum power, V_{MP} – voltage at maximum power point, I_{MP} – current at maximum power point.

The most important figure of merit characterizing solar cell is efficiency described by the following formula:

$$\eta = \frac{P_{MP}}{P_{in}} = \frac{FFV_{OC}I_{SC}}{P_{in}}, \tag{2}$$

where P_{in} – the incident light power.

All the above described parameters depend on the external conditions in which the cells operate.

Solar irradiance is the most important of environmental parameters that influences the performance of solar cells. In general the value of this parameter depends on latitude, but well known distribution of solar irradiance on the globe (Kawajiri et al. 2011) shows also the influence of altitude, vicinity of surface water, flora and local climate conditions. In the given location solar radiation changes also with different cloudy conditions, part of a day (because of air mass factor AM) and spectral effects which are affected by humidity, AM and also aerosol particles concentration. The accessible solar radiation at latitude about 50°N varies in the range of: 50–150 W/m² (winter) and 100–300 W/m² (summer) for cloudy conditions, 150–300 W/m² (winter) and 300–600 W/m² (summer) for partly cloudy conditions, 300–500 W/m² (winter) and 600–1000 W/m² (summer) for sunny clear sky. This data point to up to 20 times difference between the highest and the lowest level of radiation. Usually output power of PV modules follows the irradiation level but some kinds of them perform significantly worse at low irradiance because of the series and shunt resistance (Randall & Jacot 2003).

Another significant parameter influencing PV devices performance is wind speed that affects module temperature and indirectly its efficiency. Wind speed is influenced by both environmental and artificial factors, so wind resources differ in various locations depending on topography, neighboring objects like watery areas, mountains and buildings. They also can vary during the day or night and during the year. Altitude, which involves thermal effects, is a crucial factor determining wind speed, too. Among several formulas used for determining wind speed dependency on the altitude, $v(H)$ there is one which is widely applied in Europe and is called the logarithmic wind profile law:

$$v = v_0 \frac{ln\frac{H}{z_0}}{ln\frac{H_0}{z_0}}, \tag{3}$$

where v_0 – the speed to the reference height $H_0 = 10$ m, z_0 – the roughness coefficient length (m) which is influenced by the roughness factor for a given land type, time of a day and season. The values of wind speed usually change in the range of less than 3 m/s to 12 m/s for different roughness of the land and different altitude.

Both solar radiation level and wind speed influence the local temperature and in consequence the temperature of working PV modules. In external conditions, thermal effects are also influenced by the altitude value since the temperature in the atmosphere decreases 0.6–1⁰, depending on the humidity, for every 100 m.

Solar cells are sensitive to temperature since they are semiconductor devices. The temperature increase causes reduction of the band gap width because of the growth of the electrons energy in semiconductor, which results in lower energy needed to break the bonds. Growth of temperature value results in V_{OC} and power decrease in spite of slight I_{SC} increase.

There are numerous models describing the solar cells performance dependence on external conditions like solar irradiation, temperature and wind speed. Some of them were chosen to be applied in the presented work. The used calculation techniques were proved to be useful in a comparative study for an accurate estimation of electric performance of the given PV system (Schwingshackl et al. 2013). Different models were used to predict wind and solar irradiance effect on PV cells temperature and in consequence on its efficiency.

The empirical model by Faiman (Faiman 2008) (model 1) presents the dependence of cell temperature on in-plane irradiance, ambient temperature and wind speed in the following equation:

$$T_c = T_a + \frac{I}{U_0 + U_1 v_w},$$ (4)

where I – in-plane irradiance, T_a – ambient temperature and v – wind speed.

In the model proposed by the Mattei (Mattei et al. 2006) (model 2), STC parameters and indirect relation between cell temperature, heat exchange coefficient and wind speed are necessary to estimate cell temperature according to the formula:

$$T_c = \frac{U_{PV}(v)T_a + I\left[\tau \cdot \alpha - \eta_{STC}\left(1 - \beta_{STC}T_{STC}\right)\right]}{U_{PV}(v) + \beta_{STC} \cdot \eta_{STC} \cdot I},$$
$$U_{PV}(v) = 26.6 + 2.3v,$$
$$U_{PV}(v) = 24.1 + 2.9v,$$ (5)

where U_{PV} – heat exchange coefficient, η_{STC}, β_{STC}, T_{STC} are given for the known type of module, $\tau \cdot \alpha = 0.81$.

Another type of model (model 3), using exponential dependency, applied to estimate cell temperature is described by the following expression (Schwingshackl et al. 2013):

$$T_c = T_a + I \cdot e^{-3.473 - 0.0594 \cdot v}.$$ (6)

Two of the models (eq. 4, 5) take into account type of the PV technology since the values of U_1, U_0, β_{STC} and η_{STC} differ for various types of materials but the last one (eq. 6) does not distinguish between different PV technologies. The parameters U_1, U_0 describe heat transfer, which is influenced by convection, conduction and radiation of the cell.

The efficiency dependence on the estimated cell temperature can be obtained according to the following formula (Skoplaki & Palyvos 2009):

$$\eta = \eta_{STC}[1 - \beta_{STC}(T_c - T_{STC})].$$ (7)

In the analysis we consider two halves of the year: April—September and October—March that differ in temperature and irradiation in temperate climate. Warmer part of the year is characterized by mean day temperature value equal to 14.4°C and irradiation 600 W/m². During the colder half of the year daily mean temperature equals 1.15°C and irradiation value is about 300 W/m². Assumptions of the models are summarized in Table 1.

The dependences of cell temperature on the wind speed according to three considered models are presented in Figure 1.

There are three curves, one for each model, representing data obtained for colder and warmer part of the year. Model 1 (eq. 4) shows the strongest influence of the wind speed on temperature of solar cell. The steepest part of all the visible curves is in the range up to 10 m/s of wind speed. The increase of the wind speed to 10 m/s causes the drop of cell temperature of about

Table 1. The input data for the applied models.

T_{a1} [°C]	T_{a2} [°C]	I_1 [W/m²]	I_2 [W/m²]	U_0 [W/°Cm²]	U_1 [Ws/°Cm³]	η_{STC} [%]	β_{STC} [°C⁻¹]	T_{STC} [°C]
14.4	1.15	600	300	30.02	6.28	14	0.0045	25

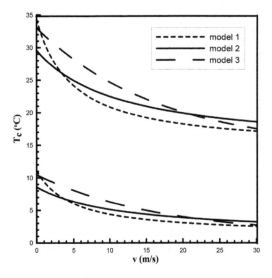

Figure 1. The dependences of poly-Si cell temperature on the wind speed according to three considered models in warm and cold part of the year.

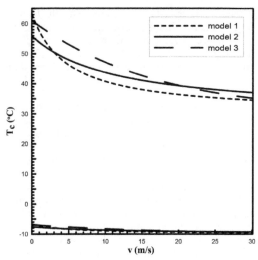

Figure 2. The dependences of poly-Si cell temperature on the wind speed according to three considered models at extreme conditions of outdoor temperature and solar irradiation.

7–14 °C depending on the model. Further increase of wind speed decreases the temperature of less than 5°C. Exponential dependency represented by model 3 shows the most uniform, almost linear changes. Analogues analysis were performed for temperature and irradiance values corresponding to extreme outdoor conditions. We choose two reference sets of data: $T_{a1} = 30°C$, $I_1 = 1000$ w/m² for hot sunny day and $T_{a2} = -10°C$, $I_2 = 100$ W/m² for cold cloudy day. All other parameters taken into account in this case are presented in Table 1.

The dependencies obtained for three used models shown in Figure 2 have similar shape as in the previous case. Cooling effect of wind is clear and results in temperature drop of more than 20°C for hot day and about 2–3°C in cold environment.

Taking into account each of three temperature models in efficiency formula (eq. 7) we obtained dependencies very similar to each other and overlapping lines shown in Figure 3. The drop rate of efficiency value with temperature in both halves of the year is the same.

The comparison of the results obtained by application of three described models with the experimental data (Schwingshackl et al. 2013) shown different root mean squared error (RMSE) for different types of PV modules technology.

In the case of poly-Si technology, considered in this work, RMSE value changes are in the range of 1.8 for model 2 to 4.5 for model 3, indicating the approach represented by eq. 5 (model 2) as the most adequate.

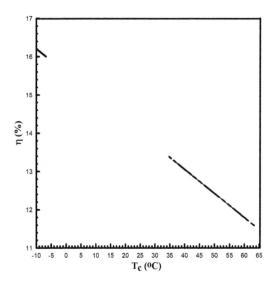

Figure 3. The changes of efficiency on the poly-Si cell temperature.

To analyze the influence of the ambient conditions on outdoor photovoltaic performance, exemplary roof-mounted grid-connected small PV installation has been chosen. The analyzed installation is located in east Spain in the city Gata de Gorgos on the north Costa Blanca in Mediterranean. Coordinates of Gata de Gorgos are 38°46'N 0°04'W and the elevation is 82 m. The system with a capacity of 11,07 kWp was installed in 2007 and consists of 54 polycrystalline modules of 205 Wp

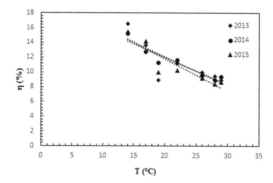

Figure 4. The influence of outdoor temperature on the efficiency of PV system in Gata de Gorgos, Spain.

each. Solar cells are oriented to south, the angle of inclination equals to 35°.

The performance analysis of described system involves estimation of various parameters that are collected by Sunny Portal, i.e. daily yield, energy and power of inverters. Data are measured 24 hours every day. One of the essential values for evaluating of PV system is its efficiency η (%), which is defined as the ratio of dc energy yield by total plane of array irradiation and the area of PV modules. The efficiency can be calculated according to the following equation:

$$\eta = \frac{E_{dc}}{H \cdot A} \cdot 100, \quad (8)$$

where E_{dc} is total dc energy yield (kWh), H is total plane of array irradiation (kWh/m^2) and A is area of the incident irradiation (m^2).

The influence of the air temperature on efficiency of the installation is shown in Figure 4. The highest values of efficiency are achieved when the outside temperature has the lowest values. The analyzed PV system had the best efficiency 17,87% when the average outdoor temperature was 14°C in year 2013 and the worst efficiency of 8,4% at temperature 28°C in 2015. According to data collected by Sunny Portal it can be seen that these dependences are analogues in the following years. The figure shows that value of system efficiency is directly related to the outdoor temperature.

4 CONCLUSION

In the planning stage of the PV investment, it is important to take into account not only the intensity of solar radiation but also other external conditions such as temperature and wind speed in the chosen location. The PV modules manufacturers provide the information on performance parameters only for Standard Test Conditions which are rarely met in natural environment but the real outdoor PV performance is influenced by different environmental variables. The decision regarding the investment in photovoltaics, especially the choice of location for the installation should be based on the theoretical predictions of PV performance in the given place. Model based analysis performed in this work shows the influence of the wind speed for the PV module temperature, which is strong for wind up to 10 m/s and less sensitive for further wind speed increase. The obtained dependences of wind cooling effect are of greater importance in the higher temperature range characteristic for warmer part of the year. The efficiency of modules exhibits the stable drop rate with the temperature growth in the entire range of temperature value. The results of models application correlate well with the real data of PV installation performance in Spain.

REFERENCES

Brundtland, G., Khalid, M., Agnelli, S. 1987. Our common future. Report of the World Commission on Environment and Development.

Cholewa, T., Pawłowski, A. 2009. Sustainable use of energy in the communal sector. Rocznik Ochrona Środowiska 11: 1165–1177.

Dubey, S., Sarvaiya, J.N., Seshadri, B. 2013. Temperature Dependent Photovoltaic (PV) Efficiency and Its Effect on PV Production in the World A Review. Energy Procedia 33: 311–321.

Faiman, D. 2008. Assessing the Outdoor Operating Temperature of Photovoltaic Modules. Prog. Photovolt: Res. Appl. 16: 307–315.

Gökmen, N., Hu, W., Hou, P., Chen, Z., Sera, D., Spataru, S. 2016. Investigation of wind speed cooling effect on PV panels in windy locations. Renewable Energy 90: 283–290. doi:10.1016/j.renene.2016.01.017

Kawajiri, K., Oozeki, T., Genchi, Y. 2011. Effect of Temperature on PV Potential in the World. Environmental Science and Technology 45: 9030–5.

Lindzen, R. 2010. Global Warming: the origin and nature of the Alleged scientific consensus. Problems of Sustainable Development 5: 13–28.

Mattei, M., Notton, G., Cristofari, C., Muselli, M., Poggi, P. 2006. Calculation of the polycrystalline PV module temperature using a simple method of energy balance. Renewable Energy 31: 553–567.

Olchowik, J.M., Jóźwik, I., Szymczuk, D., Zabielski, K., Mucha, J., Tomaszewski, R., Banaś, J., Olchowik, S., Adamczyk, J., Cieplak, T., Zdyb, A. 2004. Analysis of solar cells efficiency in hybrid solar system under conditions of south-easterly Poland. Proc. of 19th European Photovoltaic Solar Energy Conference, 7–11 June 2004, Paris, France: 3294–3296.

Olchowik, J.M., Gułkowski, S., Cieslak, K., Jóźwik, I., Banaś, J., Olchowik, S., Zdyb, A., Szymczuk, D., Adamczyk, J., Tomaszewski, R., Zabielski, K.,

Mucha, J., Cieplak, T. 2006. Comparative study of the solar modules performance in the hybrid system in south-easterly Poland during first two years of exploitation. Proc. of 21th European Photovoltaic Solar Energy Conference, 4–8 September 2006, Dresden, Germany: 3049–3050.

Pawłowski, A. 2009. Sustainable energy as a sine qua non condition for the achievement of sustainable development, Problems of Sustainable Development 4: 9–12.

Pawłowski, A. 2013. Sustainable development and globalization. Problemy Ekorozwoju 8: 5–16.

Randall, J.F. & Jacot, J. 2003. Is AM1.5 Applicable in Practice? Modelling Eight Photovoltaic Materials With Respect to Light Intensity and Two Spectra. Renewable Energy 28: 1851–1864.

Schwingshackl, C., Petitta, M., Wagner, J.E., Belluardo, G., Moser, D., Castelli, M., Zebisch, M., Tetzlaff, A. 2013. Wind effect on PV module temperature: Analysis of different techniques for an accurate estimation. Energy Procedia 40: 77–86.

Skoplaki, E., Palyvos, J.A. 2009. On the temperature dependence of photovoltaic module electrical performance: A review of efficiency/power correlations. Solar Energy 83: 614–624.

Verhelst, B., Caes, D., Vandevelde, L., Desmet, J. 2013. Prediction of yield of solar modules as a function of technological and climatic parameters. International Conference on Clean Electrical Power (ICCEP): 1–6. DOI: 10.1109/ICCEP.2013.6586956.

Environmental Engineering V – Pawłowska & Pawłowski (Eds)
© 2017 Taylor & Francis Group, London, ISBN 978-1-138-03163-0

Operational characteristics of the heat and cold storage in traction vehicles

D. Zieliński & K. Przytuła
Electrical Drives and Machines Department, Faculty of Electrical Engineering and Computer Science, Lublin University of Technology, Lublin, Poland

K. Fatyga
Faculty of Electrical Engineering and Computer Science, Lublin University of Technology, Lublin, Poland

ABSTRACT: This article presents research on minimizing power consumption by air conditioning systems in municipal buses. Proposed solution is system based on storing energy in form of latent heat. Prototype system was build and further research was conducted, testing energy storage capacity. In comparison to electrochemical energy sources prototype system proved itself to be viable solution, at the same time drastically reducing cost of purchase and operation of municipal bus.

Keywords: heat and cold storage, air conditioning system, latent heat, traction vehicle

1 INTRODUCTION

Vast development of urban areas that is observed in recent years is the main reason for which the bus would make the best means of transport for many people. Crowd and traffic jams in city centres enforce the usage of an alternative to engines propulsion systems, i.e. electric motors. This is imperative in order to avoid pollution of air due to intense production of exhaust by vehicles. Despite many advantages, buses with electrical motor as propulsion are not majority in transport fleets. The main reason is the cost of purchase which is drastically increased by price of chemical batteries. The price of batteries constitutes about 50% of the vehicle cost, with their liveliness about ten years, and average lifespan—about twenty years. What makes the case interesting, is the fact that if a battery supplied energy only to drive motors, and not air conditioning system, its size could be reduced by half in relation to the battery supplying both systems.

This article presents an alternative solution of thermal energy storage for buses, which can both store and produce heat and cold. It presents the main concept behind this solution. The aim of this system is to replace traditional ways of providing thermal comfort for passengers with innovated system, capable of the production and storage of heat and cold. The most important advantage of this solution for end user is the reduction of electrical bus purchase and operation costs, as well as an increase in the energy management efficiency.

The first part of the article presents the theoretical background, as well as the comparison between the electrochemical batteries and energy storing systems based on latent heat, while the second part focuses on the experiment done with a prototype storage system.

2 EXPLOITATION PROBLEMS, OCCURRING IN URBAN TRANSPORT VEHICLES

Modern vehicles used in transport fleets, such as URSUS T70116 electrical bus, are equipped with sets of Li-Ion batteries. The most frequent solution used by manufacturers consists in Li-Ion batteries: this is because of their high energy density per mass unit, which peaks at 150 Wh/kg (Fig 1).

In practice, their capability is much lower. In case of deep dis charge cycles, the battery may rapidly loose its capacity, and to prevent this vehicle manufacturers limit the range of battery charging, with 90% becoming fully charged, and 40% capacity becoming completely discharged. Reason after this is warranty, and requirement for 10-year lifespan. This, combined with Basic Energy Storing Capacity (BESC) at 150 Wh/kg in case of charging in range 0–100%, gives only 75 Wh/kg. Such reduced value of BESC is caused by batteries using only 50% of their energy storing capacity.

The batteries used in the vehicle Ursus are characterized by the following parameters:

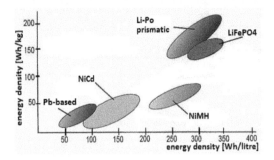

Figure 1. Examples of the energy density of the electrochemical cells.

Table 1. Parameters of batteries used in the transport vehicles.

Vehicle type	Ursus T70116
Batteries	EVC CEGTR02-1
Manufacturer	EVC-Kewelle International
Ability to store the energy	34 kWh
Capacity	53 Ah
Nominal voltage	651V
Energy density	81.92 Wh/kg
Cost	120 000 Euro

Unfortunately, because of the warranty agreement, these batteries can be discharged only in 50%, otherwise their lifespan would be greatly reduced. Manufacturers enforce this by system that detects battery charge level at 50% and sends signal that battery is discharged. Therefore, their actual energy density is only 40.96 Wh/kg. The results obtained from research done in collaboration with Municipal Transport Company in Lublin (MPK Lublin) show that the real energy density of Li-Ion batteries is 1/4 compared to one presented in articles concerning this technology. This result can be only decreased by the fact that calculations omitted aspects of battery cells ventilation, which is necessary.

3 REDUCTION OF THE OPERATING COSTS

On the basis of the research done by MPK Lublin Trolleybus Department, it was stated that half of energy stored in a vehicle is being used to cover need of passenger comfort. This creates an opportunity for alternative energy sources capable of providing heat or cold. A promising solution is an energy source that is not based on chemical transformations, as it takes place in Li-Ion batteries, but rather based on physical phase transitions

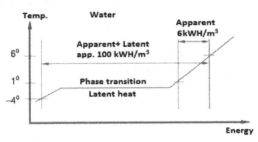

Figure 2. The possibility of energy storage in distilled water.

of medium, and storing energy in form of latent heat.

It appears, that amount of stored energy per mass unit of distilled water could be similar if not higher than the energy density of lithium-ion battery. Distilled water is able to accumulate 100 Wh/kg of energy, when its temperature is floating close to its phase transition. Out of this range, distilled water is able to store 1.16 Wh/kg per 1°C. This allows to store about 111.6 Wh/kg in the case of working cycle in temperature ranging from −4°C to 6°C. Additionally, the cost of building special storage system, which could replace half of Li-Ion batteries in single bus, is really low. The prototype version of such storage cost 12 000 PLN (~3 000Euro). Another advantage encouraging the use of this type of storage is its almost perfect invulnerability to changing temperatures, and no capacity decay over time.

4 TECHNICAL SOLUTION, PROVIDING PASSENGER THERMAL COMFORT

Air conditioning systems used in vehicles work as autonomous systems (Fotouhi et al. 2015). The status of air conditioning is based on the thermostatic controller. Its main control element is a comparator with hysteresis [2]. The system compares the value set by the driver of the vehicle with the current temperature around the passenger seat. There are a lot of disadvantages to these solutions:

* multiple startups of air conditioning system and large starting currents, reaching values 3 to 7 times higher than rated currents
* mechanical shocks, reducing the lifespan of the compressor
* inconsistent performance caused by transients
* battery lifetime being reduced due to high startup current

Furthermore, manufacturers of air conditioning systems use cheap and imprecise automatic control systems, based on mechanical devices. This approach causes a huge inertia of the control circuit [8], [9]. If the piston compressors are used, using motor drive system with converter is impossible. [10]. The main case of the limitation is poor lubrication when the motor speed is low. It causes damage to the system. For these reasons, the air conditioning system is not considered as a potential energy storage that could increase the coverage and overall efficiency of the vehicle.

5 PROTOTYPE WITH THE COLD STORAGE

The proposed solution is non-autonomous system. Main task of this system is to provide comfort for passengers, with minimal load of Lithium-Ion batteries. Moreover, the system is integrated with the main inverter and provides the ability to produce braking torque and thus the recuperation of energy. There are two hydraulic circuits in this system (Fig. 3), to store and manage heat and cold efficiently. The first circuit (cooling) is equipped with a coil compressor with a speed regulation provided by an integrated inverter. The compressor is connected with two heat exchange circuits (condenser-evaporator SK/PR) and the four-way valve Z4D.

This solution allows to change the direction of the coolant flow immediately, and thus it is possible to operate in two modes: air conditioning and heat pump. The use of thermal expansion valves and ZR1 ZR 2-based control systems utilizing electronic micro-stepping technology allows to dose the coolant to the evaporator and achieve significantly better dynamics of the whole system. In comparison to the mechanical expansion valves, the thermal expansion valves have regulation from 0 to 100%, with a time constant about 0.1 s. (Tingrui 2008). In the first circuit, R507 is used as a working medium, which operates in temperature ranging from −40°C to + 35°C. The liquid used in the system covers the full range of temperatures

Figure 4. Cartridges filled to 95% with distilled water, and prototype system to store heat or cold in the public transport vehicles.

needed to achieve thermal comfort of the passengers. The second circuit is a hydraulic system; the working medium is a water-glycol solution with a 35% concentration. It is lead to the cool storage or heat cabin KWC, installed inside the vehicle, by the PO pump and matrix controlled ZR valves. The part of the system, coupling the two circuits, is a plate heat exchanger WP. The cold storage MC is built as a thermally insulated—with polyurethane foam and polystyrene granules—tank. There are cylindrical cartridges made of HDPE plastic inside the tank, which are a heat transfer moderator (Fig. 4).

Cartridges were filled to 95% with distilled water and put in the tank, to allow free flow of the aqueous glycol solution and contact between the flowing liquid and the cartridges. The prototype storage consists of the 330 kg cartridges, which could be

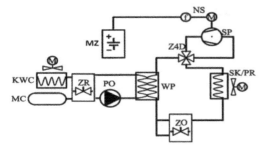

Figure 3. The prototype system to store heat or cold in the public transport vehicles.

cooled or heated to a determined temperature. The system can freeze the inserts or bring them to a boil while the vehicle is on the depot. This is determined as the cycle of normal storage charging, and it should be done during off-peak. This is especially important for the air conditioning mode when the system has the highest efficiency. When the vehicle leaves the depot and the system of thermal comfort needs to operate, a small 150 W pump and fan systems, which work inside the passenger compartment, are activated. In the case of a Ursus T70116 vehicle, it is possible to connect to electrical grid by means of pantograph traction. This feature of the Ursus vehicle gives an opportunity of charging cold storage during stops and during contacts with traction. An additional function of the system is the generation of braking torque and energy recuperation. The inverter used in the prototype system is directly connected to the main current bus. In the case of the prolonged braking or increased voltage in traction above the rated value, the system may switch to charging mode. In addition, temporary overload of the compressor is about 20 kW.

6 RESEARCH RESULTS

In order to investigate the possibility of energy storage of heat or cold, the test charging and discharging of the tank was made. One of the cartridges was equipped with three sensors PT100 and placed in the centre of the tank. The sensors are installed in order to determine the temperature distribution inside the cartridge. For this reason, the first sensor was located in the geometrical centre of the cartridge, the second sensor was installed at the inner wall of the cartridge, and the last sensor in a distance of 10 mm from the cartridge, in order to allow monitoring of the temperature of the flowing glycol. Research was carried out at constant cooling capacity, which amounted to 20 W/kg load. The temperature range was from 15°C to −15°C. The results were as follows:

The chart 4a shows the moment of transition of distilled water from liquid to solid state. The cartridge stores energy in the form of latent heat, and therefore it maintains a constant temperature for 170 min. Within this time, it may store a great amount of energy. The situation is identical in the case of full discharge Fig. 4b. Temperature contribution increases linearly until the phase transition. After 400 min, the cartridge maintains a constant temperature, whereupon the temperature rises linearly.

Analysis of the stored energy in the tank shows that the received energy is less than 10 kWh of the inserted energy. It is caused by the energy loss (imperfect insulation tank) and energy losses in the circulation pump. The average efficiency of charging and discharging tank was 80%. In comparison to the efficiency of the Li-ion battery, which is 85.5%, it is a very good result. The efficiency of the storage tank can be significantly improved by using better insulation materials. According to analysis of the amount of stored energy, the heat and cold storage can be a competitive solution. From −4C to +6C, it is possible to store about 51 kWh of energy. If the cartridge weighs 330 kg, proportionally it amounts to 151 Wh/kg, and is consistent with the theory – 160 Wh/kg. Of course, the ballast which is a water-glycol solution, supporting structures and the thermal insulation of the tank, should be subtracted from the results. The whole tank weighted 410 kg and the final results equalled 124 Wh/kg.

The analysis clearly specifies that, with respect to the municipal buses, the proposed solution involving heat/cooling tanks has the advantage over the electrochemical energy storage tanks. This advantage is visible in detail after obtaining an expanded computational analysis of results, which is currently carried out in the Department team in a similar manner to the analysis presented in the papers (Jarzyna et al. 2016; Filipek & Jarzyna 2012).

There are additional capabilities created with Computer-Aided Design (CAD) method of pilot plants. Such works are performed in a similar manner as described in (Duda et al. 2014).

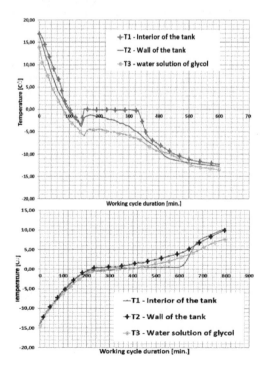

Figure 5. Cartridges temperature during work cycle: a) charging, b) discharging.

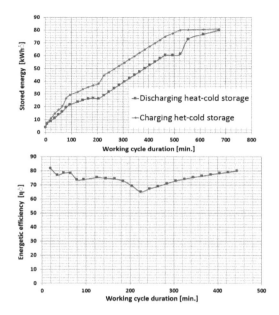

Figure 6. a) energy storage during experiment, b) energy efficiency of cold storage.

7 CONCLUSIONS

This article presented an alternate method of obtaining heat or cold energy for passenger communication vehicles. The proposed solution is based on heat/cold storage operating jointly with heat pump capable of pumping energy in two directions. The research shows that it is possible to store energy of heat or cold in the form of latent heat, in an amount comparable to charge stored in li-ion battery of same weight: for this particular case, the amount of stored energy was 124 Wh/kg, while battery capacity peaks at 150 Wh/kg. However, on the basis of the it was determined that battery capacity is at 75 Wh/kg—which is caused by manufacturers locking battery to use only part of its capacity in order to achieve longer lifespan. In that case, the heat-cold storage is able to store significantly more energy. During research, the cartridge was subcooled—this happened because the system was static, there was no stirring of cartridge, and the used container had no sharp edges. This hindered creation of crystallization nucleuses. In target, the end-user vehicles there will be stirring, due to the vehicle dynamics, as well as forced by system itself, which may improve efficiency even further. The efficiency of this solution (at 80%) is comparable to Li-Ion battery efficiency (at 85.5%), and can still be improved by using better quality insulation. Having both capacity and efficiency of Li-Ion battery and heat/cold storage system at comparable level, as well as heat/cold storage system being much cheaper allows to presume that combining both of those solutions to provide energy for electrical vehicle would drastically reduce the cost of both purchase and exploitation of such vehicle. Furthermore, the reduced cost of those vehicles may increase their popularity amongst transport fleets in companies operating in public transport, and thus increase air quality in urban areas.

REFERENCES

Cole, W.1., Powel, K. M. 1. & Edgar, T. F. in press 2012. Review of the Optimization and Advanced Control of Thermal Energy Storage Systems, *Rev. Chern. Eng.*

Filipek, P. Z. & Jarzyna, W. 2012. Financial evaluation of heat and electrical energy cogeneration in Polish household. Edited by: Pawlowski, A; Dudzinska, M. & Pawlowski, L, Conference on Environmental Engineering IV Location: Lublin, POLAND Date: SEP 03–05, *Environmental Engineering IV*: 495–498 Published: 2013.

Fotouhi, A., Auger, J., Propp, K., Longo, S. & Wild, M. 2015. A review on electric vehicle battery modelling: From Lithium-ion Toward Lithium–Sulphur, Renewable and Sustainable Energy Reviews.

Hsu Y. Y. 2007. Design and Implementation of an Air-Conditioning System with Storage Tank for Load Shifting, IEEE Transactions on Power Systems,

Jarzyna, W., Kolano, K. & Zielinski, D. 2016. Method and system of storing heat or cold in vehicles with electric propulsion. Patent Assignee POLITECHNIKA LUBELSKA.

Joybaria, M. M. & Haghighata, F. 2015. Heat and cold storage using phase change materials in domestic refrigeration systems: The state-of-the-art review, *Energy and Buildings*, 106: 111–124.

Li, H. & Jeong, S.K. 2007. The Control of Superheat and Capacity for a Variable Speed Refrigeration System Based on PI Control Logic, *International Journal of Air-Conditioning and Refrigeration* 15(2): 54–60.

Li, H., Jeong, S.K. & You, S.S. 2009. Feed forward control of capacity and superheat for a variable speed refrigeration system, *Applied Thermal Engineering* 29: 1067–1074.

Oróa, E.; de Graciaa, A., Castell,a A., Faridb, M.M. & Cabeza, L.F. 2012. Review on phase change materials (PCMs) for cold thermal energy storage applications, *Applied Energy*, 99: 513–533.

Peng, Q. & Du, Q. 2016. Progress in Heat Pump Air Conditioning Systems for Electric Vehicles—A Review, Energies—Open Access Energy Research.

Tingrui, L. 2008. Simulation and Control of Electronic Expansion Valve, *Computational Intelligence and Industrial Application*: 19–20.

Wang, Z. 2010. Experimental Study on Flow Characteristics of the Electronic Expansion Valve with Variable Condition, Power and Energy Engineering Conference (APPEEC), 28–31.

Environmental Engineering V – Pawłowska & Pawłowski (Eds)
© 2017 Taylor & Francis Group, London, ISBN 978-1-138-03163-0

Parameterisation of an electric vehicle drive for maximising the energy recovery factor

K. Kolano & M. Litwin
Electrical Drives and Machines Department, Faculty of Electrical Engineering,
Lublin University of Technology, Lublin, Poland

ABSTRACT: This article presents the results of simulations of the electric drive system in a city vehicle. The main objective of the research was to analyse the work of the recuperation of the drive. The article points out the desirability of the use of an electric drive with increased capacity to improve the recuperation performance. Attention is drawn to a significant reduction in energy consumption of a vehicle equipped with a drive system with increased power, especially during its operation in an area of considerable diversity in height. The paper points out the possibility of reducing the capacity of the chemical battery in a vehicle of this design, which will significantly reduce its production costs. For the simulation calculations an advanced ASM model of vehicle kinematics was used, prepared by dSPACE, and the electric drive model developed by the authors.

Keywords: electric vehicle, regenerative braking energy efficiency, energy consumption optimization

1 INTRODUCTION

In the situation of continuous increase in air pollution, ever more stringent emission standards are imposed on vehicle manufacturers' products. They are often so difficult to meet that more and more often the producers offer hybrid or fully electric vehicles. Vehicles with only an electrical power source are called "no-emission", as during their use no environmentally hazardous waste is produced and they do not pollute the air. This is particularly important when a large load of car traffic congests city centres, often producing smog (Clarke et al. 2010). The use of electric vehicles allows to reduce or even totally eliminate this unfavourable and unhealthy phenomenon. Such cars are also characterised by very low noise, which reduces the noise level along the routes. An electric car used in the city has another very important function, which towers over its combustion counterpart. The city, especially its centre, demands frequent acceleration and braking from car users (Siemionek & Dziubiński 2015). For each typical petrol engine car, braking can cause a loss of energy at the brakes of the vehicle in the form of heat, whereas in the case of an electric car the energy can be recycled back to the battery or to power grid through recuperation (Zielinski et al. 2015a, Zielinski et al. 2015b). In this case the energy is not lost and can be reused to drive the car. (Abousleiman & Rawashdeh 2016)

1.1 *Energy recovery in electric buses*

Recuperation, i.e. recovery of braking energy is one of the greatest benefits when using an electric car in the city. When necessary, the motor that drives the wheels of the vehicle can switch to the recovery mode of energy from a speeding vehicle (Lajunen 2013). Given the mass of the vehicle, it appears that the amount of energy that can be recovered is considerable, which reduces the average energy consumption of the vehicle and thus helps to increase its range. Another advantage is the substantial reduction in the cost of operating such a vehicle. By less frequent use of the brakes, we reduce their consumption and thus lengthen their lifespan. Recuperation can be used not only during braking, but also in other situations, such as rolling down a steep hill. We then use the potential energy of the car to recharge the batteries. The greater the diversity of terrain, the higher the profits offered in relation to the combustion counterpart. What is interesting is that in an electric car there is no concept of wear and tear of the drive unit, which can very often move smoothly from motor to generator operation, thus giving designers a wide leeway in the selection of the control algorithm of such a vehicle (Speidel S & Bräunl, 2014).

As the scope of the expected dynamics during acceleration is significantly lower than during braking, the recovery process itself is not enough to slow down the vehicle in the whole range of driver requirements and must be supported by conventional mechanical brakes. The relationship between the braking torque of the drive system and the torque generated by the braking system is determined individually for the vehicle and depends largely on the power of the drive system that can be transferred to the battery. This reduces the overall energy balance of the vehicle and invites one to analyse the possibility of using an oversized drive system, whose full potential would be used precisely in the braking process.

2 THE IDEA OF THE RESEARCH METHOD

The realities of the market confront manufacturers with the need for frequent introduction of new solutions to improve those already in use and upgrade such parameters as acceleration, the amount of energy or fuel consumed, and many others. This makes it necessary to undertake research with a high degree of complexity, which in principle is time-consuming and cost-intensive. In the initial phase of development of the automotive industry it was mainly experimental research that was used. With the development of information technology the potential of simulation studies has been noticed by vehicle manufacturers. In addition to speeding up work and reducing its cost, research teams received an asset which is always lacking during road tests, i.e. constancy of the test parameters. One has to struggle with including such factors as wind, repeatability of maneuvers, reproducibility of the running speed, pressure, and many others. These elements cannot be recreated, or it is heavily impeded, during experimental tests.

Another aspect in favour of conducting simulation tests is elimination of the need for a test track. It is necessary during the research work to eliminate at least one of the biggest variable factors, that is traffic.

Simulation methods are based on increasingly sophisticated mathematical models, subject to continuous development. Modified mathematical models are more accurate and reflect reality faithfully. They are also often combined in ever larger groups, which allows not only for mapping the dynamics of the car itself, but also its behaviour in given conditions, the external environment (surface, weather conditions). These features make the simulation method very widespread in comparative studies of energy consumption before and after modifications in the car. Such studies require very high repeatability, the same environmental and traffic conditions. Tests allow to determine the effect of even minor construction elements on energy consumption.

2.1 Brief description of the vehicle kinematics modelling environment

The simulation method used involved ASM, a very accurate mathematical model simulation software [6]. This is a package that allows not only to reproduce the vehicle dynamics in a very exact way, but also a large number of important elements of the external environment, such as temperature, pressure, surface, weather conditions or, for example, the shape of the route and terrain important for us. Each of these elements can be freely modified and adapted to the needs of the simulation. Even small parts such as the suspension have been modelled and its flexible configuration became possible.

The simulations assumed the efficiency of the drive system for engine work at 0.9. This is a typical efficiency of a modern drive system of an electric vehicle. In the process of recovery the efficiency of the engine is set to be 0.8 due to the additional losses occurring in the machine. The same values were used for the li-ion batteries. During engine work there are losses in the cells only at the resistance, but for recuperation work the voltage level is raised and ultimately thereby the amount of energy transferred to the battery is reduced.

Simulations were performed for the test track shown in (Fig. 1) and (Fig. 2), which is a reflection of the track prepared for the vehicles involved in the ShellEco-Marathon challenge in London in

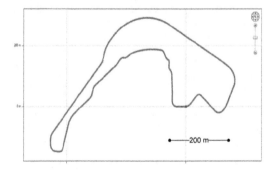

Figure 1. The route of the car in t he simulation.

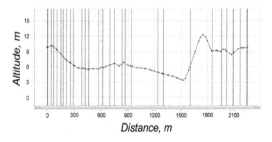

Figure 2. Altitude profile route used for the simulation.

2016. Both the shape and the relative heights are its faithful reflection.

The simulation used the data of a small city car. The parameters that were used faithfully reflect such a vehicle. The model assumed the load of two passengers weighing 80 kg each and a small luggage of 40 kg. Here are a few key parameters:

Vehicle weight:	1050 kg
Front wheels weight:	35 kg
Rear wheels weight:	32 kg
Drive:	front-wheel
Tyre width:	0.19 m
Wheel radius:	0.32 m
Drag surface:	1.94 m2
Environmental temperature:	20°C
Pressure:	1000 hPa
Engine power:	20 kW/40 kW
Engine efficiency:	0.9

3 SIMULATION RESULTS

The main motivation to start work simulation was to determine the effect of the power drive system for electric vehicle power consumption. To this end, a comparative simulation was done of a vehicle with two types of drive motors, one of whose mechanical power was twice that of the other. At the same time, the same acceleration was assumed for both systems, and the differences arise only from the amount of energy recovered during recuperation braking. The unit with the higher power rating takes into account additional losses arising from the differences in the electric size of the machines, as well as an additional mass of the drive system.

Figure 3 shows the operating principle of the braking system of an electric vehicle using the recovery effect. During the simulation it was found that the performance of most of the braking maneuvers is possible with the deceleration torque no greater than 5% of the maximum torque that can be generated by the braking system.

To the extent to which the driver performs the most deceleration maneuvers (a range of relatively small braking torques), the process of recuperation is enough to slow down the car and does not require support from a system of mechanical brakes (Abousleiman & Rawashdeh 2016). Simultaneous expansion of this range significantly reduces the need for brakes and contributes to a better balancing of the vehicle's energy.

Due to the considerable deceleration torque required it is not possible to take over the entire braking process by the drive system—it would demand a significant increase in its power and in the weight of the vehicle (Clarke et al. 2010). It is important to carry out a feasibility study of increasing the power

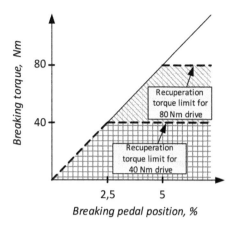

Figure 3. Dependence of the braking torque on the brake pedal position.

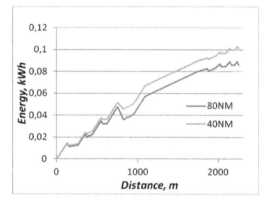

Figure 4. Dependence of the electricity consumption of a vehicle with two drive systems on flat ground.

of the drive system to the extent to which it does not increase the cost and weight while also improving the range of the vehicle.

3.1 Simulations of a vehicle travelling on flat ground

During the simulation we used the same track, but the height differences were eliminated. It would seem that only in mountainous terrain or in urban areas does it often come to regenerative operation of an electric vehicle's drive. The simulation results draw attention to the fact that even when driving outside the city and on flat ground the process of recovering kinetic energy occurs very often and has a significant impact on the overall energy balance of the vehicle. This is because the driver doing the maneuvers associated with the change of direction (turning, braking maneuvers) often uses the brake pedal.

Waveforms obtained by simulation clearly show a lower power consumption by the vehicle with a drive system of greater power that allows a greater extent of energy recovery. This difference is about 12%, which is a significant value. At the same time a greater use of the recovery effect may be imperceptible to the driver and have no negative effect on the comfort of using the vehicle (Figure 2).

The speed versus time waveform for the two drive systems shows that they differ in almost nothing from each other—the system with an oversized (heavier by 40 kg) electric drive provides no worse parameters than the standard one.

3.2 Simulations of a vehicle moving over territory with significant differences in height

Figure 6 shows the course of the amount of energy consumed in the function of the road travelled by the vehicle with two types of drive.

Also, the waveform obtained in terrain with a significant height differentiation shows a lower power consumption of the vehicle with a drive system that allows energy recovery to a greater extent during the simulation in terrain with considerable differentiation of height.

Clearly it can be seen that in terrain of considerable height variation the benefits of extending the range of recuperation work are significantly higher than those obtained over flat terrain. At the same velocity waveform (Fig. 7), an energy consumption reduced by almost 21% was obtained.

Based on the results obtained during the simulation it can clearly be stated that selection of an electric drive system taking into account only the vehicle's acceleration is improper. This procedure will only meet the requirements for the acceleration of the vehicle, but by excluding the possibility

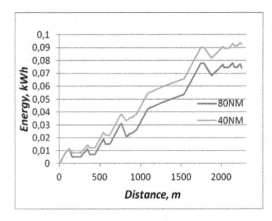

Figure 6. Dependence of the electricity consumption of a vehicle with two drive systems in terrain of considerable diversity in height.

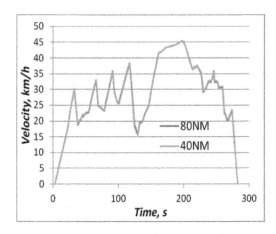

Figure 7. The course of the speed of vehicles with different drive systems.

of energy recuperation of translational motion it will entail installing more chemical batteries in the vehicle. Increasing the power of the drive system will mean that, by broadening the scope of recuperation work a greater driving range is obtained. The possibility of reducing the chemical batteries causes a significant lowering of expenses, which are disproportionately higher than the additional cost of an electric drive with increased power.

4 CONCLUSIONS

The article points to the desirability of scaling electric drive as a function of the predicted maneuvers and the shape of the route. Also, operating a vehicle in the city demands frequent braking, which means that a more powerful drive could lead to a

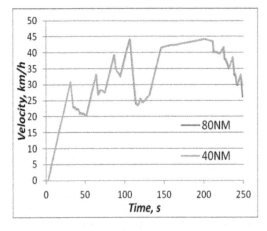

Figure 5. Speed waveforms of vehicles with different drive systems during the simulation.

reduction in energy consumption by increasing the share of energy recovered in the overall energy balance of the car. The benefits of an oversize electric propulsion system have been determined, which allows a better energy balance of the vehicle in mountainous terrain or during frequent changes of speed forced by intense maneuvers. Paradoxically, the use of an electric drive with greater power enables reduction in battery capacity while maintaining the assumed range of the vehicle.

REFERENCES

Abousleiman, R. & Rawashdeh, O. 2016. Electric vehicle modelling and energy-efficient routing using particle swarm optimization. *IET Intell. Transp. Syst.*, 10(2): 65–72.

Biczel, P. 2008. Energy Storage Systems. Chapter of Power Electronics in Smart Electrical Energy Networks Part of the series Power Systems, 269–302.

Clarke, P., Muneer, T. & Cullinane, K. 2010. Cutting vehicle emissions with regenerative braking. *Transportation Research Part D* 15: 160–167.

Cuma, M.U. & Koroglu, T. 2015. A comprehensive review on estimation strategies used in hybrid and battery electric vehicles. *Renewable and Sustainable Energy Reviews*, 42: 517–531.

Gerssen-Gondelach, S.J. & Faaij, A. 2012. Performance of batteries for electric vehicles on short and longer term. *Journal of Power Sources*, 212: 111–129.

Haddadian G., Khodayar M. & Shahidehpour M. 2015. Accelerating the Global Adoption of Electric Vehicles: Barriers and Drivers. *The Electricity Journal*, 28(10): 53–68.

Hellgren, J. 2007. Life cycle cost analysis of a car, a city bus and an intercity bus powertrain for year 2005 and 2020. *Energy Policy*, 35: 39–49.

https://www.dspace.com/en/inc/home/products/sw/automotive_simulation_models.cfm.

Lajunen, A. 2013. Energy-Optimal Velocity Profiles for Electric City Buses. 2013 IEEE International Conference on Automation Science and Engineering (CASE).

Lajunen, A. 2014. Energy consumption and cost-benefit analysis of hybrid and electric city buses. *Transportation Research, Part C* 38: 1–15.

Paska, J., Biczel, P. & Kłos, M. 2009. Hybrid power systems—An effective way of utilising primary energy sources. *Renewable Energy* 34(11): 2414–2421.

Sajadi, A., Klos, M., Biczel, P. & Biabani, M. 2012. Future perspectives of energy sector and energy market in EU. Environment and Electrical Engineering (EEEIC), 2012 11th International Conference on. 600–605.

Siemionek, E. & Dziubiński, M. 2015. Testing Energy Consumption in the Trolleybus and The Bus on a Chosen Public Transport Line in Lublin. *Advances in Science and Technology-Research Journal*, 9(26): 152–156.

Speidel, S. & Bräunl, T. 2014. Driving and charging patterns of electric vehicles for energy usage. *Renewable and Sustainable Energy Reviews*, 40: 97–110.

Wua, X., Freeseb, D., Cabrerab, A. & Kitchb, W.A. 2015. Electric vehicles energy consumption measurement and estimation. *Transportation Research, Part D: Transport and Environment*, 34: 52–67.

Zhibo, Y., Hamid, H., Tsuyoshi, K., Gnapowski S. & Hidenori A. 2014. Post-breakdown dielectric recovery characteristics of high-pressure liquid CO_2 including supercritical phase. *IEEE Transaction on Dielectrics and Electrical Insulation Society*, 21(3): 1089–1094.

Zielinski, D., Jarzyna, W. & Filipek P. 2015a. Method for testing converter network of electricity generating source, involves controlling source by first control bus and processing control signals using analog to digital converter, and transmitting signals to amplifier by control bus. Patent Number: PL407835-A1, Patent Assignee: POLITECHNIKA LUBELSKA.

Zielinski, D., Lipnicki, P. & Jarzyna W. 2015b. Synchronisation of voltage frequency converters with the grid in the presence of notching. Compel-The International *Journal for Computation and Mathematics in Electrical and Electronic Engineering*, 34(3): 657–673.

Environmental Engineering V – Pawłowska & Pawłowski (Eds)
© 2017 Taylor & Francis Group, London, ISBN 978-1-138-03163-0

Coal and biomass co-combustion process characterization using frequency analysis of flame flicker signals

A. Kotyra, W. Wójcik, D. Sawicki & K. Gromaszek
Institute of Electronics and Information Technology, Lublin University of Technology, Lublin, Poland

A. Asembay, A. Sagymbekova & A. Kozbakova
Kazakh National Research Technical University after K.I. Satpayev, Almaty, Kazakhstan

ABSTRACT: The paper presents the application of frequency analysis of time series for diagnostic purposes of biomass and coal co-combustion process in laboratory scale. The analysed time series were generated as sequences of individual image pixels of known coordinates in video recordings that were captured during combustion experiments. The resulting spatial distribution of the dominant frequency and frequency centroid were analysed for several biomass co-combustion trials where thermal power and excess air coefficient were adjusted independently.

Keywords: biomass, combustion, fuzzy methods, artificial immune systems

1 INTRODUCTION

Coal is still one of the main fuels used for both electricity and heat generation. Unfortunately, its combustion increases emissions of volatile pollutants such as Nitric Oxides (NO_X) and CO_2. For that reason, the European Union has endorsed a firm commitment that EU countries are obliged to reduce greenhouse gases by at least 20% by 2020, comparing to the 1990 level.

Dissemination of renewable fuels is considered as one of the most important means of reducing greenhouse-gas emissions, especially CO_2 for it is absorbed during plant growth and released during combustion. Therefore, it does not contribute to the greenhouse effect. Biomass co-firing with coal has the capability of reducing both nitric and sulphur oxides levels in the existing pulverized coal fired power plants. What is more, overall CO_2 emissions can be reduced because biomass is a CO_2 neutral fuel (Pawłowski & Pawłowski 2016). Moreover, substituting biomass for coal reduces SO_2 emissions, as well as NO_X due to the low sulphur and low nitrogen contents in biomass (Sami 2001). Another advantage of biomass co-firing is higher volatile contents and high reactivity of both fuel and the resulting char.

Co-firing can be adopted in the existing combustion facilities at reasonable costs in a comparatively short time. Taking the above into consideration, as well as more and more strict environmental regulations and associated penalties, power stations are seriously considering co-firing locally available biomass fuels with coal in their boilers. The existing combustion facilities can be adapted after some minor modifications. However, biomass-coal co-firing has significant drawbacks that disturb boiler operation: decreased combustion efficiency, increased slagging, and corrosion (Demirbas 2004, Pronobis 2006). Other problems are related with keeping constant physical and chemical parameters of biomass that affect performance of combustion facility and make the process difficult to maintain (Wójcik et al. 2014). Taking the above into consideration, a proper diagnostic system should be applied that would keep it at optimal operational point (Ballester & García-Armingol 2010). Diagnosis of the combustion processes and combustion facilities, due to their high complexity often utilize artificial intelligence algortihms (Shuvatov et al. 2012, Smolarz et al. 2013, Mashkov et al. 2014, Mashkov et al. 2016).

The quickest but also the most difficult mean of retrieving information of a combustion process is analysing flame radiation. It directly reflects chemical reactions that take place in the reaction zone. Comparing to chemical analysis of flue gas, optical methods (Docquier & Candel 2002) provide practically undelayed information with good spatial resolution. Combustion of pulverized fuels takes place in a turbulent flow. Local fluctuations of both fuel and gaseous reagents concentrations, as well as temperature occur. It leads to permanent local changes in the combustion process intensity that results in continuous changes in flame luminosity which can be observed as flame flicker. Combustion process affects the turbulent movement of its products and reagents determine the way the flame flicker parameters such as e.g. mean luminosity. For a given fuel mixture at constant air and fuel flow, the combustion process remains in statistical equilibrium.

Flame flicker is one of the most important observable flame parameters that is used in many industrial safety systems, mainly detecting the flame out (Lu et al. 2004). It can be also used for detecting the states of combustion process that can point to an improper operation of a burner or other maintenance issues. Although methods capable for online measurement of pulverised fuel flow in a pipe (Starke et al. 2007, Rybak et al. 2014) were developed, at the present moment in industrial conditions they are hard to be applied for economical reasons. Fuel flow magnitude is usually evaluated through indirect measurements, e.g. load of the mill and fan velocity. Therefore, the accuracy of such a measurement is low, especially in industrial conditions. Thus, it cannot be utilized to assess a combustion quality for an individual burner.

Optical methods of determining combustion process parameters are widely applied in spite of threads due to the presence of a dust (Cięszczyk et al. 2016, Komada et al. 2016, Dziubiński et al. 2016, Zyska et al. 2016). Application of flame image sequence analysis for combustion process characterization has been successfully adopted (Lu et al. 2009, González-Cencerrado 2012, Sawicki et al. 2016). However, it is generally assumed that optical access to near-burner zone of the flame is possible. In the case of front fired multi burner boilers, where burners are located in only one furnace wall, this assumption may be hard to maintain or even impossible. On the other hand, it would require a dedicated system to each burner. The paper discusses the possibility of biomass co-combustion process characterization with optical access from the opposite side of the burner mounting.

2 THE EXPERIMENTAL SET UP

All the combustion tests were done in a 0.5 MWth laboratory facility at Institute of Power Engineering, providing scaled down (1:10) combustion conditions The main part is a cylindrical combustion chamber of 0.7 m in diameter and 2.5 m long. A low-NOx swirl burner of about 0.1 m in diameter is mounted horizontally at the front wall as shown in Figure 1a. The stand is equipped with all the necessary supply systems: primary and secondary air, coal, and oil. Pulverized coal for combustion

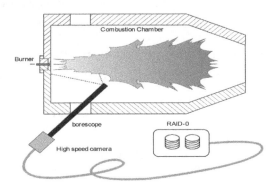

Figure 1. The borescope placement inside the combustion chamber.

is prepared in advance and dumped into the coal feeder bunker. Biomass in the form of straw is mixed with coal after passing through the feeder. Fuel mixture is delivered to the burner with the primary air. The combustion chamber has two lateral inspection openings on each side, that enabled optical access to the reaction zone. A water cooled borescope with high-speed camera attached was placed inside the combustion chamber, near the burner nozzle, as shown in Figure 1b. The borescope direction of view was 90 degrees to its axis that allowed video capture from the front of the flame at a distance of about 500 mm and at angle of 35 degrees to burner axis. The colour camera with CMOS area scan sensor was capable to acquire 500 frames per second (fps). Flame images were transferred from the interior of the combustion chamber through a 0.7 m borescope and stored in a high performance storage system.

3 THE METHODS

A sequence of N images was compacted in a single 3-dimensional data structure. The two spatial dimensions (x, y) corresponded to coordinates of each pixel of a single frame, whereas the third dimension—to a time at which the given frame was captured. The image sequence consisted of time series matched to each frame pixel of known coordinates as shown in Figure 2.

An example time series that was obtained for a given pixel coordinate x, y is presented in Figure 3. Since the images were processed as 8-bit greyscale, the pixel amplitude was between 0 and 255. Before performing the frequency analysis of the time series obtained, the mean value was subtracted:

$$a_{x,y}(t_k) = a_{x,y}(t_k) - \frac{1}{N}\sum_{k=0}^{N-1} a_{x,y}(t_k), \qquad (1)$$

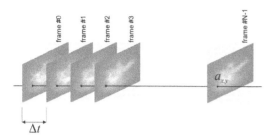

Figure 2. Data structure containing image sequence.

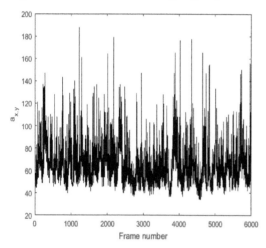

Figure 3. An example time series obtained for a given pixel coordinate x, y.

where:

$a_{x,y}$ – pixel amplitude at coordinates (x, y),
t_k – discrete time kth sample (k = 0, 1, 2...N-1),
N – number of samples.

Time series for each coordinate (x, y) in image sequence were analysed in frequency domain in order to determine spatial distribution of flame flicker. For that case, amplitude spectrum ($A_{x,y}$) can be expressed in the form of the following equation:

$$A_{x,y}(f_q) = \left|\sum_{k=0}^{N-1} a_{x,y}(t_k)\exp(-j2\pi f_q t_k)\right|, \quad (2)$$

where f_q – discrete frequency qth sample (q = 0, 1, 2...N-1).

The number of images that are captured in one second (fps) is $1/\Delta t$ where Δt denotes time resolution (distance between two neighbouring time samples). The maximum frequency of flame flickering f_{max} can be calculated on the basis of the Shannon theorem:

$$f_{max} = \frac{1}{2\Delta t} = \frac{\text{fps}}{2}. \quad (3)$$

As the number of samples in time and frequency domains equal N, the frequency resolution Δf can be expressed as:

$$\Delta f = \frac{\text{fps}}{N}. \quad (4)$$

The spectra obtained were characterized by the frequency of highest magnitude and spectral centroid that is defined as (Kua, 2010):

$$centroid_{x,y} = \sum_{k=0}^{N-1}|A_{x,y}(f_k)|f(t_k) \bigg/ \sum_{k=0}^{N-1}|A_{x,y}(f_k)| \quad (5)$$

4 EXPERIMENTS AND RESULTS DISCUSSION

During the data acquisition, the combustion process remained in stationary conditions. The single image series recording lasted 40 seconds at a rate of 150 fps, that corresponded to N = 6000 points in time domain. The fragment of an example series is shown in Figure 4.

The resolution of each captured image was 800 × 800 pixels yielding a total number of 64000 time series. According to Nyquist Sampling Theo-

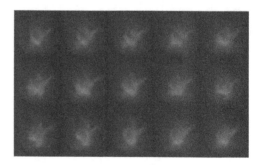

Figure 4. A fragment of an example sequence of images obtained during laboratory tests.

Table 1. Settings of combustion facility during biomass-coal co-combustion tests.

Test #	Fuel flow kg/h	P_{th} kW	λ –
1	37.2	~250	0.75
2	36.0	~250	0.65
3	38.4	~250	0.85
4	58.8	~380	0.75
5	54.0	~380	0.65
6	58.2	~380	0.85

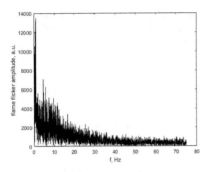

Figure 5. Amplitude spectrum obtained for an example time series.

rem, the maximum resolvable frequency is 75 Hz and following eq. (4) the frequency resolution obtained was 0.025 Hz. The amplitude spectrum obtained for an example time series is presented in Figure 5.

As it can be observed in the spectrum plot, the majority of signal power is contained within frequencies of about 1 Hz. Thus, it is necessary to provide the frequency resolution at least of the order of 0.1 Hz.

The laboratory tests were conducted for several combustion states with different settings of fuel and secondary air flows. In the first case, combustion of hard coal mixed with 10% content of biomass was investigated for two different settings of fuel flows,

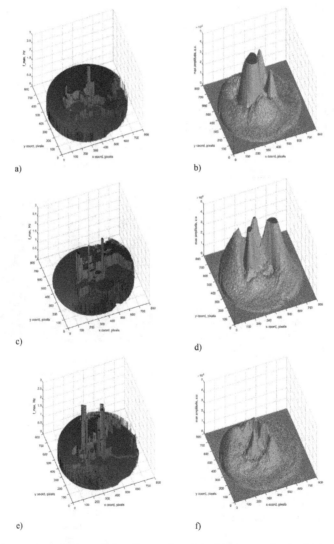

Figure 6. Spatial distributions of flame flicker frequencies of the highest amplitude obtained during combustion tests: a) test #1, c) test #2, e) test #3, spatial distributions of amplitudes of the strongest flame flicker frequency: b) test #1, d) test #2, f) test #3.

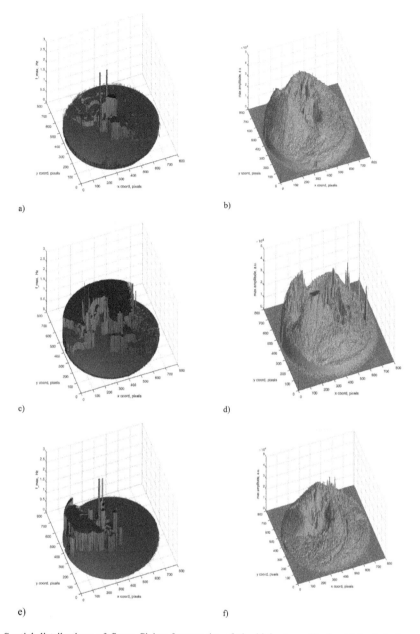

Figure 7. Spatial distributions of flame flicker frequencies of the highest amplitude obtained during combustion tests: a) test #4, c) test #5, e) test #6, spatial distributions of amplitudes of the strongest flame flicker frequency: b) test #4, d) test #5, f) test #6.

independently for three different excess air coefficients (λ) that is defined as a quotient the mass of air to combust 1 kg of fuel to mass of stoichiometric air. For that reason, secondary air flows for the same λ was different for different settings of fuel flow. The exact values of fuel flows, the resulting thermal power (P_{th}), and λ are collected in Table 1.

Spatial distributions of the maximum frequency of amplitude spectra, as well as their magnitudes are shown in Figures 6 and 7 for tests #1 – #3 (lower facility thermal load) and #4 – #6 (higher thermal load), respectively. The results obtained for frequency centroids are presented in Figures 8a– 8f.

283

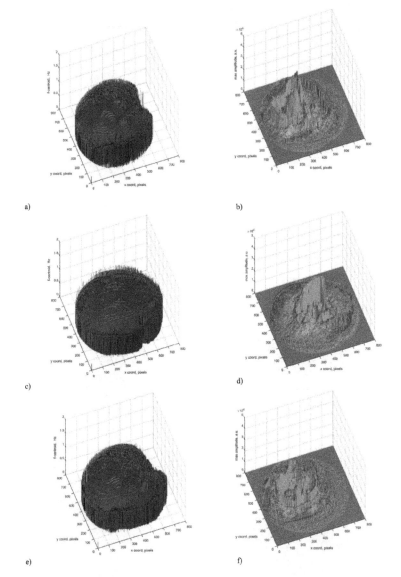

Figure 8. Spatial distributions of frequency centroids obtained during combustion tests: a) test #4, c) test #5, e) test #6, centroid magnitudes: b) test #4, d) test #5, f) test #6.

5 DISCUSSION

Due to borescope placement constraints, the burner is located in the middle of the captured images and the near-burner zone cannot be observed. The spotted flame flicker frequencies of the highest amplitudes, as it was mentioned before, are around 1 Hz. As it can be observed in Figure 5, amplitudes of flame flicker frequency drop exponentially, and above 40–45 Hz they remain on the same level.

The highest flame flicker frequencies, especially for lower thermal load are observed in the centre (Figures 6a, 6c, 6e) of images being analysed and point to burners mouth. Comparing the amplitudes of the strongest flame flicker frequency to each other, it can be noticed, that their values are higher for tests #1 and #2 (Figures 6b and 6d) when lower fuel flow and lower λ value were set.

Frequency centroid spatial distribution is much more uniform comparing to that of flicker frequencies of highest amplitude (Figures 8a, 8c, 8e). However, comparing to the amplitudes of the strongest flicker frequency, centroid magnitudes provide better localisation feature (Figures 8b, 8d, 8f).

6 CONCLUSIONS

The results obtained strongly depend on the way the images are retrieved from combustion chamber, as well as the burner type and size of combustion chamber. The camera is not located perpendicularly to burner axis as it does in majority of cases. In practice, it is often the only possible camera mounting. In the case of the power boilers with wall-mounted burners, it is the desired way of mounting for it provides optical access to all burners from a single point.

The frequency patterns obtained are more resistant to possible signal degradation due to luminous internal walls of the burner. The results obtained have shown that spatial distribution of frequency based measures of flame flicker can be applied for recognition of different states of biomass co-combustion process. Thus, they can be applied in combustion process diagnostic systems.

REFERENCES

Ballester, J. & García-Armingol, T. 2010. Diagnostic techniques for the monitoring and control of practical flames, *Prog Energy Combust*. 36: 375–411.

Cięszczyk, S. Komada, P. Akhmetova, A. & Mussabekova A. 2016. Metoda analizy widm mierzonych z wykorzystaniem spektrometrów OP-FTIR w monitorowaniu powietrza atmosferycznego oraz gazów w procesach przemysłowych. *Rocznik Ochrona Środowiska*, 18(2): 218–234.

Demirbas, A. 2004. Combustion characteristics of different biomass fuels. *Progress in Energy and Combustion Science* 30: 219–230

Docquier, N. & Candel, S. 2002. Combustion control and sensors: a review. *Prog Energy Comb Sci* 28: 107–50.

Dziubiński, G. Harasim, D. Skorupski, K. Mussabekov, K. Kalizhanova, A & Toigozhinova, A. 2016. Optymalizacja parametrów światłowodowych czujników do pomiaru temperatury. *Rocznik Ochrona Środowiska*, 18(2): 309–324.

González-Cencerrado, A. Peña, B. & Gil, A. 2012. Coal flame characterization by means of digital image processing in a semi-industrial scale PF swirl burner. *Applied Energy* 94: 375–384.

Kua, J.M.K., Thiruvaran, T., Nosratighods, M., Ambikairajah, E. & Epps, J. 2010. Investigation of spectral centroid magnitude and frequency for speaker recognition. *In Proc. Odyssey, The Speaker and Language Recognition Workshop, Brno*, Czech Republic: 34–39.

Komada, P. Cięszczyk, S. Zhirnowa, O. & Askarova, N. 2016. Optyczna metoda diagnostyki gazu syntezowego i biomasy. *Rocznik Ochrona Środowiska*, 18(2), 271–283.

Lu, G. Gilbert, G. & Yan, Y. 2009. Vision based monitoring and characterization of combustion flames. *Journal of Physics: Conference Series* 15: 194–200.

Lu, G Yan, Y. Colechin, M. & Hill, R. 2004. Monitoring of Oscillatory Characteristics of Pulverised Coal Flames through Image Processing and Spectral Analysis. *Instrumentation and Measurement Technology Conference, Como*, Italy: 1801–1805.

Mashkov, V. Smolarz, A. Lytvynenko, V. & Gromaszek, K. 2014. The problem of system fault-tolerance. *IAPGOŚ* 2014(3): 41–44.

Mashkov, V. Smolarz, A. & Lytvynenko, V. 2016. Development issues in algorithms for system level self-diagnosis. *IAPGOŚ* 2016(1): 26–28.

Pawłowski, A. & Pawłowski L. 2016. Wpływ sposobów pozyskiwania energii na realizację paradygmatów zrówmoważonego rozwoju. *Rocznik Ochrona Środowiska*, 18(2): 19–37.

Pronobis, M. 2006. The influence of biomass co-combustion on boiler fouling and efficiency. *Fuel* 85: 474–480.

Rybak, G. Chaniecki, Z. Grudzień, K. Romanowski, A. & Sankowski, D. 2014. Non-invasive methods of industrial process control. *IAPGOŚ* 2014(3): 41–45.

Sami, M., Annamalai, K. & Wooldridge, M. 2010: Cofiring of coal and biomass fuel blends. *Progress in Energy and Combustion Science* 27: 171–214.

Sawicki, D. Kotyra, A. Akhmetova, A. Imanbek, B. & Suleymenov, A. 2016. Wykorzystanie metod optycznych do klasyfikacji stanu procesu współspalania pyłu węglowego i biomasy. *Rocznik Ochrona Środowiska*, 18(2): 404–415.

Shuvatov, T. Suleimenov, B. & Komada, P. 2012. Gas turbine fault diagnostic system based on fuzzy logic. *IAPGOŚ* 2012(3): 40–42.

Smolarz, A. Lytvynenko, V. Kozhukhovskaya, O. & Gromaszek, K. 2013. Combined clonal negative selection algorithm for diagnostics of combustion in individual pc burner. *IAPGOŚ* 2013(4): 69–73.

Starke, M. Schulpin, H.J. Haug, M. & Schreiber, M. 2007. Measuring Coal Particles in the Pipe. *Power Engineering* 111: 45–48.

Wójcik, W. Gromaszek, K. Kotyra, A. & Ławicki, T. 2014. Optimal control for combustion process. *Przegląd Elektrotechniczny* 4(90): 157–160.

Zyska, T. Wójcik, W. Imanbek, B. & Zhirnowa, O. 2016. Diagnostyka stanu czujnika termoelektrycznego w procesie zgazowania biomasy. *Rocznik Ochrona Środowiska*, 18(2): 652–666.

Environmental Engineering V – Pawłowska & Pawłowski (Eds)
© 2017 Taylor & Francis Group, London, ISBN 978-1-138-03163-0

Battery-supported trolleybus traction network—a component of the municipal smart grid

W. Jarzyna & D. Zieliński
Faculty of Electrical Engineering and Computer Science, Lublin University of Technology, Lublin, Poland

P. Hołyszko
Municipal Transport Company in Lublin, Trolleybus Department, Lublin, Poland

ABSTRACT: The article presents research on minimising transmission losses in the trolleybus electric traction. The problem in question is the minimisation of these losses by applying systems with storage batteries. Different locations of such batteries in the network were analysed. The calculation of the power loss, voltage and current drops were made using the actual measurement results on a sample section of the trolleybus line. More than a sixfold reduction of the power loss was obtained for the supercapacitor battery location at the end of the analysed section. These results provide clear evidence that the retrofitting of electric traction in correctly placed storage batteries brings significant economic effects.

Keywords: trolleybus traction network, transmission losses, storage batteries system, buck and boost converter

1 INTRODUCTION

Energy efficiency of a fleet of electric trolleybus city transportation requires re-evaluations and changes in the way such traction is powered and functions. Modern technology solutions in trolleybuses (Filin, et al., 2015), (Hamacek et al. 2014), power systems of the grid, energy storage and transmission systems, allow for the introduction of innovations that improve the efficiency and reliability of the entire system (Maciołek & Drążek, 2004), (Radecki & Chudzik, 2014). The extent to which these innovative actions can contribute to improving the efficiency is the goal of this article.

Tests have been developed for selected sections of the urban trolleybus network in Lublin. There was an analysis of the existing solutions and examination of the conditions and the expected effects of changes in the structure of the trolleybus traction after the introduction of innovations, which essentially consist in equipping the traction network in electrochemical storage batteries.

The reason for taking up this work is a significant expansion of the trolleybus traction in Lublin, carried out from 2012. Apart from almost doubling the length of the line, the modernization is changing the structure of power supply. There is a departure from the central system of connections

in favour of separate power supply sections. Such actions are caused by several factors. One of them is the construction of routes very distant from the existing rectifier stations. It became a necessity to reduce transmission losses and increase operational reliability. The other factor is associated with the need for the development of energy recovery, or energy recovered during the electric braking of trolleybuses (Spichartz, & Sourkounis; 2013). It happens that in systems without batteries, and in the absence of energy reception by another trolleybus, this energy is lost at resistors (Lewandowski, 2009). Installing traction batteries, this energy can be used effectively.

In connection with the arguments quoted, the task was undertaken of examining how to improve the energy efficiency of the trolleybus traction system and determine the energy indicators. This study is the result of conceptual efforts of a team from the Lublin University of Technology and the Department of Trolleybuses at the Municipal Transport Company in Lublin. It consists of elements such as a critical review of the literature, drawing conclusions and choosing the most advantageous solution, proposing a sample topology and assessing the distribution of its structural elements, sample calculations and interpretation of results and conclusions.

2 DESCRIPTION OF THE EXISTING STRUCTURE AND REVIEW OF TECHNICAL SOLUTIONS

An electric trolleybus network is a dual system of supplying DC power to city trolleybus transport. It is connected to rectifier substations, powered by the distribution network of the three-phase alternating current. The rectifier systems used are mostly uncontrolled 12-pulse phase commutation rectifiers (Kulesza & Sikora, 2012). They allow one-way transmission of electricity from the distribution network of alternating current to electric vehicles powered from the AC mains. An example of the structure of the power section of the grid is shown in Figure 1.

The network consists of sections supplied from rectifier stations. Between the individual sections there is no galvanic connection. Each section supplies a group of vehicles, the number of which depends on the time of day. The majority of vehicles in use is equipped with modern, vector-controlled electric drives allowing energy recovery during braking (Kazmierkowski, et al., 2008). The braking energy is fed back by trolleybuses to the overhead line, from which it must be received by other trolleybuses or managed by some other receivers. Thus, if other trolleybuses are operating in the section, the recovery energy is used by them. On the other hand, in the absence of such receivers, this energy in the existing solution is lost in braking resistors.

It should be noted that the number of trolleybuses and their place on the route change dynamically. Moving trolleybuses come to a halt, during which stage the energy demand drops significantly. In addition, during braking, the power of recovery may exceed the demand by another vehicle in motion. For this reason, the traction voltage is increased, and when the 780V threshold is reached, braking is switched to resistance breaking. Examples of the effects of increased voltage with increasing the breaking current is shown in Figure 2.

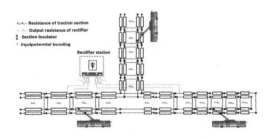

Figure 1. An example scheme of a two-track catenary consisting of two sections isolated from each other and supplied from a rectifier substation.

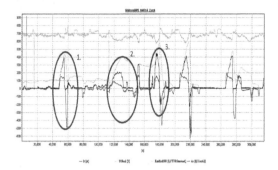

Figure 2. An example of the effect of increased voltage with the flow of braking power (traction voltage—blue, the current measured by the trolleybus pantograph—purple).

The areas circled in Figure 2 can be interpreted as follows:

- Area 1—a high negative value of the current at the pantograph means that during braking the recovery current is transmitted to the network. This current is accompanied by a voltage increase up to 780V, and after reaching this level, the recovery current decreases to zero as braking passes into resistance breaking.
- Area 2—the vehicle climbs the hill; compared to the traction's open circuit voltage of 720V, this tension significantly drops to the value of 600V.
- Area 3—starting the vehicle at the very top of the hill and braking during descent. Very heavy traction load causing a decrease in the voltage to 570V, then increase and operation with the braking resistor.

Due to the limitations of the possibilities of using recovery energy, the energy efficiency is reduced and variable dynamic states reduce the reliability and security of work. In order to limit these drawbacks, the following methods are used:

a. Changes in traffic organisation consisting in increasing the number of vehicles in use and reducing the number of passengers carried by each of these vehicles,
b. The use of bi-directional converter systems in the rectifier stations allowing return of braking energy to the grid (Hahashi & Mino, 2012),
c. Retrofitting sections with storage batteries, which draws from the network excess recuperation energy and then releases this energy in times of increased demand during starting trolleybuses (Fei et al., 2016).

Solution a) can have positive effects only on a small scale. It gives limited benefits and does not prevent the formation of transient states in the network. Due to a number of factors affecting the traffic of vehicles, it is almost impossible to schedule a mutually complementary operation of two or more vehicles.

The use of bi-directional power converters (solution b) allows to transfer excess energy to the grid. Such modern converters have a stabilizing effect on the grid voltage, limiting dangerous voltage increases, but do not guarantee the compensation of voltage drops in the catenary caused by a large momentary power consumption in distant substation points of the network.

Solution c), on the other hand, guarantees performing regenerative braking regardless of the operating status of other vehicles. Furthermore, the additional expected effect is restricting the momentary surges in energy demand caused by the need to power the rectifier stations, thereby contributing to a reduction in the power guaranteed by the grid distributor.

The discussion of the solutions presented indicates that the best approach is a traction network reconfiguration based on the construction of additional distributed storage batteries connected to the network through four-quadrant power converters (Massot-Campos et al., 2011).

Analysis of the market potential of the currently used batteries points to a division into short—and long-term energy accumulation systems.

Due to the nature of voltage interference in the catenary (Fig. 2), particular attention is focused on the accumulation systems characterized by high dynamics of energy exchange with traction. These requirements result from the need to ensure AC charging and discharging at a steepness of increase not less than 10 A/1 µs, with momentary values reaching the level of 500 A.

These criteria are met by the flywheel type of storages (Sun et al., 2016) as well as Superconductive Magnetic Energy Storage (SMES) systems (Molina & Mercado, 2011). Unfortunately, due to very high prices and the disadvantage of the requirement for cyclic maintenance of equipment, these systems are not currently used in similar solutions. In contrast, these defects are not found in modern supercapacitors and lithium-ion batteries, which are much cheaper and do not require cyclic maintenance.

Although, among up-to-date Li-ion batteries we can find suitable energy storages, particular interest are systems integrated with supercapacitors. The life of such batteries is counted in millions of cycles, and their charging and discharging power is very high. Furthermore, supercapacitors have no temperature limits in their operation, as is the case with chemical cells. For these reasons, further studies avail themselves of systems with supercapacitors.

Application of additional storage batteries requires the use of special matching converters, which among other things provide control of power flow, voltage levels and proper impedance between supercapacitor and the power grid. For the DC network, the direction of power flow is dependent on the voltage difference between the source and the receiver, wherein there must always be a condition that the source of energy has a higher voltage than the receiver. Hence the converters used are systems that increase voltage (boost converters) (Camara et al., 2010) or ones that lower it (buck converters) (Camara et al., 2010).

Taking into account these conditions of the flow of power and the fact that usually the supercapacitor battery voltages are smaller than the DC voltages, boost type converters were used for battery discharging. They allow even a tenfold increase in input voltage (Camara et al., 2010).

The situation is reversed in the case of charging supercapacitors, during which a requirement is precise regulation of voltage, reduced in relation to the mains voltage. In this case, the proper structure of the inverter is a classic buck DC/DC converter. Such converters allow the use of almost 100% of the energy potential of the source. In addition, systems with such topologies have the added advantage of the ability of parallel connection, which enables the phase shifting of signals modulating the converter keys and, consequently, a reduction of higher harmonics at the output of the battery and reduction in the dimensions of EMC filters (Przytuła, 2016). An element protecting these storage batteries from possible punctures are resistors, which contain supercapacitors in the case of exceeding the voltage at the battery's terminals.

3 EXPERIMENTAL-COMPUTATIONAL RESEARCH

The aim of the research is to determine the effects which may be brought in the catenary the application of supercapacitor storage batteries. Tests are

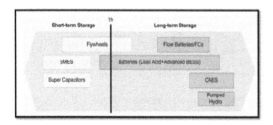

Figure 3. The division of storage batteries on the basis of their duration (Energy Storage Association, 2003).

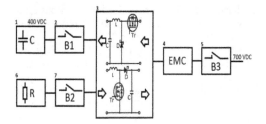

Figure 4. The block diagram showing how to attach a bank of supercapacitors to traction, with: 1—battery of supercapacitors, 3—buck and boost converters, 2,7,5—DC circuit breakers, 6—discharge resistor.

conducted mainly to find out to what extent the choice of the battery's location in the overhead line influences the energy savings achieved.

The study consists of two parts: experimental and computational. The results of the experimental phase constitute the basis for the calculation of power and voltage drops. In the experimental part, while driving along a selected section of the trolleybus route, registration was made of the momentary values of: power, current, voltage and distance as a function of time. The results obtained were used in the calculations.

To determine the losses, batteries were positioned at three points of the selected part of traction (Fig. 5a), which is powered by one side of the rectifier substations. An experimental ride of the electric vehicle was made, which lasted 300 s. In order to obtain simulation data a Solaris Trollino vehicle was used, equipped in a Skoda Blue Drive vector drive.

The computational model was designed in the Matlab Simulink program. Input values are experimental measurements, saved as text files. During calculations four cases were considered—a network without batteries and with them, situated in the rectifier station at the opposite end of the line and in the middle of it.

During the passage the vehicle was repeatedly accelerated and stopped. From the waveform registered (Figure 2), three events were selected (Fig. 5b), which served as the inputs to the computer simulation done in Matlab/SIMULINK. Then a test was conducted in order to verify the usefulness of storage batteries for traction systems. The prototype battery was used with the parameters set forth in Table 1.

An objective assessment of the considered options of the catenary can be obtained by calculating the energy consumed during the trolleybus passage analysed. This energy is defined in Figure 9 by calculating the integral of the power output curve shown in Figures 8a, b.

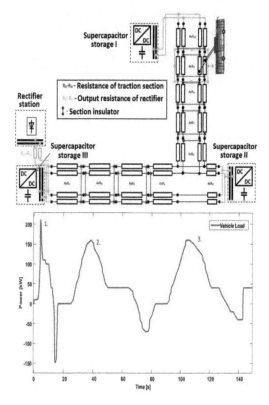

Figure 5. a) Scheme of the analysed section of the catenary with drawings of the three cases of supercapacitor battery positioning, b) selected fragments of the trolleybus ride containing cycles consisting of: starting, acceleration and braking.

Table 1. Parameters of the simulated route together with the applied batteries and converters.

Name:	Characteristic parameter:
Battery type:	Battery with supercapacitor
Supercapacitor capacity:	150 F
Max. voltage of operating supercapacitor:	400 V
Converter power:	150 kW
Converter type:	buck-boost
Permissible operating voltage of the converter:	400–850 V
Max. current of operating supercapacitor:	500 A
Operating mode of the converter:	voltage stabilisation/current limit/static charging

The calculation results are illustrated in sequence in Figures 6, 7 and 8. They show graphs of current, voltage and power as a function of time, for the selected portions of the route identified by the characteristics plotted on Figure 5b.

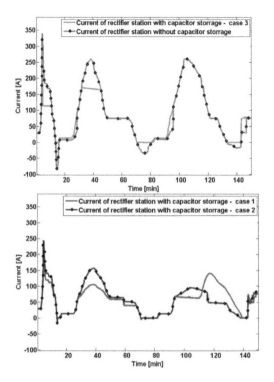

Figure 6. The current drawn from the rectifier station, located at the beginning of the experimental section of traction: case 1 – battery at the end of traction, case 2 – battery in the middle of traction, case 3 – battery in the rectifier station.

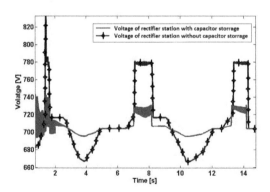

Figure 7. The voltage measured at the output of the rectifier station during experimental journey: system with supercapacitor battery.

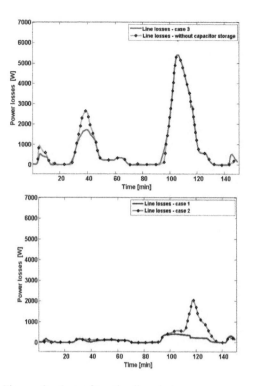

Figure 8. Loss of traction lines during passage.

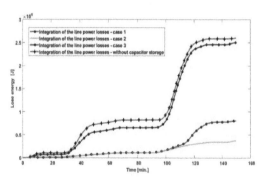

Figure 9 Graph of the integral of the power loss of the transmission line.

In assessing the amount of energy consumed to drive a trolleybus along the selected distance, it was observed that the most advantageous location of the battery is the unpowered catenary end. This is evidenced by the calculation of the relative losses compared to maximum losses occurring in the system without a battery. For the analysed cases they are as follows:

14.3%—case 1, battery placed at the end of the line opposite to the rectifier station,

31.2%—case 2, battery installed in the middle of the route,

96.1%—case 3, battery placed at the rectifier station.

It follows that the best solution is to install a storage battery in the middle of the analysed line. Such positioning decisively reduces the voltage drop on the catenary resistances. In addition, it supplements the power system most advantageously, limiting the maximum values of the current supplied to the trolleybus drive from the rectifier station.

4 CONCLUSION AND FUTURE WORKS

The obtained results show that there is a clear dependence of transmission losses in the catenary on the location of storage batteries. The lightest losses were sustained by the batteries placed in the middle of the route, and the heaviest by batteries installed in the rectifier station. The difference between the two locations is very big. In practice, this means that it does not pay off to install storage batteries in the rectifier station, as this location does not reduce transmission losses and only slightly supplements the rectifier station.

The location of the battery at the end of the line opposite to the rectifier station reduces transmission losses to the greatest extent and supplements the primary power source. Only slightly worse than this location came out the results for the placement of a battery of supercapacitors at the end of the line, opposite the rectifier station. This result indicates that the installation of several batteries in different places of the line should bring further efficiency improvements.

The research results provide a rich source of information that will be used for further analysis. Because the calculation is made using actual measurement data, the results obtained represent an important value because of their qualitative and quantitative evaluation. On their basis one can estimate global energy profits in daily cycles. This requires to extend the calculation model in order to include all the vehicle transits in the analysed section and take into account variations related to the intensity of traffic (Hamacek et al. 2014), (Maciołek & Drążek, 2004), (Radecki & Chudzik, 2014).

REFERENCES

Camara, M. B., Gualous, H. Gustin, F. & Berthon, A. 2010. DC/DC Converter Design for Supercapacitor and Battery Power Management in Hybrid Vehicle Applications—Polynomial Control Strategy, *IEEE Transactions on Industrial Electronics*, 57(2).

Energy Storage Association. 2003. *Applications of Electricity Storage*. Available from http://electricitystorage. org/technologies_applications.htm, [Aug., 2009].

Fei, L., Xuyang, L., Yajie, Z. & Zhongping, Y. 2016. Control Strategies with Dynamic Threshold Adjustment for Supercapacitor Energy Storage System Considering the Train and Substation Characteristics in Urban Rail Transit, *Energies*.

Filin, S., Filina-Dawidowicz, L. & Chmielewski, W. 2015. Present state and development prospects of wheeled vehicles powered from the catenary, *Logistyka*, 2–17(2).

Hahashi, Y. & Mino, M. 2012. High-density bidirectional rectifier for next generation 380V DC distribution system, *Power Electronics Conference and Exposition* (APEC).

Hamacek, Š., Bartłomiejczyk, M., Hrbáč, R., Mišák, S., & Stýskala V. (2014). Energy recovery effectiveness in trolleybus transport. *Electric Power Systems Research*, 112, 1–11.

Kazmierkowski, M. P., Swierczynski, D. Wojcik, P. & Janaszek, M. (2008). Direct torque controlled PWM inverter fed PMSM drive for public transport, *IEEE International Workshop on Advanced Motion Control*.

Kulesza, B. & Sikora, A. 2012. Features of economical traction 12-phase rectifier transformer, *Przegląd Elektrotechniczny* (Electrical Review), ISSN 0033–2097, R. 88 NR 8/2012.

Lewandowski, M. 2009. Model of traction-vehicle system for analysis the phenomena occurring during the braking of vehicle, *Pojazdy Szynowe*, 4: 17–21.

Maciołek, T. & Drążek, Z. 2004. Tram vehicle energy accumulator-on-board or in substation. *International Conference Speedam*, 132–137.

Massot-Campos, M., Montesinos-Miracle, D., Galceran-Arellano, S. & Rufer A. 2011.Multilevel two quadrant DC/DC converter for regenerative braking in mobile applications, *Power Electronics and Applications*.

Molina, M. G. & Mercado, P. E. (2011). Power Flow Stabilization and Control of Microgrid with Wind Generation by Superconducting Magnetic Energy Storage, *IEEE Transactions on Power Electronics* 26(3).

Przytuła, K. 2016. Przełączenia awaryjne oraz analiza stanów przejściowych wielokanałowego przekształtnika sieciowego, *Informatyka, Automatyka, Pomiary W Gospodarce I Ochronie Środowiska* - 2016, 1: 55–58.

Radecki, A. & Chudzik, P. 2014. Algorytm sterowania zasobnikiem superkondensatorowym pojazdu trakcyjnego minimalizujący przesyłowe straty mocy uwzględniający stany pracy sieci trakcyjnej. *Przegląd Elektrotechniczny*, ISSN 0033–2097, R. 90 NR 6/2014. doi:10.12915/pe.2014.06.24.

Spichartz, P. & Sourkounis, C. (2013). Measurement of braking energy recuperation in electric vehicles, *Power Electronics and Applications* (EPE),15th European Conference; 2–6 Sept. 2013.

Sun, B. Dragičević, T., Freijedo, F. D. & Vasquez, J. C. (2016). A Control Algorithm for Electric Vehicle Fast Charging Stations Equipped With Flywheel Energy Storage Systems, *IEEE Transactions on Power Electronics* 31(9).

Environmental Engineering V – Pawłowska & Pawłowski (Eds)
© 2017 Taylor & Francis Group, London, ISBN 978-1-138-03163-0

Analysis of energy consumption of public transport in Lublin

M. Dziubiński, E. Siemionek, A. Drozd & S. Kołodziej
Faculty of Mechanical Engineering, Lublin University of Technology, Lublin, Poland

W. Jarzyna
Faculty of Electrical Engineering and Computer Science, Lublin University of Technology, Lublin, Poland

ABSTRACT: The aim of this study is to analyze the energy consumption of the public transport in Lublin on selected means of transport (Solaris Trollino 12S trolleybus, Solaris Trollino 18 M and Ursus E70110 electric bus). Currently, Lublin's transportation fleet is equipped with an electric drive due to the lack of emission of greenhouse gasses in the place of exploitation, quiet drive operation and greater efficiency than the internal combustion engine, as well as its durability. Modern traction drives allow to carry out the process of regenerative braking with recuperation of energy to the trolley wires or energy storage systems. On the basis of the energy consumption analysis of three means of transportation equipped with an electric drive, we can assume that the regenerative braking process significantly contributed to the limitation of the energy consumption.

Keywords: electric bus, energy consumption, modeling, reduction of toxic substances, simulation tests

1 INTRODUCTION

The exponential development of road transport leads to a rapid depletion of the natural reserves of oil. In their evaluation of natural resources analysts use the so-called Hubbert peak theory. The most favourable scenario predicts that up to 2020 oil production in the world will grow, reaching in that year the value of 40 million barrels, after which it will go down. This means that oil resources are limited and can be lost during one lifetime. The combustion of fossil fuels, including oil, contributes to climate change caused by CO_2 emissions. Both the depletion of oil and CO_2 emissions threaten the realisation of the fundamental paradigm of sustainable development, formulated in the famous Brundtland report *Our Common Future* (1987). This paradigm dictates that meeting the needs of people living today should not deprive future generations of their ability to meet their own needs (Pawłowski 2009, 2013; Udo et al. 2010). A critical factor in achieving sustainable development is energy supply (Pawłowski 2009, Cholewa et al. 2009, Żelazna 2016), including transport needs (OECD 2010).

Within the theory of sustainable development a number of recommendations are formulated in terms of moral duty (Sztumski 2016, Cao et al. 2016). One of the recommended directions is the broad use of biofuels (Cao et al. 2016, Liu 2015). Although this trend is widely promoted, one should also note voices about the negative consequences of the use of biofuels (Pawłowski 2015). Great oppor-

tunities lie in the use of electric drives (Chłopek 2012, Polakowski 2016), in particular when batteries are charged with photovoltaic panels.

Parallel work is being done both on limiting the amount of burned petroleum products and on mitigation of emission by upgrading conventional engines. It is estimated that by 2030 road transport will reduce emissions of CO_2 by 12.5 GT, mainly as a result of upgrading efficiency by 52%, introducing electric vehicles (17%), the use of biofuels (17%) and the introduction of hydrogen as a fuel (14%) (IEA 2008). The reduction of vehicles negative impact on the environment is associated with the decrease of fuel consumption by vehicles. One of the ways of reducing fuel consumption and exhaust fumes emission is the use of hybrid drive and electric drive systems (Bartłomiejczyk & Połom 2013, Hamacek & Bartłomiejczyk 2014, Feng-chun, Liu & Zhen 2014, Wołek & Wyszomirski 2013).

The public transport of the city of Lublin uses trolley buses and electric buses (Tarkowski & Siemionek 2011, Siemionek 2013, Siemionek & Dziubiński 2015), which have the following advantages: they do not emit exhaust fumes at the site of their operation and they are characterized by smooth and quiet work of the drive.

The basic energetic quality of the vehicle directly affecting its operation cost is the consumption of electricity or fuel. The consumption is dependent on the energy used for motion equal to the work done by the vehicle driving force in the covered distance.

Vehicle motion can be divided into three phases: acceleration, steady motion and braking. The problem of energy consumption in particular phases of vehicle motion was studied by the following authors: (Siłka 1998, 2002; Chłopek 2006, 2012; Rudnicki 2008; Merkisz 2012).

Different kinds of driving tests depending on the city type are used for the purpose of studies. SORT driving tests simulate traffic conditions in a large city, an average city and in the suburbs of major cities in order to standardize procedures related to the measurement of mileage energy consumption for public transport vehicles by Union Internationale des Transports Publics (UITP).

The subject of energy consumption in a vehicle equipped with electric drive has been analysed in the paper. A part of the studies was to carry out the verification of energy consumption calculation model for the motion of a trolleybus equipped with the drive allowing for regenerative braking with the recuperation of energy to overhead lines with the use of Matlab/Simulink software. The theoretical influence of vehicle load on the value of total, mileage and unitary energy consumption of the motion has been determined on the example of trolleybuses. Changes in energy consumption in time were simulated for three SORT driving tests for the trolleybus of 12 S type loaded with the weight of 20 passengers and a driver. The analysis of changes in energy consumption was carried out in experimental studies. In the course of measurements, changes in voltage, current, speed and instantaneous acceleration values of tested vehicles were recorded.

2 THE ANALYSIS OF ENERGY CONSUMPTION IN VEHICLES EQUIPPED WITH ELECTRIC DRIVE

Energy consumption of the electric vehicle can be illustrated by the formula (1).

$$E_{EL} = E + \Delta E_s + \Delta E_p + \Delta E_j \qquad (1)$$

where: E—motion energy consumption [kJ], ΔE_s—losses connected with energy generation in the engine [kJ], ΔE_p—energy losses connected with drive transmission [kJ], ΔE_j—energy losses connected with idle running [kJ].

Power required for starting the vehicle is presented in the formula (2):

$$N_k = (F_t + F_p) \cdot v \qquad (2)$$

where: F_t—rolling resistance force [N], F_p—air resistance force [N], v—velocity [m/s]

The power on the engine shaft can be calculated from the formula (3):

$$N_s = \frac{N_k}{\eta_p} \qquad (3)$$

where: N_k—power required for starting the vehicle [kW], η_p—efficiency of drive transmission system.

The power drawn from the batteries is presented in formula (4):

$$N_{ak} = \frac{N_k}{\eta_a} \qquad (4)$$

where: N_k—power required for starting the vehicle [kW], η_a—efficiency of the batteries

The current drawn from the batteries is presented in formula (5):

$$I_{ak} = \frac{N_{ak}}{U_{ak}} \qquad (5)$$

where: N_{ak}—power drawn from the batteries [kW], U_{ak}—battery voltage [V].

The range of the vehicle equipped with electric drive is presented in formula (6):

$$Z = v \cdot t = v \cdot \frac{Q_{ak}}{I_{ak}} \qquad (6)$$

where: v—velocity [m/s], t—time [s], Q_{ak}—battery capacity [Ah], I_{ak}—current drawn from the batteries [A].

For the purpose of standardizing procedures associated with the measurement of mileage power consumption for public transport vehicles, the SORT test (Eng. Standarised On-Road Tests) drawn up by UITP (Fr. Union Internationale Transports Publics desec.) was implemented. It can be conducted in real traffic conditions, as well as in laboratory conditions with the use of undercarriage test bench. In order to adjust the profile of the cycle to the conditions in a given city, three base cycles were drawn up.

The first base cycle is SORT 1 cycle (Heavy Urban Cycle) simulating driving in the centre of a

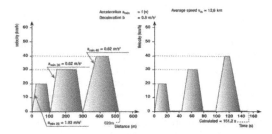

Figure 1. Sort 1 cycle.

big city (Fig. 1). Communication speed of SORT 1 cycle equals 12.6 km/h.

The second base cycle is SORT 2 cycle (Easy Urban Cycle) reflecting the conditions in an average city. Communication speed of SORT 2 cycle equals 18.6 km/h. The third base cycle is SORT 3 cycle (Easy Suburban Cycle) reflecting the conditions in smaller cities or in the suburbs of bigger cities. Communication speed of SORT 3 cycle equals 26.3 km/h.

The measurement of energy/fuel consumption is carried out separately for each of the above-mentioned cycles, according to the application of the tested bus/trolleybus. Each base cycle consists of three sections separated by a definite space of time, containing basic phases of elementary velocity profile.

3 DEVELOPING THE MODEL IN MATLAB/SIMULINK SOFTWARE

The aim of the studies is to realize the calculation model of energy consumption in the motion of the trolleybus equipped with the drive allowing for regenerative braking with recuperation of energy to overhead lines. The model consists of basic function blocks of Simulink package and four subsystems. The constant parameters used in the model are the following: vehicle mass, gravitational acceleration, rolling resistance coefficient, shape coefficient, the surface of the front wall, air density, test overall path, rotating masses coefficient, cumulative losses caused by the limited efficiency of trolleybus drive particular elements and specific parameters of given motion phases. The main system of the model "modelling.slx" is shown in Fig. 2.

The purpose of function blocks used in the subsystem (Fig. 3) is to calculate energy consumption in particular motion phases and the amount of energy recovered during regenerative braking.

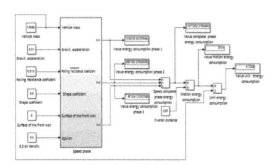

Figure 2. The diagram of the main system "modelling.slx".

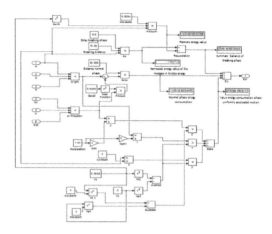

Figure 3. The diagram of the subsystem "Power consumption in cycle 1" "modeling.slx".

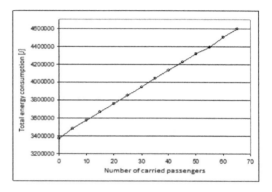

Figure 4. Diagram presenting the relationship of total energy consumption and the number of carried passengers for the trolleybus of 12 S type.

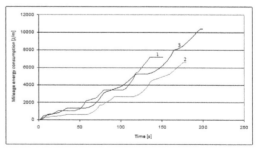

Figure 5. Changes in mileage energy consumption in time for the three SORT driving tests for the trolleybus of 12 S type loaded with the weight of 20 passengers and a driver (1—SORT 1, 2 SORT—2, 3—SORT 3).

The analysis of the obtained results showed that energy consumption increases together with the increase in the number of carried passengers. The relationship between the total energy consumption in the trolleybus motion and the number of carried passengers is presented in Figure 4.

Changes in mileage energy consumption in time for the three SORT driving tests are shown in Fig. 5. Changes in mileage energy consumption for the test SORT 1 are marked with number 1, for the test SORT 2—number 2, and for the test SORT 3—number 3.

4 EXPERIMENTAL STUDIES

In the research part, changes in energy consumption of motion in the conditions of real operation were recorded. The recorded courses of changes in the voltage and amperage of the current drawn from overhead lines in time for the 12 S trolleybus are shown in Fig. 6. Voltage is marked with number 1, and amperage with number 2. During the drive, voltage and amperage values of the drawn current were recorded every 0.3 second. Changes in the value of motion energy consumption during the tests for the trolleybus of 12 S type are shown in Fig. 7. Energy consumption changed in a manner dependent on particular phases of motion. At acceleration, motion energy consumption increased. During the phase of steady motion, the value of energy consumption remained constant, and during the braking phase it decreased due to the recuperation of electricity to overhead lines (negative current values).

The second tested vehicle was a 12-meter-long electric bus (Fig. 8). The testing of an electric bus comprised three phases of driving in urban traffic including acceleration, steady driving, braking and stopping. Depending on the intensity of particular phases of motion, the value of energy consumption and the amperage of current drawn from the battery changed.

The second tested vehicle was a 12-meter-long electric bus (Fig. 8).

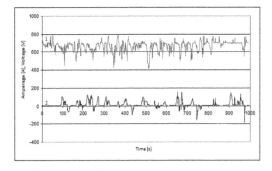

Figure 6. The recorded course of changes in voltage (1) and amperage of the drawn current (2) by the 12 S trolleybus during the second test.

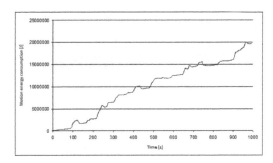

Figure 7. Changes in motion energy consumption during the test for the 12S trolleybus.

Figure 8. Electric bus.

The testing of an electric bus comprised three phases of driving in urban traffic including acceleration, steady driving, braking and stopping. Depending on the intensity of particular phases of motion, the value of energy consumption and the amperage of current drawn from the battery changed.

The recorded courses of voltage and amperage values for the current drawn from traction batteries in time for the electric bus are shown in Fig. 9. Voltage was marked with number 1, and amperage with number 2. During the drive, current voltage and amperage values of the drawn current were recorded every 0.5 second. Energy consumption increased in a manner characteristic of the urban cycle, depending on the intensity of acceleration, steady driving and braking phases (Fig. 10). As in the previous studies, amperage took on a negative value while braking, due to regenerative braking process with energy recuperation. The electric bus used traction batteries, during the tests voltage and drawn current changed their values over the course of time.

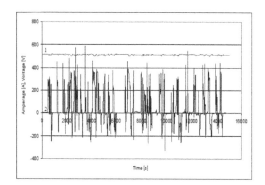

Figure 9. The recorded course of changes in voltage (1) and the amperage of the current (2) drawn from traction batteries by the electric bus during the test.

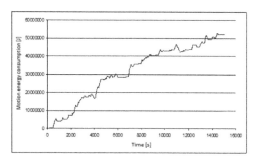

Figure 10. Changes in energy consumption values during the testing of the electric bus.

5 CONCLUSIONS

On the basis of literature analysis, as well as the carried out simulation and experimental tests, it can be concluded that the basic element directly affecting the cost of public transport vehicle operation is the consumption of fuel or electricity. Modern traction drives enable to carry out the process of regenerative braking with recuperation of energy to overhead lines or energy storing accumulators. On the basis of energy consumption analysis carried out for two means of transport equipped with electric drive, it can be stated that the process of regenerative braking significantly contributed to the decrease in energy consumption. The replacement of internal combustion engines with electric propulsion systems resulted in the reduction of toxic substances emission, reduction of noise and CO_2 emissions.

REFERENCES

Adamiec, M. & Dziubiński, M. 2009. Alkaliczne ogniwo paliwowe—aspekt sprawności, *Przegląd Elektrotechniczny*, 85(4): 182–185.

Bartłomiejczyk, M. & Połom, M. 2013. Przestrzenne aspekty efektywności hamowania odzyskowego w komunikacji trolejbusowej, *Logistyka* 6.

Brundtland, G., Khalid, M., Agnelli, S., Al-Athel, S., Chidzero, B., Fadika, L., Hauff, V., Lang, I., Shijun, M., Morindo de Botero, M., Singh, M. & Okita, S. 1987. Our Common Future ('Brundtland raport'),

Cao, Y., You, J., Shi, Y. & Hu, W. 2016. Evaluation of Automoblie Manufacturing Enterprise Competitiveness from Social Responsibility Perspective, *Problemy Ekorozowju/Problems of Sustainable Development*, 11(2): 89–98.

Cao, Y., You, J., Wang, R. & Shi, Y. 2016. Designing a Mixed Evaluation System for Green Manufacturing of Automotive Industry, *Problemy Ekorozowju/Problems of Sustainable Development*, 11(1): 73–86.

Chłopek, Z., 2012. Badanie zużycia energii przez samochód elektryczny, *Archiwum Motoryzacji* 3.

Chłopek, Z.,2006. Ocena zużycia energii przez autobusy komunikacji miejskiej, *Journal of KONES* 13(1).

Cholewa, T. & Pawłowski, A., 2009. Sustainable Use of Energy in the Communal Sector, *Problemy Ekorozwoju/Problems of Sustainable Development*, 11(2): 1165–1177.

EFA 2015, Air quality in Europe, EFA Raport, No5/2015.

Hamacek, S., Bartłomiejczyk, M., Hrbá, R., Misák, S. & Styskala, V. (2014). Energy recovery effectiveness in trolleybus transport, *Electric Power Systems Research* 07.

IEA 2012, Global transport outlook to 2050 Targets and scenarios for a low-carbon transport sector.

IEA-2008, World Energy Outlook 2008, Global transport sector emissions reduction to 2030.

Liu, H. 2015. Biofuel's Sustainable Development under the Trilemma of Energy, (2015), Environment and Economy, *Problemy Ekorozowju/Problems of Sustainable Development*, 10(1): 55–59.

Merkisz, J., Molik, P., Nowak, M. & Ziółkowski, A. 2012. Cykle jezdne pojazdów komunikacji miejskiej na przykładzie aglomeracji poznańskiej, *Logistyka* 3.

Pawłowski, A. 2009. Sustainable energy as a sine qua non condition for the achievement of sustainable development, *Problemy Ekorozowju/Problems of Sustainable Development*, 4(2): 9–12.

Pawłowski, A. 2009. Theoretical Aspects of Sustainable Development Concept, *Rocznik Ochrona Środowiska* 11(2): 985–994.

Pawłowski, A., 2013. Sustainable Development and Globalization, *Problemy Ekorozowju/Problems of Sustainable Development*, 8(2): 5–16.

Pawłowski, A. & Pawłowski, L. 2008. Sustainable development in contemporary civilization. Part 1: The environment and sustainable development, *Problemy Ekorozowju/Problems of Sustainable Development*, 3(1): 53–65.

Pawłowski, L. 2015. Where is the World heading? Social crisis created by promotion of biofuels and noadays liberal capitalism, *Rocznik Ochrona Środowiska*, 17(1): 26–39.

Polakowski, K. 2016, Samochód elektryczny a ochrona atmosfery, *Rocznik Ochrona Środowiska*, 18: 628–639.

Rudnicki T. 2008. Pojazdy z silnikami elektrycznymi, *Zeszyty Problemowe-Maszyny Elektryczne* Nr 80/2008, ISSN 0239-3646, 245–250.

Siemionek, E. & Dziubiński, M. 2015. Testing energy consumption in the trolleybus and the bus on a chosen public transport line in Lublin, *Advances in Science and Technology Research Journal* 26(9): 152–155.

Siemionek, E. 2013. Analiza energochłonności ruchu trolejbusów, *Advances in Science and Technology Research Journal*, 18(7): 81–84

Siłka, W. 1998. Analiza wpływu parametrów cyklu jezdnego na energochłonność ruchu samochodu, *Zeszyt 14 Monografia 2* PAN oddział Kraków.

Siłka, W. 2002. Teoria ruchu samochodu, *WNT*, Warszawa.

Sun, F.-C., Bin, L. & Wang, Z.-p. 2014. Analysis of Energy Consumption Characteristics of Dual-source Trolleybus, *Transportation Electrification Asia-Pacific (ITEC Asia-Pacific)*, 2014 IEEE Conference and Expo Beijing.

Sztumski, W, 2016. The Impact of Sustainable Development on the Homeostasis of the Social Environment and the Matter of Survival, *Problemy Ekorozwoju/ Problems of Sustainable Development*, 11(1): 41–47.

Tarkowski, P. & Siemionek, E., 2011. Energy consumption in the motion of the vehicle with electric propulsion, *Teka Komisji Motoryzacji i Energetyki Rolnictwa PAN*, 11c: 320–326.

Udo, V. & Pawłowski, A. 2010. Human Progress Towards Equitable Sustainable Development: A Philosophical Exploration. *Problemy Ekorozwoju/Problems of Sustainable Development* 5(1): 23–44.

Wołek, M. & Wyszomirski, 2013. *The Trolleybus as an Urban Means of Transport in the Light of the Trolley Projec. Uniwersytet Gdański*, Gdańsk.

Żelazna, A., Zdyb, A. & Pawłowski A. 2016. The influence of selected factors on PV systems environment al indicators, *Rocznik Ochrona Środowiska* 18: 722–732.

Ecological aspect of electronic ignition and electronic injection system

M. Dziubiński
Faculty of Mechanical Engineering, Lublin University of Technology, Lublin, Poland

ABSTRACT: The maintenance of vehicles generates high exhaust emissions and particular attention should be paid to the reduction of their negative impact on the natural environment. The paper presents selected solutions of electrical ignition and electrical injection in the area of exhaust emissions research conducted in real operational conditions of a vehicle. The paper described the simulation model of characteristics ignition and injection elements influence on exhaust emission from means of transport. The scope of investigations included determining environmental parameters, such as the concentration of nitrogen oxides NO_x, carbon monoxide CO and carbon dioxide CO_2 in exhaust gases. Experimental tests were carried out on the test stand and real object. On the basis of the obtained results, a diagnostic model which enables to recognize a failure can be worked out.

Keywords: exhaust fumes, injection systems, ignition systems, combustion engine, On-Board Diagnostics (OBD)

1 INTRODUCTION

A huge increase in the consumption of fossil fuels which took place in recent not only leads to the threat of depletion of resources, but also to an excessive emission of greenhouse gases. One of the important factors leading to the exhaustion of crude oil is road transport, which is also responsible for air pollution, especially in cities. Such status quo threatens the implementation of sustainable development, which was formulated for the first time in a well-known UN report Our Common Future. This report presented a holistic concept of human civilization. The main principle of sustainable development is ensuring the intra-generational equity, i.e. equal access to basic goods for all living people and avoiding polluting the environment (Udo et al. 2010, Pawłowski 2008, 2009a, 2009b, 2013). On the other hand, the principle of international equity advocates frugal and efficient use of non-renewable resources, so that future generations could have decent living conditions.

Energy suppl is one of the key problems (Sztumski 2016). Without sufficient energy supply, human civilization would not be able to develop. Meanwhile, acquisition of usable energy form fossil fuels leads to the emission of greenhouse gases which induce climatic changes, thus threatening the development of societies (Lindzen et al. 2016, Liu 2015, Żelazna et al. 2015).

Road transport is one of the factors causing a serious deterioration of air cities. Therefore, actions are taken in order to mitigate the negative impact on the natural and social environment (Cao et al. 2016 a,b, Chakraborty et al. 2016, Jarzyna et al. 2014, Duran et al. 2013).

Due to the fact that vehicles that run on oil are dominant in road transport, it is justified to conduct studies of reducing emissions of pollutants from such cars.

One of the fundamental parameters characterizing the engine operating conditions, and especially the process of compounding a mixture, is the air fuel ratio (AFR) λ. The emission of basic toxic compounds is largely determined by the value of this ratio (Fig. 1). If a mixture allowing complete fuel combustion is produced in the supply system, it is determined as stoichiometric mixture and referred to as λ = 1.0, whereas a weak mixture is determined as the value λ > 1.0, and a rich mixture λ < 1.0. Determination of the value of ratio, in addition to controlling the content of other ingre-

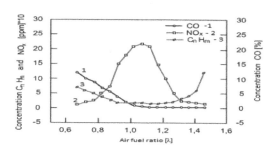

Figure 1. Content of basic toxic compounds in the exhaust fumes depending on the ratio λ.

dients, enables to conclude a proper operation of the fuel supply system. Obtaining adjustable composition of mixture, where the exhaust fumes would contain least toxic compounds, is difficult, because it leads to a compromise – it is not possible to obtain the minimum values of all exhaust constituents.

According to the standards of the European Union, a car with spark and compression ignition engine should meet the conditions of the content of toxic components in the exhaust fumes as shown in Table 1.

The introduction of the requirements of standards of On-Board Diagnostics (OBD) (Merkisz 2003, 2007, 2011, 2012, 2013, 2014) let to stricter emission control systems of toxic components. OBD systems must be able to recognize and record the wrong indications of systems associated with the emission of toxic components of exhaust gases. The occurrence of damages should be recorded and stored by the OBD system in the form of error codes.

Basic requirements of OBD include:

- evaluation of catalyst performance by examination of the content of hydrocarbons in the exhaust fumes,
- monitoring of disposal system of fumes from consumable fluids,
- control of the fuel feed system and exhaust gas recirculation system
- identification and location of the absence of combustion.

The combustion of weak and layered mixtures definitely increases the tendency of the engine to the occurrence of cycles without combustion. Due to the increasing use of combustion systems of weak mixtures, early detection of lack of combustion becomes extremely important. Lack of combustion is also undesirable because of the content of unburnt hydrocarbons in the exhaust fumes, which are burnt in the exhaust manifold and in the catalyst, possibly leading to damage. The effect of the phenomenon is also a significant increase in HC emissions (Chłopek 1999, Gronowicz 2003, Rokosch 2007).

Table 1. The emission standards of exhaust fumes for passenger cars according to the EU [g/km].

Level	Date	CO	HC	HC + NOx	NOx	PM
Diesel engine						
Euro 1 T	1992,07	2,72 (3,16)	--	0,97 (1,13)	--	0,14 (0,18)
Euro 2, IDI	1996,01	1,0	--	0,7	--	0,08
Euro 2, DI	1996,01 jedan	1,0	--	0,9	--	0,10
Euro 3	2000,01	0,64	--	0,56	0,50	0,05
Euro 4	2005,01	0,50	--	0,30	0,25	0,025
Euro 5	2009,09	0,50	--	0,23	0,18	0,005
Euro 6	2014,09	0,50	--	0,17	0,08	0,005
Spark Ignition engine						
Euro 1 T	1992,07	2,72 (3,16)	--	0,97 (1,13)	--	--
EURO 2	1996,01	2,2	--	0,5	--	--
Euro 3	2000,01	2,30	0,20	--	0,15	--
Euro 4	2005,01	1,0	0,10	--	0,08	--
Euro 5	2009,09	1,0	0,10	--	0,06	0,005
Euro 6	2014,09	1,0	0,10	--	0,06	0,005

2 MODEL OF INJECTION AND IGNITION SYSTEM AND ITS VERIFICATION

Electronic control of fuel injection system for determining the injected dose on the basis of the mass air flow and other parameters of the engine is currently being applied. The amount of fuel is adjusted to a signal received from the oxygen sensor. The signal from the oxygen sensor correlated with the information about the instantaneous position of the crankshaft, corrects fuel quantity in the cylinder. The customized adjustment of the composition of mixture for the other cylinders is performed similarly.

The model of lambda sensor operation needs to take into account not only the indications of the probe, but also the amount of fuel calculated on the basis on the inlet manifold and the amount of fuel delivered in the previous cycle (Fig. 2). The correction algorithm for indications of lambda sensor in application, is based on the equation (1) contained in:

$$m_{wtr}^{t} = m_{wtr}^{t-t_k} + t_k \left[\frac{\hat{m}_{pow}^{t}}{L_t} - k * \mathrm{sgn}(U_{\lambda}^{t-tk} - 0.45[V]) \right] \quad (1)$$

where:

$\hat{m}_{pow}^{t}$ – calculated mass of the vacuumed air
$m_{wtr}^{t-t_k}$ – the mass of fuel injected in the previous cycle
t_k – the interval of time between successive realizations of injection
L_t – theoretical request air for combustion
U_{λ} – pressure of lambda sensor

Model of ignition system contained in the application enables visualization of voltage on the secondary side of the ignition circuit (Figures 3–5).

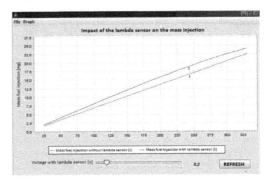

Figure 2. The window of the application can visualize the impact of the lambda probe on the mass injection for voltage 0.2 V read from probe.

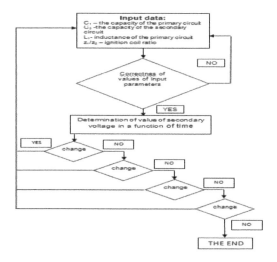

Figure 3. Functional model of the voltage on the secondary side of the ignition system.

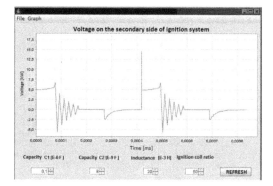

Figure 4. The voltage on the secondary side of ignition system for $C_1 = 0.1*10^{-6}$ F, $C_2 = 9 * 10^{-9}$ F, $L_1 = 20 * 10^{-3}$ H, and the ratio of the number of windings $z_2/z_1 = 500$.

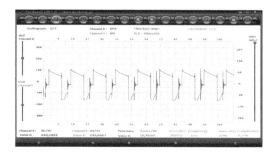

Figure 5. Oscillograph record $U = f(t)$ of the primary winding of the ignition coil for: n = 4000 rev/min.

It draws on the formula (2) on the basis of which the maximum voltage is calculated after introduction of the relevant size of the C_1, C_2, L_1 and z_1/z_2.

The maximum value of the electromotive force E_2 is:

$$E_{2max} = I_P \sqrt{\frac{L_1}{C_1\left(\frac{z_1}{z_2}\right)^2 + C_2}} \qquad (1)$$

where:

E_2 – electromotive force on the secondary side
I_P – the value of current in the primary circuit of the ignition system
L_1 – inductance of the primary circuit
C_1, C_2 – the capacity of primary and secondary ignition system
z_1, z_2 – number of turns of the primary and secondary winding of ignition system

3 EXPERIMENTAL RESEARCH ON A REAL OBJECT

As part of the work, studies of fuel track and ignition system, and their impact on toxicity were conducted. The study was conducted in the laboratory of the Department of Motor Vehicle in the

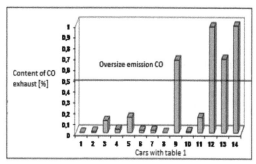

Figure 6. The graph showing the content of CO exhaust fumes of cars (2500 rev/min), because of the high emission of CO, car no. 1 was omitted.

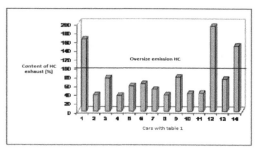

Figure 7. Content of HC in exhaust fumes of cars (2500 rev/min).

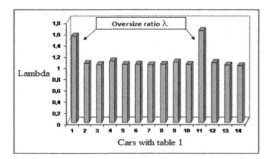

Figure 8. The figure of ratio λ for cars (idle speed).

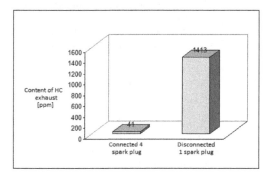

Figure 9. Comparison of hydrocarbon emissions in the absence of combustion in one cylinder at idle speed.

Lublin University of Technology. During the studies, the analysis of exhaust fumes of 14 cars was conducted. Failures of the spark plug were simulated in cars and the impact of the lack of combustion in the cylinders selected on the emission of toxic components was determined. In the absence of combustion in one of the cylinders, the content of hydrocarbons (HC) sharply rises in the exhaust fumes as it is shown in (Figures 6–9).

4 CONCLUSION

On the basis of the ignition system analysis, it can be stated that the following factors have a significant influence on the toxicity of fumes: firing ignitions, overtaking the ignition angle, degree of operational use of the spark plug, degree of operational use of spark plug wires, random damage to the ignition coil, deregulation of the ignition system, damage to the selected sensors.

On the basis of the analysis of exhaust gas components, it can be concluded that the ratio of fuel supplied to air in the mixture has the decisive influence on the pollution. The fuel and air track elements constitute factors that influence the weight ratio. These are: signal of the lambda sensor, fuel pressure (pressure regulator), capacity of fuel pump, supply voltage, the patency of the fuel hoses (inflow and outflow), the efficiency of the injectors, the patency of the fuel filter, tightness of air track, and permeability of the air filter.

The studies, involving the exhaust gas analyzer, confirmed the impact of damage of components of injection and ignition systems on the increase of hydrocarbons HC and uneven operation of engine. The analysis of the composition of the exhaust fumes for fourteen cars showed the exceeding of CO content in the exhaust fumes for 4 cars.

The proposed model of the fuel injection and ignition systems enables to interpret the impact of selected parameters on the mass of injected fuel and the voltage of the secondary ignition.

The course of characteristics of the ignition voltage and the impact of the lambda probe on the mass of injected fuel was verified on the test stand in order to develop diagnostic patterns.

REFERENCES

Adamiec, M., Dziubiński, M. 2009. Alkaliczne ogniwo paliwowe – aspekt sprawności, *Przegląd Elektrotechniczny*, 85(4): 182–185.

Chakraborty, S., Sadhu, P. K., Gosowami, U. 2016. Barriers in the Advancement of Solar Energy in Developing Countries like India, *Problemy Ekorozwoju – Problems of Sustinable Development*, 11(2): 75–80.

Cao, Y., You, J., Wang, R. & Shi, Y. 2016. Designing a Mixed Evaluation System for Green Manufacturing of Automotive Industry, *Problemy Ekorozwoju/Problems of Sustainable Development*, 11(1): 73–86.

Cao, Y., You, J., Shi, Y. & Hu, W. 2016. Evaluation of Automoblie Manufacturing Enterprise Competitiveness from Social Responsibility Perspective, *Problemy Ekorozwoju/Problems of Sustainable Development*, 11(2): 89–98.

Chłopek, Z. 1999. Modeling of exhaust emissions in the condition of traction exploitation of combustion engines, *Scientic works of the Wroclaw University of Technology, The publishing house of the Warsaw University of Technolgy*, Notebook 173, Warsaw.

Duran, J. et al. 2013. Renewable Energy and Socio-Economic Development in the European Union, *Problemy Ekorozwoju/Problems of Sustainable Development*, 8(1): 105–114.

Gronowicz, J. 2003. Protection of the environment in land transport, *The publishing house and printing agency of the Institute for Sustainable Technologies*, ISBN 83-7204-313-2, Radom.

Jarzyna, W., Augustyniak, M., Bochenski, M., et al. 2012. PD and LQR controllers applied to vibration damping of an active composite beam. *Przegląd Elekrotechniczny*, 88(10B): 128–131.

Jarzyna, W., Augustyniak, M. & Wojcik, A. 2012. Application of a fuzzy PI controller in regulation of active piezoelectric composite structure. *Przegląd Elektrotechniczny*, 88(10A): 298–301,

Jarzyna, W., Pawłowski A. & Viktarovich N. 2014. Technological development of wind energy and compliance with the requirements for sustainable development, *Problemy Ekorozowju/Problems of Sustainable Development*, 9(1): 167–177.

Lindzen, R. S. & Sloan, A. P. 2014. Global Warming and the Irrelevance of Science, *Problemy Ekorozwoju/Problems of Sustainable Development*, 11(2).

Liu, H. 2015. Biofuel's Sustainable Development under the Trilemma of Energy, 2015. Environment and Economy, *Problemy Ekorozwoju/Problems of Sustainable Development*, 10(1): 55–59.

Merkisz, J. 2003. Ecology of transport, engineering of machines, Vol. 8. Notebook 4, *Publishing agency of Wroclaw Council FSNT NOT*, ISBN 1426-708X, Wrocław.

Merkisz, J. & Mazurek, S., 2007. On-board diagnostics of publishing house of communication and transport *WKiŁ*, ISBN 978-83-206-1633-0, Warsaw.

Merkisz, J., Lijewski, P. & Fuć, P. 2011. Exhaust emissions measured under Real traffic conditions from vehicles fitted with spark ignitron and compression ignitron engines, *The Archives of Transport*, Vol. 13(2): 165–173.

Merkisz, J., Pielecha, J. & Radzimirski, S. 2012. Emission of automotive pollution in the light of new regulations of the European Union of publishing house of communication and transport *WKiŁ*, ISBN 978-83-206-1831-0, Warsaw.

Merkisz, J. & Piaseczny, L. 2013. Ecological security of a Marine combustion engine – a formal approach, *Journal of KONBIN* 1(25): 5–12.

Merkisz, J. & Pielecha, J. 2014. Emission of particulates from automotive source, *The publishing house of Poznań University of Technology*, ISBN 978-83-7775-325-5, Poznań.

Pawłowski, A. 2008. The Role of Social Sciences and Philosophy in Shaping of the Sustainable Development Concept, *Problemy Ekorozwoju/Problems of Sustainable Development*, 3(1): 7–11.

Pawłowski, A. 2009a. Sustainable Energy as a Sine Qua Non Condition for the Achievement of Sustainable Development, *Problemy Ekorozwoju/Problems of Sustainable Development*, 4(2): 9–12.

Pawłowski, A. 2009b. Theoretical Aspects of Sustainable Development Concept, *Rocznik Ochrona Środowiska*, 11(2): 985–994.

Pawłowski, A. 2013. Sustainable Development and Globalization, *Problemy Ekorozwoju/Problems of Sustainable Development*, 8(2): 5–16.

Robert Bosh Gmbh, (2004). "The technical manual – Control of spark ignition engines. Motronic systems".

Rokosch, U. 2007. Aftertreatment systems and On-board diagnostics of publishing house of communication and transport WKiŁ, ISBN 978-83-206-1657-6, Warsaw.

Siemionek, E. & Dziubiński, M. 2015. Testing energy consumption in the trolleybus and the bus on a chosen public transport line in Lublin, *Advances in Science and Technology Research Journal*, 26(9): 152–155.

Sztumski, W. 2016. The Impact of Sustainable Development on the Homeostasis of the Social Environment and the Matter of Survival, *Problemy Ekorozwoju/Problems of Sustainable Development*, 11(1): 41–47.

Udo, V. & Pawłowski, A., 2010. Human Progress Towards Equitable Sustainable Development: A Philosophical Exploration. *Problemy Ekorozwoju/Problems of Sustainable Development* 5(1): 23–44.

Żelazna, A. & Gołębiowska, J. 2015. The Measures of Sustainable Development – a Study Based on the European Monitoring of Energy-related Indicators, *Problemy Ekorozwoju/Problems of Sustainable Development*, 10(2): 169–77.

Environmental Engineering V – Pawłowska & Pawłowski (Eds)
© 2017 Taylor & Francis Group, London, ISBN 978-1-138-03163-0

Testing of exhaust emissions of vehicles combustion engines

M. Dziubiński
Faculty of Mechanical Engineering, Lublin University of Technology, Lublin, Poland

ABSTRACT: The study analyzed the systems having a direct impact on exhaust emissions, which must be checked and replaced throughout the life of the vehicle. We have analyzed exhaust-gas after treatment systems for spark ignition engines and exhaust gas after treatment system of diesel engine, which allows for the reduction of nitrogen oxides contained in the exhaust gas. The part dedicated to research, describes check-ups for engines fueled with petrol and LPG at idle and at higher speeds. The following exhaust-gas components were recorded: carbon monoxide CO, hydrocarbons HC, carbon dioxide, CO_2, nitric oxide NOx and lambda. As part of the simulation tests, registration of the composition of the exhaust gases for selected supply system damage was carried out. The results will help to develop a database of diagnostic patterns.

Keywords: exhaust emissions, lambda sensor, combustion engines, damage of injection control system, spark ignition

1 INTRODUCTION

Earth's ecosystem is limited, which means that the amount of non-renovable resources, including fossil fuels, is limited as well. It is estimated that with the current rate of crude oil consumption, the deposits will suffice for another 60–70 years (Pawłowski 2008, 2012). Present consumption of oil is extremely high. Nowadays, 2 dm³ of oil are used daily per each person living on the Earth (Armaroli and Balzani 2006). Simultaneously, combustion of fossil fuels contributes to the emission of greenhouse gases. In 1990–2014, the emission form majority of sources dropped: fuel combustion (without transport) form 62.3% to 55.1%, industrial processes and product use from 8.9% to 8.5%, waste management from 4.2% to 3.3%.

On the other hand, in the case of road transport, it increased from 14.9% to 23.2% (EEA 2014). This means that the emission of greenhouse gases from road transport is on the rise. Counteracting unfavorable climatic changes through the mitigation of greenhouse gases emissions is an essential element of ensuring sustainable development for our civilization. This idea integrates all areas of human activity (Pawłowski 2008,2009a, 2009b, 2013, Udo et al. 2010, Sztumski 2016). From the point of view of mitigating greenhouse emission, utilizing renewable fuels is the most favorable solution (Duran et al. 2013, Liu 2015, Xue et al. 2014). In the case of road transport, biofuels (Piementel 2012, Cao et al. 2016) and electric cars are widely used (Shao et al. 2014, Cao et al. 2016).

However, taking into account that majority of cars in use today are driven by combustion engines, mitigating emission form this type if engines may also contribute to a decrease in the emission of CO_2 and other pollutants.

Operating conditions of combustion engine in a vehicle motor are varied. Engine speed in the range from idle speed to work at full load, especially in the warm-up phase of the engine has a decisive effect on exhaust emissions. Acceptable values of the emissions components are continuously reduced and currently effective EURO 6 (Chłopek 2014, Czech 2014) standard obliges to use systems designed to reduce the toxicity of exhaust gases. The most important systems that reduce the toxicity of exhaust gases are: engine power control for fuel in a closed loop, exhaust gas EGR recirculation, system of additional air (AIR), pulsatile system of additional air (PAIR), draining system of fuel vapor EVAP, venting of the crankcase, and cutting off the fuel supply when braking by engine. These solutions and control systems of the toxicity of exhaust gases are used in various combinations depending on the cubic capacity of the engine and the engine control system (EEC 5 or Motronic). Fuel supply system of the engine in a closed loop is the most important measure reducing the levels of toxic compounds in exhaust gases. The fuel supply system in a closed-loop includes three-way catalyst TWC, a sensor or sensors of the level of oxygen in the exhaust gas (lambda sensor and controller). Fuel supply in a closed-loop regulates the creation of compounding alternately rich and poor;

hence, the oxygen content in the exhaust gas is changed continuously from lower to higher values (Brzozowska 2009, Chłopek 2010, Merkisz 2009). Therefore, the three-way catalyst works with conversion yields of 95%, because the NOx reduction needs oxygen deficiency. While the oxidation of HC and CO requires the presence of free oxygen. The efficiency of three-way catalyst not operating in a closed loop reaches about 70%. Fuel supply in a closed-loop regulates the formation of the mixture, allowing optimum combustion within a narrow range of lambda $\lambda = 1$, it allows a significant reduction in the untreated exhaust. Lambda sensor sends to the electronic system of engine control voltage signal of direct current. The voltage signal is between 0.1–0.9 V or 0 to 5 V, depending on the level of oxygen present in exhaust gases.

On the basis of sudden changes in the signal, voltage control unit detects the degree of enrichment of the mixture and immediately corrects the fuel injection amount such that the lambda value oscillates around the value of $\lambda = 1$. The mixture alternately enriched and lowered results in the desired conversion rate for the three-way catalyst (Fig. 1). In some vehicles equipped with engine management system and OBD system (On Board Diagnosis) there are one or two lambda sensors mounted behind the catalyst. The purpose of the lambda sensor behind the catalyst is to detect trace amounts of oxygen and nitrogen oxides in the exhaust gas after passing through the catalyst. Since the introduction of Euro 5 standard for ignition engines, using sensors of concentrations of nitrogen oxides and ammonia in the exhaust gas became a commonplace. A characteristic feature of SCR reactors (Selective Catalytic Reduction) is the existence of temperature limit of about 450°C in addition to the onset temperature of about 200°C, above which the degree of conversion is significantly reduced (Fig. 2) (Rokosch 2007, Bosch 2004, Chłopek 2012, Bielaczyc 2012). Urea, which

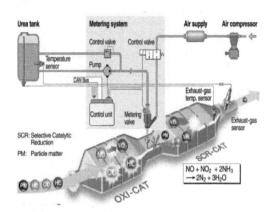

Figure 2. Exhaust-Gas after treatment system for NO_x.

is added to the exhaust, determines the course of the respective reactions and is stored in a secondary tank placed in a motor vehicle.

2 EXPERIMENTAL RESEARCH OF THE CONTENT OF TOXIC COMPONENTS IN THE ENGINE EXHAUST FUELED BY PETROL AND LPG

Tests were carried out on an engine dynamometer of Faculty of Mechanical Engineering in Technical University of Lublin. The study was conducted using diagnostic equipment of Bosch for 17 cars using petrol and LPG.

The test results are shown on the figures from Fig. 3 to Fig. 7.

When operating at idle speed, petrol-fueled engines exceeded the allowable value of CO in four cases.

When operating at higher speeds, petrol-fueled engines exceeded the limit value (point line) of CO in the three cases.

In the case of LPG, the allowable value of CO was not exceeded in any case.

The content of hydrocarbons in the exhaust gas of the tested vehicle is exceeded in eleven cases out of 17. The increase of hydrocarbon content as in cases 8,10,13 can cause misfire. For all gas-fueled engines HC value is exceeded.

In five cases out of thirteen petrol-fueled engines, lambda did not fit into the acceptable range.

For engines fueled by LPG lambda was exceeded, all engines possessed a poor mix.

As part of the experimental research, following engine malfunctions were recorded:

- Disconnected lambda sensor,
- Disconnected temperature sensor,
- Disconnected throttle potentiometer,

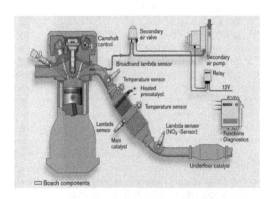

Figure 1. Exhaust-Gas after treatment—SULEV system.

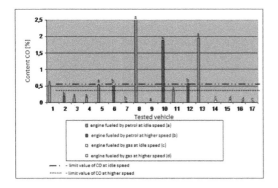

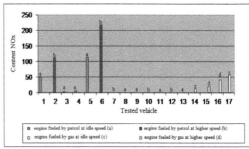

Figure 6. The content of NOx in the exhaust of vehicles tested.

Figure 3. The content of CO in exhaust gases of vehicles tested.

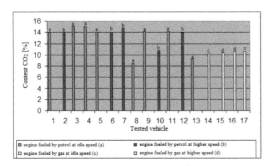

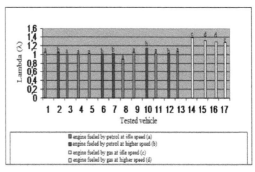

Figure 7. Lambda (λ) for vehicles tested.

Figure 4. The content of CO_2 in the exhaust gases of the tested vehicles.

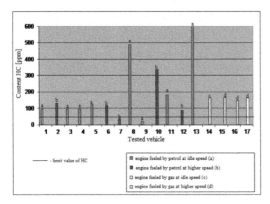

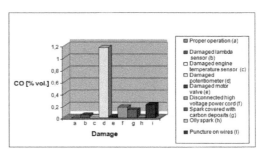

Figure 8. The content of CO in the exhaust gases for selected damages.

Figure 5. The content of HC in exhaust gases vehicles tested.

- Disconnected motor valve,
- Disconnected high voltage power cord of the first cylinder,
- Faulty spark (covered with carbon deposits)
- Faulty spark (oily)
- Puncture on high-voltage wires. The recorded measurement results are shown in Figures 8 to Fig. 12.

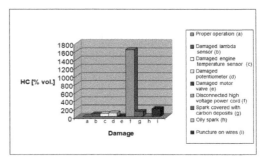

Figure 9. The content of HC in the exhaust gases for selected damages.

307

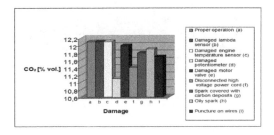

Figure 10. The content of CO_2 in exhaust gases for the selected damages.

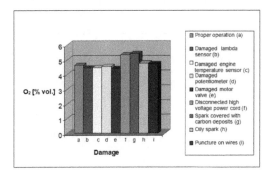

Figure 11. The content of O_2 in the flue gas with the selected damages.

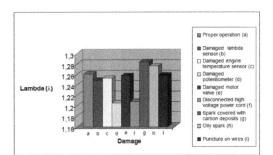

Figure 12. The content of oxygen in the exhaust gas for the selected damages.

3 CONCLUSIONS

Enrichment of the dose of fuel at warming up of a cold engine lambda 0.886 resulted in an increase of emissions of CO to 6.92% and an increase in emissions of unburned hydrocarbons to 490 ppm. Engines fueled with LPG with poor mixture of lambda 1,398 - 1, 283 resulted in a reduction of CO emissions, while all gas-fueled engines exceeded the value of HC.

On the basis of the survey it can be said that any type of damage in the control system changes the composition of the exhaust gas, exceeding the legal limit. The analysis made in the research part on the damage of fuel track sensors suggests that the most common symptom of damage is an increase in emission of HC hydrocarbon. The reason for this phenomenon is the fact that in the event of failure of any sensor the controller is in a state of emergency and the replacement value of the injection time is selected, which is due to the imprecise selection usually is not completely burned, which directly affects the growth of hydrocarbon emissions

High levels of HC, which exceed the permissible standards several times, are the result of damage to the ignition control system. The damage done to the controller and the electrical system usually leads to the exclusion of a cylinder from the work and the entire part of unburned mixture enters directly into the exhaust system. An example can be the analysis of exhaust gas while removing the high voltage wire from a single spark, for that damage the emission of hydrocarbons is increased several times.

Increased emissions of HC contributes to disturbing the work of lambda sensor from with a signal transmitted to the controller is bonded with ill-chosen mixture for subsequent cycles. As a result of the resulting conjugate, continuous reading of too rich mixture led to the generation of poor mixture to the working cylinders which leads to incorrect combustion process.

Analysis of other toxic components showed no change to the simulated damage exceeding the legal limit. However, in the case of damages of throttle potentiometer, the CO value increased exceeding emission standards with a very high level of lambda.

The recorded results will allow for the development of a database of diagnostic patterns for different types of damages of vehicles.

REFERENCES

Adamiec, M. & Dziubiński, M. 2009. Alkaliczne ogniwo paliwowe—aspekt sprawności, *Przegląd Elektrotechniczny*, 85(4): 182–185.

Armaroli, N, & Balzani, V. 2006. The Future of Energy Supply: Challenges and Opportunities, *Angew. Chem. Int. Ed.*, 45: 2–17.

Bielaczyc, P. & Szczotka, A. 2012. The Potential of Current European Light Duty CNG-fuelled Vehicles to Meet EURO 6 requirements, *Combustion Engines* 151(4).

Brzozowska, L., Brzozowski, K. & Drąg, Ł. 2009. Transport drogowy a jakość powietrza atmosferycznego, *Modelowanie komputerowe w mezoskali*, ISBN 978-83-206-1714-6, WKiŁ, Warsaw.

Cao, Y., You, J., Shi, Y. & Hu, W. 2016. Evaluation of Automobile Manufacturing Enterprise Competitiveness from Social Responsibility Perspective, *Problemy Ekorozowju/Problems of Sustainable Development*, 11(2): 89–98.

Cao, Y., You, J., Wang, R. & Shi, Y. 2016. Designing a Mixed Evaluation System for Green Manufacturing of Automotive Industry, *Problemy Ekorozowju/ Problems of Sustainable Development*, 11(1): 73–86.

Chłopek, Z., 2010. A Correlation Analysis of the Pollutant Emission from a Self Ignition Engine, *Combustion Engines*, 140(1).

Chłopek, Z., 2012. Testing of Hazards to the Environment Caused by Particulate Matter During Use of Vehicles, *Maintenance and reliability*, 14(2).

Chłopek, Z. 2014. The Assessment of the Pollutant Emission from the Self Ignition Engine in its Different Operating States, *The Archives of Transport*, 29(1).

Czech, P. 2014. Wspomaganie systemu OBD sztucznymi sieciami neuronowymi wykorzystującymi sygnały wibroakustyczne jako metoda diagnozowania uszkodzeń silników spalinowych w pojazdach, ISBN 978-83-7789-263-3, *Institute of Technology Utilization—National Research Institute in Radom*, Radom.

Dziubak, T. 2015. Techniczna eksploatacja układu zasilania powietrzem silników spalinowych pojazdów mechanicznych, ISBN978-83-7938-070-1, *Millitary University of Technology*, Warsaw.

Duran, J. et al. 2013. Renewable Energy and Socio-Economic Development in the European Union, *Problemy Ekorozwoju/Problems of Sustainable Development*, 8(1): 105–114.

EEA (2014), Air quality in Europe—2014 report.

EEA (2014), Air Pollution Fact Sheet, Poland.

IPPC 2014, Contribution of working Group III to the Fifth Assessment Report of the Intergovernmental Panel on Climate Change.

Jarzyna, W., Augustyniak, M., Bochenski, M. et al. 2012. PD and LQR controllers applied to vibration damping of an active composite beam. *Przegląd elektrotechniczny*, 88(10B): 128–131.

Jarzyna, W., Augustyniak, M. & Wojcik, A. 2012. Application of a fuzzy PI controller in regulation of active piezoelectric composite structure. *Przegląd elektrotechniczny*, 88(10A): 298–301.

Liu, H. 2015. Biofuel's Sustainable Development under the Trilemma of Energy, (2015), Environment and Economy, *Problemy Ekorozwoju/Problems of Sustainable Development*, 10(1): 55–59.

Merkisz, J., Pielecha, J., Radzimirski, S. 2009. Pragmatyczne podstawy ochrony powietrza atmosferycznego w transporcie drogowym, ISBN 978–83-7143-839-4, *Publishing Company of Poznan University of Technology*, Poznań.

OECD- 2010, Transport Outlook 2010, The Potential for Innovation.

Pawłowski, L. 2012 Czy liberalny kapitalizm i globalizacja umożliwiają realizację strategii zrównoważonego rozwoju? *Problemy Ekorozwoju/ Problems of Sustainable Development* 7(2): 7–13.

Pawłowski, A. & Pawłowski, L. 2008. Sustainable development in Contemporary civilisation, Part 1: The Environment and Sustainable Development, *Problemy Ekorozwoju/Problems of Sustainable Development*, 3(1): 53–65.

Piementel, D. 2012. Energy Production from Maize, *Problemy Ekorozwoju/Problems of Sustainable Development*, 7(2): 15–22.

Robert Bosch GmbH, 2004. Sterowanie silników o zapłonie samoczynnym, *WKiŁ*.

Robert Bosch GmbH, 2004. Sterowanie silników o zapłonie iskrowym—układy Motronic, *WKiŁ*.

Rokosch, U. 2007. Układy oczyszczania spalin i pokładowe systemy diagnostyczne samochodów OBD, *Wydawnictwa Komunikacji i Łączności WKiŁ*, ISBN 978-83-206-1657-6, Warsaw.

Shao, L., Xue, Y. & You, J. 2014. A conceptual Framework for Business Model Innovation: The Case of Electric Vehicles in China, *Problemy Ekorozwoju/ Problems of Sustainable Development*, 9(2): 27–37.

Siemionek, E. & Dziubiński, M. 2015. Testing energy consumption in the trolleybus and the bus on a chosen public transport line in Lublin, *Advances in Science and Technology Research* Journal, 26(9): 152–155.

Sztumski, W. 2016. The Impact of Sustainable Development on the Homeostasis of the Social Environment and the Matter of Survival, *Problemy Ekorozwoju/ Problems of Sustainable Development*, 11(1): 41–47.

Udo, V. & Pawłowski, A. 2010. Human Progress Towards Equitable Sustainable Development: A Philosophical Exploration. *Problemy Ekorozwoju/Problems of Sustainable Development*, 5(1): 23–44.

Xue, Y., You, J. & Shao, L. 2014. Understanding socio-technical barriers to sustainable mobility—insights from Demonstration Program of EVs in China, *Problemy Ekorozwoju/Problems of Sustainable Development*, 9(1): 29–36.

Environmental Engineering V – Pawłowska & Pawłowski (Eds)
© 2017 Taylor & Francis Group, London, ISBN 978-1-138-03163-0

Artificial intelligence methods in diagnostics of coal-biomass blends co-combustion in pulverised coal burners

A. Smolarz, W. Wójcik, K. Gromaszek & P. Komada
Faculty of Electrical Engineering and Computer Science, Lublin University of Technology, Lublin, Poland

V.I. Lytvynenko
Kherson National Technical University, Kherson, Ukraine

N. Mussabekov, L. Yesmakhanova & A. Toigozhinova
Kazakh National Research Technical University after K.I. Satpaev, Almaty, Kazakhstan

ABSTRACT: The paper presents technologies being developed in the Institute of Electronics and Information Technologies at Lublin University of Technology. They use optical sensors and artificial intelligence methods for process supervision and diagnostics. Research is aimed to develop a system allowing a parametric evaluation of the quality of pulverized coal burner operation in order to decrease its environmental impact. The article shows results for coal-shredded straw blends, yet the methodology may be applied for other types of blends.

Keywords: biomass, combustion, fuzzy methods, artificial immune systems

1 INTRODUCTION

It seems that in spite of growing share of other types of energy sources, burning various types of fuels will remain the main source of energy throughout the next decades. Unfortunately, it will also remain the greatest source of atmospheric pollution. On 23rd January 2008, the European Commission put forward a far-reaching package of proposals where it commits itself to reduce its overall emissions to at least 20% below 1990 levels by 2020. It has also set the target of increasing the share of renewables in energy use to 20% by 2020. The latter commitment results in search of new technologies that partially or entirely make use of renewable energy sources. This is also a case of combustion technologies where alternative fuels obtained from renewable sources are used. In the case of pulverized coal burners, biomass co-combustion may be applied. The efficient and clean combustion of those fuels poses a number of technical challenges. In general, alternative fuels are characterised by low to very-low calorific values and by fluctuating properties (among different batches, or along the time in a continuous process). The variability of this type of fuels can bring the system to off-design operation and cause increased pollutant emissions, lower efficiency or flame stability problems. Therefore,

permanent supervision and optimization becomes an issue that should be addressed in order to guarantee the reliability of a practical system.

The paper presents the technologies being developed in the Institute of Electronics and Information Technologies at Lublin University of Technology. They use optical sensors and artificial intelligence methods for process supervision and diagnostics. Research is aimed at developing a system allowing parametric evaluation of the quality of pulverized coal burner operation. The information about the status of the device is useful only when it has a form understandable to the operator or automatic control system and diagnostics. In the article, we analyse the case when improper operation of the burner consists in too high or too low excess air coefficient, diagnostics will therefore be relied on to detect three states.

There are many classification methods, possible to use in this case, starting from the Fisher method (LDA) to a currently popular method of Support Vector Machine (SVM) or neural networks. Due to the highly nonlinear nature of dependencies and lack of an analytical model, the artificial intelligence methods were used. Two examples will be shown, i.e. the fuzzy networks and a relatively new class of classification methods—artificial immunology algorithms.

311

2 TESTS—METHODOLOGY, FACILITY AND MEASUREMENTS

Combustion of pulverized coal was examined through optical methods, which were based on analysis of wide spectrum radiation emitted by the flame. The analysis also takes into account spatial features of such radiation source. Combustion of pulverized coal in the power burner takes place in a turbulent flow. In its each point, local fluctuations of both fuel and gaseous reagents concentrations, as well as temperature occur. It leads to permanent local changes in combustion process intensity, which result in continuous changes in flame luminosity that can be observed as flame flicker. As combustion process affects the turbulent movement of its products and reagents, it determines the way the flame flicker parameters such as e.g. mean luminosity and luminosity frequency spectrum. A number of combustion supervision and flame-fault protection systems use information contained within flame flicker. The multichannel fibre-optic flame monitoring system developed at Lublin University of Technology belongs to this class of solutions, but additionally it allows observation of selected areas of the flame (Wójcik 2001, Wójcik 2004).

Experiments were conducted on test rig located in the Institute of Power Engineering in Warsaw. It is a combustion chamber with a single pulverized coal swirl burner made in 1:10 scale in relation to a low-emission industrial burner. This object was chosen because of its ability to perform experiments with a single burner, and its good instrumentation. All measured quantities are visualized and recorded by the data acquisition system. Sampling period is 1s. The combustion chamber is equipped with the above-mentioned optical fibre probe which allows observation of five different areas of the flame.

The experiment begins with bringing the chamber to the proper temperature. When the temperature stabilizes, series of measurements are performed with changing air and fuel flows. During an individual measurement the amounts of fuel and air are kept constant. A single measurement lasts approximately 300 seconds. Such measurement method is used in order to eliminate the impact of the transport delay of gas analysers. It is assumed that during the measurement the conditions are fixed and the emission values stabilized. The tests were conducted at three different thermal loads, for pure pulverised coal and 10% blend with biomass (shredded straw). The amount of secondary air was being changed in order to achieve the air excess corresponding to normal operation, too high and too low conditions. Voltage signals corresponding to the instantaneous brightness of the flame of the areas observed by individual optical

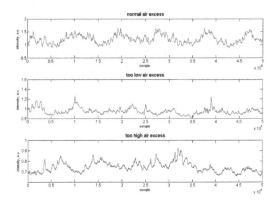

Figure 1. Example measurements corresponding to normal, too low and too high air excess ratio.

fibres were sampled at the rate of 8 KS/s and saved by a dedicated system. Figure 1 shows example measurements corresponding to normal, too low and too high air excess ratio.

3 NEURO-FUZZY ALGORITHMS

As first, the diagnostic capability of the combustion process is shown on the example of fuzzy neural network. The combination of neural networks with fuzzy logic has many benefits, especially where traditional methods and solutions do not give good results or to use them for specific tasks would be too time-consuming or costly. This method usually yields better results than SVM and classical neural networks (Xu 2012); besides, by removing the last layer of the network fuzzy, information can be obtained and used for warning when symptoms of malfunction (or failure) appear.

Neuro-fuzzy is one of the concepts of artificial intelligence that refers to combinations of artificial neural networks and fuzzy logic, proposed by J. S. R. Jang (Jang 1991). Neuro-fuzzy hybridization results in a hybrid intelligent system that synergizes these two techniques by combining the connectionist structure of neural networks and the human-like reasoning style of fuzzy systems through the use of fuzzy sets and a linguistic model consisting of a set of IF-THEN fuzzy rules. The main advantage of neuro-fuzzy systems is that they are universal approximators with the ability to solicit interpretable IF-THEN rules, yet it involves two contradictory requirements in fuzzy modeling: interpretability versus accuracy. The approach presented in the article focuses on the latter one, implementing the Takagi-Sugeno-Kang (TSK) model. The general idea of a neuro-fuzzy model is depicted in Figure 3. The structure of the model is determined before training and does not change. Training takes the form of adjusting the fuzzy membership functions parameters, in order to minimise the model error.

The Matlab fuzzy logic toolbox and its ANFIS (adaptive neuro-fuzzy inference system) tool were used to design neuro-fuzzy models. The tool allows the construction and training of the Sugeno models using the methods typical for neural networks, e.g. error backpropagation. For this purpose the fuzzy model is converted into an equivalent neural network with the structure of a multilayer perceptron. Due to the strategy of measurements the data were grouped in distinct centres so the fuzzy model structure was generated by subtractive clustering.

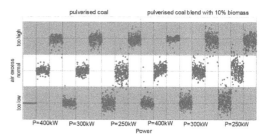

Figure 2. Fuzzy classification of the excess air ratio for different burner output and two types of fuel.

4 NEURO-FUZZY CLASSIFICATION

After completing the measurements the stored time series from each fibre was filtered in order to obtain the same time basis as for process parameters (1 sample per second) and to obtain the following parameters calculated over a period of 1 second: the average intensity value, intensity variance, number of mean value crossings and number of zero crossings (changes of sign) of the signal derivative. Such a choice of parameters was made on the basis of previous studies (Smolarz 1999).

As an example, results for a mixture consisting in 90% of the pulverized coal and 10% biomass (shredded straw) are shown. The mixture was prepared prior to the combustion test.

Because both the PCA analysis and orthogonality analysis did not demonstrate the possibility to omit any of the features, all 30 were used. Due to the methodology of research in which the series of tests were made during which constant conditions were kept, measurements were clearly grouped. Therefore, subtractive clustering method was used to determine the number of membership functions. The set of features of a total of 2700 measurements has been divided into a learning part and testing part, 80% and 20% respectively. Using 31 membership functions a classification error of 0.21% for the training set and 1.85% for the test set was obtained. The network was also trained using measurements of both the pure pulverized coal, as well as mixtures with the biomass. A total of 18 variants of burner operation were included with 3 power levels, 3 excess air levels, and 2 types of fuel. The resulting classifier was more versatile. For 80% to 20% learning/testing set division the classification error was 0.8% and 1.3%, respectively. For 60% to 40% division, the error has risen to 1.7% and 5.8%, respectively. This is partly due to the fact that some input was out of training range—neural networks do not have the ability to extrapolate. Leaving the fuzzy outputs Fig. 2 enables to notice a rising uncertainty in the low-power rich-biomass region.

On the basis of the analysis it can be concluded that method presented above is suitable for detecting deviation of the excess air ratio λ of ± 0.1 from the correct value. In the case of coal and biomass co-firing increase in sensitivity may be difficult, especially at a low load of the burner. Method performance is also confirmed by the test in which type of fuel (coal or blend) is being recognised on the basis of optical signals with the accuracy of 99%.

5 ARTIFICIAL IMMUNE ALGORITHMS

In the 1990s, Artificial Immune System (AIS) emerged as a new computational research filed inspired by the simulation of biological behavior of Natural Immune System (NIS). The NIS is a very complex biological network with rapid and effective mechanisms for defending the body against a specific foreign body material or pathogenic material called antigen.

The Artificial Immune Systems, as defined by de Castro and Timmis (de Castro 2003) are: "Adaptive systems inspired by theoretical immunology and observed immune functions, principles and models, which are applied to problem solving". However, AIS are one of many types of algorithms inspired by biological systems, such as neural networks, evolutionary algorithms and swarm intelligence. There are many different types of algorithms within AIS and the research to date has focused primarily on the theories of immune networks, clonal selection and negative selection. These theories have been abstracted into various algorithms and applied to a wide variety of application areas, such as anomaly detection, pattern recognition, learning and robotics.

5.1 Negative selection algorithm

The negative selection of T-cells is responsible for eliminating the T-cells whose receptors are capable of binding with self-peptides presented by self-MHC molecules. This process guarantees that the

T-cells that leave the thymus do not recognize any self-cell or molecule. Forrest et al. (Forrest 1994) proposed a change detection algorithm inspired by the negative selection of T-cells within the thymus. This procedure was named *negative selection algorithm* and was originally applied in computational security. A single type of immune cell was modelled: T-cells were represented as bit strings of length L. The negative selection algorithm of Forrest and collaborators is simple. Given a set of self-peptides, named self-set **S**, the T-cell receptors will have to be tested for their capability of binding the self-peptides. If a T-cell recognizes a self-peptide—it is discarded, otherwise it is selected as an immune-competent cell and enters the available repertoire **A**.

The idea of negative selection algorithm is to generate a set of detectors in a complementary set of **N** and then to use these detectors for binary classification as "Self" or "Non-Self". Formally, the negative selection algorithm can be represented as (Lytvynenko 2006, Lytvynenko 2008):

$$NegAlg = (\Sigma^L, L, \mathbf{S}, \mathbf{N}, r, n, s, pr)$$

where Σ^L denotes shape-space; L is receptor length; **S** is "Self" detector set; **N** is "Non-Self" detector set; r denotes cross-reactive threshold; n is total number of appointed detectors; s is detector set size; pr denotes rule matching rows in adjacent positions.

The negative selection algorithm can be summarized as follows (Lytvynenko 2008) (Fig. 3):

- *Initialization:* randomly generate strings and place them in a set **P** of immature T-cells, assuming all the molecules (receptors and self-peptides) are represented as binary strings of the same length **L**.
- *Affinity evaluation:* determine the affinity of all T-cells in V with all elements of the self-set **S**.
- *Generation of the available repertoire:* if the affinity of an immature T—cell with at least one self-peptide is greater than or equal to a give cross reactive threshold, then the T-cell recognizes this self-peptide and has to be eliminated (negative selection); else the T-cell is introduced into the available repertoire **A**.

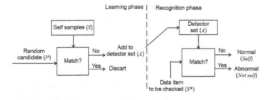

Figure 3. Negative selection algorithm.

The process of generating the available repertoire in the negative selection algorithm was termed learning phase. The algorithm is also composed of a monitoring phase. In the monitoring phase, a set **S*** of protected strings is matched against the elements of the available repertoire **A**. The set **S*** might be the own set **S**, a completely new set, or composed of elements of **S**. If recognition occurs, then a non-self pattern (string) is detected.

It is well known, that the algorithm of Negative Selection (NS) has some restrictions and limitations (de Castro 2003). It is not appropriate, for example, when the number of self samples is small and sparse.

Some limitations of the binary-string representation in NS algorithms are as follows:

- binary matching rules are not able to capture the semantics of some complex self/non-self spaces,
- it is difficult to extract meaningful domain knowledge,
- in some cases, a large number of detectors are needed to guarantee better coverage (detection rate),
- it is difficult to integrate the NS algorithm with other immune algorithms,
- the crisp boundary of "self" and "non-self" may be very hard to define.

In real-valued representation the detectors are represented by hyper-shapes in n-dimensional space. The algorithms use geometrical spaces and heuristics to distribute detectors in the non-self space.

Some limitations of the real-valued representation in NS algorithms are:

- the issue of holes in certain geometrical shapes, and may need multi-shaped detectors,
- curse of dimensionality,
- the estimation of coverage,
- the selection of distance measure.

During our experiments it has been established that generation of set of detectors at training phase occurs casually. This happens because it is impossible to define in advance the minimum necessary quantity of detectors which will provide the maximum quality of recognition. The increase in quantity of detectors leads to a delay of a phase of recognition, and its reduction—to deterioration of work of algorithm, since the probability of formation of the "cavities" which are areas in space of "Non-self" that are not distinguished by any of detectors increases. Thus, a problem of the given research is working out of an advanced method of generation of the detectors, capable of adaptive selection of their options, quantity and an arrangement.

5.2 Clonal selection algorithm

Presently, the algorithm CLONALG exists in two forms: (1) for optimization problems solving, and (2) for solving problems of classification and pattern recognition. Basic clonal selection algorithm (de Castro)., named CLONALG, works as in Fig. 4.

Formally, the algorithm of clonal selection can be represented as (Lytvynenko 2008):

$$CLONALG = (P^l, G^k, l, k, m_{Ab}, \delta\delta, f, I, \tau, G, AB, S, C, M, n, d)$$

where P^l is space of search (space of forms); G_k is space representation; l is the length of vector of attributes (dimension of space of search); k is the length of antibody receptor; m_{Ab} is dimension of population of antibodies; δ is the expression function; f is the affinity function; I is the function of initialization of the initial population of antibodies; τ is the condition of completion of algorithm work; AG is the subset of antigens; AB is the population of antibodies; S is the operator of selection; C is the operator of cloning; M is the mutation operator; n is the number of the best antibodies selected for cloning; d is the number of the worst antibodies subjected to substitution for new ones. The algorithm used in our experiments contains adaptive antibody described in (Smolarz 2013).

5.3 Combined clonal and negative selection algorithm

The next issue to be addressed is the fact that the negative selection algorithm is intended only to solve the problems of anomaly detection and binary classification. Only a small number of works describe the use of negative selection algorithm for multi-class classification problems (de

---

1. Initialization; randomly initialize a repertoire (population) of attribute strings (immune cells).
2. Population loop .for each antigen, do:
   2.1. Selection ; select those cells whose affinities with the antigen are greater.
   2.2. Reproduction and genetic variation: generate copies of the immune cells: the better each cell recognizes the antigen, the more copies are produced. Mutate (perform variations) in each cell inversely proportional to their affinity: the higher the affinity, the smaller the mutation rote.
   2.3. Affinity evaluation: evaluate the affinity of each mutated cell with the antigen.
3. Cycle: repeat Step 2 until a given convergence criterion is met.

Figure 4. Standard clonal selection algorithm.

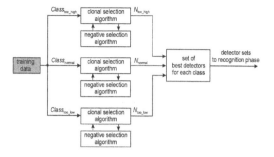

Figure 5. Synthesis of multi-class classifiers—learning phase.

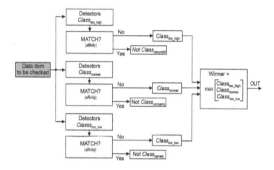

Figure 6. Synthesis of multi-class classifiers—classification phase.

Castro 2003). The classifier presented in this paper is based on the hybridization process of negative selection with clonal selection, and was designed to solve problems of classification to many classes. The concept of classification is used in terms of supervised learning, which allows categorizing objects into known groups using training set prepared beforehand. The proposed Combined Clonal Negative Selection Algorithm for Multi-Class problem classification) consists of a group of n elements responsible for assigning patterns to corresponding n classes. Synthesis of such classifier designed for the problem of detection of excess air diagnostics is shown in Figures 5 and 6, for learning phase and classification phase respectively.

6 ARTIFICIAL IMMUNE CLASSIFICATION

In this particular case, the selected coefficients of Discrete Wavelet Transform (DWT) (Wójcik 2003) within the local signal intensity of the flame radiation were chosen as the features of the flame. The negative clonal selection was used as classification algorithm. There are three classes to be recognised; hence, the classifier contained three subsets of detectors (Fig. 4). The measurement data was processed with DWT (Daubechies 6) in the windows of length of 16,384 samples. Subsequently, the statistical parameters (maximum value, minimum and average and standard deviation) of the most significant transform coefficients—D1, D2, D3, D4 and A1 – were calculated to give the vector of 20 features for each class. A set of features was randomly divided into learning and testing subsets by 30% / 70% and 70% / 30% out of 2400 data points.

In order to avoid the bias associated with the random sampling of the training data, the k-fold cross-validation was also performed. In k-fold cross-validation, the data is partitioned into k subsets of approximately equal size. Training and testing the algorithm is performed k times. Each time, one of the k subsets is used as the test set and the other k–1 subsets are put together to form a training set. Thus, k different test results exist for the algorithm. However, these k results are used to estimate performance measures for the classification system.

The common performance measures used in diagnostics are accuracy, sensitivity and specificity. Accuracy expresses the ability of the classifier to produce accurate diagnosis. The ability of the model to identify the occurrence of a target class accurately is determined by sensitivity. Specificity is determined by the ability of the algorithm to separate the target class. The accuracy can be expressed as:

$$Accuracy(Z) = \frac{\sum_{i=1}^{|z|} Assess(z_i)}{|Z|},$$

$$Assess(z) = \begin{cases} 1, & if\ classify(z) = z.c \\ 0, & otherwise \end{cases}$$

where: z denotes the patterns in testing set to be classified, $z.c$ is the class of pattern z, $classify(z)$ returns the classification of z by classification algorithm. For sensitivity and specificity analysis, the following equations can be used:

$$Sensitivity = \frac{TP}{TP+FN}, \quad Specificity = \frac{TN}{TN+FP}$$

where: TP, TN, FP and FN denote respectively true positive, true negative, false positive and false negative classification.

Classification tests using negative clonal selection algorithm were made according to the algorithm shown in Fig. 5. Table 1 contains the results of performance analysis. Average accuracy was about 98.99%. Classification accuracy obtained using fuzzy networks (TSK) was about 96.4%. Normalised execution time of both algorithms was similar.

Table 1. Performance measures for negative clonal selection algorithm.

Learning set/testing set distribution	Accuracy	Sensitivity	Specificity
40/60	98.95	99.25	99.10
60/40	99.18	99.20	99.45
80/20 5-fold cross-validation	98.85	98.75	99.25
Mean	98.99	99.07	99.27

7 CONCLUSIONS AND REMARKS

Optical signal can be used for diagnostics of an individual burner. The optical signal is the fastest and provides a selective way of getting information about the quality of combustion. Its interpretation, however, poses many difficulties.

The studies, described in the article, confirm that in order to obtain the information about correct range of a pulverized coal burner air excess ratio, the estimate calculated on the basis of immediate optical signals can be used instead of the delayed signals from the gas analyzers. The use of neuro-fuzzy models allows to obtain diagnostic signals of satisfactory accuracy and time, which allows application in control systems. The method has the disadvantage of neural network techniques. Although it has sufficient accuracy within the trained input space, it is unable to extrapolate, leading to erroneous results in the case of combustion conditions in which the inputs to the network are out of boundaries that were previously trained.

The modified negative selection procedure that uses optimization, as well as the artificial immune network for optimization parameters detectors was developed. A distinctive feature of this procedure is a modification of the learning process, through which the adaptive selection of settings, as well as the number and location of detectors is implemented.

Experimental studies have shown high efficiency of the proposed procedure, which is evident in its stability through adaptive value of cross-reactive threshold; optimality due to the adaptive immune network configuration size, i.e. the number of required detectors; accuracy by reducing the number and size of the created "cavities".

Classification accuracy of the negative clonal selection algorithm was better than the one of fuzzy (TSK) algorithm when applied to the problem of detection of anomalies in air excess ratio using optical system. Considering similar computational complexity of the above-mentioned algorithms, the advantage of the former one is clear. The negative clonal selection algorithm can then be used for diagnostics of correct co-firing of pulverised coal blends with biomass in individual PC burner.

ACKNOWLEDGEMENTS

A part of the research leading to these results has received funding from the European Union's Research Fund for Coal and Steel (RFCS) research programme under contract n°RFCR-CT-2008-00009 - SMARTBURN.

REFERENCES

De Castro, L. N. & Timmis, J. I. 2003. Artificial immune systems as a novel soft computing paradigm. *Soft Computing Journal* 7: 526–544.
Forrest, S. 1994. Self-nonself discrimination in a computer. *Proc. of the 1994 Ieee Symposium on Research in Security and Privacy.*
Jang, J.-S. R. 1991. Rule extraction using generalized neural networks. *Proc. of the 4th IFSA World Congress, in the Volume for Artificial Intelligence*: 82–86.
Lytvynenko, V. I. 2006. Immune classifier for binary classification tasks—practical implementation. *System technologies* 5(46): 113–126.
Lytvynenko, V. I. 2008. Comparative experimental study of a modified negative selection algorithm and clonal selection algorithm negative for solving classification.

Vestnik Kherson National Technical University, 4(33): 7–14.
Smolarz, A., Wójcik, W., Kotyra, A., Wojciechowski, C. & Komada, P. 1999. Fibre optic monitoring system. *Proceedings of SPIE* 4239: 129–132.
Smolarz, A., Lytvynenko, V., Kozhukhovskaya, O. & Gromaszek, K. 2013. Combined clonal negative selection algorithm for diagnostics of combustion in individual PC burner. *IAPGOS* 4: 69–73.
Wójcik, W., Surtel, W., Smolarz, A., Kotyra, A. & Komada, P. 2001. Optical fiber system for combustion quality analysis in power boilers. *Proceedings of SPIE* 4425: 517–522.
Wójcik, W., Kotyra, A., Komada, P., Przyłucki, S., Smolarz, A., Golec, T. 2003. The methods of choosing the proper wavelet for analyzing the signals of the flame monitoring system. *Proceedings of SPIE* 5124: 226–231.
Wojcik, W., Golec, T., Kotyra, A., Smolarz, A. & Komada P. 2004. Concept of application of signals from fiber optic system for flame monitoring to control separate pulverized coal burner. *Proc. SPIE* 5484: 427.
Wójcik, W., Kotyra, A., Smolarz, A. & Gromaszek, K. 2011. Modern methods of monitoring and controlling combustion of solid fuels in order to reduce its environmental impact. *Rocznik Ochrona Srodowiska* 13(2): 1559–1576
Wojcik, W., Gromaszek, K., Kotyra, A. & Ławicki, T. 2012. Pulverized coal combustion boiler efficient control. Przeglad Elektrotechniczny 88(11b): 316–319.
Xu, L., Tan, C., Li, X., Cheng, Y. & Li, X. 2012. Fuel-type identification using joint probability density arbiter and soft-computing techniques. *IEEE Transactions on Instrumentation and Measurement* 61(2): 286–296.

Environmental Engineering V – Pawłowska & Pawłowski (Eds)
© 2017 Taylor & Francis Group, London, ISBN 978-1-138-03163-0

Combustion process diagnosis and control using optical methods

K. Gromaszek, A. Kotyra & W. Wójcik
Faculty of Electrical Engineering and Computer Science, Lublin University of Technology, Lublin, Poland

B. Imanbek, A. Asembay, Y. Orakbayev & A. Kalizhanova
Kazakh National Technical University after K.I. Satpaev, Almaty, Kazakhstan

ABSTRACT: The paper discusses the conditions for diagnosis of combustion process using additional signals from non-invasive optical methods. This approach allows to obtain additional information about the ongoing process in an non-invasive way. The radiation emitted by the flame is a reflection of the chemical and physical reactions, occurring in the combustion process. The changes of the flame position in space results in shape fluctuations that can be interpreted as disruption of the balance results. The authors assumed that under certain conditions the shape of the flame can be an important indicator. Considering adequate emission flame spectrum, it is possible to determine the content appropriate process parameters. The optical methods based control seems to be particularly important for effective combustion and co-combustion, considering both economical and ecological factors in sense of industrial ecology.

Keywords: combustion diagnosis, optical methods, control algorithm, industrial ecology

1 INTRODUCTION

Various goods, services, and systems can be examined from an industrial ecology perspective, energy is probably the most fundamental and intricate in its characteristics. One definition of industrial ecology is "the network of all industrial processes as they may interact with each other and live off each other, not only in the economic sense but also in the sense of the direct use of each other's energy and material wastes" (Ausubel, 1992). There are several arguments for switching from coal-fired generation to gas generation. This is mainly due to lower emissions per kilowatt-hour generated. The discussion of many potential energy resources and delivery systems, including renewables and high-tech systems still remains. As element of industrial ecology policy, European Union implemented regulations for countries to reduce amount of emissions. It is a challenge, especially for countries, where energetics is coal based (Mukhanov, 2012), (Smolarz, 2013), (Zyska, 2016).

It is axiomatic that the aim of industrial activity is to satisfy human needs; hence, energy has a special place in industrial ecology. The coal is still the main fuel used in electricity generation around the world and it contains impurities that increase pollutant emissions significantly new combustion techniques are developed e.g. air staging, reburning and flue gas circulation (Li, 2003). In countries

with coal-based energetics there are long-term and short-term solutions. Long-term solutions cover increasing renewables systems and developing innovative technologies based on new energy sources. The short term solutions may involve soft indirect methods e.g. process optimization using alternative tools and approaches (Komada, 2016), (Sawicki 2016).

Additionally, fossil fuel depletion forces the use of renewable fuels such as biomass, where biomass is milled and burned simultaneously with coal in existing power stations. However, low-emission combustion techniques, including biomass co-combustion have negative effects: directly—influence on process control stability/efficiency and indirectly on combustion installations via increased corrosion or boiler slagging (Hein, 1998). These effects can be minimized using additional information about the process. Proper combustion monitoring (diagnosis) system ought to be applied (Koshymbaev, 2014), (Kotyra, 2010), (Pawłowski, 2016), (Shuvatov, 2012).

The efficiency of pulverized fuel depends on different parameters. Recirculation vortexes that lengthen the paths of the coal grains passing through the flame to minimize generation of thermal oxides of nitrogen (NO_x) are often applied, as pulverized coal combustion low-emission techniques. In order to make combustion of pulverized coal more efficient and clean, it is necessary to measure its key parameters. The information taken at the output (exhaust gas collector) is delayed and averaged. Although in (Fristrom 1995), several combustion diagnostic direct techniques are presented the most of them are impossible to utilize under industrial conditions or they are expensive. It creates new opportunities to supplement existing systems with fast and minimally invasive optical methods. Such approach allows to use image processing based information in process control system. This paper focuses on using optical methods usage in combustion and biomass co-combustion process.

2 MATERIAL AND METHODS

Radiation emitted by the flame is a reflection of chemical reactions and physical processes occurring in the combustion process. Both optical and acoustic diagnostics belong to the most important methods that allow the non-invasive way to obtain non-delayed and spatially selective additional information about the ongoing combustion process.

The spectrum of flames in the visible emission, provides information to determine the content of the air-fuel ratio, the quantity of heat release and temperature. Among optical methods, image processing based approach seems to be particularly important (Cięszczyk, 2016), (Dziubiński, 2016).

The still and apparent position of flame is the result of dynamic equilibrium between the local flame propagation speed and the speed of the incoming fuel mixture. Changes of the flame front position in space are seen as the flame shape fluctuations that are disruption of the balance results. This allows to assume that the shape of a flame can be an indicator of the combustion process, occurring under certain conditions (Kotyra, 2010).

Research of control system solution used data from the process of pulverized coal combustion and biomass co-combustion. They were conducted in 0.5 MWth laboratory combustion rig at the Institute of Power Engineering in Warsaw. This unit simulates the scaled down (10:1) combustion conditions of a full-scale swirl burner fired with pulverized coal with biomass added. The test stand comprises a horizontal layout consisting of a cylindrical combustion chamber 0.7 m in diameter and 2.5 m long, as shown in Fig. 1a.

A model of a low-NOx swirl burner about 0.1 m in diameter is mounted at the front wall. The stand is equipped with all necessary supply systems; primary and secondary air, coal, and oil. A mixture of pulverized coal and biomass is prepared in advance and dumped into the coal feeder bunker when the combustion chamber achieves stable conditions (Mashkov, 2014), (Mashkov, 2016).

The combustion test consisted of the following steps. First, the combustion chamber was warmed up by burning oil. When the temperature had risen sufficiently, the feeding device was started and the air-fuel mixture was delivered to the burner, simultaneously with the oil. After reaching the proper temperature level, the oil supply was switched off (Kotyra 2010). In the next stage pulverized coal was included into the process. Afterwards, several tests with different proportions of biomass in relation to the pulverized coal were conducted.

Measuring the physical-chemical quantities in such complex plant is a potential problem. For dynamic processes, measured values are

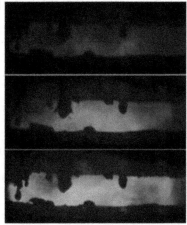

Figure 1. View of the laboratory combustion stand with camera and fiber-optic probe installed (a) and combustion sequences (b).

generally delayed and distorted. Measurements contain errors due to measuring conditions (dirt, slag, dust deposition). Moreover, they can provide indirect measurements or averaged because of the difficulty or inability to install the sensor at a particular location of the device.

In the proposed solution, a classical approach is supplemented with information about the image parameters flame, registered with a fast CCD camera or fiber-optic probe (see Fig. 1b).

2.1 Process modeling

Usually, traditional control systems of power plants are based on a number of regulation loops and feedforward compensators that contribute to maintaining the main process variables within reasonable values. With a few modifications, these structures can be adopted to the modern, demand-driven energy systems. Introduction of distributed generation and smart-grid systems require plants with enhanced load-following capability. Facing the occurrence of sudden load changes, control system emergency procedures or safety features ought to keep avoiding potentially dangerous behaviour. The dynamic performance can be improved significantly when using multi-variable control techniques instead of the classic SISO loops. A multi-variable, Multiple-Input, Multiple Output (MIMO) control scheme would reduce potentially dangerous events and unnecessary (redundant, tentative) emergency procedures. Both the power reliability and plant efficiency would therefore be increased.

The Nonlinear Autoregressive network with exogenous inputs (NARX) is a recurrent dynamic network, with feedback connections enclosing several layers of the network. The NARX model is based on the linear ARX model, which is commonly used in time-series modeling. The defining equation for the NARX model is as follows:

$$y(t) = f(y(t-1), y(t-2), \ldots, y(t-n_y), u(t-1), u(t-2), \ldots, u(t-n_u)), \quad (1)$$

where the next value of the dependent output signal is regressed on previous values of the output signal and previous values of an independent (exogenous) input signal. The NARX model can be implemented using a feedforward neural network to approximate the function. A diagram of the resulting network is shown in the Fig. 2a, where a two-layer feedforward network is used for the approximation. This implementation also allows for a vector ARX model, where the input and output can be multidimensional.

The output of the NARX network can be considered as an estimate of the output of modeled

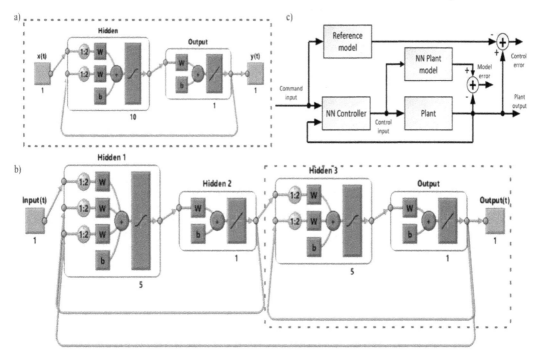

Figure 2. The structure of the neural network used in the plant identification phase (a), the structure of MRAC network (b) and MRAC control scheme (c).

nonlinear dynamic system (Suresh, 2005). The output is feedback to the input of the feedforward neural network as part of the standard NARX architecture. Regarding the fact that the true output is available during the training of the network, it is possible to create a series-parallel architecture (Narendra, 1991), in which the true output is used instead of feeding back the estimated output. This has two advantages. The first is that the input to the feedforward network is more accurate. The second is that the resulting network has a purely feedforward architecture, and static backpropagation can be used for training.

The custom architecture used for further analyses is the Model Reference Adaptive Control (MRAC) system. Such model reference control architecture has two subnetworks. One subnetwork is the model of the plant to be controlled. The other subnetwork is the controller. Obtaining the trained NARX plant model, it is possible to create the total MRAC system and insert the NARX model inside and then add the feedback connections to the feedforward network. The next stage of training was focused on training of controller subnetwork (White, 1996). In order to the closed-loop MRAC system to respond in the same way as the reference model (used to generate data), the weights from the trained plant model network ought to be inserted into the appropriate location of the MRAC system. Then to achieve plant an initial input of zero, the output weights of the controller network were set to zero.

The final MRAC network is presented in the Fig. 2b, where layer 3 and layer 4 (output) make up the plant model subnetwork. Layer 1 and layer 2 make up the controller.

3 RESULTS AND DISCUSSION

As a result, the analyses highlighted the relationship between the parameters that describe the variation of the flame and the temperature of the exhaust gas in the chamber, or the amount of air flow in the secondary factor. Thus, if the temperature is slowly varying size, having an inert nature, the synthesis of the controller can be used quick-picture (actually a parameter or group of the image parameters). Primary air is used mainly for delivering pulverized coal to the burner nozzle, while secondary air is used for regulation purposes. Input parameters, such as the coal-biomass mixture and air flows, were changed several times during the tests, in order to create various combustion states.

Due to the incomplete knowledge of the control object or unexpected changes in the system performance with fixed parameters classic algorithms can be inappropriate. Adaptive control methods

would solve these problems. In turn, the required knowledge of object is achieved by artificial neural network modeling (Arabas, 1998), (Calise 1996).

3.1 Identification of a model for control

In optimal control algorithms, a reduced order model of the plant is usually applied. For the most implementations such models are generally linear. The choice of a specific control algorithm often imposes the model structure that needs to be implemented.

In general, physical-based models allow the designer to better understand the plant dynamics and their structure and parameterization may be related to both geometric and physical characteristic of the plant. Physical based models expose the process non-linearities and approach to their description. Their main purpose is to provide accurate simulation of plant behavior over wide range of operating conditions. The simulation approach significantly simplifies the work of the control designer, allowing to avoid expensive and time-consuming in-field testing. The design cycle is reduced and various alternative configurations can be tested without impact on the plant.

In order to ensure both control algorithm and simulator convergence, the reduced order model ought to be used. It is usually achieved by linearization of the non-linear equations, in order to obtain a set of linear state-space. The procedure is complicated for multivariable system with a number of set variables with implicit relations. For SISO feedback control loops inside system structure, the linearization is more complex. In such a case, designers use black-box identification methods of multivariable model from available, acquired data sets. Multivariable models contain large number of parameters whose estimation and selection are affected by excessive computational load and numerical errors. Regarding the MIMO systems, they are often presented in state-space form. It is caused by the access to the information on the dynamics in a limited number of parameters on one hand. However, on the other hand, there are subspace model identification (SMI) methods, that lead to the reliable models when the number of states, inputs and outputs is high. The training of the MRAC system took much longer that the training of the NARX plant model regarding to the fact, that the network is recurrent and dynamic backpropagation was used. After the network was trained, it was tested by applying a test input to the MRAC network. There were two MRAC systems designed and compared. The first one used non-optic, measurement based set of input vectors, quantitatively describing the flow of secondary air, fuel expense, respectively, and vectors describing exhaust temperature in the

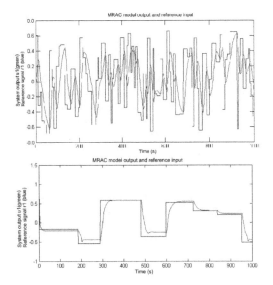

Figure 3. The result of MRAC system response to the system reference input: (a) without additional information from optical signals and (b) with flame image descriptor contour length vector.

chamber, recorded in the first measurement point, respectively. The second scheme used secondary air flow control signal and selected flame image descriptors based on Otsu's method, i.e. flame surface area and contour length.

Figure 3 shows system response to the system reference input in both cases: with classic measurements (a) and in the case of flame image descriptor contour length vector (b).

In Fig. 3, the plant model output does follow the reference input with the correct critically damped response, even though the input sequence was not the same as the input sequence in the training data. The steady state response is not perfect for each step, but this could be improved with a larger training set and perhaps more hidden neurons. From the obtained results of the proposed neural adaptive controls, it can be concluded that: control signals are bound, abrupt changes of system parameters involve a sudden changes of amplitudes of command laws and the outputs of the controlled system.

4 CONCLUSION

The aim of this paper is to propose model reference control scheme for complex combustion control. Two approaches of the neural adaptive control have been designed, developed and tested successfully. The considered scheme takes into account the multi-input multi-output nature of the combustion process.

The proposed algorithm uses artificial neural network capabilities to prevent situation in which information from the vision system became unreliable or significantly delayed. Using additional information optical signals based descriptors vectors for flame surface area and contour length proved system tracking properties.

Regarding to the conducted simulation test the following results were achieved: The increment of the prediction horizon n allows better performance since a greater prediction of the future error is possible. While using temperature values, its error weight must be high regarding to the fact the classical temperature regulation is slow and responsible for overall performance. High control horizon values return undesired oscillations.

Simulated biomass co-combustion results ought to be treated carefully due to the fact that its irregular consistence causes varying process parameters. Undoubtedly, better results can be obtained using biogas.

Further works ought to be concentrated on the more precise robustness verification of the proposed controller.

REFERENCES

Arabas, J., Domański, P.D. & Świrski, K. 1998. Optymalizacja kotła pyłowego i obniżenie emisji NOx przy wykorzystaniu sieci neuronowych - metody soft computing, Energetyka, 1: 21–33.

Ausubel, J.H. 1992. Industrial Ecology: Reflections on a colloquium. Proceedings of the National Academy of Sciences of USA, 89:879–884.

Calise, A. J. 1996. Neural Networks in Non-Linear Aircraft Flight Control, IEEE Aerospace and Electronic Systems Magazine, 11(7): 5–10.

Cięszczyk, S., Komada, P., Akhmetova, A. & Mussabekova A. 2016. Metoda analizy widm mierzonych z wykorzystaniem spektrometrów OP-FTIR w monitorowaniu powietrza atmosferycznego oraz gazów w procesach przemysłowych, Rocznik Ochrona Środowiska, 18: 218–234.

Dziubiński, G., Harasim, D., Skorupski, K., Mussabekov, K., Kalizhanova, A. & Toigozhinova, A. 2016. Optymalizacja parametrów światłowodowych czujników do pomiaru temperatury, Rocznik Ochrona Środowiska, 18: 309–324.

Fristrom, R.M. 1995. Flame structure and processes, London, Oxford University Press.

Hein, K.R.G. & Bemtgen, J.M. 1998. EU clean coal technology – co-combustion of coal and biomass, Fuel Processing Technology, 159–169.

Komada, P., Cięszczyk, S., Zhirnova O. & Askarova N. 2016. Optyczna metoda diagnostyki gazu syntezowego z biomasy, Rocznik Ochrona Środowiska, 18: 271–283.

Koshymbaev, S., Shegebaeva, Z. & Wójcik, W. 2014. Definition of the objects of multivariable control of technological process of smelting industry on the basis of optimization model, IAPGOŚ, 1: 18–20.

Kotyra, A., Wójcik, W. & Golec, T. 2010. Environmental Engineering III, Chapter 84. Assessment of the combustion of biomass and pulverized coal by combining principal component analysis and image processing techniques, CRC Press, 575–579.

Li, Z.Q. & Jin, Y. 2003. Numerical simulation of pulverized coal combustion and NO formation, Chemical Engineering Science, 58, 1: 5161–5171.

Mashkov, V., Smolarz, A. & Lytvynenko, V., Gromaszek, K. 2014. The problem of system fault-tolerance, IAPGOŚ, 4: 41–44.

Mashkov, V., Smolarz, A. & Lytvynenko, V. 2016. Development issues in algorithms for system level self-diagnosis, IAPGOŚ, 6: 26–28.

Mukhanov, B., Suleimenov, A. & Komada P. 2012. Control system elaboration for phosphorite charge pelletizing process, IAPGOŚ, 3: 25–27.

Narendra, K. S. & Parthasarathy, K. 1991. Learning Automata Approach to Hierarchical Multiobjective Analysis, IEEE Transactions on Systems, Man and Cybernetics, 20(1): 263–272.

Pawłowski A. & Pawłowski L. 2016. Wpływ sposobów pozyskiwania energii na realizację paradygmatów zrównoważonego rozwoju. Rocznik Ochrona Środowiska, 18:19–37.

Sawicki, D., Kotyra, A., Akhmetova, A., Imanbek, B. & Suleymenov, A. 2016. Wykorzystanie metod optycznych do klasyfikacji stanu procesu współspalania pyłu węglowego i biomasy, Rocznik Ochrona Środowiska, 18: 404–415.

Smolarz, A., Lytvynenko, V. & Kozhukhovskaya, O. 2013. Combined clonal negative selection algorithm for diagnostics of combustion in individual PC burner, IAPGOŚ, 4: 69–73

Suresh, S., Kannan, N., Omkar, S. N. & Mani V. 2005. Nonlinear Lateral Command Control Using Neural Network for F-16 Aircraft, Proc. 2005 American Control Conference, 2658–2663.

Shuvatov, T. & Suleimenov, B. & Komada, P. 2012. Gas turbine fault diagnostic system based on fuzzy logic, IAPGOŚ, 40–42.

White, D. A. & Sofge, D. A. 1993. Handbook of Intelligent Control: Neural, Fuzzy, and Adaptive Approaches, New York: Van Nostrand and Reinhold.

Zyska, T., Wójcik, W., Imanbek, B. & Zhirnova, O. 2016. Diagnostyka stanu czujnika termoelektrycznego w procesie zgazowania biomasy, Rocznik Ochrona Środowiska, 18:652–666.

Author index

Adamiak, K. 137
Asembay, A. 279, 319
Askarova, N. 155

Bąk, J. 101
Banasik, K. 211
Baran, S. 169
Bartoszek, L. 83
Bergier, T. 71, 95
Bik-Małodzińska, M. 169
Boczoń, A. 25
Boryczko, K. 15, 53
Bozkurt, D. 65
Brandyk, A. 25
Bukowska-Belniak, B. 33

Cel, W. 231, 239, 257
Chojniak, J. 115
Chomczyńska, M. 181
Chuchro, M. 33
Cichoń, T. 59
Cimochowicz-Rybicka, M. 123
Czech, T. 137
Czechowska-Kosacka, A. 231,
239

Diatczyk, J. 65
Dorgeloh, E. 115
Drewnowski, J. 187
Drozd, A. 293
Duda, A. 257
Dziubiński, M. 293, 299, 305

Ejhed, H. 115

Fatyga, K. 267
Fijałkowska, D. 161

Górka, J. 123
Gromaszek, K. 279, 311, 319
Gronowska-Szneler, M.A. 217
Gruca-Rokosz, R. 83
Gruszczyński, S. 71
Guz, E. 249
Guz, Ł. 203, 249

Hegedusova, B. 115
Hernandez De Vega, C. 187
Hołyszko, P. 287

Imanbek, B. 319

Jałowiecki, Ł. 115
Janowska, B. 161, 239
Jarzyna, W. 287, 293
Jaworek, A. 137

Kaleta, J. 1
Kalizhanova, A. 319
Kiczko, A. 25
Kida, M. 1
Klepacz-Smółka, A. 145
Kolano, K. 273
Kołodziej, S. 293
Komada, P. 311
Koszelnik, P. 1, 83
Kotyra, A. 279, 319
Kowalczyk, A. 89
Kowalewski, Z. 71
Kozbakova, A. 279
Krajewski, A. 211
Krawczak, E. 261
Królikowska, J. 59, 101
Krupa, A. 137
Kwiatkowski, M. 65

Łagód, G. 187, 203
Ledakowicz, S. 145
Lenik, Z. 257
Leśniak, A. 33
Litwin, M. 273
Lupa, M. 33
Lytvynenko, V.I. 155, 311

Majewski, G. 25
Marchewicz, A. 137
Masłoń, A. 109
Mussabekov, N. 311
Myszura, M. 169

Orakbayev, Y. 319
Osypenko, V.P. 155
Osypenko, V.V. 155

Papciak, D. 1
Pasierb, A. 15
Pawłat, J. 65
Pawłowska, M. 169
Pawłowski, A. 169
Paździor, K. 145
Piaskowski, K. 131, 197
Piecuch, T. 89

Piegdoń, I. 7
Piekarski, J. 89
Piotrowicz, A. 249
Płaza, G. 115
Porretta-Tomaszewska, P. 25
Przytuła, K. 267
Puszkarewicz, A. 1

Rak, J. 53

Sagymbekova, A. 279
Sawicki, D. 279
Sawicki, J.M. 217
Schmidt, R. 197
Siemionek, E. 293
Sikorska, A.E. 211
Smolarz, A. 311
Sobczuk, H. 203
Sobczyk, A.T. 137
Styszko, L. 161
Suchorab, Z. 203
Świderska-Dąbrowska, R. 131,
197
Szaja, A. 187
Szeląg, B. 43
Szostek, K. 33
Szpak, D. 7
Szymański, K. 239

Tchórzewska-Cieślak, B. 1, 7
Terebun, P. 65
Toigozhinova, A. 311
Tomaszek, J.A. 109

Wasilewicz, M. 211
Wassilkowska, A. 101
Werle, S. 225
Wesołowska, S. 169
Wiśniewski, K. 187
Włodyka-Bergier, A. 71, 95
Wójcik, W. 155, 279, 311, 319
Wrębiak, J. 145
Wróbel, M. 25

Yesmakhanova, L. 311

Żaba, T. 101
Zdyb, A. 261
Zhassandykyzy, M. 155
Zieliński, D. 267, 287
Żukowska, G. 169